Planning Canadian Communities

An Introduction to the Principles, Practice, and Participants

Fifth Edition

Gerald Hodge

David L.A. Gordon

Australia Canada Mexico Singapore Spain United Kingdom United States

THOMSON
NELSON

Planning Canadian Communities: An Introduction to the Principles, Practice, and Participants, Fifth Edition
by Gerald Hodge and David L.A. Gordon

Associate Vice President Editorial Director:
Evelyn Veitch

Editor-in-Chief, Higher Education:
Anne Williams

Executive Editor:
Paul Fam

Marketing Manager:
Sean Chamberland

Developmental Editor:
Heather Parker

Photo Researcher/ Permissions Coordinator:
Nicola Winstanley

Content Production Manager:
Wendy Yano

Copy Editor:
Elizabeth Phinney

Proofreader:
Carol J. Anderson

Indexer:
Gillian Watts

Manufacturing Coordinator:
Pauline Long

Design Director:
Ken Phipps

Interior Design:
Katherine Strain

Cover Design:
Peter Papayanakis

Cover Images:
Top: City of Vancouver; Middle: © Toronto Star/First Light; Bottom: © Masterfile (Royalty-Free Div.); Background: Courtesy of The City of Calgary—The Bridges redevelopment project.

Compositor:
International Typesetting and Composition

Printer:
Thomson/West

Printed and bound in the United States
2 3 4 10 09 08 07

For more information contact Nelson, 1120 Birchmount Road, Toronto, Ontario, M1K 5G4. Or you can visit our Internet site at http://www.nelson.com

Library and Archives Canada Cataloguing in Publication Data

Hodge, Gerald
Planning Canadian Communities: an introduction to the principles, practice, and participants/Gerald Hodge, David L. A. Gordon.— 5th ed.

Includes bibliographical references and index
ISBN-13: 978-0-17-625242-7
ISBN-10: 0-17-625242-8

1. City planning—Canada—Textbooks. I. Gordon, David L. A., 1954– II. Title.

HT169.C2H62 2007 307.1′20971
C2006-905970-5

In memory of
Jane Jacobs, 1916–2006,
and to
Sarah Jane Rudder Gordon
for their inspiration.

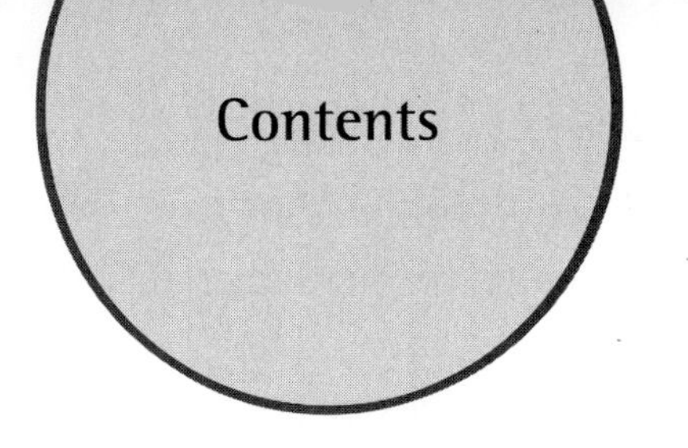

Contents

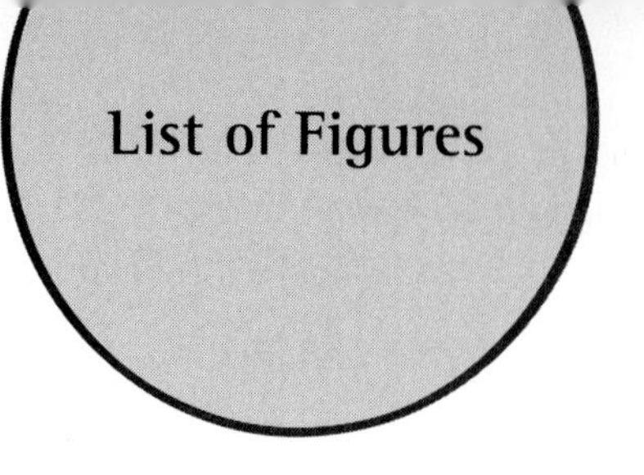

List of Figures

Preface

With this edition, *Planning Canadian Communities* begins its 21st year of analysis of the public endeavour of community planning in the cities, towns, and regions of Canada. The past two decades have been a time of immense learning, both for the field of community planning and for the authors. The task of community planning has become broader and more complex, as has the task reflecting upon it between these covers. One constant in this edition, as with its predecessors, is that it remains a personal view of how community planning started, how it works today, and who participates in it, albeit now of two authors. Occasionally within this perspective, we offer our views of its shortcomings and its potential for further development.

Another constant is our belief that the *community plan* is an essential component in Canadian planning practice. A plan cannot, of course, be realistically attained and maintained without the use of effective and fully participatory processes; but processes without a plan to achieve are little more than political self-indulgence. Successful community planning is, therefore, a blend of both *plan* and *process*. Yet another constant is our view is that there is a *Canadian* mode of community planning. In the hybrid tradition of this country, British and American planning approaches have combined with our own needs and cultures to produce a distinctive and often-envied style and approach. A primary aim of this book is, thus, to examine community planning in diverse settings across this country and to reveal its Canadian-ness. In pursuit of this goal, we try to make the often bewildering array of activities and institutions that occur under the rubric of community planning more understandable to students of planning, practising professionals, and, not least, citizens.

We are aware that it is a tall order to facilitate a dialogue about the best way to build ecologically, economically, physically, and socially sound communities. And it is a task that has become no less easy in the 20 years since this book was first published (and promises not to become any easier in the future). Still, presenting these topics through the structure comprising community planning's *principles, practice,* and *participants* seems viable. And to aid in this, the fifth edition has been extensively revised. The chapters dealing with the history and background of community planning have been completely revised and arranged chronologically to reflect several recurring themes in the 19th and 20th centuries. The chapters dealing with planning practice have been revised to better reflect current perspectives in the field. Likewise, the chapters dealing with planning's participants have been broadened to reflect not only the diversity of people in our communities but also the most recent thinking regarding this key component. As well, references to the literature have almost doubled to reflect the burgeoning Canadian writing on community planning, and a new set of "case studies," Planning Issues, which depict planning problems being dealt with in communities across Canada, are integrated with the revised and updated text. More than a score of new illustrations have been added, along with an updated selection of relevant websites to enhance the scope of each chapter. There is now a website devoted to the book where additional websites and illustrations and a comprehensive bibliography are available: www.planningcanadiancommunities.ca.

In the end, community planning cannot be done from a book or a website. But it is our hope that these pages may foster a greater appreciation of the need for visions—plans—of what Canadian communities can become, plans that engage and involve the community in their preparation.

Acknowledgments

A substantial revision of a book, such as occurred this time around, reveals again how much depends upon the contributions of others. They helped us build on the strong foundations created by the four previous editors, in order: Peter Milroy, Dave Ward, Avivah Wargon, and Karina Ten Veldhuis. Many other readers, reviewers, students, and colleagues contributed along the way, not the least being the late Kent Gerecke. To all of them we offer our thanks for their support and thoughtful comments, and apologize for not being able mention each one.

For this edition, we are grateful for the editorial guidance of Anthony Rezek, Katherine Goodes, and Heather Parker. Many of Andrew Milos's original graphics remain, and Kelly McNicol and Xian Zhang have revised some and added others in his format. Our appreciation extends especially to the Canadian Institute of Planners, who made their collection of award-winning plans available and to Patrick Condon (UBC), Andres Duany (Miami), and Avi Friedman (McGill) who provided additional images. Throughout, Jo-Anne Rudachuk responded with forbearance and good humour when faced with yet another draft to process, and our research assistants, Tasha Elliott, Dayna Lafferty, and Raktim Mitra, were there when needed with timely and thoughtful contributions. Many Canadian planners advised us on provincial planning legislation, practice, and terminology for which we are indebted, especially to: Michèle Bertol (Iqaluit), Neil Craik (UNB), Steven Croft (Halifax), Howard Epstein (Dalhousie), Jane Glenn (McGill), Jill Grant (Dalhousie), Felix Hoehn (U. of Saskatchewan), Steven Horn (Yukon), Barb Jeffrey (York Region), Eran Kaplinsky (U. of Toronto), Andrea Law (Vancouver), Clarissa Lo (Iqaluit), Byron Miller (U. of Calgary), Elaine Mitchell (St. John's), Nalini Naidoo (Dillon Yellowknife), Ken O'Brien (St. John's), J.F. Schommer (Regina), Joy St. John (Concordia), Shelley Steel (Calgary), George Stetkiewicz (Yukon). Not least, our thanks to the librarians at the Hornby Island Branch of the Vancouver Island Regional Library for facilitating many interlibrary loans and to John Meligrana for bringing the Queen's University Library to the west coast.

The Social Sciences and Humanities Research Council of Canada and a Fulbright Fellowship supported David Gordon's planning history research, which also benefited from comments by Eugenie Birch (Penn.), Raphael Fischler (McGill), and Larry Vale (MIT). Our colleagues at Queen's University, in particular Susan Hendler, Hok-Lin Leung, and Mohammad Qadeer, who provided new insights on feminist approaches, diversity, planning theory, and multiculturalism, were always generous in their support. Other colleagues provided valuable suggestions in the course of preparing this edition, including Chris Fullerton (Brock), Mark Seasons (Waterloo), and anonymous reviewers from Alberta, Brock, Guelph, Langara College, and McMaster; to the latter, we may not have written exactly the book you wanted, but your suggestions did influence our revisions. And, as any teacher will attest, our students are a constant source of new insights and ideas. We can only apologize that it sometimes takes a few years to answer some of the best questions from the classroom. Of course, no one named above is responsible for the book's contents or its errors; that responsibility remains with us.

Finally, as every author knows, there are not sufficient words or ways to acknowledge the companionship, patience, and support of one's partner when so much attention turns toward a book. To ours, Sharron Milstein and Katherine Rudder, very special thanks.

Gerald Hodge
Hornby Island, British Columbia

David L.A. Gordon
Kingston, Ontario

August 2006

Cover Images

Front top: Vancouver's urban waterfront in False Creek North.
Courtesy: City of Vancouver

Front middle: Suburban development on agricultural land.

Front bottom: Careful design allows public use of environmentally sensitive areas.

Background Image (Front and Back Covers): Aerial View of "The Bridges"—a transit-oriented development in Calgary's inner suburbs that features green buildings, good urban design and some affordable housing. Courtesy: City of Calgary and Canadian Institute of Planners.

Gerald Hodge is one of Canada's foremost community and regional planners. Dr. Hodge has been involved in planning education and research for more than 40 years. From 1973 to 1986, he was Director of the School of Urban and Regional Planning at Queen's University and has also taught planning at the University of Toronto and UBC. Among his many publications are *Planning Canadian Regions* (with I.M. Robinson) and *Towns and Villages in Canada* (with M.A. Qadeer). Now retired to Hornby Island, B.C., he continues to write on planning matters and to stay involved in local planning issues. He holds a Ph.D. from MIT and an MCP from the University of California at Berkeley.

David Gordon teaches planning at Queen's University. Prior to joining the School of Urban and Regional Planning, he practised in the public and private sector for 15 years, twice sharing the Canadian Institute of Planners' National Award of Distinction. His publications on planning history and urban redevelopment include *Planning Twentieth Century Capital Cities* and *Battery Park City: Politics and Planning on the New York Waterfront*. He has also taught urban design and development at the University of Toronto and Ryerson University, and holds an M.Pl. from Queen's and a D.Des. from Harvard.

PART ONE

1

The Roots of Canadian Community Planning

INTRODUCTION

Both the shape of our communities today and the methods we now use for planning them are a legacy of the past. Communities do not just grow, as Thomas Adams has noted. Their form and functioning are the result of countless decisions made over the generations of their life, decisions that have embodied the needs, experience, values, and aspirations of their builders. Each generation builds upon the experience and physical structures of those that have gone before. Builders may accept and refine what is there, sometimes building and rebuilding in a dramatic way. They may achieve livable communities or they may not. They will do so in response to real problems and possible solutions rendered by someone with varying degrees of forethought, or what we now call planning.

Similarly, the public activity we in Canada call community planning evolved alongside changing community problems and perceptions of those problems by citizens, professionals, and governments. It works with a built environment that is the result of accumulated decisions upon which it will add yet another layer reflecting long-standing ideals and contemporary ideas and premises regarding the physical form of communities, the social needs of residents, and the institutional means for achieving them. The next five chapters describe the broad evolution of community-building abroad and in Canada and, especially, the gradual refinement of community-planning approaches over the past century and a half. In other words, they show the depth of community planning's roots and what nurtured it into the form this practice has taken today.

The image above is from the Woodlands Rezoning *project, New Westminister, British Columbia, which received the Canadian Institute of Planners' Award for Planning Excellence, Category of Implementation, 2004.*

1

Chapter One

The Need for Community Planning

Cities do not grow—all of them are planned.

Thomas Adams, 1922

Community planning in Canada is now well into its second one hundred years, and its value to communities is acknowledged across the country. Communities of all sizes and in all regions have their own plans. Whether in the Gwich'in Settlement Area in the Northwest Territories, Ucluelet on the B.C. coast, St. Stephen in New Brunswick, Victoria, Winnipeg, or Montreal, few would take issue with the view that planning meets some important needs of their communities. In spite of periodic confrontations over the building of new shopping centres, high-rise apartments, and highways, or the demolition of historic buildings, or the failure to conserve open space, the regard for planning and making plans is strong. Even in these contentious situations, the essential debate is not about the need for planning, but for *better* planning—not *whether* but *how* it should be done.

The activity of community planning has come to be strongly valued over and above its immediate achievements: it has become, to use a sociological term, "institutionalized." There exists in every province and territory some form of legislation that both sanctions the notion of planning and specifies its format for that region's communities. From coast to coast it is considered a normal and necessary public function. However, the fact that we have a plethora of governmental frameworks and legislative means in Canada to facilitate community planning does not explain *why* a community needs planning. The sets of planning procedures and legalistic steps, which may sometimes seem overwhelming, are not in themselves community planning. They are means, not ends.

The real need for community planning arises because *people* in a community wish to improve their environment. This need has not always been so widely accepted, nor have concerns over community environments always been the same, as we shall see in the next four chapters. But then, the concerns raised by communities and the approaches used in their resolution are always changing. The planners for today's cities and towns work within community contexts in which populations have grown, aged, and become more culturally diverse. Moreover, citizens have become insistent in their desire to be a part of the community-planning process. This is making for a more demanding climate for today's planners, but it highlights a question that always pervades the practice of community planning, as it does all chapters of this book:

- ***What generates the need for planning community environments?***

The Essential Raison D'être

In order to appreciate the raison d'être for the planning that abounds in most communities, let us begin by focusing on the sorts of actions a number of Canadian communities have taken in their pursuit of planning. This will provide a basis for deriving some guiding principles about the need for community planning.[1]

- ***Halifax, Vancouver, Ottawa,*** and ***Montreal*** have each developed planning regulations to control the location of tall downtown buildings so as not to spoil important public views. In Halifax, planners sought to protect the traditional view of the harbour from Citadel Hill; Vancouver planners valued views of the mountains; in Ottawa, it was to retain views of the Parliament Buildings; while in Montreal, there was concern over losing views of Mount Royal.
- ***Natuashish,*** on the mainland of Labrador, is a brand-new community, which the Mushuau Innu began planning in 1994 to give them a site where they could revitalize their culture, economy, and community; it has now been completed.
- The 80-kilometre stretch of the Saskatchewan River centred on ***Saskatoon*** is covered by a 100-year master plan to achieve a balance between conservation of the environmentally fragile river banks and the desire for recreation and urban uses; the multilevel Meewasin Valley Authority is charged with implementing the plan.
- The channel of the Little River in ***Windsor,*** Ontario, used to be a tire-filled wasteland; now it is rehabilitated and part of the city's plan for a connected greenway for the whole city.
- ***Abbotsford,*** on Vancouver's eastern fringe, has a new neighbourhood, Auguston, which follows the design principles of New Urbanism, with mixed-housing types, garages in back lanes, front porches, and walkways from the local commercial area. The ***Calgary*** neighbourhood of McKenzie Towne also follows these principles.
- The town of ***Markham,*** on Toronto's northeastern edge, took seriously the challenge to "restore biodiversity" and produced a Plan for the Environment with a vision for protecting, restoring, and linking up natural features that they are now using to guide their broader community planning.
- ***Red Deer,*** Alberta, recently completed a Downtown Action Plan to provide a long-term vision for development of its central area. It included several major initiatives: a promenade on the central axis of 48th Street, links to the Riverlands Area, and a general policy to put "Pedestrians First."
- ***Fermont,*** a resource town of 5000 in Quebec's subarctic region, has a unique weatherproof downtown core, while in ***Tumbler Ridge*** in B.C., ***Leaf Rapids*** in Manitoba, and ***Fort McMurray*** in Alberta, the planners also paid attention to the climate and the need to keep things compact for easy access in all seasons. And ***Iqaluit,*** in Nunavut, has adopted designs for sustainable Arctic subdivisions.

- Mixed-use projects are being planned in Canada's largest cities to rehabilitate run-down areas: the Benny Farm in ***Montreal***, the Woodward's project in ***Vancouver***, and Regent Park in ***Toronto***.

What do all of these planning initiatives—the New Urbanism neighbourhood in Calgary, the restoration and conservation of river banks in Saskatoon and Windsor, the biodiversity plan in Markham, and the protected views in Halifax—have in common? For one, each of these planning efforts deals with various elements of a community's **physical or built environment**, the built and/or natural environment. Each community also undertook to make a plan for the purpose of achieving a goal desired by its citizens. In short, the answer to our question is: **community planning is about attaining a *preferred* future built and natural environment.**

More specifically, the community preferences we see expressed in these examples arise from either, or both, of two basic needs. The situations of Halifax and Saskatoon can provide helpful insights regarding these needs. Halifax is responding to an impending **problem**—new development that will affect historic views. Saskatoon's Meewasin Valley Authority is pursuing broader **aspirations**—to secure a stable riverbank environment for the enjoyment and use of all citizens well into the future. A look back at the other examples will show that both problems or aspirations, or both, regarding a community's built and natural environments guided the planning activity. Thus, we can state that there are two principal reasons that a community is stimulated to undertake community planning:

1. **A community may wish to solve some problems associated with its development;** and/or
2. **A community may wish to achieve some preferred form of development.**

These two needs are often intertwined. For example, a community faced with the need to solve a particular development problem (e.g., the cleanup of a river course, as with Windsor), may decide to reassess its overall plan for parks and greenways. Likewise, plans that aim for a preferred future almost always encounter latent development problems that need a solution.

The Need to Solve Development Problems

Of the two basic reasons noted above, a community's need to solve present or future problems in its physical environment is probably the one most people think of first. The problematic situations that spark planning action reflect the concerns and conditions of the time. In the 1990s, for example, actual and potential deterioration of the natural environment were prominent planning concerns. In the 1980s, growth management and affordable housing occupied planners' attention. And in the 1970s, the concerns were very often about the effects that new, large-scale projects—e.g., apartment-building complexes, expressways, second airports, shopping centres—might have on existing community areas. The 1960s began with major concerns over physical deterioration in communities; urban renewal became the hoped-for solution, until the "bulldozer approach" used to achieve it was called into question. Canadian urban communities in the two decades after World War II were concerned with how to accommodate burgeoning populations—how to provide them with housing, public utilities, schools, and parks, as well as how to cope with their infatuation with the automobile. The problem of housing for the urban poor concerned communities in the 1950s, as it did 50 years before and still does today.

Every decade has its characteristic problems, as well as new versions of past problems, because cities and towns are always changing. They may grow in population numbers and the area they occupy, decline in their economic well-being, or diversify in the cultures among residents. Even when there is no quantitative growth, there is change through the population, buildings, and infrastructure that goes on continually. The development of the physical environment of a community, new or old, can be overtaken by new economic trends, new technology, new attitudes, and new activities. Whatever the source of change, every community must confront development problems at one time or another—traffic congestion, housing affordability, deteriorating natural environments, and neighbourhood stability, to name just a few.

Growth and Development Problems

Although the content of the planning "agenda" may differ among communities, a common feature of all is the need for planning to deal with problems associated with growth and development. The Vancouver area is currently experiencing population growth of about 40 000 persons per year, for example, while many smaller communities in rural regions of B.C. and elsewhere are declining in population numbers. Other places face a growth in the number of automobiles in their downtown areas while still others lose jobs as factories close. In all these instances a **quantitative change**, either positive or negative, occurs (and sometimes both as one change affects another facet of the community). Thus, when we talk about "growth" in a community, it refers to either increases or decreases in the size of population, the number of structures, the traffic, the space required for the community, and so on.

"Development," by contrast, refers generally to **qualitative change** in the community, such as the opening up of new areas, the replacement of one land use with another, a shift in the cultural mix of the population, or the emergence of new modes of building and doing business. A typical development situation these days is found in the proposal for a "big-box" retail complex on the outskirts of a town or city; this often has major implications for a community's established downtown area as well as for its road network. An analogy from early in the 20th century would be the cities and towns that were trying to accommodate an influx of new population while coping with the spread of disease and extensive fires (see Chapter 4). Housing, then, was in short supply and often inadequate, as were water and sewage systems, where they existed at all. Both examples indicate that growth and development may generate problems that affect the whole community. Further, both situations not only posed threats to parts of the established community but also caused uncertainty about the future form and condition of the overall community.

Both growth and development are always occurring in a community from a variety of sources. So, is planning needed just because there is growth and development? The answer to this is that the need for planning depends upon whether the growth and development give rise to actual or anticipated problems *for* the community or lessens its achievement of desired future conditions.

The Public Interest

It was stated above that planning is concerned with a community attaining preferred future conditions in its built and natural environments. The community's *preferences* are thus the prime consideration when seeking a solution to growth and development problems—that is, in making a plan. Indeed, it is the "sum" of these community preferences that validates any proposed plan for they are considered to reflect the "public good" or to be in the **public interest**. In other words, community planning aims to promote the interest of the entire public—of all citizens—and thereby give social legitimacy to plans.

As reasonable as this principle sounds, obtaining community agreement about what constitutes the public interest is not a simple matter. In 19th-century Canada, decades passed before the public health and safety problems that plagued its cities were recognized and acted upon (see Chapter 3). If this seems incredible to us today, we need only consider that it is more than four decades since Rachel Carson, in her book *Silent Spring*, alerted us to the dangers of environmental degradation and pollution, problems that continue to plague us. In part, the task of identifying the public interest is made more complex by the fact that our communities are composites of privately owned properties, structures, and services—not to mention vehicles—and publicly provided facilities, programs, and means of communication. All the diverse interests implied by this are involved in maintaining and developing communities, and their preferences do not always coincide. Much of community planning, therefore, involves trying to reconcile private and public preferences into acceptable community preferences to solve problems caused by growth and development. This is not always accomplished smoothly or quickly, and there is probably no reason to expect that either agreement or equanimity will always prevail in planning issues.

On the private side of the community, those whose interests lie in maintaining a stable, tranquil setting of, for example, a neighbourhood, may clash with those who wish to pursue long-held development objectives and rights to change their properties and structures. Or there may be clashes between private interests and the public good, such as proposals for development in what the community considers inappropriate locations. Halifax's concern that new development might block historic views is a typical case, and so is the concern in many communities over allowing housing developments in flood-prone areas. Sometimes clashes arise over proposals by public bodies to accommodate growth or new technologies, such as building an expressway or a second airport; and sometimes there are clashes between different public bodies and their interests and aspirations, as in where to locate dump sites for toxic waste. Meanwhile, in all such clashes, each side tends to advocate its position as being in the public interest.

How can this be? And how can planners "sum" the various preferences, especially with our communities accommodating more and more diverse interests? It is coming to be seen more and more that the public interest is not monolithic and neutral. Our cities and towns are made up of multiple publics, each with a potentially different view of the public good.[2] The one-time scenario of a solid cornerstone of the public interest, with the planner tending and interpreting it in the realm of physical development, is shifting in the face of the increasing diversity of interests in today's cities and towns. The public interest is coming to be seen as something that *becomes known* through collaborative planning processes of consensus-building.[3] The planner's task, then, is to ensure that these processes are as inclusive as possible and that all perspectives on what constitutes the public good are aired, including those held by the planner.[4] In this way, something more appropriately termed the **common interest** can emerge and sanction planning actions. (The changing definition and role of the public interest in planning are discussed more fully in Chapters 7 and 15.)

External Effects

It is fair to say that most of the problems of growth and development that planning deals with arise from the **external effects** that an instance of development has upon other facets of the built and natural environment. These external effects, which are often called "overspill effects" or "externalities," may be localized or they may affect the entire community and even beyond. For example, traffic congestion results from traffic destined for (or leaving) one or a set of land uses (e.g., a stadium or a downtown area) exceeding the capacity of the means of moving the traffic, and the effects (e.g., noise, gridlock, pollution) then spill over into adjacent areas.

Problems can arise from the development of a new professional office building near a low-density residential area: it may generate traffic and parking needs that can be only met on adjacent streets. A school board's decision to close a neighbourhood school can have external effects when parents have to worry that their children will have to walk farther to school and perhaps cross dangerous streets as well. Or the design of a new subdivision's storm-drainage system will need to be checked to make sure that it doesn't cause pollution in a nearby stream that feeds into the community's water supply.

The idea of anticipating externalities accompanying development proposals and projects has come to be a central part of community-planning practice. Partly owing to greater understanding about how the various parts of a community function and relate to one another, community planning now takes into consideration the **potential** effects of a development project on a neighbourhood, the community as a whole, or even its region. Planners will examine plans for a proposed highway linking the city centre and the suburbs for its effects on the property values, traffic patterns, and air quality of the areas it passes through, for example. A good deal of community planning, therefore, is as much about trying to grasp the potential external effects of development as it is about responding to the direct and immediate effects.

The need to grasp potential problems arising from development proposals has brought two other considerations into planning. The first is the recognition that external effects may not be known from our previous experience—they may have to be estimated or possibly just guessed at. Moreover, we are beginning to realize that many effects of development may not even be known until well into the future, as has been the case with many pollution problems. A second consideration is the role of scientific and technical knowledge in understanding the complexity of a community's growth and change. The result is to make the professional planner's task increasingly demanding (and some would say nigh impossible).[5]

The need to predict any problems associated with development proposals and other changes in the community has, for example, introduced into the planner's milieu the need to consider several sophisticated notions. One has to do with the costs and benefits associated with proposed developments, sometimes phrased as "who benefits and who pays"? For almost all development projects have external effects beyond the subject property. Another is the "precautionary principle" whereby the external effects of a project could possibly be harmful, but cannot be fully established. Yet another is the notion of making a "risk assessment" of a project's external effects becoming manifest. These and other planning analyses are discussed more fully in Chapters 7 and 14.

To recapitulate, community planning deals not just with solving problems but also with solving them so that the community turns out to be as good as, or better than, it was expected to be. Even in the solving of problems, a community tries through its planning to achieve some of its ideals. In this concern for ideals, we see a connection with the other major reason for community planning—the desire to achieve some improvement in the form of the community.

The Need to Achieve an Improved Environment

Community planning is concerned with more than solving the problems posed by current development. Goals, ideal aims, are important as well. There is a tradition of idealism that influences much of what is done in the name of community planning. Probably the most quoted aphorism on the subject of idealism is that of Daniel Burnham—"make no little plans." Burnham was the co-author of the path-breaking 1909 plan for Chicago and its region.

> Make no little plans; they have no magic to stir men's blood and probably themselves will not be realized. Make big plans: aim high in hope and work,

remembering that a noble, logical diagram, once recorded, will never die, but long after we are gone will be a living thing, asserting itself with ever growing insistency.

An intrinsic part of most community-planning efforts involves the consideration of ideal situations and the aspirations of people for their community. It is a tradition with deep roots in the history of city-building. The plans for ancient cities in Greece, Rome, China, the Middle East, and the Americas reveal a concern for the proper location of various functions of the city. The sites of religious buildings and public areas, and even the geographical orientation of the streets, often were considered symbolic and thus important. Cities themselves were seen as symbols of a society's aspiration to achieve progress and human betterment. As Aristotle noted, "A Citie is a perfect and absolute assembly or communion of many townes or streets in one."[6] Through succeeding eras, the city as a reflection of human aspirations continued to be a major theme in philosophy and culture, although the images changed from the religious to the secular as societies changed.

This tradition is still a powerful one in community planning. The sentiments expressed by Burnham encourage not just boldness but also encourage the community to express its hopes and ideals in its plans, rather than merely deal with solutions to temporary problems or "choose among lesser evils," as David Riesman has stated.[7] Indeed, Gordon Stephenson, a British planner who worked and taught in Canada in the 1950s, unequivocally observes: "I do not believe we can make worthy plans without having ideal conceptions...."[8] And there are many examples in Canada where the ideals of the plan-makers for human betterment are central to the community plan.

Nowhere is this more striking than on our resource frontier. Kapuskasing, Ontario, and Temiskaming, Quebec (see Figure 4.9, page 83), both planned by Thomas Adams, one of the great planners associated with Canada, and built in the 1920s, are two such places. Kitimat, B.C. shown in the accompanying photograph (see Figure 1.1), and largely designed by the acclaimed American planner Clarence Stein in the 1950s, is another instance of planning to achieve ideal living conditions for an entire community. Both these planners were well acquainted with the efforts of Ebenezer Howard in England to promote, and eventually build, self-contained Garden Cities (see Chapter 3). The latter concept of an ideal community became an extremely powerful model for those concerned with planning and building suburban communities in Canada, from Strathcona Park near Edmonton to Ajax near Toronto.

Other Canadian community plans, while equally idealistic in concept, have striven to improve the environment on a less sweeping scale. One of the most notable is the 1915 plan for Ottawa, which aimed to achieve a monumental quality for the nation's capital.[9] Of the same genre is the Wascana Centre area around the legislative buildings in Regina. More recently, the planning of Harbourfront in Toronto and False Creek South in Vancouver reflects the aspirations of these communities to create a new but lasting image in their environments. In these examples, something new is designed and built. Rooted in this same theme are the historical conservation programs undertaken by some communities, such as Niagara-on-the-Lake, Ontario, and St. Andrews, New Brunswick, where the desire is to preserve surroundings of special value for future generations.

Whether the planning involves the design for some special new environment or the refurbishing of highly valued older environments, the thrust is essentially the same: creating an image, or vision, of a possible community environment. At its most dramatic, this facet of planning in modern times has produced plans that have resulted in, for example, the impressive capital cities of Brasilia in Brazil (see Figure 1.2) and Chandigarh in India. Brasilia is even more successful than its planners hoped and now houses two million people, compared to the half million originally envisioned.[10] But even less spectacular planning situations also reflect a human need to strive for improved community environments. There is a utopian element in most community planning, as there has

Figure 1.1 First Phase of a New Canadian Community: Kitimat, B.C., 1957

American planner Clarence Stein was commissioned to design a city in northern British Columbia, ultimately for 50 000 people, based on the "neighbourhood unit" principle. The economic base of the community is a huge aluminum smelter.

Source: Reprinted with the permission of Alcan Inc.

been since the first communities were planned. But these visions *are* realizable. According to an old adage, "A planner is someone with his head in the clouds and his feet planted firmly on the ground."

The heady experience of planning a completely new town is not available to most community planners. Rather, it is planning for rebuilding, and refurbishing, and revitalizing a preexisting community that is the most common planning experience today. Further, the motivation for an improved community environment nowadays is often less in terms of the appearance of a city or town and more in its "livability" and "sustainability." The active engagement with caring for the natural environment of a community characterizes much planning that is currently being done across the country from Markham to Calgary. This is not to say that these goals are achieved easily, for today's planning milieu involves a wider array of participants and a tug-and-pull of views in reconciling them. This reality of competing goals and objectives in community plan-making will be evident throughout the remainder of the book.

Putting the "Community" into Canadian Planning

The term "community planning," used throughout this book, requires some explanation. Although one does not encounter the term as often nowadays, it is peculiarly Canadian and especially appropriate to describe the activity of planning living environments in our variously sized settlements. "Community planning" entered the Canadian lexicon

Figure 1.2 Plan for a New Capital City: Brasilia, 1957

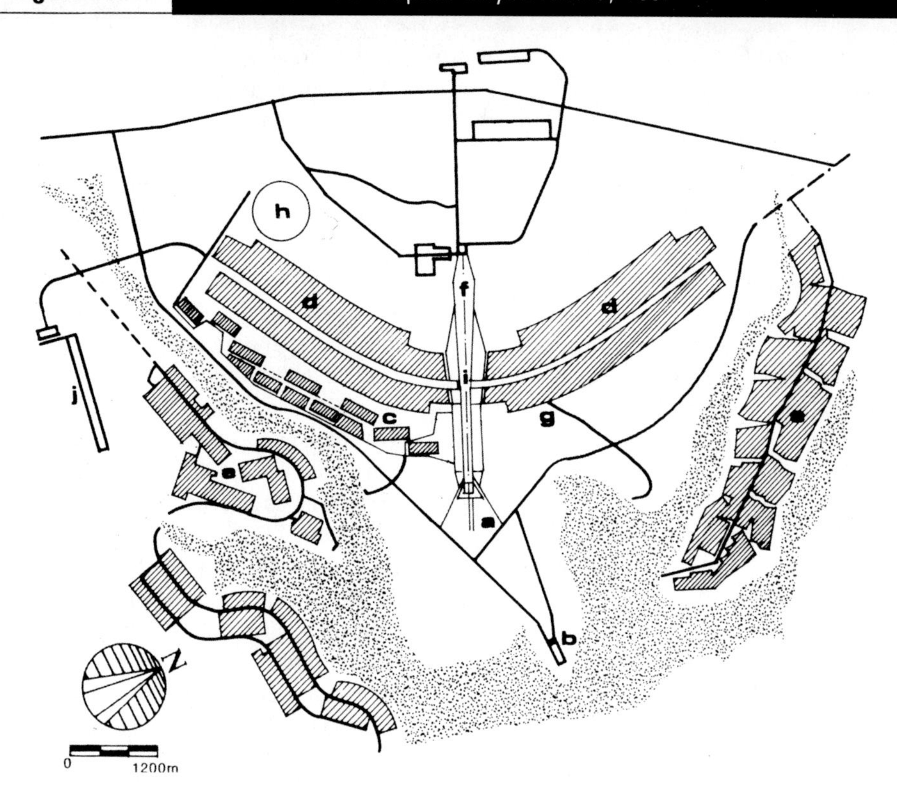

The idea for an inland capital for Brazil dated from its independence in 1822. A competition for a plan was held in 1956, and the curved cruciform plan of Lucio Costa was chosen. Fine buildings were designed by architect Oscar Niemeyer, and the first phase of the highly symbolic city was completed in 1960. Key: (a) main square, (b) President's residence, (c) foreign embassies, (d, e) residential areas, (f) recreation area, (g) university, (h) cemetery, (i) main traffic interchange, (j) airport.

not long after World War II, when the Community Planning Association of Canada (CPAC) was formed. In the United States, the comparable term is "city planning," which has been in use since shortly after 1900. The British have used the term "town planning" since the late 19th century. In the academic world in Canada and the United States, "urban planning" has been the accepted term since 1960.

There is no record available to explain why the Community Planning Association chose "community planning" for use in Canada; the term "town planning" was in widespread use from the turn of the century. Provinces had their town-planning acts, cities their town-planning commissions, and the professional planners had their town-planning institute. The first significant use of the term "community planning" appeared in a report to the Canadian government by a committee giving advice on the problems the country would face when World War II ended. The 1944 Advisory Committee on Reconstruction prepared a report entitled *Housing and Community Planning*[11]—which came to be known as the "Curtis Report" after its principal author, Professor C.H. Curtis of Queen's University. It argued that town planning had two distinct but complementary meanings. Town planning, they felt, should encompass not only the "rational physical organization" of a city but also the concept of "better community living."[12] Perhaps this was the stimulus to adopt the term "community planning" when CPAC met two years later, for many of the founding members of Canada's first (and only) nationwide citizen organization in planning had also been members of the Curtis Committee.

Planning Issue 1.1 One Island, One Park—Maybe

The Montreal Gazette
April 17, 2006

One Island, One Park—Maybe
Michelle Lalonde

There are precious few places left on Montreal Island where you can close your eyes, smell nothing but moist earth and fresh air and enjoy the rat-a-tat of woodpeckers, a chorus of wood frogs, or the trickle of spring water over moss-covered rocks. The forests and wetlands around Riviere a l'Orme in the West Island harbour some of those places, and conservationists say it's time Montrealers opened their eyes to the fact that natural spaces like these are disappearing at a shocking rate.

Since 1990, an estimated 1,000 hectares of forests have been stripped from the island to make way for private development. This city now has the lowest percentage of protected natural areas of any North American urban centre.

Only six per cent of the island's land mass is unspoiled wilderness—woods, shoreline and wetlands—and half of that is unprotected by any law or statute. Much of the unprotected half is owned by private developers.

An alliance of island conservation groups called the Green Coalition has been trying desperately to stop this steady erosion of green space. For the past two years, the coalition has been working with the provincial government on a project to give what's left of Montreal's natural spaces the legal status of a provincial park.

In fact, before he was turfed out of cabinet six weeks ago, former environment minister Thomas Mulcair was preparing to create a new provincial park on the island, starting at the shores of the Lake of Two Mountains and the Riviere a l'Orme, but eventually taking in all ecologically valuable land in the entire archipelago of Montreal. This includes the islands of Montreal and Laval, as well as dozens of other islands around them.

In an interview with *The Gazette* last week, Mulcair said the provincial park idea was "moving along apace" when he was forced out of the environment portfolio.

"The idea was that we take everything in and around the Montreal archipelago that is worth saving," and consolidate it into what the Quebec government refers to as a "national" park, or what is called a provincial park in other Canadian provinces, he said.

It would be Quebec's first provincial park in an urban setting, and the first that would be made up of many unconnected pieces of land. Mulcair's enthusiasm for the plan had conservation groups rejoicing.

It is worth noting that, prior to World War I, Thomas Adams urged Canadians to use the term "town planning" in preference to the American term "city planning."[13] He contended that the American term gave too much emphasis to the physical side of city building and did not seem to acknowledge the human side of city life. He felt that the British term "town planning" was more encompassing of the social, physical, and environmental qualities of human settlements. His preference is understandable, as he had begun his career in planning with the founders of the English Garden City movement, who advocated building new towns that integrated human and physical factors. This latter dichotomy has for a considerable time been a source of debate among the professional planners and others involved in planning for our communities. The physical factors are the facets of a community that can be planned, regulated, and shaped most easily, so the practice of community planning quite naturally gravitates in this direction. Every so often, planners are accused of emphasizing physical factors at the expense of human factors in their planning solutions. While the emphasis ebbs and flows between the human and the physical factors, our planning efforts seldom comprise only one. Thus, community planning has merit in that it brings the human or community aspects clearly into the picture.

But the new environment minister, Claude Béchard, has been silent on the project, as his government deals with the fallout from its controversial plan to sell off parts of another provincial park at Mount Orford.

If you ask David Fletcher of the Green Coalition why we should save urban forests like Anse a l'Orme from development, his blue eyes light up. Not only are these places a joy to hike through, he explains, but they provide habitat for migratory birds and many rare and endangered species. The islands in the Lake of Two Mountains, which include Montreal, Laval and many smaller islands, enjoy a micro-climate that allow certain rare plant and animal species to flourish.

For 30 years as a primary school teacher in Pierrefonds, Fletcher organized field trips so he could show children first-hand how everything—from the deer to the plants to the birds to the bivalve mollusks in the marsh—is interdependent. Some of those kids are now active in the local conservation movement, he said.

"You have to educate their viscera and their hearts.... People have to place a value on these places, emotionally as well as in their pocketbooks."

That's why the Green Coalition was pleased—but far from satisfied—in March 2004 when Montreal Mayor Gerald Tremblay's administration identified 10 "ecoterritories" on the island, including the Riviere a l'Orme eco-corridor.

And the ecoterritory designation does not automatically stop development. It means that developers will have to go before a special city committee to see if their plans can better accommodate environmental concerns.

"It's just a name that makes people think that these places are protected, which is simply not the case," said Sylvia Oljemark, co-founder of the Green Coalition.

In an interview with *The Gazette* last week, an Environment Department official said the park project is not dead, but he tried to lower expectations raised by Mulcair.

"For the moment, the Two Mountains Park project is just a concept, an idea to better plan and integrate conservation actions," in the Greater Montreal region, said Patrick Beauchesne, acting director of ecological heritage and parks with the Quebec Environment Department. "We are reflecting on it."

But Mulcair said the park project was already well advanced when he left the department.

"If you don't have that over-arching vision of what needs to be done from Lake of Two Mountains right to the other end of the island and the other islands beyond, you just won't be able to build for the future."

Source: Excerpted from Michelle Lalonde, "One Island, One Park—maybe." *Montreal Gazette*, April 17, 2006, A1. Reprinted with permission.

"Community planning" seems to be the most appropriate term to use for two other reasons. First, in the Canadian setting, a town is usually seen as a small-size settlement and a city, by contrast, a large one. However, Canadian settlements of all sizes are involved in planning, as are many rural areas that are not clusters of populations. But all may rightfully be called communities. Second, "community planning" conveys the idea that modern planning is an activity undertaken by the community and involving all who live in it. The terms city planning and town planning both suggest a technical activity dominated by professional planners. Humphrey Carver, one of the founders of CPAC, says that one of the main aims in establishing the association was to create "a framework of discussion in which laymen, professionals, and politicians could meet on equal terms to talk about their aspirations for Canadian cities."[14] The polite phrase "a framework for discussion" evolved into the more vigorous one of "citizen participation" as a way to describe the involvement of more than just planners and politicians in planning activity. Even though the involvement of community members is not uniformly practised, the question is not whether it should happen, but how and when.

"Community" planning signifies the importance of the aspiration that the community should be doing the planning.

Planning Challenges for the New Century

Community planning is a task that evolves as society evolves. Community planners therefore need to be aware of changes occurring in their communitie and/or in the larger society that contribute to them. With each instance of change or set of changes come challenges for the planner. Some of the challenges that today's planners face are new and others have only recently emerged, yet each must be met (or met again). To set the stage for the chapters ahead, it will be helpful to review briefly the major types of challenges already on Canadian planners' thresholds.

For a decade or more, since about the mid-1990s, the array of both planning issues and planning solutions being propounded has burgeoned. Some of these stem from the never-quite-achieved aim of full citizen participation. Others stem from the long-accepted but seldom-realized goal of environmental protection. Still others are relatively recent and derive from the accelerating cultural diversity of communities large and small. The emergence of globalization influences provides another challenge for planners, as do the needs of First Nations communities. But it is not just problems that are being highlighted: solutions are, too, for more effective collaborations among planning stakeholders and physical designs to attain more effective and satisfying neighbourhood living are being advocated.

This is a vibrant period for community planning. The challenges listed below could be seen as obstacles, but they also constitute opportunities for all who are engaged in community planning to expand the scope and enrich the substance of this vital public activity.

- **Citizen participation** in deciding planning issues would seem to be in accord with the basic value of democracy that planners have espoused since the beginning, yet it is far from being wholeheartedly the practice in Canadian community planning.[15] Participation and participants are central to Chapters 11 and 12.
- **Ecological planning**, or incorporating planning for the natural environment into our physical planning, is still a worthy if elusive goal after a hundred years and needs to become an integral part of the design of communities henceforth.[16]
- **Redeeming place**, or planning communities in which the inhabitants' communitarian, ecological, and aesthetic needs are integral to the final outcome, by re-creating neighbourhoods and stopping "mindless sprawl," is central to the urging of both New Urbanism and bioregionalism advocates.[17]
- **Cultural diversity**, where planning meets multiculturalism, has risen high on the planning agendas of many communities, especially the "gateway" cities such as Vancouver, Montreal, and Toronto, and challenges planners to move beyond simple awareness and sensitivity in their planning.[18]
- **Population aging**, already a prominent feature of community life, will dramatically increase in all types of communities in all parts of the country and will demand a sensitivity of plan-makers to facilitate the continuing independence of seniors to be able to carry out their daily activities and get around in the community.[19]
- **Safety and security** in communities in terms of both crime and the needs of physically impaired persons and the elderly, as basic as they seem, are not something to which most plans and plan-makers have been attentive, as recent literature reminds us.[20]
- **Globalization** is already having an impact on planning methods, research, and practice and is bound to have more with the increasing interconnectedness of people, places, institutions, and economies, with implications for sustainable development, communications, and immigration, among others.[21]

Structure and Resources of the Book

The basic premise of this book is that community planning in Canada, or elsewhere for that matter, is best understood by looking at, first, why and what is pursued in community planning; second, how community planning is done; and, third, who gets involved in a community's planning. These facets of community planning—the principles, practice, and participants—provide the focus for this book.

The Roots of Canadian Planning

The remaining chapters in Part One examine the evolution of Canadian planning from earliest settlement to the present time. Community planning in Canada has roots in long-standing ideals, ideas, and principles about how a community should develop and how it should promote the public's interest in the process of development. The next four chapters consider the legacies of the **physical forms** we find in Canadian communities today. Chapter 2 looks at how cities began and what influenced the forms planners gave them up through the 18th century. The foundations that began to be laid down in the 19th century for Canadian communities, especially regarding concerns over their appearance, living conditions, functioning, and natural environments are broached in Chapter 3. The emergence of new planning concepts and professional practice to further these latter concerns in the period 1900–45 is the focus of Chapter 4. The planning concepts and processes we know today and how they have evolved since 1945 are examined in Chapter 5.

Community Plan-Making in Canada

Part Two describes the practice of community plan-making in Canada at the present time. When community planning goes beyond ideal concepts and social concerns, a mode of practice develops to deal with elements of the environment being planned and to provide the organizational arrangements to conduct public planning. Contemporary community planning in Canada has well-developed modes of practice such that planning activities differ little from one part of the country to the other. The five chapters in this part describe, progressively, the planning for urban communities, regional and metropolitan communities, and small towns in Canada. The **built environment** and the way planners and others view it initiates the discussion (Chapter 6); this is followed, in Chapter 7, by a description of the steps and studies necessary in **making a community plan.** The **scope and role of a community plan,** the centrepiece of planning practice, are described for an **urban community** in Chapter 8. The final two chapters in this section focus on plan-making in two contrasting spatial settings: first, in large communities, metropolitan areas and non-urban regions (Chapter 9), and in quite small ones, such as towns and villages (Chapter 10).

Participants and Participation in Community Plan-Making

Part Three identifies the participants involved in making a community plan. Community planning obliges us to use collective decision-making processes because of the wide array of individuals and groups who have an interest in the future of the community. Procedures are usually conducted within a governmental framework and may be both formal and informal in nature. They, in turn, define who may participate in planning decisions and how and when decisions need to be made. Two chapters describe the texture and process of community planning decision making. First, in Chapter 11, there is an overview of the sequence of decisions responding to development initiatives and in making a plan, including the main actors in these initiatives. Second, Chapter 12 provides an in-depth view of the main participants—from planners to politicians to developers and citizens—and the dynamics of their interaction.

Tools for Implementing Community Plans

Most planners would argue that, while it is essential to prepare a plan, that is only the beginning. It is equally important to design and apply appropriate *tools* for guiding the decisions of participants so that the plan will be achieved or implemented. The chapters in Part Four examine the tools and approaches used in implementing the community plan. These follow two trajectories: the first, described in Chapter 13, are those involved in regulating land uses in a community, such as zoning and other land-use controls that guide the decisions of private landowners; the second are those initiatives that can be taken by the public policy-makers of the community in budgeting for public facilities, building regulations, and other development incentives as examined in Chapter 14.

Anticipating the Future for Community Planning

Community planning is a task that evolves as society evolves. Cities and towns continue to develop, informed by both their past and their future. New problems arise along with new versions of old problems. Thus, the true progress of community planning is as much a measure of how it deals with existing problems as how well it anticipates future conditions. The final chapter discusses a number of salient issues that will challenge Canadian community plan-makers in the future, including issues still in need of resolution, such as citizen participation and environmental protection. It also discusses issues that will become more central in the years ahead, such as sustainability, the population's aging, and its increasingly multicultural character, as well as the future of the suburbs.

Learning Resources

In addition to the main text, supplementary resources are provided with each chapter to expand the scope of learning about community planning. They proceed increasingly outward, from guiding questions within the text to stand-alone cases of contemporary planning issues in Canadian communities and to planning resources available on the Internet.

Guiding Questions. As one proceeds through the book, two leading questions are provided in each chapter. The first, at the outset of the chapter, points the general direction in which the chapter will unfold, while the second, at the chapter's end, reflects on the foregoing discussion and its relation to material to come in the next and succeeding chapters.

Planning Issues. Community planning should always be seen as a part of the human dynamic. It does not take place in books but in council chambers, neighbourhood halls, and sometimes in the streets. It is intimately involved with the politics of the community. Some of the drama associated with community planning is introduced here by a series of case studies, called Planning Issues, which depict contemporary planning stories from across Canada. Although these are necessarily "snapshots" at a particular point in time, and some may now be resolved, they are indicative of the planning issues that communities like yours and mine encounter.

Internet Resources. The Internet is already rich in resources that can be used for further study in community planning as well as in actual community-planning practice. Each chapter thus concludes with a selection of websites relevant to the subject of the chapter.

Highlights of Canadian Community Planning

The evolution of community planning in Canada may be quickly grasped in the chronological listing of important events and achievements from the 1890s to the present time provided in the appendix.

To reiterate, community planning, essentially, is about providing an appropriate physical foundation for community life. This is where our discussion begins in the next chapter. As you proceed, keep in mind the following question:

- *What community needs have planners found, time and again, they must provide for in the physical foundation of cities and towns?*

Endnotes

1. An older, but still useful source of Canadian community planning initiatives (and one which originally suggested this list) is Canadian National Committee for Habitat, *The Canadian Settlements Sampler* (Ottawa: Community Planning Press, 1976), especially 14–43. A current source of successful planning activities is the website of the Canadian Institute of Planners, available online at www.cip-icu.ca.
2. This is well argued in, among others, Leonie Sandercock, *Toward Cosmopolis: Planning for Multicultural Cities* (London: John Wiley & Sons, 1997).
3. Judith E. Innes, "Planning through Consensus Building," *Journal of the American Planning Association* 62:4 (Autumn 1996), 460–472.
4. Jill Grant, "Rethinking the Public Interest as a Planning Concept," *Plan Canada* 45:2 (2005), 48–50.
5. Sandercock, *Toward Cosmopolis.*
6. From Aristotle's *Politics,* in a 1598 translation, as cited in Helen Rosenau, *The Ideal City* (London: Studio Vista, 1974), 12.
7. David Riesman, "Some Observations on Community Plans and Utopia," *Yale Law Journal* 57 (December 1947).
8. Gordon Stephenson, "Some Thoughts on the Planning of Metropolitan Regions," *Papers of the Regional Science Association* 4 (1958), 27–38.
9. The story of this remarkable plan is described in David L.A. Gordon, "A City Beautiful Plan for Canada's Capital: Edward Bennett and the 1915 Plan for Ottawa and Hull," *Planning Perspectives* 13 (1998), 275–300. It and its antecedents are elaborated in David Gordon, ed., *Planning Twentieth-Century Capital Cities* (London: Routledge 2006).
10. Joachim Domingo Roriz, "Brasilia: A National Project for Development and Integration," *Plan Canada* 40:3 (April–May 2000), 16–17.
11. Canada, Advisory Committee on Reconstruction, *Housing and Community Planning,* vol. 4 (Ottawa:King's Printer, March 1944), Report of the Subcommittee.
12. Ibid., 178.
13. Thomas Adams, "What Town Planning Really Means," *The Canadian Municipal Journal* 10 (July 1914).
14. Humphrey Carver, *Compassionate Landscape* (Toronto: University of Toronto Press, 1975), 90.
15. Among many recent Canadian articles on citizen participation are: Noel Keough, "Calgary's Citizen-Led Community Sustainability Indicators Project," *Plan Canada* 43:1 (January–March 2003), 35–36; John Blakney, "Citizen Bane," *Plan Canada* 37:3 (May 1997), 12–17; and Beth Sanders, "View From The Forks: Coming to Terms with Perceptions of Public Participation," *Plan Canada* 38:2 (March 1998), 30–32.
16. John Newton, "Exploring Linkages Between Community Planning and Natural Hazard Mitigation in Ontario," *Plan Canada* 44:1 (January–March 2004), 22–24; and Mary-Ellen Tyler, "Ecological Planning in an Age of Myth-Information," *Plan Canada* 40:1 (December 1999–January 2000), 20–21.
17. Kirsty MacDonald, "Turning Alleys into Assets," *Plan Canada* 44:2 (April–June 2004), 48–51; and Andrea Gabor and Frank Lewinberg, "New Urbanism," *Plan Canada* 37:4 (July 1997), 12–17.
18. Mohammad Qadeer, "Dealing with Ethnic Enclaves Demands Sensitivity and Pragmatism," *Ontario Planning Journal* 20:1 (2005), 10–11; and Sandeep Kumar and Bonica Leung, "Formation of an Ethnic Enclave: Process and Motivations," *Plan Canada* 45:2 (Summer 2005), 43–45.
19. Gerald Hodge, *The Seniors' Surge: The Geography of Aging in Canada* (Vancouver: UBC Press, forthcoming 2007); and Mary Catherine Mehak, "New Urbanism and Aging in Place," *Plan Canada* 42:1 (January–March 2002), 21–23.
20. Gerda Wekerle, "From Eyes on the Street to Safe Cities," *Places* 13:1 (Winter 2000), 44–49; Rob Imrie, "Barriered and Bounded Places and the Spatialities of Disability," *Urban Studies* 38:2 (February 2001), 231–237; and Hester Parr and Ruth Butler, "New Geographies of Illness, Impairment and Disability," in H. Parr and R. Butler, eds., *Mind and Body Spaces* (London: Routledge 1999), 1–24.
21. Manuel Castells, "Planning in the Information Age," *Plan Canada* 40:1 (December 1999–January 2000), 18–19; and Farokh Afshar and Keith Pezzoli, "Integrating Globalization and Planning," *Journal of Planning Education and Research* 20 (Spring 2001), 277–280.

Internet Resources

There are several websites that are important in themselves and also act as gateway sites for community planners to an array of links to other relevant sites. The list begins with Canadian sites and continues with those in the United States. URL addresses are current as of the year of publication; however, they may change. (Readers are reminded that, in some cases, they may need to revert to the main address (before the "/" slashes) and then follow appropriate links to the desired site.)

Chapter-Relevant Sites

Planning Canadian Communities
www.planningcanadiancommunities.ca

The Canadian Institute of Planners
www.cip-icu.ca

The Canadian Urban Institute
www.canurb.com

Canada Mortgage and Housing Corporation
www.cmhc-schl.gc.ca

United Nations Habitat
www.unhabitat.org

International Society of City and Regional Planners (ISoCaRP)
www.isocarp.org

Cyburbia—The Planning Community
www.cyburbia.org

2

Chapter Two

The Beginnings of Today's Cities

Such is the tenacity of these simple geometric forms, the circle, the straight line, and the right angle. They survive because of their adaptability.

Hans Blumenfeld, 1943

The communities we know today are part of a heritage of city- and town-building knowledge stretching back several millennia. From Mesopotamia and the Nile Valley to China, India, Greece, Rome, and Renaissance Europe there is vast experience in community planning available. Our geographic and historic circumstances may differ, but we can learn from the achievements and problems of past planners and the form and function of the communities they designed. These European precedents can be found in the form of the 17th- and 18th-century Canadian communities, while our 19th-century cities began to benefit from planning concepts aimed at righting the ills of the emerging industrial city. In short, those involved in planning a Canadian community today are heirs of planning decisions and planning ideas of the past. Drawing from this repertoire of planning thought and action, consider the following question:

- ***What are the major physical development issues in community building that planners of the past faced and planners today still face?***

Perennial Planning Issues

Over the 7000 years since humankind began designing communities, six important issues have recurred for planners. These issues need to be addressed either when building a wholly new community or expanding an existing one.[1] They are fundamental to much of what this book is about and need to be reviewed now. The queries that accompany each are key ones for planners to consider, but there will always be others, depending upon the community and the era:

1. **The selection of the site.** Will it be a hilltop or valley, an island or cape, a harbour, or some other transportation advantage? How will it fit with its natural environment?
2. **The function (or purpose) of the community.** What needs is the settlement intended to satisfy? Is it to provide protection, facilitate commerce, act as a political or religious focal point?
3. **The form of the community**. What are the aspirations of the populace, the aesthetic considerations of the culture, the functional needs to be served?
4. **The allocation of land uses.** Where will the people live, do business, govern, worship, congregate for public activities?
5. **The need for connection**. How will people and goods circulate and be exchanged among various land uses?
6. **Accommodating growth and change**. Should the community be more intensively developed within its present confines or be extended on the periphery if it needs to grow?

Although planners must acknowledge all six of these issues, the importance they accord to each will vary depending upon a number of things: the community's values, the human needs, the nature of the site, the technology available (especially for building and circulation), and the economic and other functions. Once decisions are made and carried out in any of these issue areas, the physical shape and arrangement of the community will be influenced through all succeeding periods. Consider these points in regard to the various historical cases of city building and planning described throughout the remainder of this chapter and in those that follow.

Cities of the Ancient World

Our best archeological research about the beginning of cities shows that substantial settlements were being built about 5000 BCE. There were village settlements long before this, and villages still abound, but these simple groupings of more or less equal social units are not cities. As eminent Canadian planner Hans Blumenfeld has noted, "only where the plurality of social units of a villages is combined with the social and functional differentiation found in the castle can we talk of a city."[2] Thus, cities emerge when a society begins to distinguish such social needs as defence, promotion of worship, or symbolizing political control in a region, and then combines these functions with the need to house a large population nearby. The effort to combine these functions at a particular location constitutes the earliest community planning.

The archeological record identifies five major clusters of ancient cities:

- Mesopotamia from ca. 5000 BCE to perhaps 1800 BCE
- The Nile Valley from ca. 4500 BCE to perhaps 2100 BCE
- The Indus Valley from ca. 2300 BCE to 1800 BCE
- China from ca. 2500 BCE to today
- Central and South America from ca. 1000 BCE to AD 1500

While the archeological research is fairly clear about the location and age of these ancient cities, there is no consensus about *why* the villages evolved into these early urban civilizations. The conventional view is that more efficient agricultural methods allowed some settlements to support activities other than subsistence farming through their surplus production.[3] And all of these civilizations did have substantial agricultural production, from Mesopotamian wheat to Aztec corn. But this "agriculture-first" view of urban civilization has been challenged by new theories suggesting the importance of economic creativity in urban development.[4] We will encounter these ideas again in later chapters that discuss planning post-industrial Canadian cities.

Cities of Mesopotamia

The cities of the Tigris and Euphrates Valleys, especially Babylon, provide good examples of planning in ancient times. Babylon (see Figure 2.1) was located in the Euphrates Valley (about 80 kilometres south of present-day Baghdad, Iraq). Its location was a "cape" in the river, which probably provided a good site for defence, transportation, and water supply. The river bisected the city and a wall encircled both sides enclosing an area of about 2 square kilometres. Powerful emperors lived in Babylon, and during their tenure the city was embellished in many ways. The most famous emperor, Nebuchadnezzar (625–551 BCE), as well as building the renowned Hanging Gardens, was said by the Greek writer Herodotus to have paved the great processional avenue with "limestone flags."[5]

Of the dwellings of common folk (the population of Babylon is reputed to have reached close to 10 000 at this time), little can be said. They would have occupied the bulk of the land in the city, but the residential areas seem not to have had any regular layout. Dwellings were arranged within blocks of land perhaps a hectare in size. The housing blocks were not regular in size or shape, and thus the spaces between them, which served as streets, were irregular.

Two factors account for the amorphous form of the interior of these cities. The first is that people moved around mostly on foot (wagons or carts were an exception). The second factor is that the commerce and manufacturing, and even a good deal of the agriculture, were conducted within the dwellings or land in the housing blocks. Streets were not required to provide easy connections throughout the city for the exchange of goods or for people to get to jobs. This general pattern of cities, comprising dominant public areas, a few major roads leading to monuments and gates, and rather undifferentiated residential areas all within a city wall, persisted until the Middle Ages in Europe.

The Nile Valley

Ancient Egyptian cities were quite different from the walled Mesopotamian fortresses. The Nile Valley was dotted with cities constructed as monuments to the Pharaohs; the desert, mountains, and the Egyptian army were the defences against invaders. Although a temple city such as Memphis was settled for 1500 years, some royal capitals, such as Akhetatan (1369–1354 BCE), might be occupied for only 15 years while the Pharaoh's monuments were built. These cities had the semiformal planning of a long-term construction camp, although some of the workers' villages were laid out in a tight grid pattern. These disappeared after the death of the Pharaoh, leaving the monuments to be maintained by priests.[6]

The Indus Valley

Urban civilization on the Indian subcontinent appears to have emerged in the Indus River basin almost 4200 years ago. Cities such as Harappa and Mohenjo Daro were laid out with their important streets in a basic grid, aligned in a north–south orientation (see Figure 2.2). Almost all housing was in the form of narrow, attached dwellings; detached houses were quite rare in ancient cities around the world. These cities held up to 30 000 people at their peak, but were mysteriously abandoned after about 500 years. But ancient Indian architectural manuals continued to promote strong geometric forms, based on Mandala, which influenced the plans of future cities.[7]

Figure 2.1 Babylon, 6th Century BCE

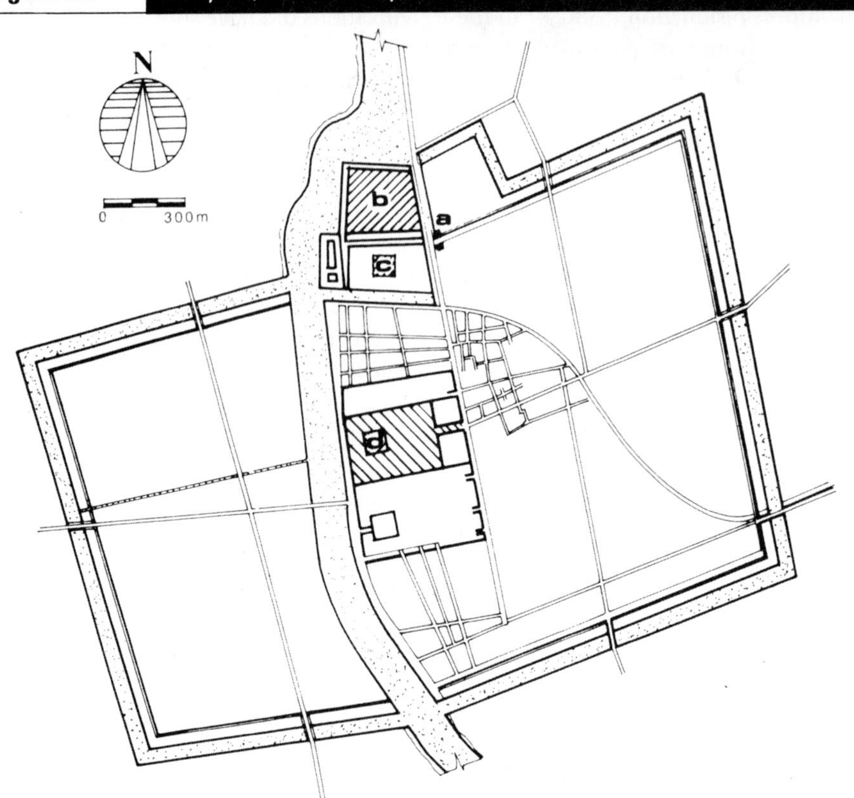

In the time of Nebuchadnezzar, Babylon spanned the Euphrates River, and its walls and moat encompassed an area of about 3 km^2. Its population was around 10 000. Key: (a) main gate, (b) fort, (c) Hanging Gardens, (d) temple.

China

Ancient Chinese community planning dates back as far as the Mesopotamian period. Danish planner Steen Eiler Rasmussen tells us that when Emperor Kublai Khan built his capital in 1268, he chose a location where "five different towns had existed, one after the other, over a period of thousands of years; each one perfectly rectangular and each one oriented exactly north–south and east–west."[8] This precise orientation follows the ancient philosophical principle of *feng shui*, "wind and water," which promotes harmony between various pairs of physical factors within a geographic framework related to the sun. Early Chinese cities had fortified walls with a main gate on the south side. There was usually a great processional road leading from the gate directly north to the entrance of the ruler's palace or to a temple. Gates and main streets often led in the other three directions. The interior of the city was subdivided into square blocks of land with great families often segregated into special city districts with their own walls.[9] Lanes between housing blocks were narrow and not arranged in a uniform pattern. Housing blocks tended to be square (see Figure 2.3).

The Americas

On the other side of the world, the Aztecs, Mayans, and Inca were building elaborate cities in Central and South America, in places like the Yucatan, Central Mexico, Andean Bolivia, and Peru. The approach of the Incas to planning their communities is best seen

Figure 2.2 Mohenjo Daro, India, ca. 2000 BCE

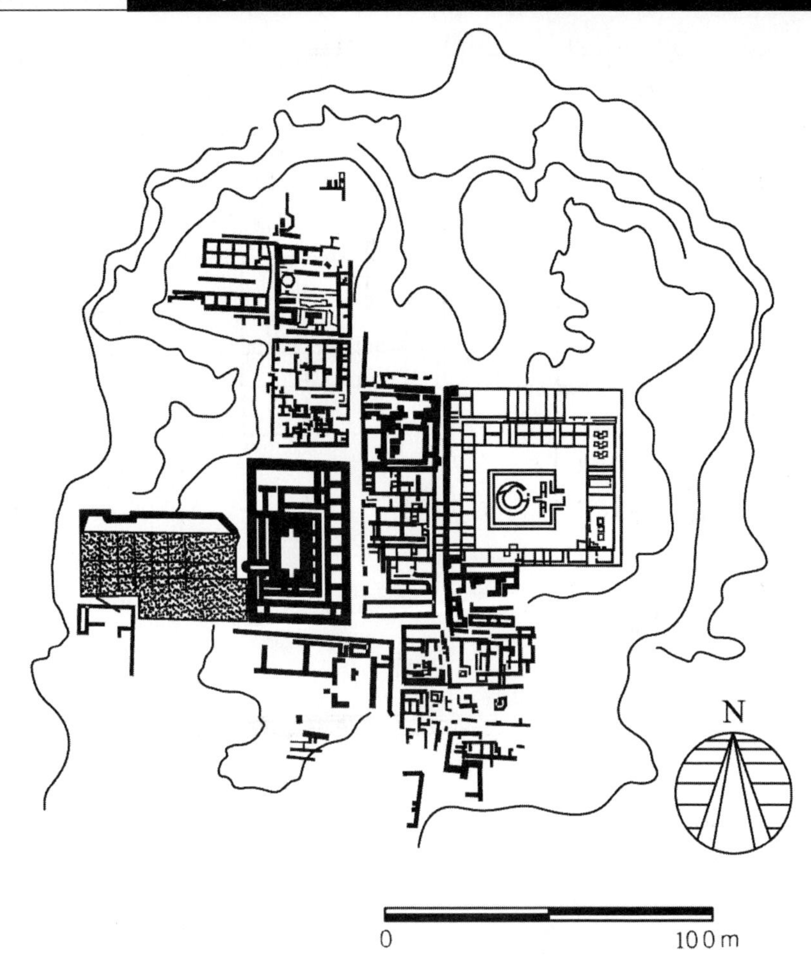

The regular geometric layout and orientation of the ancient cities of the Indus Valley are probably the earliest examples of deliberate urban planning for an entire community. Both the citadel, shown here, and the lower town were based on a grid of streets aligned north–south, probably for religious reasons.

in Cuzco, Peru. As American planner Francis Violich describes it, the Incan plan for Cuzco was divided into four sections, with a main plaza at the junction of the four main roads, each of which led to the four great regions of the Empire. Minor plazas were located elsewhere within the city walls and "each city block was an allotment for one city family; its distance from the centre of the city depended upon the degree of relationship to the Inca ruler."[10] Some of the more remote Incan cities, like Macchu Picchu, occupied mountainous sites, quite different from the fertile valleys of the Asian civilizations. But the Aztec, Mayan, and Incan cities were largely destroyed during the Spanish conquest of the 16th century, so these ancient urban forms had little influence upon North American urban planning following European settlement.

Greek and Roman City Planning

European city development was extensively influenced by the urban civilizations of ancient Greece and Rome. Sir Peter Hall credits the golden age of Athens (500–400 BCE) with establishing the philosophic, scientific, and artistic foundations of Western

Figure 2.3 Khara-Khoto, Early Chinese Colony

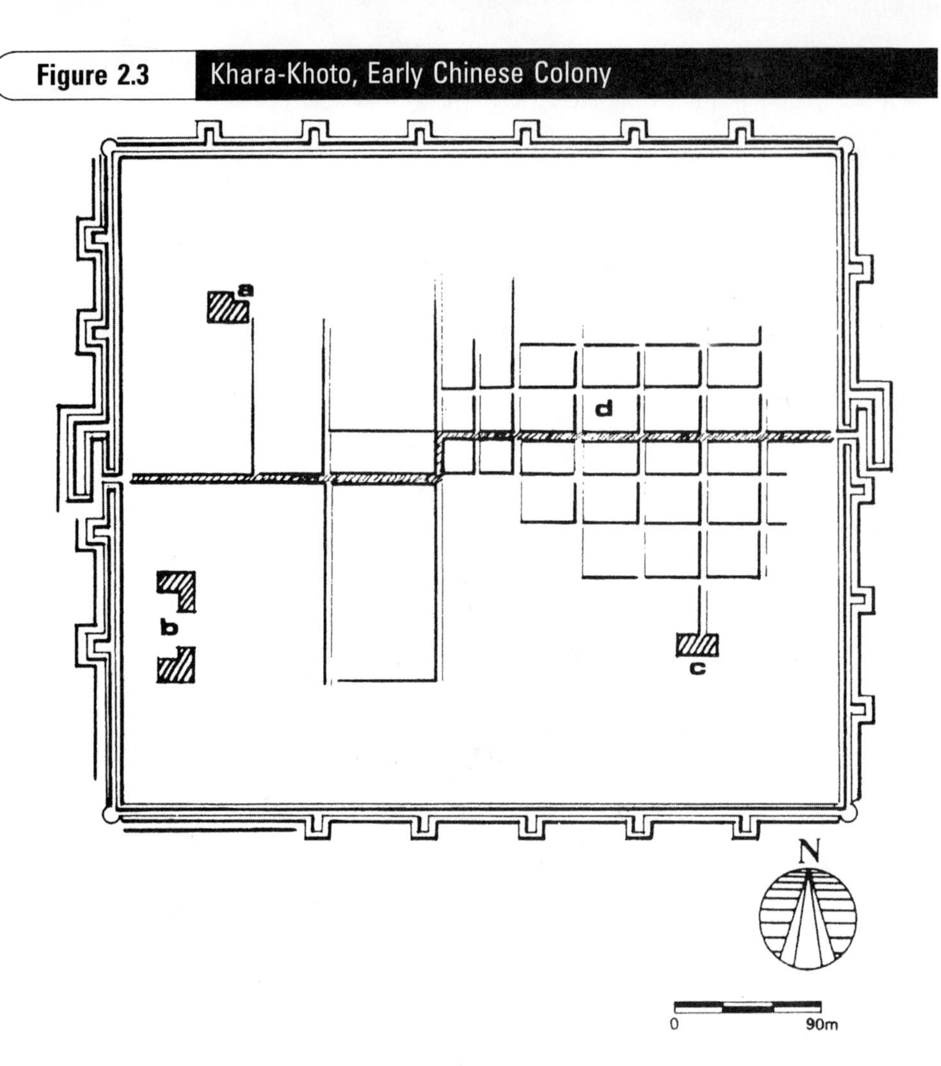

This "planted town" in Central Asia employs the planning geometry used in ancient cities in China, with walls and roads oriented precisely north–south and east–west. Roads from the gates are offset to prevent evil spirits from passing through. The town walls enclosed about 0.2 square kilometres and held a population of 4000–5000. Key: (a) tomb, (b) temple, (c) palace, (d) residential compounds.

culture.[11] The ancient Greeks were systematic planners of new communities and the Romans were extraordinary engineers, pioneering urban infrastructure techniques that would allow their imperial capital to grow to the largest city the world had ever known.[12]

The Greek Planning Tradition

The experience of the ancient Greeks is important in the history of city development and planning. Greek architects and builders, especially Hippodamus, are usually credited with originating the **gridiron** street pattern, although this is not entirely accurate, as we shall see. However, Greek architects and builders probably should be credited with the first comprehensive community plans.

In the Greek community plan, the allocation of land to the individual dwelling was the common denominator. Rectangular blocks of land were arranged to allow the dwellings within them to be oriented to the southern sun, a necessity for houses lacking central heating in the more northerly areas occupied by the Greeks. The location of public buildings and spaces determined the location of major streets. The *agora*, or marketplace, was usually built on one side of a main street, although not necessarily in the geographical centre of the city. Main streets also served the temple, the *pnyx*, an open-air podium where citizens met to consider affairs of state, the theatre, and the stadium.[13] It should be noted that main streets were often only 7 to 9 metres wide and local streets were the 3- to 5-metre spaces left between the blocks of housing. Thus, the impression of gridiron street plans for Greek cities is only partly true; the rectangular blocks of land given over to housing, rather than the streets, determined the geometric pattern. Most towns were surrounded by protective walls.

The "classic city" of the Greeks, as these plans are often called, was built by the score around the shores of the eastern Mediterranean, into Asia Minor, and on the north coast of Africa. When the Greek colonizers sought sites for their new cities, defence and access to the sea were important; thus, they chose such sites as craggy headlands on the ocean. This meant superimposing the geometrical Hippodamian street pattern on rugged sites, with the result that numerous streets were so steep as to be built only as steps. The pattern of streets may resemble today's gridiron, but it is worth remembering that most were not thoroughfares for vehicles. The plan for the town of Priene (see Figure 2.4) is a good example of classic Greek city building.

Towns and Cities of the Romans

From the 2nd century BCE, Rome succeeded Greece as the dominant European power. Over the next 500 years, the Romans extended their control from Britain to Algiers, and from Constantinople to the Danube Valley. In this period, they made two contributions to the development of community planning. The first was with the plans for colonial towns. The second was in the planning for the large cities that developed on the Italian peninsula.

In order to sustain power in distant lands, garrison towns were established and connected with roads to facilitate movement of the Roman legions. Thus, the old saying "All roads lead to Rome" aptly captures the extent of Roman domination. Hundreds of today's towns and cities of Western Europe owe their location to the sites the Romans chose for their garrisons. The Roman term *castrum*, meaning a military encampment, is found as "chester" in English town names; all such places originated from Roman camps.

The Roman garrison towns (or *coloniae*) were frequently used to deploy discharged soldiers from the huge armies of the time; all followed a similar pattern. Often rectangular in shape, they were surrounded by walls and bisected by two main streets that met at or near the centre. At this location was the forum. House blocks, or *insulae*, of 0.5 to 1.5 hectares, either square or oblong, were fairly uniform throughout the area enclosed by the walls. The two main streets were usually oriented so that one ran north–south and the other east–west; it has been conjectured that the latter street pointed toward the spot on the horizon where the sun rose on an important ceremonial day for the town. Although size varied among such towns, they were usually less than 2.6 square kilometres in size and often less than 20 hectares. The population of the larger garrison towns seldom exceeded 10 000 (see Figure 2.5).

Such home cities of the Romans as Naples, Pompeii, and Rome grew from early villages without benefit of the deliberate, overall plan used in building their colonial towns. These domestic Roman cities also represent the earliest known examples of planning and managing the development of cities of considerable size. Pompeii, a small city, reached a population of 20 000 by 79 CE; Ostia, the port city of Rome, is estimated at 50 000 by the 3rd century CE, and Rome itself reached upward of 1.5 million about the same time. To be able to house populations of this size requires methods of

Figure 2.4 Greek City: Priene, 4th Century BCE

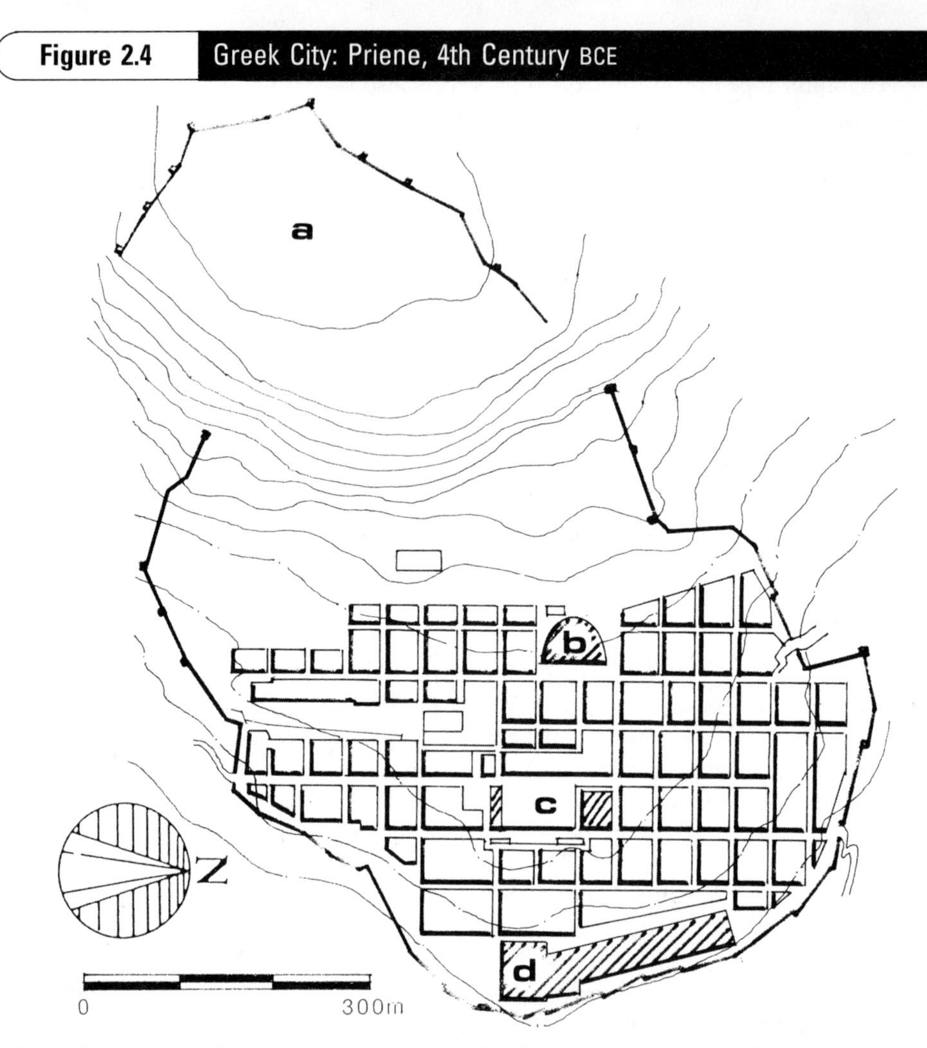

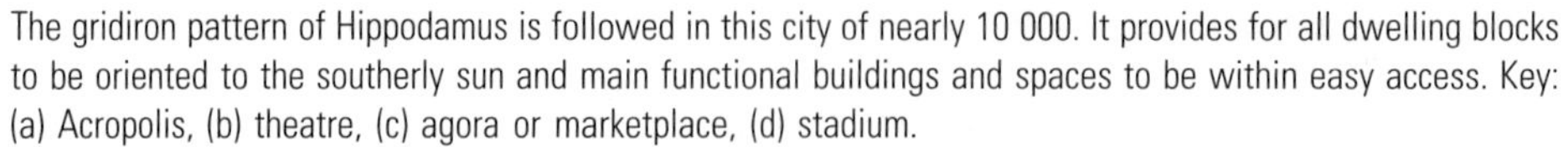
The gridiron pattern of Hippodamus is followed in this city of nearly 10 000. It provides for all dwelling blocks to be oriented to the southerly sun and main functional buildings and spaces to be within easy access. Key: (a) Acropolis, (b) theatre, (c) agora or marketplace, (d) stadium.

distributing large amounts of water, providing for drainage and sewage, and moving people and goods, as well as taking into account shopping and cultural needs.

The Romans solved the technical problems created by their growing cities in unprecedented, and often dramatic, ways. Large aqueducts were built to supply water; Rome needed 14 aqueducts. Underground sewers were constructed that are still considered feats of engineering, and the 29 main highways leading to other parts of the Empire were, in most cases, paved.[14] Canadian planner Robert McCabe's studies show that the Romans provided for the shopping needs of citizens in what corresponds to our present-day shopping centres. In conjunction with a city's forum, a macellum was constructed, a structure for many shops, often on two or more levels. In Rome, the shopping centre portion of the main forum of the city had about 170 shops and covered almost 18 600 square metres.[15] Such a centre compares to the shopping centres built in many Canadian communities. Smaller planned shopping centres were also built in the various regions (districts) of Rome to provide for "neighbourhood" shopping needs.

The large Roman cities were, undoubtedly, complex communities, in many ways rivalling modern cities. This can be grasped by comparing the population densities. Pompeii

Figure 2.5 Ancient Roman Garrison: Silchester, England

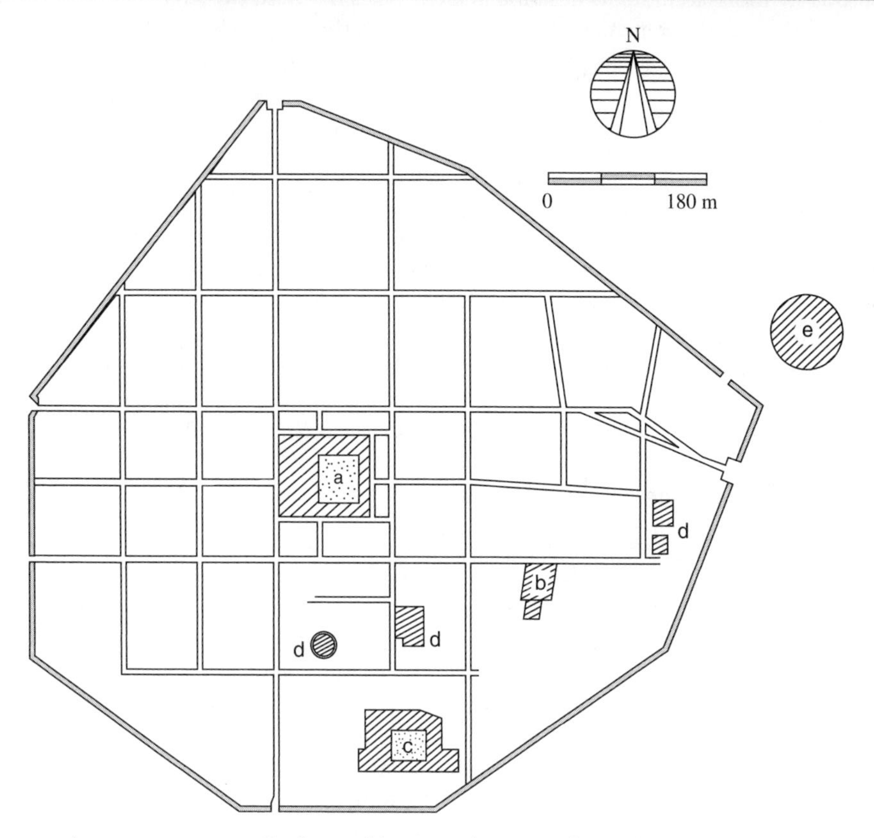

Roman garrison towns were usually dissected by two main streets, the *cardo* running north–south and the *decumanus* running east–west; the forum, with its shops, temples, and public buildings was near their intersection. The population probably did not exceed 5000. Key: (a) forum, (b) baths, (c) inn, (d) temple, (e) amphitheatre.

occupied about 85 hectares and had 20 000 residents; Rome occupied about 26 square kilometres and had, possibly, 1.5 million residents. These represent densities of around 50 000 persons per square kilometre. Toronto, in its central area, has densities under 8000 persons per square kilometre and densities of less than half that in its suburbs.

Ancient Cities in Summary

Cities in the ancient empires were established much like the "new towns" of modern times; they were often built where no previous settlement had been, or perhaps only a very small one. To quote Rasmussen again:

> [When] faced with the necessity of creating a new town in a strange place they must build it according to a preconceived plan or it will end in chaos. And that plan must, of necessity, be a very simple one, easily laid out, so that everyone with the least possible trouble can quickly discover what he has to do.[16]

The resulting physical form of cities until the 14th century shows two fundamental characteristics. First, land uses are arranged according to a "block plan," to employ Blumenfeld's term; that is, **blocks** of land devoted to public and residential uses, rather than streets, dictated the form.[17] Second, a surrounding wall often limited the community's growth and sometimes created a radial pattern to main streets providing access to gates, temples, and markets. This is the first evidence of the **radial-concentric** physical form. It is probably more accurate to describe ancient cities by these two characteristics than to compare them with later cities, where streets and traffic are much more crucial elements.

Origins of the Modern European City

It is important not to see these examples of city-building in the past as static concepts, or simply as maps. They were functioning communities that existed over long periods of time, and changed significantly during their history: shifting the allocations of land, adding new buildings, making new road connections, even expanding in many cases. They were "continuing" cities, to use the key word from the title of Vance's book.[18] The evolution that occurs in city form and composition should become even more evident as we proceed to describe more modern instances in this section. It is a necessary perspective when one considers not only where a city has come from but also what it might become, for you can be sure it will continue to change in the future.

From the Dark Ages to the Middle Ages

The several hundred years after the fall of Rome in the fifth century, the period called the Dark Ages, were characterized by a general social formlessness and a lack of large settlements in Europe. The political stability and protection provided by the empires of the Greeks and Romans ceased, and large cities could not be maintained; neither their supply of food nor their trade and other basic needs could be assured. The settlement pattern of Europe reverted to one of villages.

As strong kings, princes, and bishops emerged to provide security over larger regions, from the 9th century onward, European town life greatly expanded. Many towns and cities began from modest roots in the feudal countryside where, often, some dwellings and a market had been established at the gate of a castle or monastery, leading to the extension of the castle walls. Many new towns were also established in this period.

The urban historian A.E.J. Morris classifies the towns resulting from the urban expansion of the Middle Ages as being either **organic growth towns** or **new towns.**[19] The former were those places that grew on the sites of old Roman towns, on the sites of fortified feudal villages, or on the sites of agricultural villages. London, Cologne, and Paris grew from such roots and greatly expanded in this period. The physical form of organic growth towns is partly conveyed by the name—that is, they generally lack any overall pattern. Except for a few main roads, the streets meander, and blocks of land are irregular in shape, as the plan of Carcassonne shows (See Figure 2.6). One still finds a residue of this amorphous medieval development at the centre of many large European cities; there are even a few examples in North America, such as the "Lower Town" in Quebec City.

Morris subdivides the new towns of the Middle Ages into two types: **bastides** and **planted towns**. Both were used to consolidate territorial control by a king or ruling house. Bastides, as the name suggests, were bastions, fortified towns built to a predetermined plan. They employed a clear pattern for streets and blocks of land within rectangular or circular walls. Bastides were built mostly in the 13th century, in France, Wales, and England by Edward I, and by other rulers in Germany and Bohemia. Planted towns were new towns developed to promote trade as well as protect territory; New Brandenburg (See Figure 2.7) is an example of one such town. Cities such as Londonderry in Ireland and Berne in Switzerland owe their existence to this new form of town development.

Moreover, bastides and planted towns would prove to be models for cities in the New World a century or so later. When bastides were first built, there was a need to attract settlers to these regions, and this was done in part by providing the newcomers

Figure 2.6 13th-Century Carcassonne, France

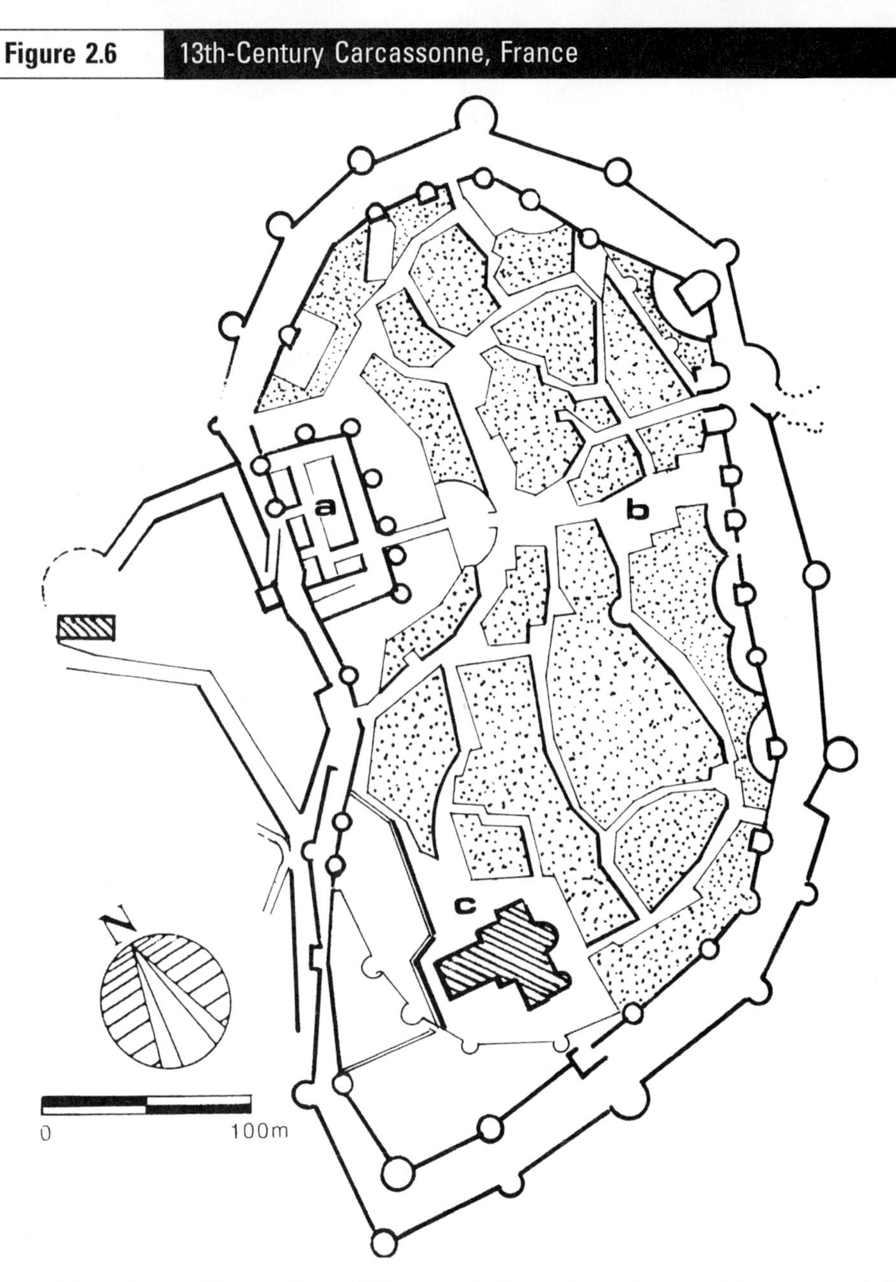

A typical "organic growth" town of the Middle Ages: the few main streets connect the gates, castle (a), church (c), and market (b) as needed, and the overall pattern is irregular.

with plots of land inside and outside the walls. Land-use allocation was much more egalitarian in the bastides than in earlier settlements. Their general gridiron pattern facilitated this, and the English, in particular, would come to use this model in their North American colonies.[20]

Whether new towns or those developed from earlier settlements, medieval towns were alike in many respects. The component parts were the wall and its gates, the marketplace, often with a market hall and commercial buildings, the church, and, sometimes, the castle. Main streets connected the main buildings and the market with the gates, and joined main areas together. The rest of the land within the wall was allocated in blocks for residences and gardens. When land was needed for, say, a market or church

Figure 2.7 Fortified German Town: New Brandenburg, 1248

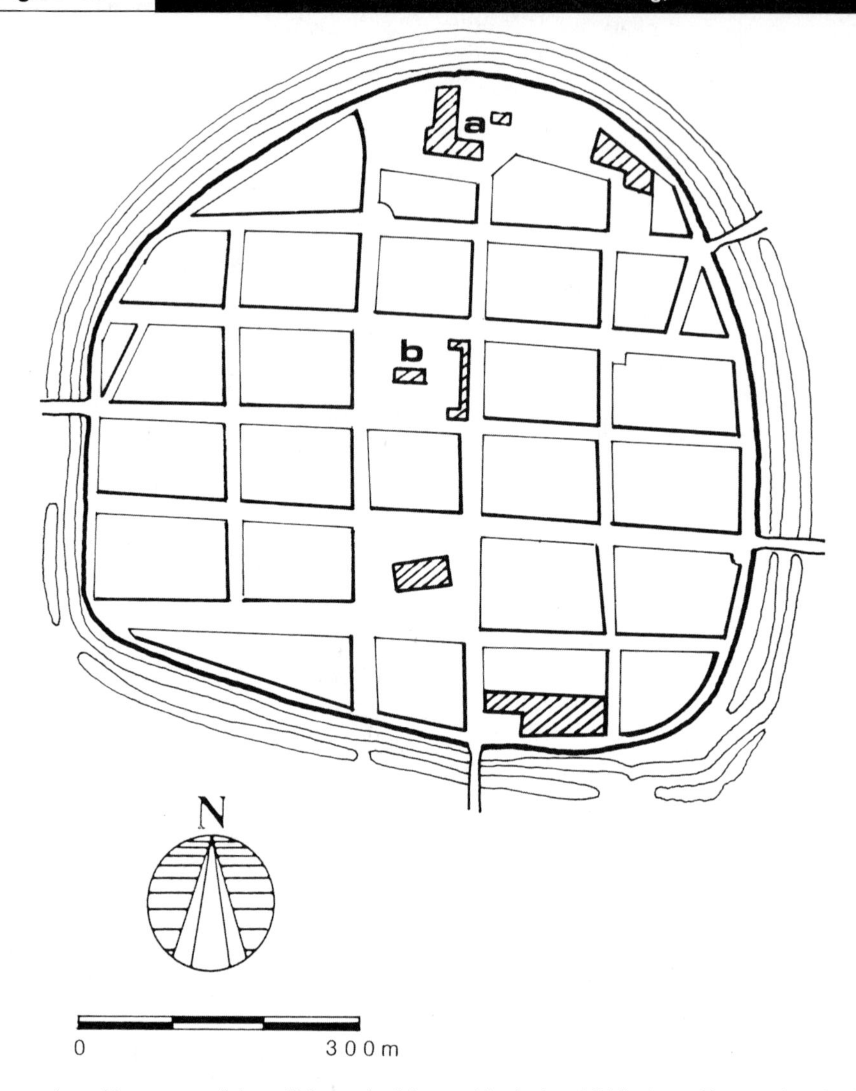

Towns such as this one were "planted" for territorial control in the late Middle Ages. They employed a fairly regular gridiron street pattern. Key: (a) church, (b) market square.

square, either a block of land was left vacant or a street was widened at that point. Most streets were narrow, for movement was largely on foot.

During the town-building of the Middle Ages, material well-being also improved. Tradesmen, artisans, and merchants in many towns became prosperous and powerful enough to challenge the local rulers and gain citizenship. Especially in northern Europe, they often built guildhalls and a town hall adjacent to the market area of the town to signify the importance of the "burghers," or citizens of the "burg." In the long run of history, these steps to allocate the space of the city among king, church, and citizens were decisive to the development of a modern, segmented, urban way of life.

Renaissance and Baroque Cities: The Importance of Design

The period from 1400 to 1800, which encompasses the Renaissance and Baroque eras, is an influential one in regard to the form of cities today, for this is the period when the compact settlements of medieval Europe were dramatically extended, refined, and restructured. The city became an object of design, a means to express the aesthetic and functional aims of the period. These aims, in short, were a desire for order and discipline and a desire to impress. They can be seen in the designs of the painting and architecture of the time. In turn, they led to the designs for such special features as the gardens of Versailles, the grand boulevards of Paris, the piazzas of Rome, and the Georgian squares of London. The colonial cities that Europeans took to the Americas and Asia show the imprint of Renaissance/Baroque city design, as we see below.

The major achievements in city-building in this period came after 1600. They were stimulated by the confluence of five factors: (1) the aesthetic theory and concepts coming from a revival of interest in the classical art forms of Rome and Greece; (2) the invention of printing and improvements in the production of paper; (3) the growth of wheeled traffic as a result of replacing the solid wheel with a lighter one made of a separate rim, spokes, and hub; (4) the invention of gunpowder, which rendered medieval fortifications obsolete; and (5) the accumulation of immense, autocratic powers by the heads of some nation-states and city-states.

The cumulative effect of these five factors was that cities were literally opened up by new avenues, public squares, and new residential districts. Moreover, the planning was a self-conscious undertaking, with certain design principles at the forefront: symmetry, coherence, perspective, and monumentality. The informal and somewhat ad hoc arrangements of the medieval town gave way to a "preoccupation," as Morris calls it, "with making a balanced composition" when making a city plan.[21]

Most of the planning activity was devoted to restructuring existing cities. The planners, who were mainly architects and engineers, employed three design components:[22] the main straight avenue, the gridiron pattern for local streets in new districts, and enclosed spaces. As Lewis Mumford says:

> The avenue is the most important symbol and the main fact about the baroque city. Not always was it possible to design a whole new city in the baroque mode; but in the layout of half a dozen new avenues, or in a new quarter, its character could be re-defined.[23]

The main avenue, often radiating from a monument or public square, was a functional necessity as much as a Renaissance/Baroque design concept (see Figure 2.8), for in this period the volume of wheeled traffic increased tremendously. There was resistance to this new mode of travel, as there would be later to railroads and urban expressways in cities, but the advantages were obvious for the flow of goods, people, and military equipment. The advent of the straight street for moving traffic, both on the main avenues and in new districts, marks the beginning of what Blumenfeld describes as the "street plan" for cities.

The public square was just as important in Renaissance/Baroque city design as the straight street. Squares were also conceived on an axial basis, sometimes provided by a main avenue leading to the square. In addition, squares were enclosed spaces; that is, they were designed as three-dimensional spaces to be surrounded by buildings or other landscape features providing an architectural harmony. Even main streets with their uniform façades took on this same sense of enclosure. This quality of architectural coherence, so often lauded by visitors to Europe's cities, probably best epitomizes the notion of urbanity in a city, a quality still sought in the building of present-day civic centres.

The cities of the 1400–1800 period still needed to be bastions of defence, but owing to the perfection of gunpowder and the cannon new means of fortification had to be found. Countless attempts were made to create an ideally defensible and livable city

Figure 2.8 18th-Century Karlsruhe, Germany

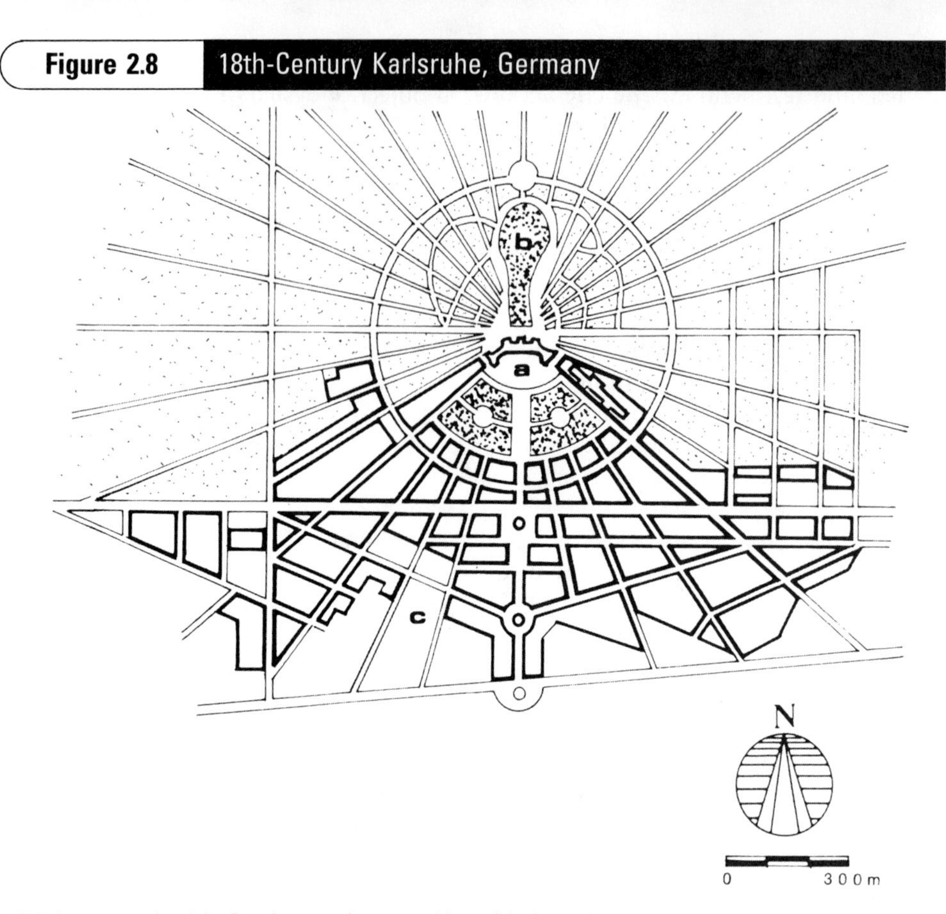

This is an example of the Renaissance city as an object of design, using straight and diagonal boulevards leading to focal points, in this case, the palace (a) and the royal gardens (b). The remainder of the town (c) had a modified gridiron street pattern.

through the design of intricate systems of bastions and citadels. Plans for fortified towns, often based on the concepts of French military engineer Sebastian Vauban, were printed and widely available throughout Europe. Fortified cities influenced the modifications made to existing cities and the plans for several entirely new ones that have since become major cities: St. Petersburg in Russia, Mannheim in Germany, and Gothenburg in Sweden.

The Imprint of the Renaissance in the Americas

It was during the Renaissance, often called the Age of Discovery, that the systematic exploration and settlement of the Americas began. The initial European settlements were cast in the mould of city forms conceived in the home countries of the explorers and settlers. Defence of territorial claims was paramount at first, and the earliest settlements adopted the patterns of fortified towns in Europe, with a wall and citadel or main battery. Within the walls one finds the Renaissance street patterns and public squares. The plan for Louisbourg, Nova Scotia, provides an excellent illustration of the early French conception of new town-building in the New World. New Amsterdam, the original Dutch settlement at New York, was also built on similar principles.

The Spanish, the first and most assiduous colonizers, approached the planning of towns according to a set of written rules entitled The Laws of the Indies (sometimes called "America's first planning legislation"),[24] which covered everything from the selection of sites, the building of ramparts, and the specifications for the main plaza, to the location of principal streets and important buildings, and the size of lots.

The gridiron arrangement chosen by the French for Montreal, which faces on and parallels the St. Lawrence River, followed the bastide model; it is believed to have influenced such other French outposts in the Mississippi Valley as St. Louis, Mobile, and New Orleans.[25] The influence of Renaissance planning ideas through English settlement in North America is best seen in the towns established from New York to Georgia. These towns employed the gridiron layout of streets combined with public squares, following the pattern set by the new, fashionable residential districts of English cities in the mid-17th century.[26]

Two important examples are the plans for Philadelphia (See Figure 2.9) in 1682 and Savannah in 1735, which carry forth the concept for an entire city. The plan William Penn chose for Philadelphia was a gridiron of streets in which the two major streets crossed near the centre and formed a public square, and in which each quadrant had its own square. James Oglethorpe's plan for Savannah made more refined use of the Georgian square by providing one in each ward of 40 houses; the wards were bounded by roads of the main gridiron and can be likened in concept to the "superblocks" used in 20th-century planning.

The last direct Renaissance influence on North American cities came from the plan Major Pierre Charles L'Enfant prepared for Washington in 1791 (See Figure 2.10). L'Enfant was a French émigré engineer and artist who served under George Washington, and who had grown up at the court of Versailles. In remarkably short order, estimated at less than six months, he developed an adaptation of Baroque monumental vistas, city squares, and diagonal streets superimposed on a grid of local streets. Despite such mundane impediments as awkward intersections and building lots (caused by the radial avenues crossing the local grid) the plan is noteworthy for its grand scale and its assiduous application of Baroque concepts to an entire city. Plans for several other cities employing a similar radial-grid pattern—Detroit, Buffalo, and Indianapolis—followed shortly thereafter, but only small parts of them were ever completed.

The L'Enfant plan nearly did not come to fruition. Its implementation depended on the sale of private building lots to pay the costs of the new public buildings, avenues, and squares. Washington went through a long and fitful period when land either did not sell, or was not built upon, or, when built upon, encroached on public areas. Construction even took place on the Mall linking the Capitol and the Washington Monument, and a railway was authorized to cross it as well. For much of the 19th century, Washington was referred to as "a plan without a city."

In contrast to city development in Baroque Europe, the communities of North America were promoted as places where colonists would have a large measure of freedom to develop their own land. Thus, the control that rulers of European states were able to impose over civic structures in their home cities did not exist to the same degree in their colonies. Renaissance designs could be brought to North America, but could not always be achieved.

Contributions of 19th-Century Idealist Planners

Even as designs for cities in Europe were elaborated, these cities began to expand precipitously as the Industrial Revolution came into its own. The burden of this expansion was felt in inadequate housing, water supply, and sewage disposal; in short, in massive slums. The congestion, squalor, and filth made the lives of the lower class unbearable and threatened the quality of life for other sections of society. The Industrial Revolution also broke down the centuries-old connection between the town and countryside. The persistence of these conditions led to many proposals to create new urban

Figure 2.9 Philadelphia: Penn's Plan of 1682

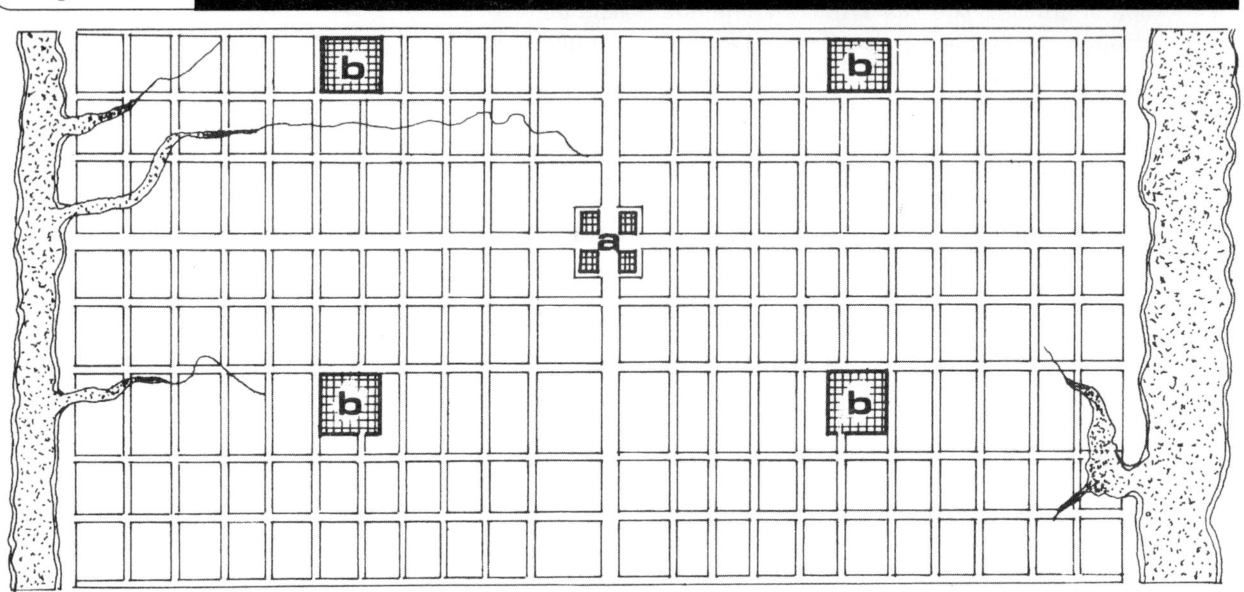

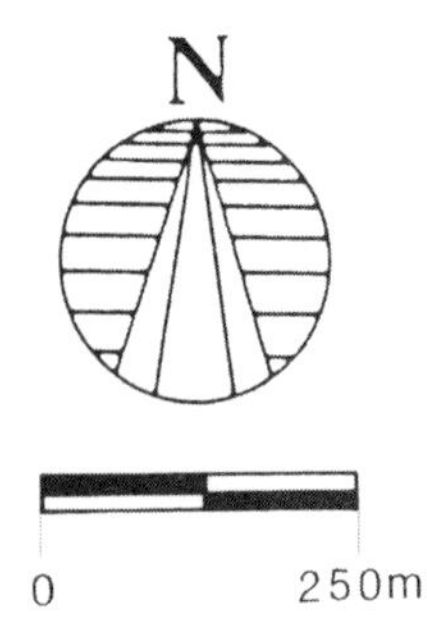

0 250m

William Penn chose a plan reflecting English Renaissance city design, using the regular grid block and Georgian squares (a, b). The main cross streets were 100 feet (30 metres) wide, and stretched about two miles (3.2 kilometres) east and west and one mile (1.6 kilometres) north and south. The central square (a) was 4 hectares in size.

environments. Philanthropic efforts from the ruling class, political initiatives from radicals and socialists, and ideas from utopian thinkers all combined to suggest ways to intervene in the deplorable physical conditions. Importantly, they all carried their own ideals and values about the conduct of community life as well. Many of these ideals remain firmly embedded in approaches to community planning today.

Robert Owen and the Utopians

One such idealistic scheme was that of **Robert Owen**, a rich English industrialist, who presented a plan for a cooperative community combining industry and agriculture in 1816. His proposal for New Lanark was for a settlement of 1200 people covering about 480 hectares of agricultural land. Dwellings were grouped around a large open square, three sides of which would be taken up by residences for couples and children of less than three years of age; the fourth side would contain the young people's dormitories, the infirmary, and guest accommodation. The central area would have such public buildings as schools, a library, a communal restaurant, and clubs, along with other areas for recreation. Around the perimeter would be gardens, farm buildings, and industrial units. To complement his physical plan, Owen proposed a cooperative community structure with much emphasis on education and on blending work in the factory with study and leisure.

Figure 2.10 L'Enfant's Plan for Washington, 1791

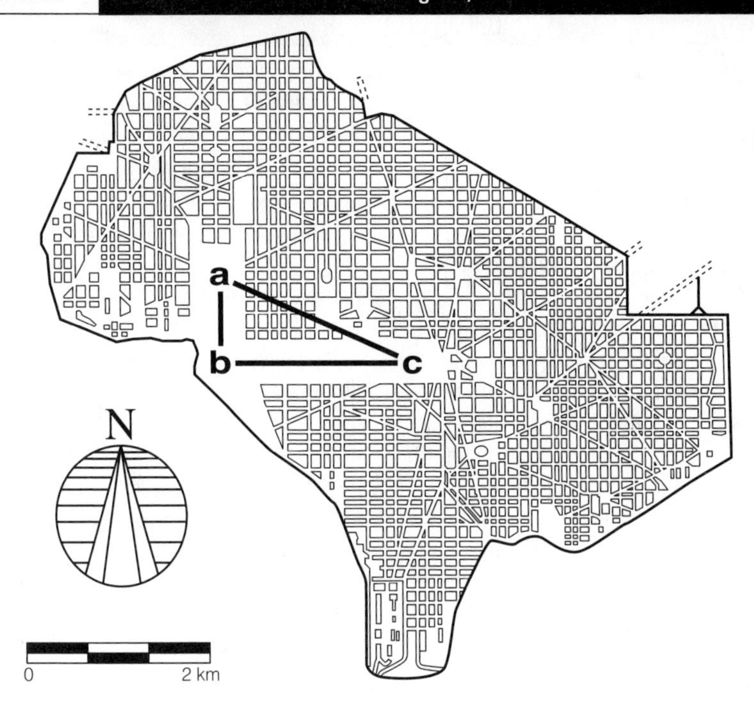

Pierre L'Enfant adopted the French Renaissance traditions for his grand plan for the government buildings and monuments of Washington. A system of radial streets focuses on important points and is superimposed on a gridiron. His "federal triangle" links (a) the President's (White) House, (b) the Washington Monument, and (c) the Congressional (Capitol) Building.

Owen did not succeed in gaining support from authorities to build a prototype of his new town in Britain. He tried to develop a similar community in the United States in 1825—New Harmony, Indiana—but had to give that up in a few years. Despite these disappointments, Owen's ideas were very influential. His **parallelogram system**, named for the rectangular form of the town he propounded, came to be widely known, much cited, and imitated.

Charles Fourier, a French utopian writer of the same era, proposed a community of about 1600 people carefully chosen according to their ages, who would be housed in a single building, the Phalanstery, surrounded by over 2 square kilometres of land. All accommodation would be communal to concentrate human relationships and achieve the "universal harmony" he propounded. Between 1830 and 1850, there were at least 50 attempts to create Phalanstery in different countries. The connection with the pastoral landscape is also important, as planning historian Leonardo Benevolo notes about Fourier's plan:

> The land shall be provided with a fine stream of water, it shall be intersected by hills and adapted to varied cultivation; it should be contiguous with a forest and not far removed from a large city, but sufficiently so to escape intruders.[27]

Godin, a French admirer of Fourier's ideas, adapted the idea to the city and industrial production. His proposals for Familistery promoted the family unit, which would be housed in its own apartment, but with communal amenities; a workers' cooperative would control the housing and the factory. **Cabet**, another source of French utopian thought, proposed a city, called Icaria, of metropolitan scale. Cabet conceived of a physical form that would separate traffic from pedestrian walkways, put factories on the outskirts, and have two circumferential boulevards. A prototype of Icaria was begun in Illinois in 1849 by Cabet and several hundred followers, but never grew very large.

Planning Issue 2.1 Downtowns Seen as Vital: Strip Malls Can't Replace Public Squares

The Calgary Herald
July 9, 2005

Downtowns Seen as Vital: Strip Malls Can't Replace Public Squares
Avi Friedman

SHERWOOD PARK—Sherwood Park is like many cities in the oil-rich province of Alberta—it's booming, with people flocking from across Canada to work in its refineries and petroleum-processing plants that tower in the distance and spit fire into the sky.

When I visited this city of 50,000 northeast of Edmonton on a cold December night, it was in the midst of pondering its future expansion and managing its growth. The town planner who drove me to town offered to show me around, and we drove through sprawling suburban neighbourhoods with rows of huge homes and oversized garages. We also passed through a wide boulevard lined with big-box stores surrounded by vast empty parking lots.

"Let's drive through downtown for a bit," I suggested.

"We don't have any," she replied.

"Where do your young people congregate on a Saturday night? Where do Canada Day and Saint Patrick's Day parades take place?" I asked.

"We mostly meet in malls, and our parades pass through the boulevard that you just saw. Otherwise, we drive to downtown Edmonton," she said.

Her answer puzzled me. I wondered whether suburban towns really needed downtowns. Could the town hall, sports arena and medical clinics be scattered throughout the city, or should they be concentrated in a central area? Are main streets, with their shops and restaurants, needed? Can malls serve as gathering and meeting places, or should towns invest in public squares that are used only occasionally?

European cities exemplify the evolution of Western urban history and hold the answers. Built on the ruins of ancient Roman towns, they often had two main arteries crossing their heart. The intersection was the place where civic and religious institutions were, with a church, government house and a market commonly framing a public square.

The proliferation of private car ownership changed all that, and the rapid development of suburban communities turned them, years later, into edge cities. Unlike older cities whose organic evolution took centuries, new towns did not develop from a core and, in fact, did not need one.

Leaders of new suburban towns failed, unfortunately, to recognize that the core is not only a place of commerce, but a social centre. In old cities, the centre was the place one went to see and be seen. It was the town's living room and knowledge-exchange place. Citizens went there to hear important announcements, greet visitors and see soldiers on their way to and from wars.

Mass communication replaced the town's public square, and residents learned more about events from their city's newspapers, television and the Internet than they learned from each other. Yet, most suburban towns lost a unique opportunity to bring people together. The places and the circumstances that could inspire communal ownership no longer exist.

A strip mall or a row of big boxes cannot replace a public square framed by tall buildings with stores at their feet, along with trees, benches and perhaps an amphitheatre.

I told some of this to the people of Sherwood Park in a town-wide consultation. I suggested that rather than explore growth, they look into their hearts, and that it is never too late to build tomorrow's legacy.

I left not knowing whether someone took note of my message. Several months later, I was mailed a newspaper clipping—it had an article and a drawing of Sherwood Park's proposed town square.

Avi Friedman teaches architecture at McGill University.

Source: Excerpted from Avi Friedman, "Downtown Seen as Vital: Strip Malls Can't Replace Public Squares," *Calgary Herald*, July 9, 2005, J11. Reprinted with permission from the author.

Two British ideal town plans of the mid-19th century sought to bring order to chaotic city environments through the design of the physical form alone and did not attempt to restructure social relations. Hygeia, planned by **Benjamin Richardson**, and Victoria, by **John Buckingham**, both insisted on the need for fresh air, light, and water, and on a regular and uniform design to bring order to the lives of city dwellers. Buckingham's plan is the best known, with its series of concentric squares, each containing rows of buildings for houses, workshops, and so forth. Public buildings were at the centre, as were the houses for the wealthiest inhabitants; factories and other industrial activities were located away from the centre. It was in many ways a plan to help achieve efficient production for factories and provide full employment and amicable surroundings for workers.

Although most of these ideal towns were never built, the efforts of the idealists were important in the development of planning concepts in two ways: (1) they were the first attempts at community planning that considered the sources of employment *and* the social structure of the inhabitants; and (2) they represented the beginning of the development of technical and geometric skills applied to community planning. The widespread enthusiasm garnered by these proposals indicated the increasing willingness of people to accept technical professional advice in improving communities.

Model Industrial Towns

The proposals of the utopians, in stressing the connection between improved living conditions for industrial workers and social harmony, stimulated many industrial owners in the second half of the 19th century to build "model" communities. The aims were more paternalistic than philanthropic, concerned as they were with enhancing workers' efficiency. Nevertheless, the plans for model communities provided further evidence of the ability to conceive of technical solutions to some of the urban problems of the time. In 1846, the industrial village of Bessbrook was begun for linen mill workers near Newry, Ireland. In 1853, Titus Salt started building Saltaire, a model town for the 3000 workers at his textile mill near Bradford in England. Similar communities were built in France, Belgium, Holland, and Italy. The Krupp family of industrialists built four workers' settlements around Essen in Germany between 1863 and 1875.

Several new British industrial towns built toward the end of the 19th century are also worthy of note. Bourneville, near Birmingham, was built by chocolate manufacturer George Cadbury in 1879. Residential areas were broken into groups through the use of intervening parks and playgrounds; the land on which the town was built was deeded to the community and has remained under town ownership to this day. Cocoa manufacturer Joseph Rowntree built Earswick near York in 1905; it too was made a community trust. The architects Barry Parker and Raymond Unwin planned Earswick as they had Port Sunlight for Lever Brothers in 1886. In these towns, Parker and Unwin began to design housing areas in large blocks with their own interior gardens and play spaces. These design values are mirrored in post–World War II subdivisions right across Canada.

Patrick Geddes

One of the most important 19th-century contributors to planning ideas that are still cogent today is Patrick Geddes. Born in 1854, Geddes trained as a botanist with Thomas Huxley and thereby gained his enduring interest in ecological principles. His concern over the abysmal living conditions in British cities led him to try and apply these principles to the organization of cities. His means for doing this were pragmatic, intellectual, and professional. In the 1880s, he moved into one of Edinburgh's filthiest slum tenements and proceeded to renovate it as a way of helping the inhabitants to see how they could improve their own living quarters. In 1892, he established his Outlook Tower in Edinburgh, which allowed one to view the city and to view an exhibition of Geddes' ideas about the interrelationships of social, economic, and physical features of a city.

As one biographer has said, "Geddes cannot be limited to a single discipline or profession, he must be described in a hyphenated manner, as a planner-teacher-sociologist-political economist-botanist-activist."[28] Because of these prodigious interests, Geddes brought a new approach to the task of community planning. He appreciated that human life and its individual and collective environments, both natural and manufactured, are all intertwined, and that improving cities means understanding them and planning for that reality rather than substituting another urban form. Geddes was a practical idealist:

> Eutopia, then, lies in the city around us: and it must be planned and realised, here or nowhere, by us as its citizens—each a citizen of both the actual and the ideal city seen increasingly as one.[29]

Geddes pioneered his ideas through various forms—exhibitions, lectures, writing—and through planning practices. In the early part of the 20th century, Geddes practised planning in Britain and in India. He saw community planning, in many ways, as a learning experience, both for the community and the planner. Geddes is often credited with the dictum "No plan before survey." He stressed the importance of knowing a town's geography, history, economy, social conditions, and means of transportation and communication. Not only was he the first planner to specify what should be in the "town survey," but he also advocated that its results—maps, charts, photographs, and so on—be put on public display. Canadian planner Kent Gerecke cites Geddes' "integrative linking" as perhaps the contribution most missing from today's planning.[30]

Geddes was among the first to give thoughtful consideration to what came to be called "urban renewal" in the 1950s. He saw no merit in bulldozing slum areas to the ground because the displaced people would simply be forced to relocate. His concept of **conservative surgery** required that efforts be made to remove as few buildings as possible and to repair and modify existing structures. Human disruption would thus be minimized, and the physical fabric of the city maintained. Geddes was also the first to recognize that as cities spread out they often grew together into a "conurbation." He cautioned against the effects on the natural landscape, agricultural resources, and rural communities that would result from simply letting cities grow outward. Patrick Geddes is truly the first regional planner (see also Chapter 9).

Reflections

The ideas for new physical forms for communities and for alterations to existing forms, as traced above, represent technical design solutions to meet the needs of a community. Underlying these solutions is the aim of meeting the social needs of a community so that the daily activities of people, businesses, institutions, and governments can be conducted with relative ease. There must be adequate shelter, potable water, and means of disposing of waste; provisions for safety and security; ways of getting around; and venues for social occasions. From the time of the earliest city-builders, these needs were recognized and met with varying degrees of success. With the advent of the Industrial Revolution, communities began growing very quickly in both population size and industrial activity. Not only did it become difficult to meet these basic needs, but they were often overlooked or disregarded. The idealist planners were stimulated by these situations. They made a lasting contribution by clearly stating that community planning had *social* components and that planners could and should address them.

By the middle of the 19th century, the broad strokes of the future community-planning profession were becoming clear. For one, planning would need to go beyond a community's physical form and provide for **living areas**, **working areas** (for industry, shopping, business), **transportation**, and **public facilities** within the overall design. For another, it would need to recognize the natural environment as integral to a community's well-being. And, not least, it would need to devise the means by which a pluralist society could carry out such planning. The next three chapters trace the evolution of planners' and communities' sometimes fitful responses to this challenge over the ensuing century and a half, bringing us to the current community planning ethos and practice in Canada. In every endeavour, the past remains important because it is a road map to both the present and the future. Thus, a question to keep in mind regarding this and the next few chapters is:

- ***How can we use the planning lessons of the past to better serve today's needs of a community's residents?***

Endnotes

1. An excellent background on the issues surrounding city-building throughout history is found in James W. Vance, *The Continuing City* (Baltimore: Johns Hopkins University Press, 1990), 24ff. Perhaps the best chronological account of the physical planning of cities is A.E.J. Morris, *History of Urban Form* (New York: John Wiley, 1979). Spiro Kostof's *The City Shaped: City Patterns and Meanings Through History* (Boston: Little Brown, 1991) and *The City Assembled: Elements of Urban Form Through History* (Boston: Little Brown, 1992) address selected urban form issues in greater depth.
2. Hans Blumenfeld, "Form and Function in Urban Communities," in Hans Blumenfeld, *The Modern Metropolis*, edited by Paul D. Spreiregen (Montreal: Harvest House, 1967), 4.
3. Adam Smith, *The Wealth of Nations,* 1775; and Lewis Mumford, *The City in History* (New York: Harcourt Brace and World, 1961).
4. Jane Jacobs, *The Economy of Cities* (New York: Random House, 1969) and *Cities and the Wealth of Nations* (New York: Random House, 1984); also Peter Hall, *Cities and Civilization* (London: Weidenfield and Nicolson, 1998).
5. F. Haverfield, *Ancient Town Planning* (Oxford: Clarendon Press, 1913), 25.
6. H.W. Fairman, "Town Planning in Pharaonic Egypt," *Town Planning Review* (April 1949); Mumford, *City in History*, Ch. 7; Morris, *History of Urban Form*, 12–14.
7. Morris, *History of Urban Form*, 14–16, 255.
8. Steen Eiler Rasmussen, *Towns and Buildings* (Cambridge, MA: MIT Press, 1949), 8.
9. Thomas Adams, *Outline of Town and City Planning* (New York: Russell Sage Foundation, 1935), 47.
10. Francis Violich, *Cities of Latin America* (New York: Reinhold Publishing, 1944), 22ff.
11. Hall, *Cities and Civilization*, Ch. 2.
12. Ibid., Ch. 22.
13. R.E. Wycherly, *How the Greeks Built Cities* (London: Macmillan, 1962).
14. Robert W. McCabe, "Shops and Shopping in Ancient Rome," *Plan Canada* 19 (September–December 1979), 183–199.
15. Ibid.
16. Rasmussen, *Towns and Buildings*, 8.
17. Blumenfeld, "Form and Function," 24.
18. Vance, *Continuing City*.
19. Morris, *History of Urban Form*, 66.
20. Vance, *Continuing City*, 200ff.
21. Morris, *History of Urban Form*, 125.
22. Ibid., 125ff.
23. Mumford, *City in History*, 367.
24. John Reps, *Town Planning in Frontier America* (Princeton, NJ: Princeton University Press, 1969), 41.
25. Ibid., 82.
26. Vance, *Continuing City*, 234ff.
27. As quoted in Leonardo Benevolo, *The Origins of Modern Town Planning* (London: Routledge & Kegan Paul, 1967), 59.
28. Marshall Stalley, ed., *Patrick Geddes: Spokesman for Man and the Environment* (New Brunswick, NJ: Rutgers University Press, 1972), xii; and Helen Meller, *Patrick Geddes: Social Evolutionist and City Planner* (London: Routledge, 1990).
29. Stalley, *Patrick Geddes*, 112.
30. Kent Gerecke, "Patrick Geddes, a Message for Today!" *City Magazine* 10:3 (Winter 1988), 27–35.

Internet Resources

Chapter-Relevant Sites

Planning Canadian Communities
www.planningcanadiancommunities.ca

Mohenjo-Daro: Indus Valley civilization
www.mohenjodaro.org

Mesopotamia
www.mesopotamia.co.uk

City water resources in Rome, 800bce–2000ad
www.iath.virginia.edu/waters

Robert Owen and New Lanark:
www.robert-owen.com

Urban planning 1794-1918 compiled by John Reps
www.library.cornell.edu/Reps/DOCS

3

Chapter Three

Foundations of Canadian Community Planning

The development of [Canadian] towns has been chaotic, and tens of thousands of so-called houses have been thrown together, which must, sooner or later, be condemned for sanitary reasons. As for town planning, there has been none.

Dr. Charles A. Hodgetts, 1912

The foundations of Canadian communities and their planning are melded from an array of experiences and ideas of the late 19th century and even prior to that time. European settlers used physical forms familiar to them to start many of our cities and towns, often locating them where Aboriginal settlements had been for centuries before. Economic and population growth along with changes in technology led, in turn, to problems of ugly streetscapes, squalid living conditions, lack of open spaces, and inadequate public utilities. Each of these concerns generated community outcry and active movements urging that they be dealt with. Not infrequently, these were in competition with one another for public backing. Such backing was often not forthcoming from local governments, which were themselves underdeveloped.

Community planning emerged because of these concerns and also within the throes of their solutions. Indeed, it evolved alongside changing urban problems and community perceptions of them. The period just before and after 1900 was a vital and formative period for community planning and merits reflection on the following question:

- *How was contemporary community planning shaped by factors prominent in 19th-century Canada?*

Aboriginal Settlements

The location and design of the Aboriginal communities in Canada had little influence upon European settlement in the 18th and 19th centuries when compared to Mexico, whose capital city is located on the site of the ancient Aztec city of Tenochtitlán. Across vast stretches of the western and northern parts of Canada, the Aboriginal peoples were nomadic hunter–gatherers, rather than city-builders. Their shelters were often portable tents or huts and domed snow houses that were renewed each season.[1]

The two small areas that had more permanent Aboriginal communities were the southern Ontario lowlands, extending to Montreal, and along the Pacific coast. The Huron and Iroquois peoples were successful farmers, raising corn and living in rectangular, barrel-roofed houses (longhouses) in small villages. Most settlements were fortified with log palisades, due to continuing warfare, and the villages were often destroyed or moved (see Figure 3.1 left). In comparison, the well-built plank houses in the unfortified fishing villages of the Pacific coast might last for decades. Some of the villages of the Salish, Bella Coola, Haida, Kwakiutl, and Nootka peoples were still intact on their coastal sites in the late 19th century (see Figure 3.1 right).

Figure 3.1 Aboriginal Settlements in Central and Western North America

Iroquois villages (left) consisted of a palisade of logs surrounding a collection of longhouses (ca. 14th century). Skidegate Village (right) of the Haida Nation in the Queen Charlotte Islands, 1878.

Source: (left) Artist's rendition of the Draper Site, by Ivan Kocsis, copyright Museum of Ontario Archaeology, London; (right) George M. Dawson/Library and National Archives of Canada/PA-037756.

If the early European settlers were clever, they paid close attention to the location of Aboriginal camps that offered clean water, shelter from winds, and access to river transportation and wildlife. Almost every European settlement prior to the railroad was sited near an Aboriginal camp for these environmental reasons, including Toronto and Kingston. Further, many place names have Aboriginal roots: Canada, Quebec, Toronto, and Ottawa (indeed, Toronto meant "meeting place" to First Nations people of the area). But the design and construction of European settlements was little informed by Aboriginal practices. The expansion of farms and towns across the southern half of the country displaced entire Aboriginal communities onto reserves, often with disastrous social consequences.[2] Current planning for First Nations settlements is discussed in Chapter 10.

Plans for Frontier Communities

The locations and the layouts used in the fortresses, trading posts, and railway towns from European settlement are still prominent features in their present city forms. Moreover, all but a few of the places that have become cities were developed initially on the basis of a plan. From the towns of the French regime along the St. Lawrence and of the British in the Maritimes, to those established later in Ontario, British Columbia, and the Prairie provinces, plans were usually made before or at the time of building of a community. A brief survey will help show the forms that characterized the early community plans as each region came to be settled.

French Canada

The earliest planned communities in Canada were those established by the French in the early 17th century. Three on the St. Lawrence River have become major cities today: Quebec City (1608), Trois-Rivières (1634), and Montreal (1642). Quebec City and Montreal are the most distinctive. Quebec's Old City has the character of a medieval organic growth town. The form of Montreal is more like that of a planted town, or bastide, such as those developed for settling new territories in France and other parts of Europe in the Middle Ages (see Figure 2.6, page 29).[3]

Quebec City's Lower Town, built by Champlain on a narrow river terrace, assumed a generally rectilinear pattern of narrow streets in which one block was left open for a public square, Place Royale, onto which fronted the town's church. The Upper Town developed outward from the fort and the cathedral. The *place d'armes* in front of the governor's fort gave an orientation to one set of rectangular housing blocks, while the *grande place* in front of the cathedral provided a different orientation to another set of housing blocks. Housing development and fortifications expanded hand in hand as the Upper Town grew, with the walls, in the 1720s, reaching the extent we see today. In the late 18th century, in an attempt to give some overall order, two parallel streets were run through the town, each to a main gate in the wall.[4]

Montreal's Old City was laid out in a more regular form than Quebec City's. It is shaped, as a visitor in 1721 remarked, like a "long rectangle."[5] It extends for about 1 kilometre along the face of a ridge that slopes down to the St. Lawrence River. Walls eventually enclosed the town, but these are now gone. In 1672, two major streets that ran the length of the settlement were surveyed. Other streets, running up the hill between the two main streets, were not regularly spaced. The resulting pattern is an irregular, but now picturesque, gridiron, with many of the original public squares still remaining. Recently discovered manuscripts suggest that Montreal's original plan also may have derived from Champlain.[6]

The most interesting of the early French community plans, and among the few truly original Canadian plans, are those for three villages 10 kilometres northwest of Quebec City that were laid out by Jean Talon in 1667. In an effort to settle the rural population in villages (as in France) rather than in long, narrow lots strung out along the St. Lawrence River, Talon developed three villages: Charlesbourg, Bourg Royal, and L'Auvergne. The village design is based on the idea of farm lots radiating out from a central square surrounded by a road, on the outside of which would be located the village church, flour mill, shops, and so forth. Each village was about 2.5 kilometres on a side, and the central square, called the *traite-quarre*, about 275 metres on a side.[7] A main road was projected to bisect the area in each direction to link up with the other towns. There were 10 farms of 35 *arpents* (about 14 hectares) in each quadrant. Charlesbourg is now part of the urbanized area of metropolitan Quebec City, but its radial pattern can still be discerned. Figure 3.2 shows two of these villages.

Figure 3.2 Two of the First Canadian Planned Towns, 1667

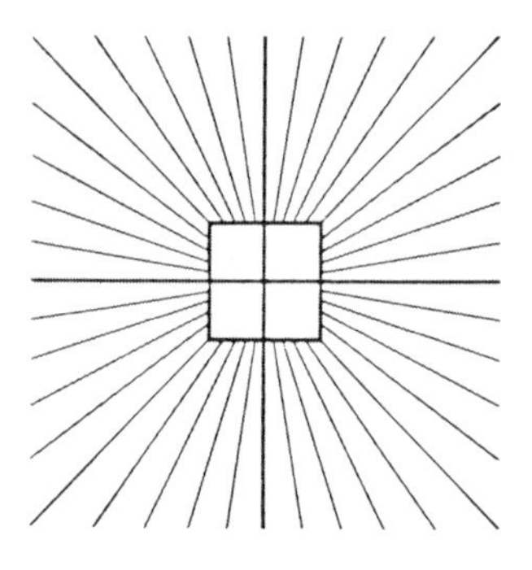

Jean Talon's radial plan for two villages north of Quebec City: Charlesbourg and Bourg Royal. A central square at the intersection of roads to neighbouring villages was reserved for a church, cemetery, and flour mill. Settlers' houses were arranged around the square on each of the 40 triangular farms. Each village was about 6 square kilometres. The aerial photo clearly shows the original radial pattern in these two towns. Both are now part of the Quebec City Urban Community.

Source: Aerial photograph © 2006. Produced under licence from Her Majesty the Queen in Right of Canada, with permission of Natural Resources Canada.

Louisbourg, the imposing fortress town on Cape Breton Island, is another major example of French community design, but here the form and concept are completely "imported." French army engineers constructed a fortified city in the tradition of Sebastian Vauban (see Figure 3.3). There were four major bastions, 5 kilometres of ramparts, many demi-bastions, and outworks surrounding the town. It was meant for 4000 persons and was laid out as a regular gridiron. The town lasted a scant 40 years; its fortifications were demolished by the British in 1760. The restoration of Louisbourg according to its original plan, when complete, will provide the best example of a late-Renaissance fortified town outside Europe.

Atlantic Canada

St. John's, Newfoundland, is the oldest community in Canada. In 1583, Sir Humphrey Gilbert formally established the British claim to the area. No plan exists for the founding of St. John's; it seems rather to have evolved from a fishing village oriented to the shoreline. The pattern that emerged gradually for the town, and that still dominates the city's central area, was of two streets that parallel the irregular curve of the shore. One, near the water's edge and called Lower Path, is now the main business street, Water Street; the other, further up the slope and called Upper Path, is also a major street today.

When the British began to establish permanent settlement in other parts of the Atlantic region, in the mid-18th century, communities were developed according to plans. These were almost invariably a regular gridiron, rectangular in shape, and carefully surveyed. Typically, British colonial towns of this period were laid out relative to a survey baseline along the harbour but sufficiently inland to provide an uninterrupted street

Figure 3.3 Louisbourg, Cape Breton Island, 1720

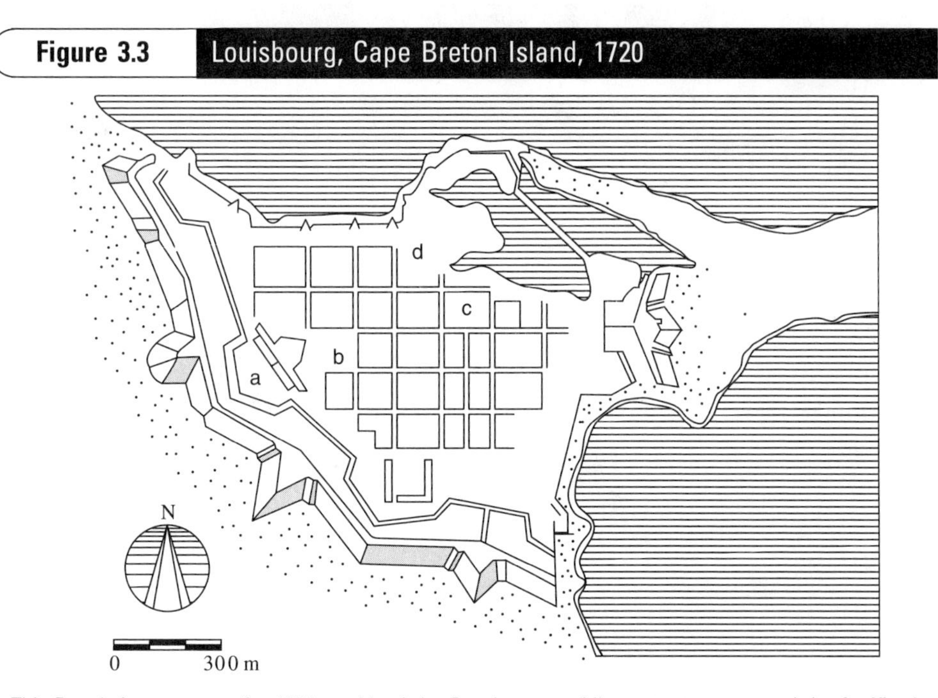

This French fortress town for 4000 combined the Renaissance gridiron street pattern and the fortification designed for the era of cannons and gunpowder. It is currently being restored. Key: (a) main bastion, (b) square, (c) hospital, (d) market.

on which other street lines could be established. This was the fashion in the original town sites for Charlottetown (1768) (see Figure 3.4), Saint John (1783), and even on the steep hillsides of Lunenburg (1753) and Halifax. These plans allowed space for a church, a governor's residence, barracks and parade ground, cemetery, and warehouses, and were usually surrounded by a palisade. In the 1749 plan for Halifax, Britain's primary North Atlantic naval station, five forts were located on the perimeter of the town. The central fort remains today on Citadel Hill, overlooking downtown and the harbour.

Although the early plans for towns in this region generally follow the simplest of surveyor's layouts, they were based on the notion of providing a pool of relatively equal building sites. This was aimed at attracting settlers from Europe, many of whom would not normally be able to have their own land. The gridiron pattern of streets and lots fills this function most effectively and would come to be the most widely used city form as Canada spread westward in the 18th and 19th centuries.

One other plan merits attention. In 1785, the Governor of Cape Breton, a well-known cartographer, Colonel des Barres, prepared a plan for Sydney that is undoubtedly derived from the forms of Georgian England.[8] It is also similar to the plan for Annapolis, Maryland, with circular plazas and axial streets. Actually, the des Barres plan was a regional plan in that five satellite communities were envisioned outside Sydney, linked by boulevards radiating from the central circular plaza.

Upper Canada

Settlement of what is now Ontario began in the years following the American Revolutionary War. There were ambitious plans for the development of this unsettled area,

Figure 3.4 Charlottetown, 1768

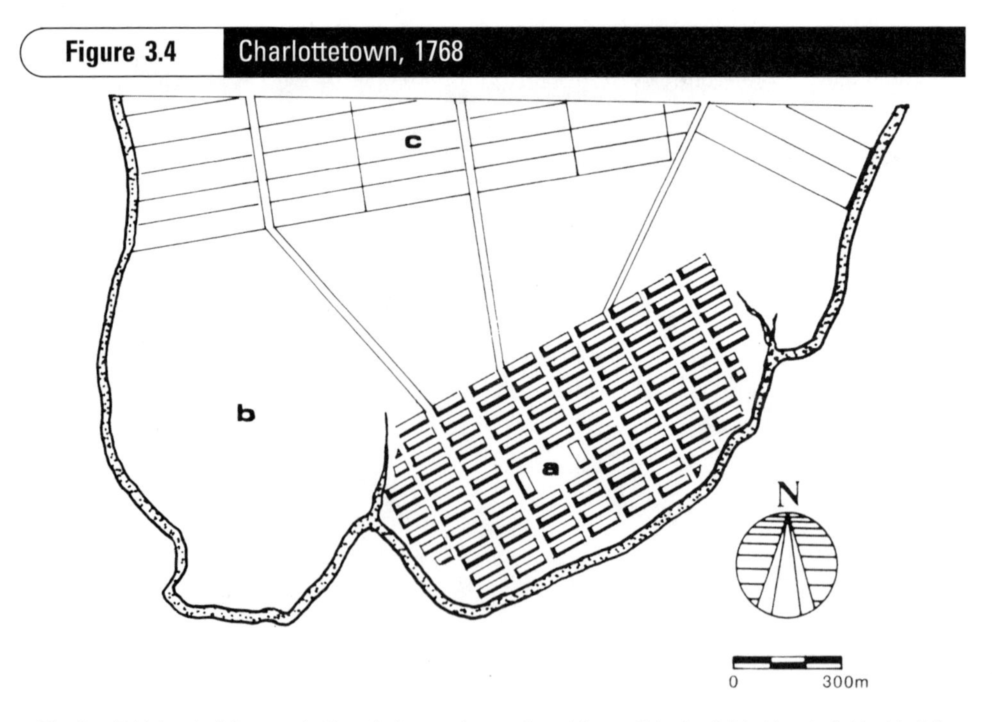

The first British colonial towns in Canada featured a regular gridiron of blocks divided into individual building lots for settlers. The grid typically started from a survey baseline along the harbour. The small blocks in the original waterfront grid have retained their historic ambience and charm. Key: (a) central square, (b) common, (c) farm lots.

which included a system of townships each averaging 100 square miles (260 square kilometres) and each containing a town site. The towns were to be a mile square (2.59 square kilometres) divided into one-acre (0.4-hectare) "town lots" and were to include space for streets, church, market, and defence works. A large military reserve of land was surrounded by the town, which was surrounded, in turn, by a grid of 10-hectare "park lots" and 80-hectare "farm lots." Although the concept of "mile square" towns was contemplated on maps prepared in Britain, such as the original plan for Toronto (see Figure 3.5), none of the five major towns that were first developed—Kingston, Toronto, London, Hamilton, and Ottawa—followed this design.[9]

Kingston was first surveyed for a town adjacent to the small military garrison in 1784 and followed the gridiron pattern being used in towns in the Maritimes at about the

Figure 3.5 Original Plan for Toronto, 1788

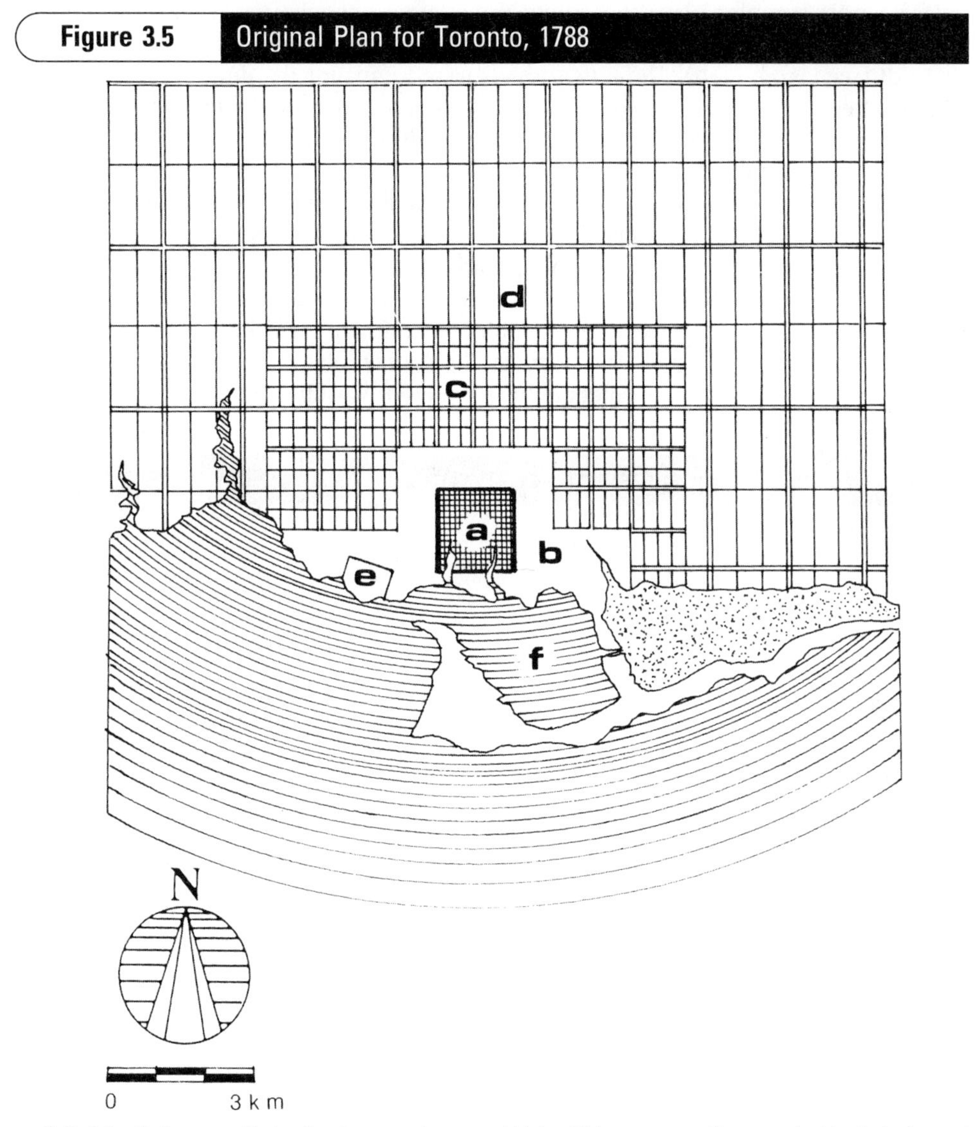

Called the "mile square" plan for the central town grid (a), which was one mile on each side, it designated five public squares and a wide surrounding common (b). This part of the plan was never implemented, but the subdivision of medium-sized "park lots" (c) and the larger farms (d) did materialize. The old Garrison Reserve (e) to the west of the town became Canada's first public park in 1848. A less ambitious town grew up beside the well-protected harbour (f).

same time. Its gridiron was "bent" to follow the shoreline of Lake Ontario. The triangular space between the two grids was reserved for public uses: a market, church, jail, and battery. Other land was reserved for gardens, a hospital, and a school, while large parcels on the outskirts were granted to the clergy, members of government, and other prominent people. This approach to town layout is the same as that of other major places developed around 1800 in Ontario. The original survey lines, moreover, continue to provide the basis of expansion. Today's arterial streets in Toronto were the original road allowances that bounded groups of six 80-hectare farm lots in Simcoe's 1793 survey.

The early Ontario gridiron town plans had no aesthetic pretensions. Their aims were primarily functional: to provide for orderly development, the equitable distribution of land, and basic public land needs at the lowest possible cost of surveying the lots. Since most of the lots were to be given away, there was no revenue to offset the costs of more complex plans that might provide public amenities. Two community plans broke with this colonial tradition: Guelph in 1827 (see Figure 3.6a) and Goderich in 1829 (see Figure 3.6b). These two towns were to anchor the development of the 400 000 hectares of land, the Huron Tract, in western Ontario. The private developer of these towns, the Canada Company, had an incentive to create attractive plans, since they wished to sell town lots. Plans for both of these places, usually credited to John Galt, adopted the idea of a radial pattern of streets converging on a town market.

Guelph was arranged in a fan shape, outward from a market beside a river; Goderich was arranged around an octagonal central market square. As in Renaissance designs, a gridiron was used for local streets. John Galt is known to have been influenced by the town-planning approaches used by the Holland Company, which acquired rights to develop upper New York state.[10] The latter development was supervised by Joseph Ellicott, whose brother had succeeded L'Enfant as the planner for Washington. Thus, the plans for

Figure 3.6 Town Plans

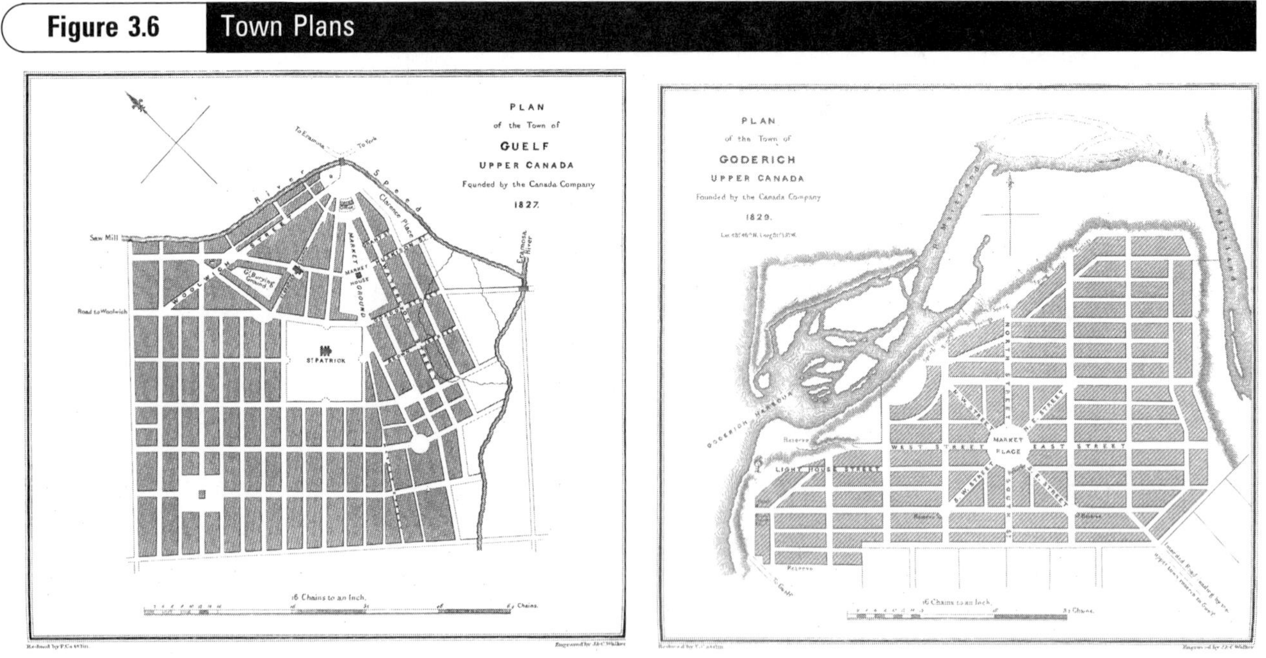

a) Guelph, Ontario, 1827 b) Goderich, Ontario, 1829

These two towns were planned by John Galt for the Canada Company to encourage settlement in western Ontario. Guelph shows the influence of the radial street pattern used in the plan for Washington, D.C.; note the central church and wedge-shaped market. The octagonal central market square in Goderich is now occupied by the courthouse.

Source: Joseph Bouchette, *The British Dominions in North America* (London: Longmans, 1831).

Guelph and Goderich are descended more from the L'Enfant concept than from Georgian England.[11] Other towns developed in the Huron Tract reverted to a common gridiron.

Western Canada

The growth of cities and towns in western Canada is, by and large, a product of national expansion rather than of colonial development from Europe.[12] There were, of course, some pre-Confederation antecedents. Many places owe their choice of site to an early fur trading post, such as Edmonton, Winnipeg, and Victoria. And British Army engineers around 1860 laid out gridiron town sites for several communities in British Columbia, including New Westminster and Vancouver, while that region was still a British colony. However, it was the establishment of railway links with eastern Canada, especially in the 1880s, that gave the stimulus to widespread town development throughout western Canada (see Figure 3.7).

Figure 3.7 New Chaplin, Saskatchewan, 1907

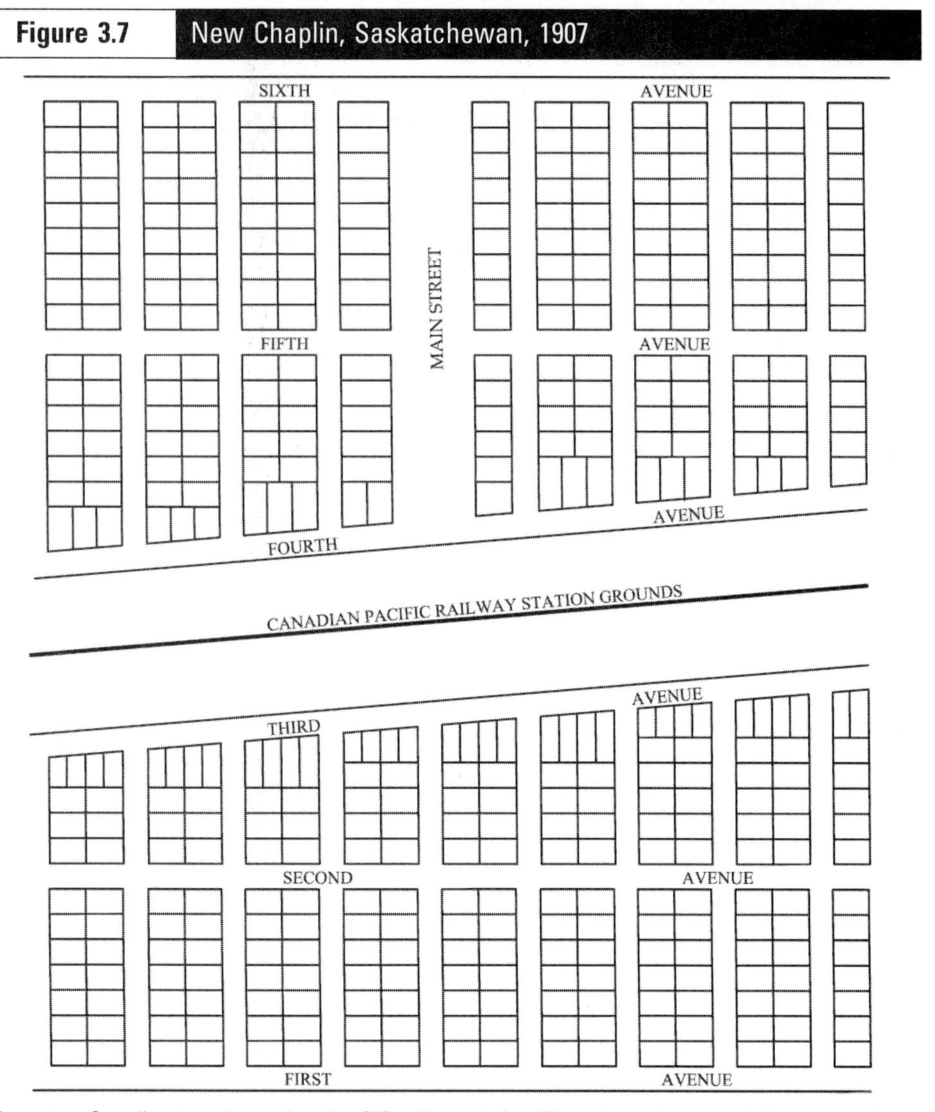

A western Canadian town focused on the CPR railway station. These town sites were laid out according to standard rules for land surveyors. Thomas Adams criticized them: "The rectangular plans, with their unnecessarily wide main streets and lanes, have not led to good results. . . . The chief advantage of the rectangular plan is, unfortunately, that its uniform lot sizes and dimensions assist speculation in land."

Source: Redrawn from Thomas Adams, *Rural Planning and Development* (Ottawa: Commission of Conservation, 1917), Figure 16.

In barely 20 years, from 1871 to 1891, almost all the places that have become major cities and towns of the region were laid out in conjunction with the advance of the transcontinental railroad and its branches. Very often laid out by railway engineers, the plans followed a common gridiron, whether for small communities or large and regardless of the topography of the site. These were plans for the sale and distribution of land. To draw again from Blumenfeld: "The right angle and straight line, convenient for the division of land, are equally convenient for the erection of buildings, [and] for the laying of pipes and rails."[13] The gridiron also suited the ebullient and egalitarian spirit that characterized the settlement of western Canada: ample land, equal parcels, similarity of sites.

The gridiron does not depend on the location of an important church, a fort, or marketplace for its beginnings. But it does depend on the ownership of land in the grid and, crucial to western Canada's communities, on where the railway station and freight yards are situated. Since the railway companies were granted land as an incentive to build in the West, they chose much of their land where they planned to build railway stations and could benefit from the sale of town lots. The initial town sites for Saskatoon, Regina, Calgary, and Vancouver did not coincide with the railway terminus. But it was the railroad's gridiron town site that became the dominant commercial centre and the location for public buildings.

In Edmonton, the presence of a 1200-hectare tract of land granted to the Hudson's Bay Company immediately west of the original town site influenced the development of its central district. Winnipeg's pattern owes much to the routes followed by fur traders north and south along the Red River and west along the Assiniboine, and to the long, narrow Métis farm lots that fronted on the rivers. Nevertheless, in western communities, regardless of rivers or rival town sites, the overall gridiron gradually reconciled many differences, and often the original survey lines still govern the direction of land development.

Early Problems of Urban Growth in Canada

The genesis of Canadian community planning began early in the 19th century, when problems associated with urban growth first became a concern. Canadian towns began to grow into cities as a result of the general industrial and commercial vigour of the 19th century. Canada was able to share in the supply of raw materials demanded by the burgeoning industrial structure of Europe, especially Britain. This meant economic expansion, population growth, and physical development for Canadian communities. The population in British North America between 1815 and 1865 grew from 0.5 million to 3.5 million people. Although most people still lived in the countryside, the major towns started to grow into cities in this period: Toronto grew from little more than 1000 people to about 50 000; Kingston grew from 6000 to 17 000; Montreal grew from 15 000 to 110 000; and Halifax grew from 15 000 to 25 000.

But sharing in the benefits of the Industrial Revolution also brought a share of the burdens. In particular, the problems of rapid urbanization that began to afflict British communities in conjunction with the Industrial Revolution soon materialized in Canada's towns and cities. Here are descriptions of two urban situations in the first half of the 19th century. The first describes conditions in the new colonial town of York in 1832 (which would be renamed Toronto two years later) while the second, which differs little, describes Glasgow, Scotland, in 1840.

> Stagnant pools of water, green as leek, and emitting deadly exhalations, are to be met with in every corner of the town—yards and cellars send forth a stench from rotten vegetables sufficient almost of itself to produce a plague and the state of the

> bay, from which a large proportion of inhabitants are supplied with water, is horrible.[14]

> In many houses there is scarcely any ventilation; dunghills lie in the vicinity of the dwellings; and from the extremely defective sewerage, filth of every kind accumulates.[15]

Disease

Sanitary deficiencies were common in towns and cities of the time on both sides of the Atlantic. Moreover, those problems defied easy solution throughout most of the 19th century. There was inadequate technical understanding about the construction of a public sewer system. It was also not understood how subsoil conditions could allow outdoor privies to contaminate wells for drinking water, and it was considered normal that individual houses and businesses should provide their own water supply and waste disposal. Combine these factors with both the lack of public resources to construct sanitary and water supply systems and the lack of authority to compel private property owners to connect to such systems, and one has the setting for tragedy.

Virulent disease often spread within and between communities. Cholera, carried by immigrants from the British Isles, spread in 1832 from near Quebec City all the way to London, Ontario, and up the Ottawa Valley in a matter of two months, for example. Typhus hit many cities, including Ottawa and Kingston, in 1847; this epidemic helped spur on public health measures by provincial governments.

Although the public and governmental concern was there, the knowledge about the origin and spread of disease was not. Until the 1870s, when microbes were first isolated as the cause of many diseases, the prevailing wisdom (sometimes referred to as the "filth theory") attributed disease to the accumulation of wastes, human and otherwise; it was commonly believed that the air was poisonous in the vicinity of accumulated wastes. Impure water was also thought to cause disease, but it was not understood until near the end of the century that waterborne microbes might contaminate a community's water supply.

Water Supply

Efforts to bring piped water to all buildings and then to secure sources of pure water were very protracted in many communities. It took Winnipeg from 1882, when a public water and sewer system was started, until 1906 to pass a bylaw requiring all buildings to be connected.[16] Ottawa began a public water system in 1872, after debating it from 1855, but it was 1915 before a pure supply was obtained.[17] It took serious typhoid outbreaks in both of these cities to resolve the problem of water and sewer services.

The immediacy of the problems of disease and pollution was associated in people's minds with the growth of towns and cities. Slowly but surely it became clear that the solutions to these problems required community-wide action. So compelling became this concern that, well into the 20th century, obtaining "healthful conditions" was a prime objective of those who propounded community-planning remedies for towns and cities. And the importance of pure water supplies continues today, as the recent events in Walkerton, Ontario, and other centres show.

Fire

Another cornerstone for community planning came out of the concern over public safety from fire. Large fires were all too common in Canadian communities throughout the 19th century (see Figure 3.8). The extent of this menace can be seen in the following list of major conflagrations in larger Canadian cities:[18]

Halifax	1750, 1861
Saint John	1837, 1841, 1845, 1877
Fredericton	1825
Quebec City	1815, 1834, 1845, 1862, 1865, 1866, 1876
Montreal	1765, 1768, 1803, 1849, 1852, 1901
Kingston	1847, 1854, 1857, 1890
Toronto	1849, 1890, 1895, 1904
Ottawa	1874, 1900, 1903
Vancouver	1888
New Westminster	1898

The control of fire in towns and cities, however, was difficult for two reasons. First, the technology of firefighting was still rudimentary, lacking the necessary vehicles, hoses, and so forth, and fire hydrants had yet to be invented. Second, and even after equipment for firefighting was available, most Canadian communities lacked ample supplies of water. There are many reports of fires raging out of control because of either insufficient water or inadequate water pressure. Thus, the public's concerns over both health and fire losses were centred to a large degree on the supply of water.

Figure 3.8 Fires in Early Cities: Ottawa, 1900

Frequent major fires occurred in Canadian cities in the early years of the 20th century before the introduction of adequate public water supply systems and firefighting equipment. These residents fled a fire that destroyed half of Hull, and jumped the river to burn 14 percent of Ottawa.

Source: Library and Archives Canada/PA 1200334.

The frequent loss of buildings in fires also caused fire insurance rates to climb dramatically in this period. Local businesses became increasingly worried over the waste and the cost of fires, which, in turn, prompted concern over the ways in which cities and towns were being built and maintained. The quality of building construction came in for scrutiny: Were buildings safe for their occupants? Would they deter the spread of fire to other buildings? These questions, considered straightforward today, were in the past not deemed appropriate: the construction of buildings was the private business of the builder or owner. As late as 1920, no province had a uniform building code and municipal building-inspection practices were also relatively new. The threat of fire is no longer the concern in city-building that it once was, but the issue of civic safety is no less present. Natural disasters such as floods and earthquakes are ever-present dangers to communities, as the floods in the Saguenay region of Quebec and in Manitoba in the mid-1990s or New Orleans in 2005 attest. These latter kinds of concerns have spawned a sub-discipline within community planning called "disaster planning."[19] (See Figure 3.8.)

Slums

As Canadian cities grew more populous in the latter part of the 19th century, attention turned increasingly to the quality of housing. The new populations in cities came almost entirely from Europe, and most of these were poor people. It was known that cholera epidemics were attributable to newly arrived immigrants. It was also observed that many immigrant workers lived in crowded, unsanitary, rudimentary dwellings. Tenuous and often unfair assumptions were made that the poor and their neighbourhoods were the prime sources of disease and fire (not to speak of moral turpitude). As one writer

Figure 3.9 Canadian Slum Housing, 1912

Wooden tenements in an eastern Canadian city, showing crowded rear lots, a narrow entrance from the street, and flimsy construction. Rear lots were seldom served by water and sewerage.

Source: City of Toronto Archives, Fonds 1244, Item 679. Reprinted with permission.

notes, "bad plumbing and crowded housing were more easily fixed on as the culprit in the spread of disease and crime than the complete framework of poverty."[20]

There were extensive slums in Canadian cities of this period. A penetrating study by Herbert Ames of a working-class district in Montreal in 1896 provides details of housing conditions.[21] In the area, 2 or 3 square kilometres just to the east of Windsor Station, over 37 000 people lived in 8300 tenement-type apartments. These densities of, respectively, 14 300 persons per square kilometre and 33 dwelling units per hectare do not seem extremely high by the standards of today's high-density, high-rise housing projects. However, these conditions were for an entire district, not just a housing project, and most of the buildings were only two and three storeys high! Along with the narrow streets and almost no open public space, the result was very high density by any standards. Furthermore, half of the dwellings had outdoor privies. Ames noted that some other wards in Montreal had densities two and three times higher than the area he studied (see, for example, Figure 3.9, page 53).

Other Canadian communities suffered similar problems. In Winnipeg, for example, in 1884 the overall density of population in the built-up area was close to 4000 per square kilometre.[22] By 1900, about half of Winnipeg's houses still had outdoor privies; add to this the widespread use of horses for transportation and the need to dispose of animal wastes. The density of development just after mid-century (1860) in Toronto had already reached over 4800 persons per square kilometre.[23] There was concern over Toronto slums from as early as 1873, and, in 1884, the Toronto Tenement Building Association was established to build houses and apartments with "modern conveniences and sanitary appliances" for working-class populations. These overall density figures are for the entire city, but the poor usually lived in neighbourhoods at two and three times the average density. The consequent crowding often led to illness and social tension. By the end of the century, efforts were under way in many cities to deal with slum conditions through public health measures, improved building practices and codes, and even new housing built specifically for workers (see Chapter 4).

Railways, Streetcars, and Urban Form

Canadian cities, both western and eastern, were significantly affected by the introduction of railways. The location of the passenger station was important to communities, but more important over the long run was the location of the marshalling yards and freight terminals, for they consumed large amounts of space. The freight yards and terminals were usually adjacent to the passenger station and, hence, next to the downtown area. In addition, the coming of railways to Canadian communities coincided with extensive industrialization in the country. Industrial firms were encouraged to locate near the freight terminals or along rail lines.

Railroad development, undoubtedly a boon to the economic development of communities, strongly affected the pattern of their growth. Typically, the railway passenger station "anchored" one side of the downtown commercial area, and on the other were the freight yards and industrial area and, frequently, the port. Housing development spread out from this core, but not evenly, for the freight yards proved to be both a physical and psychological barrier to residential growth.

Housing for the more affluent population was located away from the rail lines and industrial areas. Housing for poorer segments of the population was left to the land adjacent to industry and railways. The large, generally linear area occupied by freight yards and industries was also difficult for city development to cross without expensive underpasses or bridges. In inland cities, when development did succeed in leapfrogging the railway, it often resulted in the establishment of lower-income residential districts on, so to speak, "the other side of the tracks." In cities with harbours, the railway–industrial development usually spread along the waterfront, creating a barrier between the community and its natural marine asset (see Figure 3.10).

The street railway was also introduced to Canadian communities in the late 19th century. This form of rail transport, especially after electrification, dramatically influenced the pattern of residential development. Prior to street railways, both the dependence on foot travel and poor roads tended to limit development to within 1.6 kilometres of the central area. In the 1870s, the horse-drawn trolley permitted development to occur up to 5 kilometres from downtown and the jobs of the central industrial area. The electrification of streetcars around 1890 saw the extension of lines 10 kilometres and more from downtown stores and jobs. The norm was usually a half-hour trip—a norm that persists today in North American communities.

The streetcar lines followed streets that radiated outward from the central area of the community and were a great stimulus for land development adjacent to those routes. As a result, the overall pattern of the city assumed a finger-like shape along transit routes. The oldest housing outside the central area is found in areas served earliest by streetcar lines. A strip of retail and commercial establishments often served these extended new neighbourhoods. Almost always the gridiron formed the base of development; it was simply extended as the streetcar lines were built (see Figure 3.11).

Figure 3.10 Impact of Railways on the Centre of the City: Toronto, 1870

The advent of railways was associated with manufacturing and commercial developments, which were in the centre of most cities by the late 19th century. The marshalling yards for freight trains consumed large amounts of city space and, in port cities like Toronto and Vancouver, they cut off the harbour from the rest of the city and hampered its use by citizens.

Source: Toronto Public Library (TRL): T31103.

Figure 3.11 Effect of Transportation Modes on the Form of the City: Toronto

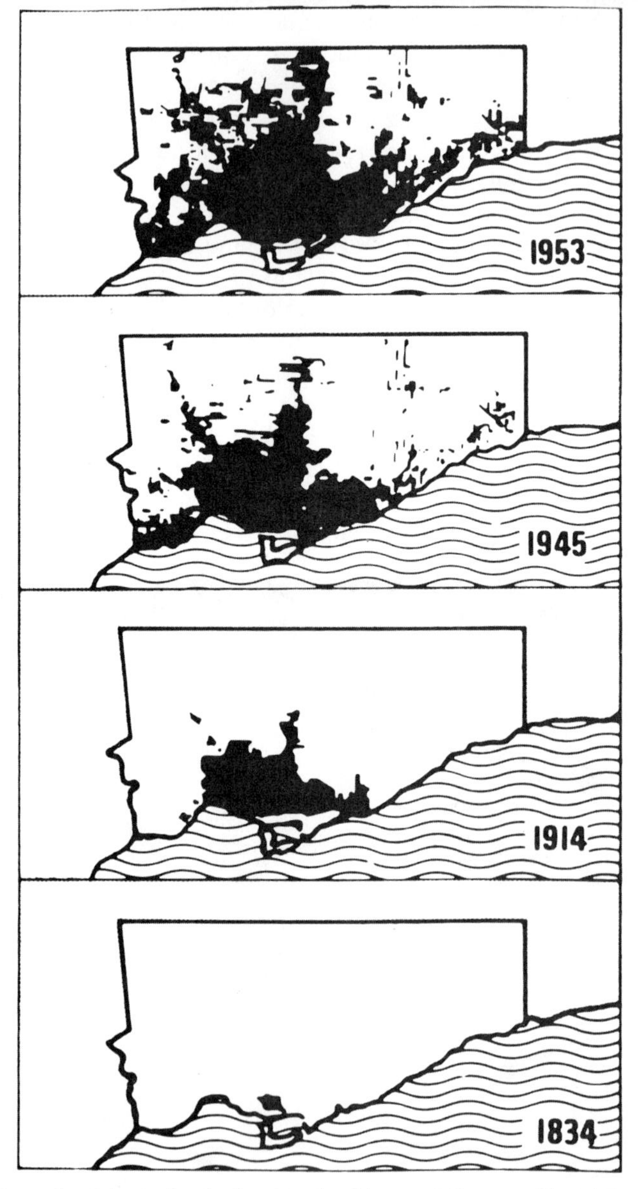

Toronto's form of growth relates to the dominant mode of transportation prevailing at the time. The city of 1914 with electrified streetcars is not only 20 times larger than in horse-and-foot-traffic times, but also shows the growth adjacent to the radial streetcar routes, which is even more exaggerated by 1945. However, within less than a decade, the proliferation of the private automobile made for a much more dispersed form.

Source: Metropolitan Toronto Planning Department.

Emergence of Modern Planning Concepts

The rapid industrialization occurring in the 19th century (beginning around 1800 in Britain, 1840 in the U.S., and 1880 in Canada) centred on cities and brought with it new people, new wealth, and new problems. Toronto's population grew from just over 56 000 in 1871 to nearly 522 000 only 50 years later. At first, no cities on either side of the Atlantic were able to cope with the demand for housing, the need for transportation, and the provision of basic water supply and sewage-disposal services. Later, when technology and social conscience caught up with these needs, the result was cities that might best be described as "cluttered": smoky industrial districts, unpaved roads, mean and crowded working-class housing, half-finished suburbs, and a plethora of new electric poles and overhead cables.

During the late 19th and early 20th centuries, in most large Canadian communities the emerging physical patterns of cities prompted concerns not unlike those receiving attention in Britain and the United States at the same time.[24] In response to these concerns, a coterie of professionals emerged who gradually codified their ideas and experience about the best physical form for communities. The Canadian planning profession developed at this time and, with its British and American counterparts, generated a rich array of planning concepts, which continue to be drawn upon.

The planning concerns for the burgeoning cities just over one hundred years ago were rooted in at least four different perceptions of city problems. One viewed the major problem as the **shabby appearance of cities**; another viewed it as the **deterioration of living conditions,** others were concerned over the **loss of the natural environment**, or **inefficiency and waste**. From each sprang different planning concepts that might produce better communities. Out of the concern over the appearance of cities came the notion of the City Beautiful and the redesign of major streets and public areas in existing cities. Out of the concern over living conditions came the notion of Garden Cities, wholly new communities designed to allow new patterns of living in less congested surroundings. Concern over the natural environment led to the Parks Movement, which built public open spaces and preserved natural areas. Concerns about city efficiency led to proposals for infrastructure systems such as electric streetcars, sewers, and paved roads. Adherents of each approach often became part of competing planning movements, whose Canadian origins are discussed below (see Figure 3.12).

The imprint of each approach is to be found in Canadian community planning of the first half of the 20th century, as described in detail in Chapter 4. During this period, each approach evolved into more refined planning ideas that included garden suburbs,

Figure 3.12 Evolution of Community-Planning Ideas, 1890–1900

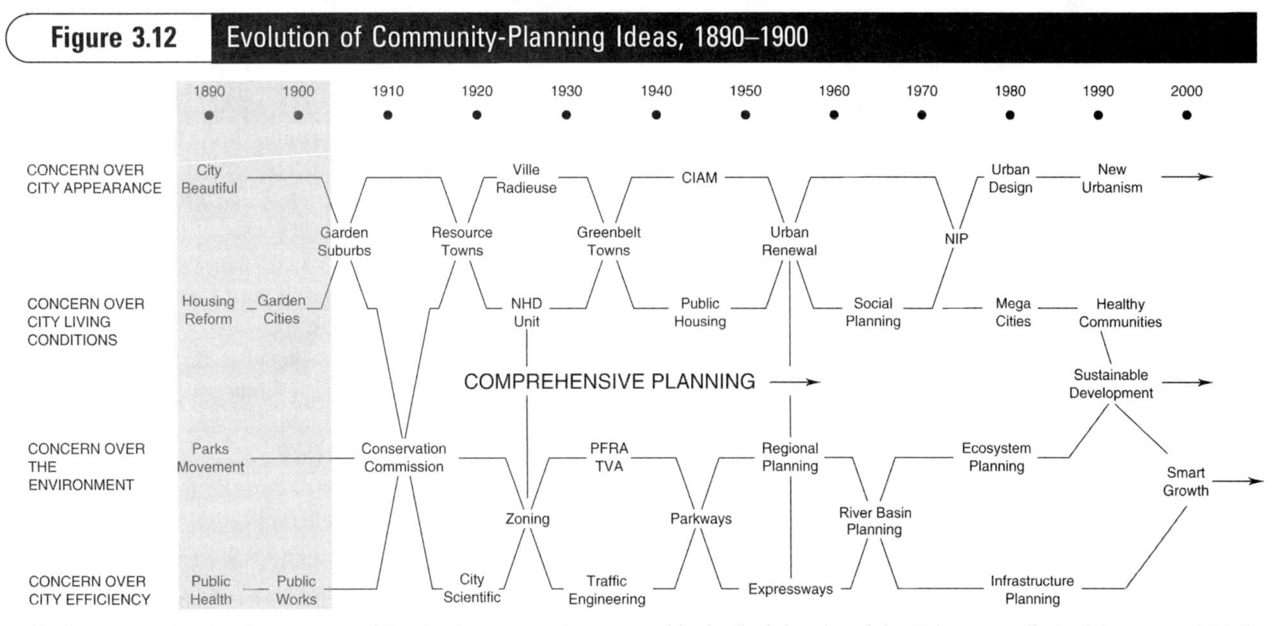

Modern community planning grew out of four basic concerns that emerged in the final decades of the 19th century. Each of these were initially articulated around broad social movements: city appearance (City Beautiful), city living (Garden Cities), the environment (Parks), and city efficiency (Public Works).

Planning Issue 3.1 Better Suburban Planning Needed to Curb Obesity, Experts Find

The Globe and Mail
October 12, 2005

Better Suburban Planning Needed to Curb Obesity, Experts Find
André Picard

It's time to shift the focus from blaming individuals for being fat to understanding how the environment we live in discourages healthy living, a scientific think tank has concluded.

"We need to look well beyond getting more gym classes for kids and better food in school cafeterias," said Diane Finegood, scientific director of the Institute of Nutrition, Metabolism and Diabetes.

"We need to understand how we can fundamentally alter the environment so the healthy choice is the easy choice."

Dr. Finegood said that means rethinking the way public policy is created and implemented in a broad range of areas, including urban planning, transportation, education, agriculture and taxation.

But the immediate focus needs to be on infrastructure, including urban design and planning, she said.

"At a time when we're about to embark on major investments in infrastructure, we need to make sure we do things differently," Dr. Finegood said.

"We can't allow ourselves to repeat the mistakes of the past."

Stephen Samis, director of health policy at the Heart and Stroke Foundation of Canada, agreed. He noted that when governments began establishing zoning bylaws, they had three main goals: to ensure the health, the safety and the welfare of citizens.

"Unfortunately, the focus today is almost exclusively on safety. We need to get back to those other basic tenets, health and welfare, in the way we regulate our built environment."

He said there are several factors that should come in to play when there is development, including building sidewalks, ensuring roads are interconnected to facilitate walking and public transport, building walking and cycling paths, ensuring mixed-use development so people are not forced to drive to stores, and even financing for new housing.

"Suburbs are a proven thing, so financing is easy, but financing smart, healthy development is a lot more difficult," Dr. Samis said. "That's not smart if you look at the impact of suburbs."

Canadian researcher Larry Frank has shown that suburbanites are about 35 percent more likely to be obese than their urban counterparts, in large part because they spend so much time in their cars.

Dr. Finegood and Dr. Samis led a think tank that met last week in Toronto to establish research priorities for tackling the obesity epidemic. The meeting involved more than 100 experts from a wide array of specialties, including medicine, public health, the environment, urban planning, economics, agriculture, the food industry and consumer groups.

While there was broad consensus on the need to address suburban sprawl as a health issue, the expert group was not able to agree on how to deal with the obesity problem through economic incentives and disincentives.

"Clearly this is a sensitive issue, and politically charged," Dr. Finegood said. "But the reality for us as scientists is that the evidence is poor: We don't know what works and doesn't work because economic incentives haven't been properly evaluated."

"Bringing a diverse group together to look at what we know and what we need to know about tackling obesity is a Canadian first," Dr. Samis said. "But it's just a first step. We're keen to start kicking some of this research out the door."

According to Statistics Canada, more than 59 percent of Canadian adults are of unhealthy weight. The total includes 23 percent who are obese (meaning more than 30 percent or more of their body weight is fat) and 36 percent who are overweight (25 percent or more fat).

---

the neighbourhood unit, and greenbelt towns. The late-20th-century legacy of each approach is examined in Chapter 5; we also see that current planning ideas such as Smart Growth, sustainable development, and New Urbanism have roots that can be traced back to the late 19th and early 20th centuries.

Concern over City Appearance: The City Beautiful Movement

In 1893, Chicago hosted the World's Columbian Exposition. The design of the grounds and buildings of this fair is credited with stimulating a surge of concern for the design of cities in North America over the ensuing 40 years. The site on the shores of Lake Michigan was selected and laid out by Boston landscape architect Frederick Law Olmsted. A strong team American designers, led by Chicago architect Daniel Burnham, conceived of a setting of buildings, avenues, statues, canals, and lagoons in true Baroque fashion. Planning historian Mel Scott calls it a "temporary wonderland of grand perspectives and cross axes ... shimmering lagoons and monumental palaces ... an enthralling amalgam of classic Greece, imperial Rome, and Bourbon Paris."[25]

The need for the beautification of cities had been stirring in the United States for at least two decades before the Chicago World's Fair. It showed up in municipal art societies, civic improvement commissions, and in the attention given to both private and public landscape architecture, notably in the development of public parks and civic centres.[26] The Chicago fair strengthened this aesthetic effort by providing design principles that could (and subsequently did) govern the design of city halls, public libraries, banks, railroad stations, civic centres, boulevards, and university campuses. Burnham was involved with refurbishing L'Enfant's plan for Washington in 1902. His 1908 plan for Chicago (prepared with Edward H. Bennett) is considered the benchmark of City Beautiful plans, containing schemes for new diagonal avenues, civic plazas, public buildings, and a series of parks along the lakeshore, and proposals for a network of highways for the entire metropolitan region of some 10 000 square kilometres and a chain of forest preserves and parkways on the periphery.[27]

The impact of the Chicago exposition was not lost on Canadian architects, engineers, and surveyors (who would soon make up the fledgling planning profession). They too were active in campaigning against the squalor and the ugly environments that were developing in Canadian communities in the period of rapid urbanization prior to World War I. Montreal architect A.T. Taylor complained (in 1893) after viewing the fair:

> The average modern city is not planned—like topsy, it just grows, and we are only allowed to touch with the finger of beauty a spot here and there. One longs for the days of Pericles or Caesar, or even those of the First Empire, when cities were laid out with beauty and effect, and were exquisite settings for noble gems of architecture.[28]

The Renaissance design principles of symmetry, coherence, and monumentality were revived by the Chicago fair, and Canadian "planning advocates" promoted them widely with governments, chambers of commerce, and corporations. Beautification proposals often arose from local groups such as civic improvement societies or business organizations, aided by the local architects' association.[29] In Vancouver, leadership came from the Vancouver Beautiful Association;[30] in Montreal and Calgary, railway companies were closely involved. These organizations initially drew membership from city elites, but debate on beautification efforts often drew in business people, and then, later, the new professional engineers, architects, and planners joined them.[31]

Concern over Housing and Living Conditions: The Garden City Concept

The Garden City concept emerged from concerns about terrible living conditions in late-19th-century British industrial cities, documented by reformers such as Charles Booth and revolutionaries such as Friedrich Engels.[32] But Britain was just the leading edge of the Industrial Revolution, and the overcrowding, disease, and poverty of London and Manchester had arrived in North America by the late 19th century. Early reactions to poor living conditions included the Utopian communities and settlements built by industrial philanthropists such as Lever's Port Sunlight and Cadbury's Bourneville, described in Chapter 2. The university settlement houses led by Jane Addams in Chicago and Henrietta Barnett in London were early attempts to improve social conditions in inner-city neighbourhoods. The Garden City concept, espoused by Ebenezer Howard, proposed, however, to deal with poor city living conditions at both the local and metropolitan scales (see Chapter 9).

In today's terminology, the Garden City would be called either a "satellite town" or "new town." Like its modern versions, the Garden City concept aimed at affecting the physical form of communities in two ways: first, it would disperse the population and industry of a large city into smaller concentrations, and second, it would create community living environments in the periphery of large cities that were more amenable than those of the city.

Ebenezer Howard, the British originator of the Garden City idea in 1898, was not an architect or surveyor but a court reporter and, as one biographer emphasizes, an inventor.[33] He presented his ideas for new towns as general diagrams rather than as plans for a particular community and location. The main dimensions of Howard's Garden City were:

1. A population of about 30 000;
2. A built-up town of 1000 acres (400 hectares);
3. An agricultural greenbelt of 5000 acres (2000 hectares) surrounding the town;
4. Provision of land for industry and commerce to supply employment to the residents:
5. An arrangement of land uses to promote convenience and reduce conflict; and
6. A means of rapid communication between the central city and the Garden City.

An important feature of the Garden City's development was that all the land, including the greenbelt, would be owned by a single limited dividend corporation and held in trust for both investors and the residents. The corporation would build infrastructure, homes, and industrial parks, leasing houses to residents and land to employers.

In the Garden City that Howard visualized, each house would have its own garden; each neighbourhood, its own area for schools, playgrounds, gardens, and churches; and the whole town its surrounding "garden" or agricultural estate, as Howard termed it. There would be a strong town centre with a town hall, concert and lecture hall, theatre, library, museum, and hospital, as well as a large public garden and ample room for shops. Each neighbourhood ("ward") would be bounded by major avenues and would house about one-sixth of the population, or about 5000 people, in individual or group housing and be only about 1.5 square kilometres in greatest extent. An innovative, but often overlooked, part of Howard's concept was for an internal "belt of green" 130 metres wide, a Grand Avenue (three times wider than University Avenue in Toronto), to separate the area for factories, warehouses, and so on from the residential area.

Howard's ideas were turned into reality with the building of two Garden Cities north of London. The first was Letchworth, about 56 kilometres north of London, which was designed by Barry Parker and Raymond Unwin and started in 1903. It has now reached close to its intended population.[34] The second, Welwyn, was started by Howard in 1919 about 27 kilometres from London.[35] Thomas Adams was associated with the building of

Letchworth as secretary of the group of investors who initiated the first Garden City. A decade later, Adams came to Canada as the Town Planning Advisor to the Canadian Commission of Conservation (see Chapter 4). The significance of the Garden City concept is to demonstrate, as Adams remarked 30 years after Letchworth, "the advantage of planning communities in all their features from the beginning."[36] Garden City ideas influenced city plans around the world over well into the next century. By 1950, greenbelts and satellite towns were important components of plans for metropolitan Moscow (1935), London (1946), and Stockholm (1950).[37]

Concern over the Environment: The Beginnings of the Parks Movement

Many of today's major public parks in Canadian cities came into existence in the late 19th century. Most were not part of the initial planning of our communities. The first plans for communities in eastern Canada provided for market squares, church squares, and military parade squares, but not public recreation areas; plans for communities in western Canada usually provided for neither squares nor parks. Indeed, the idea of space being set aside for public recreation only began to take hold in Britain in the 1840s.

Some of the earliest Canadian landscaped open spaces emerged from a serious public health problem—where to bury the dead. The early pioneers were usually buried in the yards of their original places of worship, but these soon filled up. The typhus and cholera epidemics of the mid-19th century filled every available open space in the affected towns with mass graves. The solution to this problem in several cities was the construction of a shared, non-denominational cemetery on rural lands, some distance from the edge of the built-up area. It was a surprising example of ecumenical cooperation during a period when religious rivalries were strong.

The idea followed precedents from Père Lachaise cemetery in Paris (1804) and especially Mt. Auburn Cemetery (1831) just outside Boston. Rural cemeteries were built outside Kingston (Cataraqui, 1850), Montreal (Mount Royal, 1852), Ottawa (Beechwood, 1873) and Toronto (Mount Pleasant, 1876). The grounds of the cemeteries were beautifully landscaped according to plans designed by gardeners or civil engineers—the landscape architecture profession had not yet emerged. Regular visits to the graves of family members became a social activity, rather like a picnic. The next step was to provide a public park for similar recreation, without the trip to the rural cemetery.[38]

Canada's first public parks to be supported by municipal funding were established less than a decade after those in Britain and around the same time as those in U.S. cities.[39] A major stimulus was the transfer of land reserves held by the British central government, usually for military purposes, to the local government. The Garrison Reserve (now Exhibition Park) in Toronto came into existence this way in 1848, as did Kingston's City Park in 1852, Hamilton's Gore Park in the same year, Halifax's Point Pleasant Park in 1866, the Toronto Islands in 1867, London's Victoria Park in 1869, Montreal's Isle Ste. Hélène (the site of Expo 67) in 1874, and Vancouver's Stanley Park in 1886. All of these park areas were on the periphery of their respective communities at the time. There was no concept of neighbourhood parks until around the turn of the century, when various provinces introduced legislation to promote local parks.

As Toronto's parks chairman said in 1859, public parks offered "breathing spaces where citizens might stroll, drive, or sit to enjoy the open air."[40] The outlook reflected a desire to create a natural setting, often along with an appreciation of horticulture. The Public Garden in Halifax is an example of the latter; Mount Royal Park in Montreal, an example of the former. Mount Royal Park, purchased by the city in 1872, is also significant among Canadian urban parks because its basic layout (see Figure 3.13) was done

Figure 3.13 New Mount Royal Park, Montreal, 1877, design by Frederick Law Olmsted

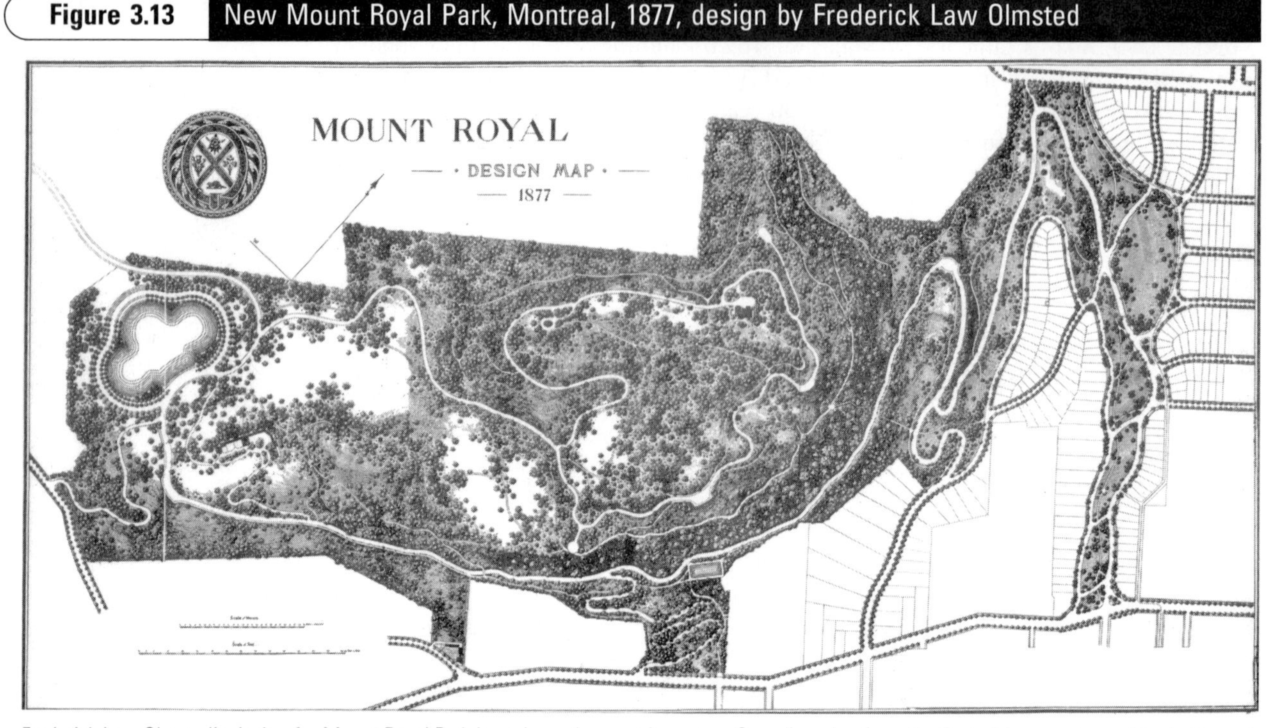

Frederick Law Olmsted's design for Mount Royal Park is perhaps the most important Canadian plan from the Parks Movement era. Olmsted's work is now cherished and recent repairs to the park have attempted to maintain his original vision.

Source: Frederick Law Olmsted, *Mount Royal* (New York: G.P. Putnam, 1881).

by Frederick Law Olmsted, who had designed Central Park in New York City. He urged Montreal to "bring out the latent loveliness of [the] mountain beauty" and retain the wilderness and sense of seclusion."[41] Olmsted's firm went on to prepare plans for parks at Niagara Falls, for Stanley Park in Vancouver, and for Rockwood Park in Saint John.

Concern over City Functioning: The City Efficient/Scientific

Fire, disease, and transportation problems of the 19th-century city were slowly addressed by retrofitting the urban areas with physical infrastructure, and regulating the use of property and construction of buildings. The need for infrastructure became apparent when the private provision of services began to fail in more crowded cities. Water of dubious quality was delivered in barrels to those who could afford it; human waste was thrown in pits on the streets; firefighting teams would watch your house burn if you had not insured with their company; city roads were almost impassable during the muddy seasons and the horse-drawn streetcars were often stuck.

As urban areas grew in overall size and density, the need for city-wide systems of infrastructure was gradually acknowledged by politicians. Planning for these systems fell into the hands of a new profession—civil engineering—that emerged at that time. Roman innovations such as aqueducts, gravity sewers, and streets paved with stone blocks were revived in Paris, New York, London, and Boston after conditions became intolerable. New technical innovations quickly followed: piped water supply delivered by steam power, coal gas for light and heat, electric streetcars and streetlights.

But the pace of technical innovation greatly outstripped the ability of the cities to implement them. Most new utilities were "natural monopolies": the most efficient water system was a single supplier of piped water under pressure for both drinking and firefighting, for example. And the part-time, volunteer local governments originally set up for rural settlement often lacked the sophistication and financial resources to develop or maintain these systems, or even manage private companies who offered to install the system if granted a monopoly contract. They tried to address building problems with a patchwork quilt of regulations. A disastrous fire might lead to regulations prohibiting thatched roofs, requiring masonry for new construction and limiting building height to the tallest ladder of the engine company. Dreadful smells from animal-rendering factories or coal-gas plants might lead to bylaws controlling their use or location. But each city's or town's regulations were different, and all were constrained by a local government system designed for rural communities.

So, advocates for planning the efficient growth of a city using the latest scientific and engineering advances were often involved in attempts to reform local government. The great technical successes of the engineers in the early 20th century often had to wait for the institutional innovations of the legislators and lawyers.

Establishing a Local Government System

Local governments were not established in Canada until well after the original colonial town sites were surveyed and settled. In many cases this took over a hundred years, and the process was often arduous. The colonial governments of Britain and France were usually loath to give up their control, especially to the local population. After the establishment of Upper and Lower Canada, progress began to be made toward local self-government. There was, however, little experience in doing so and no nationwide system to guide it. With Confederation in 1867, each province began to establish its own arrangements for local government, frequently learning from one another as their settlement patterns warranted it. As Plunkett and Betts conclude, "by the beginning of the 20th century, all ten provinces in Canada had established essentially similar systems of local government."[42]

The achievement of local government on a broad basis in Canada marks an important threshold for community planning. This institutional foundation was essential for community planning to become established and effective. To appreciate better this vital nexus, it is helpful to review the process of establishing local government in this country.

Precursors of Local Government

For the first 225 years of European settlement in what is now Canada, until the 1830s, few communities had anything resembling local government. This institution literally had to be invented. But why was it sought? The answer to that question is that any clustering of people to form a community will generate needs that are common to the setting, and some way must be found to satisfy these needs. In the early period of settlement, the needs are usually quite rudimentary. Roads and streets, a quay or dock may be needed, as well as a school and the means to fight fires. Some of these needs require construction, some require organization, and some continue to evolve into the indefinite future. All require resources—money, time, labour.

Deciding upon the needs of a community is relatively easy. Deciding upon how to mobilize the necessary resources is more difficult because it involves reconciling viewpoints on how best to provide for community needs. Much of the history of local government institutions in Canada is concerned with determining the best way to structure the decision-making process: Who should be involved? How much power should they have?

The earliest settlements were under French or British colonial rule, and the power to govern them did not lie in community hands. However, there are many instances in which the early settlers felt that neither the governments in Europe nor the colonial administrators adequately perceived the needs of their communities. Efforts to promote local self-rule began in Quebec communities in the mid-1600s. Citizens elected a few representatives to discuss with the French governor the needs of the communities. These efforts were short-lived because the central government disapproved, and it was not until the early 1800s that further steps were taken to allow municipal institutions in Quebec. The settlers of the Maritime provinces and Upper Canada (Ontario) were mainly from the newly independent United States. They brought experience of local government with them, and soon sought a voice in their community's affairs; but there was little response from the British colonial government, with one notable exception. Saint John's 5000 settlers from New England obtained a municipal charter in 1785, the first in Canada. It was 47 years before the second was issued.

Some communities in the British colonies had local government in the form of the courts of quarter session. This ancient institution consisted essentially of "magistrates" appointed by the colony's governor who attended court in each district four times a year for both judicial and legislative purposes. They were responsible for maintaining order and settling minor lawsuits, but they could also establish regulations for moving animals, license various businesses, and appoint minor officials to maintain roads.[43] Counties were delineated at this time, and a large centre within each served as the seat of the courts of quarter session. These occasional and centrally controlled mechanisms were not acclaimed by all settlers. Many, especially in Ontario, wanted locally elected councils empowered to deal with community needs. In 1812, the citizens of Kingston agitated for a charter that would allow a municipal council to pass its own bylaws and regulations.[44]

Various changes were made to meet these challenges. Local magistrates were appointed to administer "police towns," but this proved no more satisfactory than quarter sessions. The Public School Act in Ontario in 1816—the first such act in Canada—proved to be a breakthrough, however, as it allowed a community to erect a school, hire a teacher, and elect school trustees to raise taxes and administer the school. Not only did this confer the right of local election, it also established the still-prevailing separation of education from other local functions in a community. In the 1830s and 1840s, local efforts to form municipalities began to succeed. Brockville was the first in Ontario in 1832, followed by York (Toronto), Kingston, and Hamilton; in Quebec, Montreal and Quebec City obtained charters in 1832; and in the Maritimes, Halifax was incorporated in 1841, Fredericton in 1848.

The Emergence of Local Government

During the 1840s, considerable effort was made to develop a sound system of local government in both Upper and Lower Canada. Ontario's Municipal Act of 1849, often referred to as the Baldwin Act after its originator, proved to be workable and enduring. Its format was for two classes of local government: (1) cities, towns, and villages for urban communities and townships for rural communities; and (2) counties that comprised the local municipalities. Sources of tax revenue were specified, as were the form of elections and the composition of local councils and their duties. With the exception of new local government arrangements designed for metropolitan areas (see Chapter 9), local government forms are not substantially different today than those specified in the early acts.

Two important contextual elements in the development of local government in Canada concern the provincial government's role. The first derives from the fact that, except for the Prairie provinces and northern territories, all the others began as colonies governed from abroad. The tendency of British colonial administration was for strong central control. If municipal institutions were allowed, the central government would ultimately be responsible, a model still found in Britain. Thus, as the Canadian provinces were emerging, there was only grudging acceptance of local self-rule. Self-rule was, and continues to be, hemmed in by various provincial audits, mandatory approvals, and supervision of local affairs. Witness the provincial hegemony regarding local governments in, for example, the Ontario government's imposed amalgamations in Toronto, Ottawa, Hamilton, and Kingston in the late 1990s, or similar actions in Quebec and British Columbia.[45]

The second element concerns the constitutional setting for local government that was established in the British North America Act (BNA Act) of 1867. That original constitution did not provide communities with the right to a local government, but only specified that the responsibility for establishing municipal institutions lay with the provincial governments. Section 92 of the BNA Act defines this prerogative. (The 1982 Constitution carries forward the same section.) It is common to find municipalities referred to as "creatures" of the provincial government. While this superior role of the province is true in other federal states, such as the United States and Australia, what is significant in Canada is the alacrity with which provincial governments perform their role. The development both of local government and, subsequently, of community-planning institutions, bears this stamp of provincial paternalism.[46]

The local government systems that were created by the provinces in the latter half of the 19th century were limited in their roles and in their sources of revenue. Local governments were conceived primarily in terms of providing such services as roads, water, and fire protection to residences and places of business. This is often described as "providing services to property." In turn, local governments were restricted to taxes applied to private property for their revenue. It is not too difficult to appreciate how this constraint has forced municipalities to emphasize property ownership in their bylaws and regulations, as well as how it colours their outlook on the community. As a consequence, property owners and those who desired to develop property came to feel that their demands on the local government were deserving of attention. They, after all, were the primary source of municipal revenues. Lastly, municipalities were limited in their ability to provide social services. Education, health, and welfare services were prerogatives retained by the province, to be delegated—if at all—to such special agencies at the local level as school boards.

The Reform Epidemic

The beginning decades of local government were not auspicious. Inexperience with locally run institutions, combined with a bias toward property ownership and owners, was responsible for many cases of inefficiency, uneconomic practices, and even corruption. By 1900, and the dramatic upsurge in urban development, many local governments appeared unable to cope with the new demands. They were faced with responsibilities for traffic control, parks, housing, and community planning, the extent of which had not been contemplated by those who framed the various municipal acts. Inter-municipal problems arose in transportation, water supply, and protective services, because residences spread into suburbs. And financial resources were often strained to the limit when large bond issues were needed to finance public works projects.

A municipal-reform movement arose primarily from within business and professional circles, since this would likely benefit their interests. They contended that a local government's main task was to provide services efficiently and economically. City growth translated for them into technical problems in engineering and fiscal management. Goldwin Smith, prominent among those involved in the reform movement in Toronto at the turn of the century, stated: "A city is simply a densely populated district in need of a specially skilled administration."[47] Good government for a municipality was thus cast in terms of a business management model. This became a powerful viewpoint, and one that persists today in many communities.

The reformers were also generally suspicious of popularly elected local governments (even though, until recent decades, only property owners could vote in most civic elections). They argued for boards and commissions with specific functions to perform and with an appointed membership. Ostensibly, the intention was, as Vancouver's Mayor Bethune said at the time, to attract "the services of bright able men who have not the time to serve on Council."[48] Thus, across the country many such agencies were formed, with elected representatives often excluded from serving. Montreal acquired a Park Commission; Vancouver, a Water Works Commission; and Fort William–Port Arthur (Thunder Bay), a Public Utilities Commission. These joined school boards, library boards, and police commissions in a plethora of quasi-autonomous bodies, most of which had independent budgets and their own technical and professional staff. Community planning and planners would have to vie for their own position within these organizational arrangements as Canada entered the 20th century.

Reflections

It seems paradoxical in some ways that this turbulent period in urban development and civic affairs was the milieu in which community planning emerged. Yet without this struggle to solve urban problems, debate the merits of various approaches, and strengthen local government, community planning appeared as not much more than idealistic rhetoric. This was no longer the time when kings and colonial masters decided on the form of cities and towns and on how they were governed. The Industrial Revolution broke those bonds, first in Europe and then in North America, where the ethos of pluralist society and local self-government was prominent. (The latter was fostered in considerable measure by the colonial model of making land available under individual ownership to settlers.) Thus, as the various problems of city growth and development arose, there was the desire for broad debate about solutions.

Community planning, to this day, does not grow and develop independently of real urban problems or their solutions as perceived by citizens, professionals, and governments. So, in the arduous period of wrestling with how to attain urban betterment of the 19th century, those who championed a comprehensive community planning were being informed, were learning, of the perceptions of the time. These views would, in turn, need to be thrashed out by a pluralist society (however defined at the time) and be implemented. But this required effective local institutions that had, literally, to be invented in many instances. Therefore, this was a period of enormous social learning, not least for the advocates of community planning.

By the beginning of the 20th century, community planning had achieved a sufficiently stable footing to start making a difference in the form, living conditions, and functioning of Canadian cities and towns, as we shall see in Chapters 4 and 5. But note that this unfolding future was not without its own urban problems and competing views of their solution. Thus, keep in mind the question:

- ***What parallels are there for community planning in the 20th century from the experience of the previous century?***

Endnotes

1. Marc Denhez, *The Canadian Home: From Cave to Electronic Cocoon* (Toronto: Dundurn Press, 1994).
2. R. Cole Harris, *The Resettlement of British Columbia: Essays on Colonialism and Geographical Change* (Vancouver, BC: UBC Press, 1997).
3. James W. Vance, *The Continuing City* (Baltimore: Johns Hopkins University Press, 1990), 200ff.
4. Peter Moogk, *Building a House in New France* (Toronto: McClelland and Stewart, 1971), 14.
5. As quoted in Moogk, *Building a House*. This is in reference to the settlement that evolved to the north of the original French village of Ville Marie.
6. John Reps, *Town Planning in Frontier America* (Princeton, NJ: Princeton University Press, 1969), 82; and Phyllis Lambert and Alan Stewart, *Opening the Gates of Eighteenth-Century Montréal* (Montreal: Canadian Centre for Architecture, 1992); Jean-Claude Marsan, *Montreal in Evolution: Historical Analysis of the Development of Montreal's Architecture and Urban Environment* (Montreal and Kingston: McGill-Queen's University Press, 1990).
7. E. Deville, "Radial Hamlet Settlement Schemes," *Plan Canada* 15 (March 1975), 44 (reprinted from *Conservation of Life* [April 1918]).
8. Michael Hugo-Brunt, "The Origin of Colonial Settlements in the Maritimes," in L.O. Gertler, ed., *Planning the Canadian Environment* (Montreal: Harvest House, 1968), 42–83.
9. L. Gentilcore, ed., *Ontario's History in Maps* (Toronto: University of Toronto Press, 1984).
10. Clarence Karr, *The Canada Land Company: The Early Years,* Research Publication No. 3 (Ottawa: Ontario Historical Society, 1974), 24ff. Also Gilbert Stelter, "Guelph and the Early Canadian Town Planning Tradition," *Ontario History* 77 (June 1985), 83–106.
11. John Reps, *Town Planning,* 350, makes a similar point regarding the origins of the radial pattern of Buffalo, New York, which was laid out for the Holland Company by Ellicott.
12. Larry D. McCann and Peter J. Smith, "Canada Becomes Urban: Cities and Urbanization in Historical Perspective," in Trudi Bunting and Pierre Filion, eds., *Canadian Cities in Transition* (Toronto: Oxford University Press, 1991), 69–99.
13. Hans Blumenfeld, *The Modern Metropolis* (Montreal: Harvest House, 1967), 27.
14. As quoted in Charles M. Godfrey, *The Cholera Epidemics in Upper Canada 1832–1866* (Toronto: Secombe House, 1968), 20.
15. As quoted in William Ashworth, *The Genesis of Modern British Town Planning* (London: Routledge & Kegan Paul, 1954), 49.
16. Alan Artibise, *Winnipeg: A Social History of Urban Growth, 1874–1914* (Montreal and Kingston: McGill-Queen's University Press, 1975), 232ff.
17. John H. Taylor, "Fire, Disease and Water in Ottawa," *Urban History Review* 8 (June 1979), 7–37.
18. J. Grove Smith, *Fire Waste in Canada* (Ottawa: Commission of Conservation, 1918), 277–289.
19. John Whittow, "Disaster Impact and the Built Environment," *Built Environment* 21:2/3 (1996), 81–88; and Lawrence J. Vale and Thomas J. Campanella, eds., *The Resilient City: How Modern Cities Recover from Disaster* (New York: Oxford University Press, 2005).
20. Shirley Spragge, "A Confluence of Interests: Housing and Reform in Toronto, 1900–1920," in A. Artibise and G. Stelter, eds., *The Usable Urban Past* (Toronto: Carleton Library, 1979), 247–267.
21. Herbert B. Ames, *The City Below the Hill* (Montreal: Bishop Engraving, 1897), 27–47.
22. Artibise, *Winnipeg,* 150ff.
23. Peter G. Goheen, *Victorian Toronto 1850 to 1900,* Research Paper 127 (Chicago: University of Chicago, Department of Geography, 1970), 84.
24. For an example of the breadth of concerns, see Paul Rutherford, ed., *Saving the Canadian City: The First Phase, 1880–1920* (Toronto: University of Toronto Press, 1974).
25. Mel Scott, *American City Planning Since 1890* (Berkeley: University of California Press, 1969), 33.
26. William H. Wilson, *The City Beautiful Movement* (Baltimore: Johns Hopkins University Press, 1989).
27. Daniel H. Burnham and Edward H. Bennett, *Plan of Chicago* (Chicago: Commercial Club, 1909).
28. As quoted in Walter Van Nus, "The Fate of City Beautiful Thought in Canada, 1893–1930," in G. Stelter and A. Artibise, eds., *The Canadian City: Essays in Urban History* (Toronto: Macmillan, 1979), 162–185.
29. For the role of architects, see Harold Kalman, *A History of Canadian Architecture,* vol. 2 (Toronto: Oxford University Press, 1994), 649–659.
30. Lance Berelowitz, *Dream City: Vancouver and the Global Imagination* (Vancouver, BC: Douglas & McIntyre, 2005), Ch. 5.
31. Gilbert Stelter, "Rethinking the Significance of the City Beautiful Idea," in Robert Freestone, ed., *Urban Planning in a Changing World: The Twentieth Century Experience* (London: Spon, 2000), 98–117.
32. See Peter Hall, *Cities of Tomorrow* (London: Basil Blackwell, 2002), Ch. 1; and Friedrich Engels, *The Condition of the Working Class in England in 1844,* W.O. Henderson and Witt Challoner, trans. (London: Blackwell, 1958/1845).
33. F.J. Osborn makes the point in his "Preface" to Ebenezer Howard, *Garden Cities of Tomorrow* (London: Faber and Faber, 1946). Howard's original 1898 work has been reprinted as *Tomorrow: A Peaceful Path to Real Reform* (London: Routledge, 2003) with commentary by Peter Hall, Dennis Hardy, and Colin Ward.
34. Mervyn Miller, *Letchworth: The First Garden City* (Chichester, UK: Phillimore, 2002).
35. Maurice De Soissons, *Welwyn Garden City: A Town Designed for Healthy Living* (Cambridge, UK: Publications for Companies, 1988).
36. Thomas Adams, *Outline of Town and City Planning* (New York: Russell Sage Foundation, 1935), 275.
37. See Stephen Ward, ed., *The Garden City: Past, Present and Future* (London: Spon, 1992); Peter Hall and Colin Ward, *Sociable Cities: The Legacy of Ebenezer Howard* (Chichester, UK: Wiley, 1998); and Stephen Ward, *Planning the Twentieth-Century City: The Advanced Capitalist World* (New York: Wiley, 2002).
38. David Schuyler, *The New Urban Landscape: The Redefinition of City Form in Nineteenth-Century America* (Baltimore: Johns Hopkins University Press, 1986); and J.R. Wright, *Urban Parks in Ontario, Part I: Origins to 1860* (Toronto: Ontario Ministry of Tourism and Recreation, 1984); and McKendry Jennifer, "The Role of Cataraqui Cemetery in the Rural Cemetery Movement," *Historic Kingston* 44 (1996), 3–9, 52.
39. J.R. Wright, *Urban Parks in Ontario, Part II: The Public Park Movement, 1860–1914* (Toronto: Ontario Ministry of Tourism and Recreation, 1984). The author provided the original stimulus for this section.
40. As quoted in Elsie Marie McFarland, *The Development of Public Recreation in Canada,* (Toronto: Canadian Parks/Recreation Association, 1974), 14.
41. Frederick Law Olmsted, *Mount Royal* (New York: G.P. Putnam's Sons, 1881), 64. For the Olmsted firm's work in Canada, see Nancy Pollack-Ellwand, "The Olmsted Firm in Canada: A Correction of the Record," *Planning Perspectives* 21 (July 2006), 277–310.
42. T.J. Plunkett and G.M. Betts, *The Management of Canadian Urban Government* (Kingston: Queen's University Institute of Local Government, 1978), 58.

43. Ibid., 48.

44. Ibid., 50.

45. John Meligrana, ed., *Redrawing Local Government Boundaries: An International Study of Politics, Procedures and Decisions* (Vancouver, BC: UBC Press, 2004).

46. Ibid.

47. As quoted in Weaver, *Shaping the Canadian City*, 72.

48. Ibid., 70.

Internet Resources

Chapter-Relevant Sites

Planning Canadian Communities
www.planningcanadiancommunities.ca

Canadian Urban History
www.uoguelph.ca/history/urban.html

The City Beautiful Movement
xroads.virginia.edu/~cap/CITYBEAUTIFUL/city.html

The Plan of Chicago
www.encyclopedia.chicagohistory.org/pages/10537.html

Garden Cities of To-Morrow
www.library.cornell.edu/Reps/DOCS/howard.htm

National Association for Olmsted Parks
www.olmsted.org

The City Scientific
www.library.cornell.edu/Reps/DOCS/ford_13.htm

4

Chapter Four

Pioneering Community Planning in Canada, 1900–1945

You may ask, is it reasonable ... to make plans for generations in the distant future? We have only to study the history of older cities, and note at what enormous cost they have overcome the lack of provision for their growth, to realize that the future prosperity and beauty of the city depends ... upon the ability to look ahead.

Frederick Todd, 1903

It is widely accepted today that community planning aims to improve the quality of daily life in our cities, towns, and regions. But any such public activity does not come into being either quickly or independently of its context. There has to be acknowledged, first, that one or more problems is affecting community well-being, and then a desire to find a solution to it. The latter, importantly, depends on a sufficient body of people being convinced that the activity can contribute significantly to the welfare and prosperity of city and town dwellers. As the 20th century opened, a succession of major problems affected Canadian cities and towns, from their disparate appearance to their inefficient functioning and their unsanitary housing. Each of these sought a response in the new activity of community planning and, in turn, influenced its development through the first half of the century.

The period 1900–1945 saw the pioneering of the ideas and practice of community planning in Canada. It consisted of communities taking some halting steps to assume responsibility for their planning problems and to draw upon technical and professional assistance in this regard. It was also a tumultuous period, with further dramatic growth of cities and advances in transportation, both of which brought new problems. Nevertheless, this period is among the most important in the evolution of community planning. Within it we see the development of the first locally sponsored plans in Kitchener and Toronto, the path-breaking work of the Commission of Conservation, the emergence of provincial planning legislation, and the emergence of a planning profession. In this context a key question to consider is:

- ***How and why did we come to use the planning tools and institutions to which we are so accustomed today?***

The Impact of 20th-Century Urban Problems

The 20th century presented new problems for Canadian cities and towns. The first dozen years of this century constituted a period of unparalleled prosperity, from the wheat fields of Saskatchewan to the wharves of Vancouver and Saint John. Immigrants poured into Canada ostensibly "to open up the West," but most of them ended up in the nation's cities. Urban-based industries thrived on the new inexpensive labour supply. In addition, technological solutions provided cities with infrastructure, both below ground (water and sewerage) and above ground (electric lines, paved roads), fairly rapidly and inexpensively. Probably most important in this period, however, was the advent of powered urban rail transportation—the electric streetcar and the commuter railroad—and slightly later, the self-propelled bus.

Problems of Excessive Subdivision

Urban growth was so highly valued by communities that it was pursued through aggressive "boosterism." Ample supplies of land were deemed essential to accommodate the hoped-for growth. No thorough study has ever been made of the amount of land that was subdivided in anticipation of urban growth in the decade or so before World War I. However, it is possible to glean its extent in several cities from concerned accounts of the problems raised by over-subdivision. Here are a few examples comparing the actual population just prior to 1914 with the population that could be accommodated on the already subdivided land:

Calgary	50 000/770 000
Ottawa-Hull	123 000/1 600 000
Edmonton	40 000/500 000
Saskatoon	12 000/750 000
Vancouver	115 000/750 000

Abetting this surge in land speculation were dramatic increases in the coverage of street railways. Figures for Toronto and Vancouver convey something of the picture that was occurring in other cities. In 1880, Toronto had 30 kilometres of street railways, and only 10 years later had 110 kilometres, while Vancouver went from 26 kilometres in 1900 to 165 kilometres by 1914. The impact of such expansion can be appreciated from the fact that each linear kilometre of streetcar line could serve about 2 square kilometres of residential land or, in pre-1914 densities, potentially 10 000 persons. Not until the automobile freeways of the 1950s would Canadian cities again witness such an impact on this scale.

And, like the freeways later, the streetcar lines brought potential development but, frequently, not actual building. Land was surveyed by the hundreds of hectares, and subdivision plans were registered in land titles offices. There were, of course, not enough people heading for Canada's cities to come close, in most cases, to absorbing the amount of new building lots. Severe problems arose for cities in the short run because of the premature subdivision. As Walter Van Nus notes:

> A developer's desire to extract the maximum number of lots ... often led him to ignore the location and/or width of projected or existing streets nearby, if by doing so he could squeeze more lots out of the property.[1]

This lack of simple coordination of the extensions to cities often required, at a later time, expensive road relocation by the municipalities. In addition, the subdivision development was usually scattered, and this meant extra costs in providing municipal services such as water and sewer lines, roads, sidewalks, and street lighting. Because intervening undeveloped land did not pay its share of the servicing costs, the taxes had to be raised from the new building lots. And, frequently, the land that came into actual development was unsuitable for building—topographically too steep, many rock outcroppings, or poorly drained—and this not only led to immediate problems for communities but often was also a source of problems that many decades later would be termed "urban sprawl."

Cities that espoused growth found themselves on the horns of a dilemma. First were the enormous costs of servicing the new suburbs. Those that demurred in making these expenditures might find the building going to an adjacent municipality. Second, people seeking a building lot were finding land values and taxes much higher along streetcar routes and accordingly sought land beyond the end of the streetcar line. This tended to spread the costs further afield, but not necessarily the benefits. Third, most cities were faced with large housing shortages, which the mere subdivision of land on the fringe did little to alleviate. Land prices and rents rose sharply in Canadian cities in the several years before World War I, especially in and around downtown areas. Poor people and manufacturing firms traditionally located near the centre suffered most. Many firms moved to suburban locations, not infrequently followed by the shacktowns of their workers.

During this period of crises in city-building, a crucial ideological issue had to be resolved. The boom reflected the ethos of economic progress for many, while bringing debilitating circumstances for most others. In whose interest should reforms be made? The view that came to prevail, to quote historian John Weaver, was that "civic resources should assist the endeavours of those who do most for the material growth of the community, namely business and real estate interests."[2]

New Dimensions to Urban Problems

These new urban problems had several significant impacts for the then-fledgling planning profession.[3] They created an awareness of the importance of land subdivision, the developers who undertook it, and the problems of efficiency often associated with it.

Land Subdivision. The unprecedented scale of pre-1914 subdivision brought home the importance of this process. When subdividing a larger piece of property into house lots or other small parcels of land, the new pieces of property acquire independent status, and the community is obliged to honour this whenever they should be built upon. A community's future development pattern and its quality are thereby constrained. On the one hand, there are such technical aspects as road alignment, drainage, lot size, and ease of providing water and sewer lines in which the community has a rightful interest. On the other hand, there is the issue of the degree to which a community should intervene in the development of privately owned property.

The Land Developer. The act of subdividing land in a speculative way involved a relatively new actor in the planning and development of communities—the land developer. The surge of growth created a demand for city land so large that it attracted many more people and groups to seek profit in providing space for newcomers. They joined such old land grantees as the Canadian Pacific Railway, the British American Land Company, and the Hudson's Bay Company; the latter frequently influenced the location of new municipal development by their decisions. On a smaller scale, but equally pervasive, were a coterie of individual landowners (large and small), their agents, and brokers, all anxious to share in the potential profit of the new growth. This was the period for the beginning of a real estate industry. The views of land development interests, the real estate industry, and homeowners all focused upon the primacy of private property. Their concerns held sway in most municipal councils at that time, much as they seem to do today.

Efficiency. The unfettered land development in the early 20th century led to unconnected streets and poorly drained sites, among other inefficiencies. Faced with this large-scale, rapid land development, the tendency of planners to prepare physical designs for new development seemed often to be out of tune with the vigorous, widely supported process of land speculation. Planners thus began to promote planning as a means of obtaining efficiency in city development.

Planning Responses to Early Urban Problems

As communities tackled the problems of disease, pure water supplies, fire, sanitation, slums, and excessive subdivision, their experience gave rise to two important realizations. First, it became evident that these problems resulted from the pace of growth and development of cities. Second, and more slowly, came the realization that the solutions to these problems lay in better coordination, regulation, and physical arrangement of the overall development of cities. The issues of public health, fire safety, and adequate housing became the first social goals of community planning. These issues would be found, time and again, enshrined in the preambles of planning legislation proclaiming that the bylaw or plan was aimed at "improving the health, safety, and public welfare" of the community.

Further, five areas of public concern about city and town development had gathered considerable momentum as the 19th century came to an end. Four of these grew out of substantive issues that have already been identified (see Figure 4.1). The fifth grew out of a widespread dismay over the capability of local government officials to deal adequately with the needs of a community:

1. the city's appearance
2. urban living conditions and housing
3. the state of the natural environment
4. concern over efficient city functioning
5. reform of local government

Figure 4.1 Evolution of Community Planning Ideas, 1900–1945

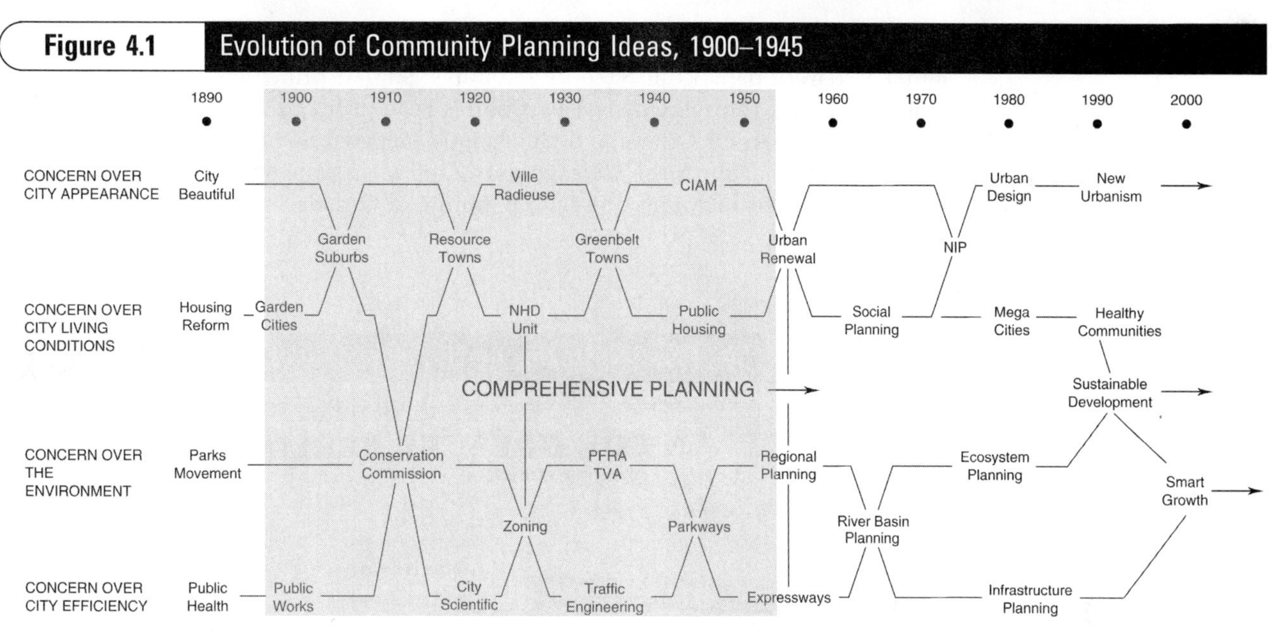

The first half of the 20th century saw the initial planning concerns of the late 1800s put into practice and refined, blended with others, and professionalized. Architects, for example, are often involved in plans concerned with city appearance; social workers and housing advocates are key allies when addressing city living conditions; landscape architects and ecologists help address environmental concerns; and engineers, public-health advocates, and lawyers are involved in plans to improve city efficiency. The foundations of contemporary planning approaches were laid in this period.

Source: David L.A. Gordon

An active movement developed around each of these concerns among members of the public, various professions, newspapers, and many public officials who lobbied for the mitigation of urban problems, the beginnings of which were described in Chapter 3. Often the concern focused on a special-interest organization. There was, of course, overlap in the interests of each; indeed, as can happen in a small country, many of the same people were involved in two or more of these crusades for improvement in Canadian city life.

The community (town) planning movement came out of the confluence of these five streams of concern during the first dozen years of the 20th century. The proponents of each shared many objectives, but they also brought with them differences in values, professional outlook, skills, and jurisdictional focus.

Concern over City Appearance

The burgeoning City Beautiful movement, so profoundly influenced by the 1893 Chicago World's Fair, lent considerable momentum to addressing concerns over the appearance of Canadian cities in the early 20th century. But beautification as an objective began to falter after World War I as new concerns over city efficiency, the environment, and living conditions gained prominence, and the Beaux Arts basis for City Beautiful aesthetics was attacked by the Modern design movement of the 1920s and 1930s. By the Depression of the 1930s, any proposal to make cities more attractive seemed like a waste of scarce public resources.

The Peak of the City Beautiful, 1900–1915

In 1906, the first citywide planning proposal in Canada to emanate from City Beautiful approaches was undertaken for Toronto by the Ontario Association of Architects and the Toronto Guild of Civic Art. It contained plans for a series of diagonal streets and a system of parks connected by parkways. Similar principles guided the plans made for Berlin (now Kitchener) by Charles Leavitt, Jr., in 1914 (see Figure 4.2), and the 1915 plan for Ottawa and Hull jointly prepared by Edward Bennett, Daniel Burnham's associate on the Chicago plan, and Canadian Arthur Bunnell, whose name would recur in connection with planning in Canada through the next several decades.[4]

Figure 4.2 City Plan for Greater Berlin (Kitchener-Waterloo, Ontario), 1914

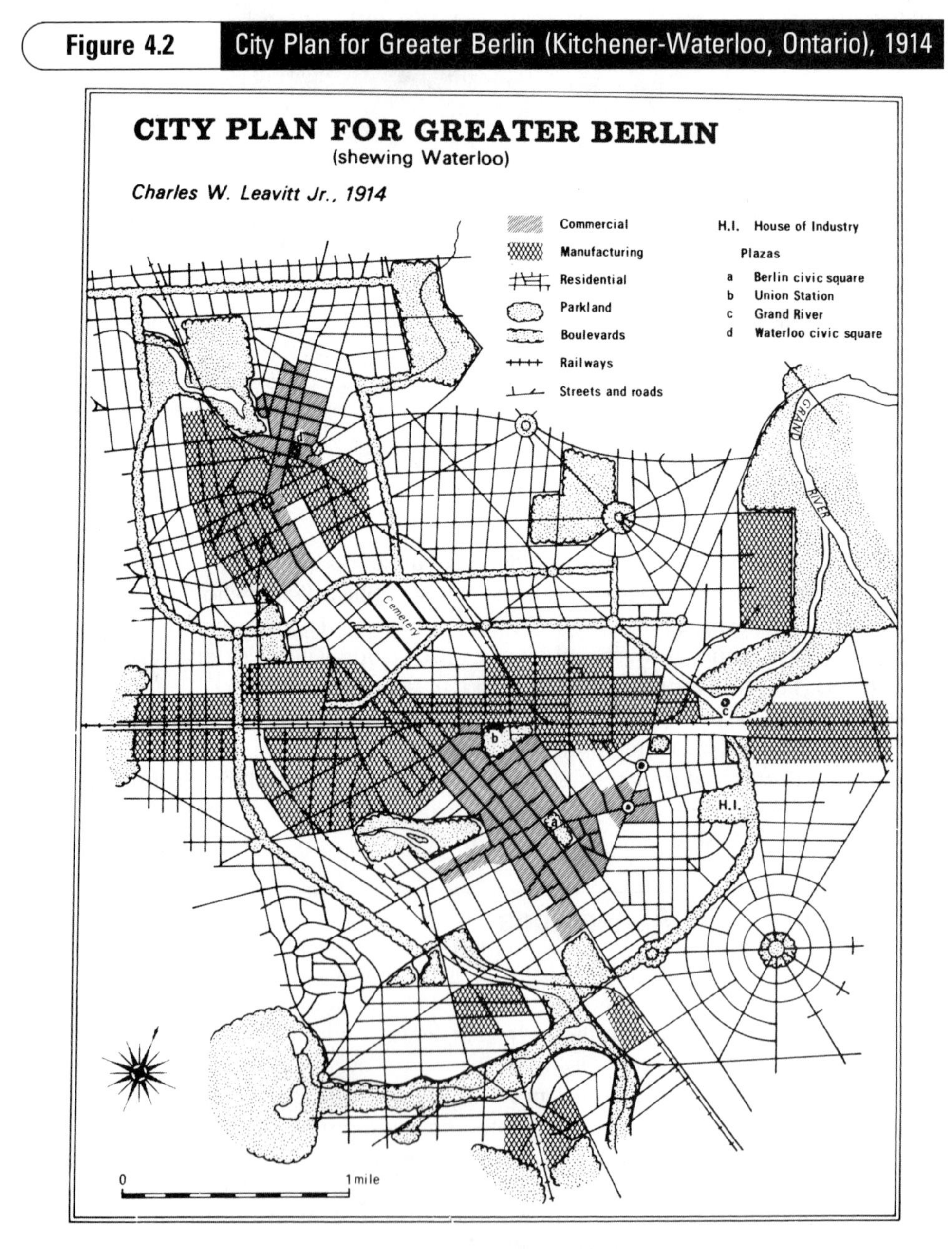

This plan by American planner Charles Leavitt, Jr., employs all the City Beautiful design devices: diagonal avenues leading to city squares, circular streets, and parkways. Kitchener was named Berlin prior to World War I.

Plans prepared in the City Beautiful style also influenced the development of some provincial capitals, especially Regina. English landscape architect Thomas Mawson's 1914 plan for the lands surrounding the new Saskatchewan legislative building influenced all future designs for the Wascana Centre, although the new City Hall was not built until 1978.[5]

The achievement of civic grandeur was an aim of City Beautiful planning.[6] This was often expressed in plans of civic centres, with monumental public buildings grouped around a public square and a broad tree-lined avenue leading to it as in the one proposed for Edmonton (Figure 4.3). Others were proposed (but never built) for Calgary by British landscape architect Thomas Mawson in 1914[7] and for Vancouver by American planner Harland Bartholomew in 1929.[8] Canadian Horace Seymour collaborated on the latter plan and also included City Beautiful elements in other plans he made at the time.[9] In 1921, another respected Canadian planner, Noulan Cauchon, prepared a plan for Hamilton's civic centre in the same tradition, and the monumental quality we now see in Toronto's University Avenue, which leads to the provincial legislative buildings, was cast in a 1929 redevelopment plan for that city.[10] One might add to this list of City Beautiful projects the former city of Maisonneuve and the many railway stations built in a grand manner in Canadian cities in this period.[11]

The **City Beautiful** movement was later commonly described in condescending terms—as "mere adornment," as having failed to address the "real problems" of city housing and sanitation, as extravagant.[12] A number of important factors are missed in these debates. First, the classic mode that characterized many of the architecture and design concepts was the design style of the time, and it was a style that favoured adornment of buildings and public places. Second, the design style was rooted in powerful aesthetic principles that had endured from Renaissance times: symmetry, coherence, perspective, and monumentality. Even those not professing City Beautiful tenets used design elements that drew upon these principles. Ebenezer Howard had his Grand Avenue and radial streets leading to the town centre; Thomas Adams's plan for several resource towns employed a central tree-lined boulevard around which to organize the community. As recently as 1979, the plan for a new civic centre in Calgary was criticized on the same grounds as Mawson's City Beautiful centre half a century earlier. Third, regardless of design style, City Beautiful planners had correctly identified most of the main elements of a community's physical form with which a planner needed to work: the street pattern, the public buildings, and the parks—the elements under public control.[13]

The Rise of Modernism and Le Corbusier

The Beaux Arts roots of City Beautiful were attacked in the clamourous times following World War I. The Modernist movement in literature, art, and architecture declared "the end of history," and models of Greek and Roman buildings were consigned to the basement storerooms of architectural academies. At the influential Bauhaus School of Weimar, Germany, students in art, architecture, and city planning started with common studios that focused on the design possibilities of modern materials, rather than the historic precedents from the past. Planners were given a clean slate—the "*tabula rasa*"—and asked to imagine the perfect modern city.

The results were often shocking, sometimes brilliant, and a striking departure from the classical compositions of the City Beautiful or the small-scale arts and crafts humanism of Garden City projects. Perhaps the most influential ideas for cities came from the Swiss-born architect Le Corbusier in the 1920s, who declared that a house is a machine to live in. Instead of small clusters of mostly single-family houses, each with its neighbourhood park, Le Corbusier envisioned the city as a huge park where 60-storey office towers and 10-storey apartment buildings were woven in zigzag form across landscaped space in his Contemporary City (Ville Contemporaine)

Figure 4.3 Proposed Civic Centre for Edmonton, 1915

A typical proposal for a grandiose civic centre by Thomas Mawson, similar to those in other Canadian cities in this period, when Edmonton had barely 40 000 people. The Minneapolis landscape architects Morell and Nichol prepared similar proposals for Saskatoon. The aim of such City Beautiful designs was to bring a sense of civic grandeur to otherwise mundane cities.

Source: Glenbow Archives, Calgary, Alberta, NC–6–161.

concept (see Figure 4.4). Hardly more than 5 percent of the ground would be covered, and many buildings constructed on stilts would allow the park space to flow underneath. Le Corbusier promoted skyscrapers and the possibilities they gave for "concentrating" the population without "congesting" it.[14] His concept had people living within a park at very high densities—3000 persons per hectare—but also promised to save them considerable time in horizontal travel compared to a low-density, spread-out city.

Figure 4.4 Le Corbusier's View of the Contemporary City, 1922

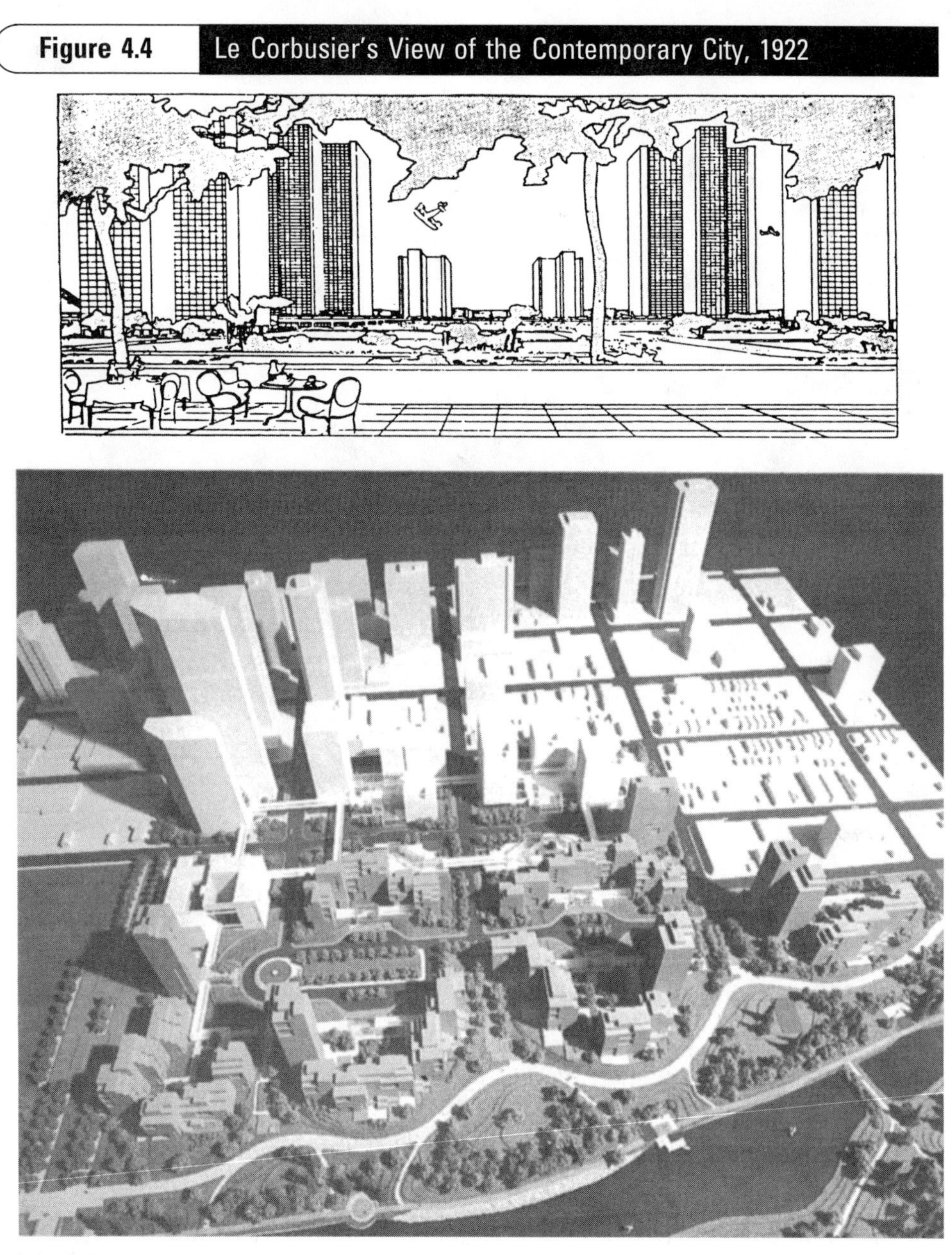

In Le Corbusier's eyes, modern technology allowed cities to avoid congestion and to enjoy open space at their very centres. Tall towers are arranged within a park-like setting, with very little of the ground area covered by buildings (top). His concepts are still very influential, as is evident in the 1979 proposals for Calgary's downtown (bottom).

Source: Le Corbusier, "Le Corbusier's View of the Contemporary City, 1992." © FLC (Paris)/SODRAC (Montreal) 2002 (top); Calgary Planning Department (bottom).

Although the Contemporary City was designed as the ideal solution for building a modern city for three million people on a "green field" site, Le Corbusier also believed that his principles should be applied in the reconstruction of existing European cities such as Paris, his adopted home. In his 1925 Plan Voisin, he proposed demolition of most of the historic centre of Paris, north of the Seine, to build a modern city centre with high-rise office buildings, avenues for fast-moving traffic, and apartment blocks (see Figure 4.5). Since Paris was considered the epitome of urban sophistication and

Figure 4.5 Le Corbusier's Plan Voisin for Central Paris, 1925

Le Corbusier's proposal was to demolish the streets and buildings in central Paris, leaving only the Louvre. The vacant land was to be rebuilt with the elements of the Ville Contemporaine—tall office blocks, mid-rise apartments, and expressways. Paris rejected this plan, but a surprising number of cities followed the formula during the 1950s and 1960s.

Source: © FLC (Paris)/SODRAC (2006).

beauty at the time, the Plan Voisin was greeted with consternation, but the uproar made Le Corbusier famous.

The CIAM and the Athens Charter

By the mid-1920s, Le Corbusier made links with the leaders of the German design schools and other European architects and planners. He became a central figure in the CIAM (Congrès Internationaux d'Architecture Moderne), which discussed the new theories for building and planning.[15] Although Le Corbusier was not successful in building his early plans, other CIAM members built demonstration projects in Frankfurt, Amsterdam, and Vienna.

The CIAM continued to refine its planning theories during the economic depression of the 1930s, when little building took place in Europe or North America. Some Modern architects and planners tried their hand in the Soviet Union, while others prepared a framework for planning, named after the 1933 CIAM Athens conference. The Athens Charter recommended that urban planning be divided into four functions: dwelling, work, recreation, and transportation. It also recommended rigorous separation of land uses and their connection by high-speed transportation, making use of automobiles on exclusive roadways, electric railroads in subways, and aircraft.[16] The Modern planners regarded the historical city as inefficient and obsolete, and were ready with a well-elaborated theory for planning a new kind of city in the reconstruction after World War II.

Concern over Housing and Living Conditions

In the early 20th century, the concern over urban living conditions continued, and the first steps were made in housing reform to provide inexpensive homes for working-class families. Some Garden City principles were incorporated into garden suburbs and planned resource towns in the Canadian north. And a new set of residential planning principles—the neighbourhood unit—was tested in the Radburn and Greenbelt "new towns" in the United States.

The Housing Reform Movement

Those involved with public health matters were among the first to try to arouse public concern over housing conditions, especially for the poor. They urged sanitary improvements and reductions in density. The initial arguments were humanitarian in nature, and, as broader support was sought, they were augmented later with the plea that poor housing promoted disease and caused major costs for the nation's businesses. Industrialists and businessmen had joined the debate over housing in the late 1890s. From their perspective, poor housing, which facilitated disease among workers, also led to absenteeism and lowered productivity. Moreover, much of the poor housing occupied by workers was expensive, and high rents usually meant pressure for more wages, the industrialists unabashedly stated.[17]

But obtaining better housing for workers was difficult. With the dramatic growth of Canadian cities in the early part of the 20th century, the cost of suburban land was pushed up and inner-city land prices stayed high as downtown areas burgeoned. Municipally provided (public) housing was considered but did not receive much support until the 1920s. Tenant cooperatives were also proposed following the urgings of Henry Vivian, a British MP who toured Canada in 1910 to promote the idea of a partnership of tenants subscribing the capital for a housing development. The most notable such project was Cité Jardin in Rosemont.[18] Also in Montreal, industrialist Herbert Ames built a model tenement project called Diamond Court soon after the turn of the century to demonstrate how low-income housing could be improved.

Limited-dividend housing (providing a fixed rate of return on the investment) built with private capital was favoured mostly by affluent members of the community. It tapped their philanthropic spirit and reduced any "socialist" tendencies to have the municipality provide housing. A popular phrase of the time was Thomas Roden's "philanthropy and five percent," which referred to the dividend that investors would receive in building worker housing. In 1907, Roden helped form the Toronto Housing Company, which eventually built several hundred dwellings. This experience is important in several ways. First, there was local government involvement in the projects, for although the company was started by private investors, its bonds were guaranteed up to 85 percent by the city government. Ontario passed legislation allowing this in 1913, following the lead of Nova Scotia, which included the option in its 1912 Town Planning Act (the second in the country). Second, the projects were designed as row houses surrounding ample courtyards in the tradition of Hampstead Garden Suburb housing (see Figure 4.6). In this way, both high density and amenity were achieved, and to this day, the projects retain a distinctive, pleasing character. And, third, organizers of the company, most notably G. Frank Beer and Sir Edmund Osler, were also prominent in arguing for community-wide planning. They campaigned widely to provide playgrounds and better transportation, as well as to reduce the congestion of the population. Beer, in his remarks to the 1914 conference of the Commission of Conservation in Ottawa, linked "city planning" to conservation: "The conservation of life and desirable living conditions . . . are inseparable."[19]

The Garden Suburb

A hybrid of Garden City and City Beautiful approaches was the Garden Suburb, so called because it employed the generous residential environment of the newly developed

Figure 4.6 Housing Project for Workers: Toronto, 1913

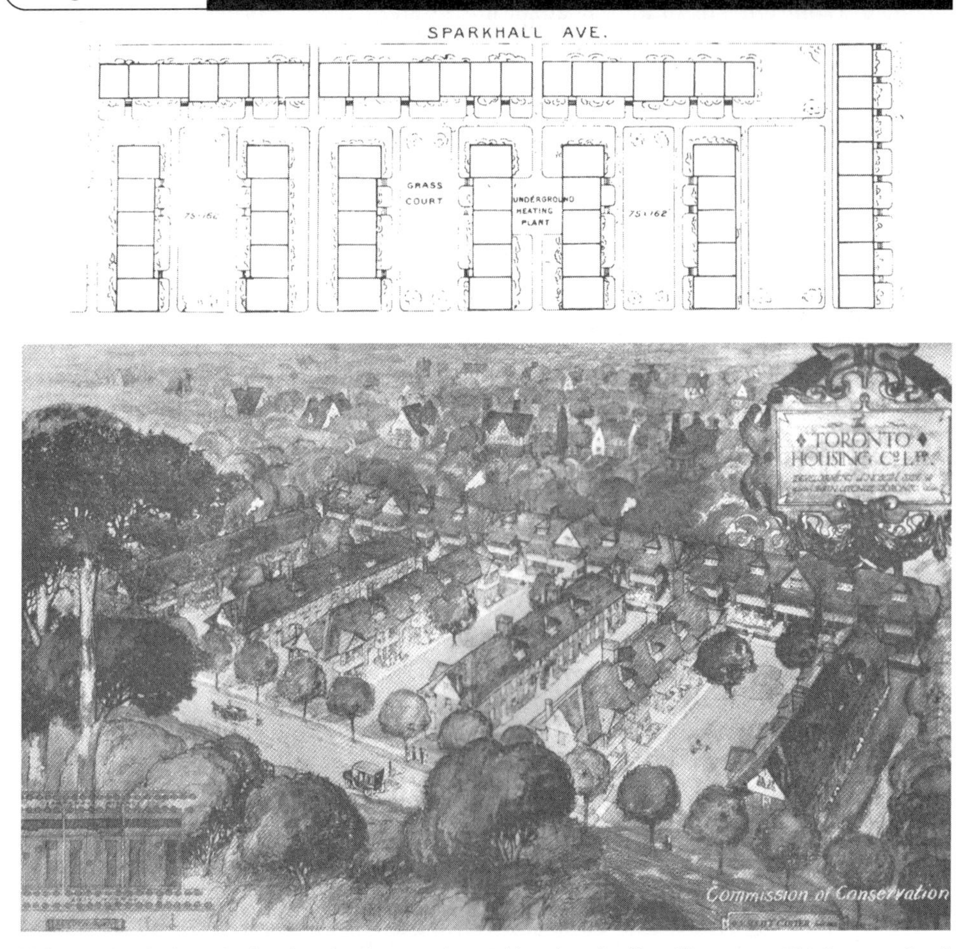

Plan and sketch view of a housing development for working-class families. There were 204 "cottage flats" (one- to four-bedroom row houses). The Toronto Housing Company (a limited-dividend corporation) built the project, which is still in use as a non-profit housing cooperative.

Source: Plan of devasted area

Letchworth Garden City, and was usually located just beyond the built-up urban area. Further, as with City Beautiful ideas, it broke with the standard gridiron pattern of streets, often termed at the time the "monotonous grid."

The forerunner of the Garden Suburb approach is Hampstead Garden Suburb in London, England, which was designed by Raymond Unwin, co-designer of Letchworth. Its curving streets fitted to the topography, and its parks and open space gave the inspiration for dozens of such residential areas across North America as well as Europe.[20] The most important Garden Suburb experiment in the U.S. was Forest Hills Gardens in New York City, planned by Frederick Law Olmsted, Jr., and developed by the Russell Sage Foundation.[21] The notable Garden Suburb projects in Canada are Oak Bay in Victoria,[22] Shaughnessy Heights in Vancouver (see Figure 4.7), Mount Royal in Calgary, Tuxedo Park in Winnipeg, Leaside and Forest Hill in Toronto, Lindenlea in Ottawa,[23] and the Town of Mont-Royal in Montreal.[24] All of these were begun before 1920; many of them were built on land owned by a railway company and the designers were usually major landscape architecture firms such as the Olmsteds or Frederick Todd.

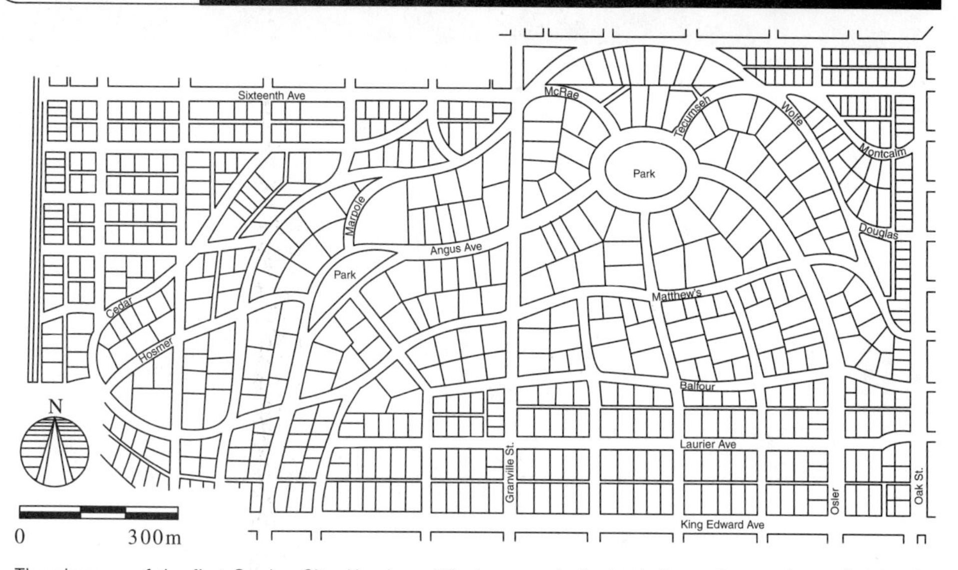

The planners of the first Garden City, Unwin and Parker, popularized this form of street layout in planning new suburbs. It was used in the design of such districts for wealthy Canadian homeowners as Mount Royal in Calgary, Leaside in Toronto, and the Town of Mont-Royal in Montreal, as well as this one in Vancouver.The gridiron was discarded by Frederick Todd in favour of curving streets, which respected the hilly topography.

Thomas Adams's 1917 plan for the rebuilding of the Richmond District in Halifax is a notable Garden Suburb design. The district originally had a common gridiron pattern, even though it was on a steep hillside, and was devastated by an enormous munitions ship explosion and fire. Adams's proposed curving streets conformed to the topography, and his plan was largely applied in the reconstruction.[25] The adjacent Hydrostone neighbourhood provided affordable housing made of fireproof concrete block ("Hydrostone"). Adams' simple design alternated service alleys and residential streets with wide central boulevards. The alleys hide the service poles and garages, while the tree-lined boulevards provide a small park for every block. The Hydrostone neighbourhood has had enduring appeal, and was the first area designated as a National Historic Site for its planning principles (see Figure 4.8).

Most Canadian Garden Suburbs were never intended to provide housing for the mass of people in the community. Their generous—even by today's standards—design features meant high prices were attached to the building lots. Yet the high design standards have proved a crucial factor in the persistence of such areas as Shaughnessy Heights and the Town of Mont-Royal as favoured residential districts for three-quarters of a century while many adjacent areas have deteriorated. Finally, the Garden Suburb approach provided the stimulus to later community designers of resource towns and metropolitan suburbs alike in Canada.

Planned Communities for Resource Development

The development of dozens of new towns on the Canadian resource frontier paralleled the expansion of cities in the early part of the 20th century. The planning of these resource settlements ranged in sophistication from a simple grid survey appended to the site for the mine, mill, or smelter, to conscious attempts to create attractive, healthful communities. Possibly the first such planned resource town was Nanaimo, British Columbia, with an interesting "cobweb" street design (dating from about 1880) of radials and

Figure 4.8 A Garden Suburb with Affordable Housing: The Richmond Distric and Hydrostone Project, Halifax

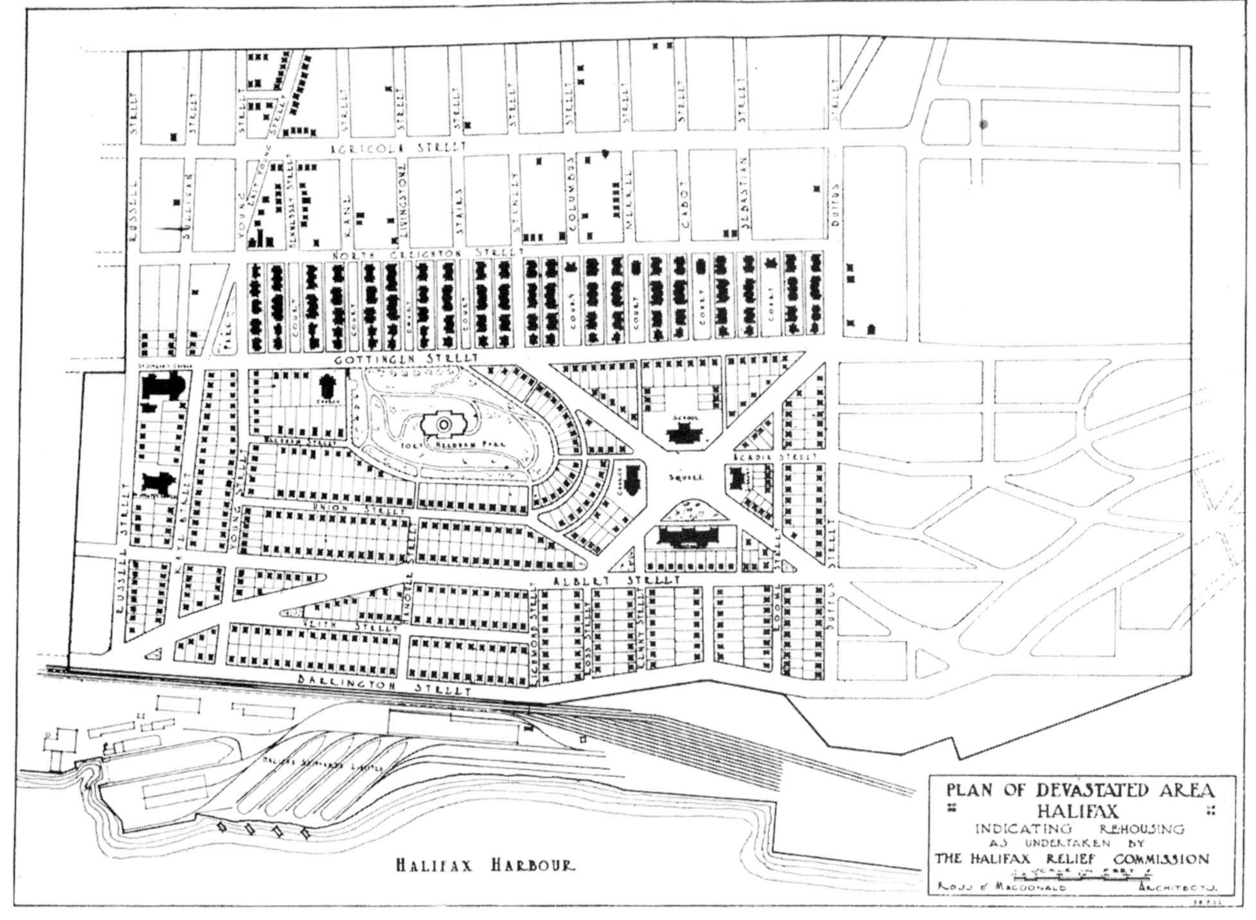

Thomas Adams's 1917 plan for schools at the central square, and a memorial on top of the hill at the central park. The Hydrostone neighbourhood is the regular grid of short blocks, north of the park.

Source: "Plan of Devasted Area Halifax indicating rehousing as undertaken by the Halifax Relief Commission." Ross & MacDonald, architects, in Construction, vol. xii, no. 10 (Toronto, October 1999), p. 295; NSARM, Halifax Relief Commission, MG 26, Series R, no. 1717.29.

circumferentials. In 1904, a plan of grandiose proportions was prepared for Prince Rupert, also in British Columbia, by prominent U.S. landscape architects. Its two main avenues and its circles, crescents, and public sites were meant to be the centre of a city of 100 000 people, rivalling Vancouver for Pacific trade.[26] Although this dream was not realized, the original planned layout is still evident in the present community. On a more modest scale were the plans for three pulp and paper mill towns: Iroquois Falls (1915) and Kapuskasing (1921) in Ontario, and Temiskaming (1917) in Quebec.

The plan for Temiskaming (see Figure 4.9) is noteworthy both because it was designed by Thomas Adams and because it shows the influence of Garden City planning principles. The resource frontier town, of course, gave the opportunity of planning for an entire community from the ground up and employing such ideas as the greenbelt, the separation of conflicting uses, street patterns fitting the contour of the land, and ample land for housing. These towns could demonstrate the importance of careful, overall planning that planning efforts in already built-up cities couldn't achieve. Thomas Adams said, in reference to his plan for Temiskaming:

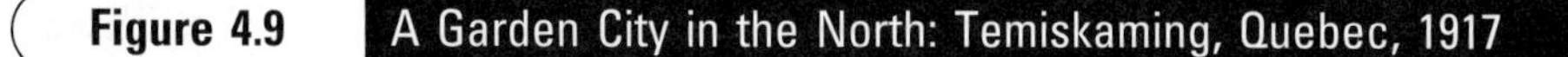

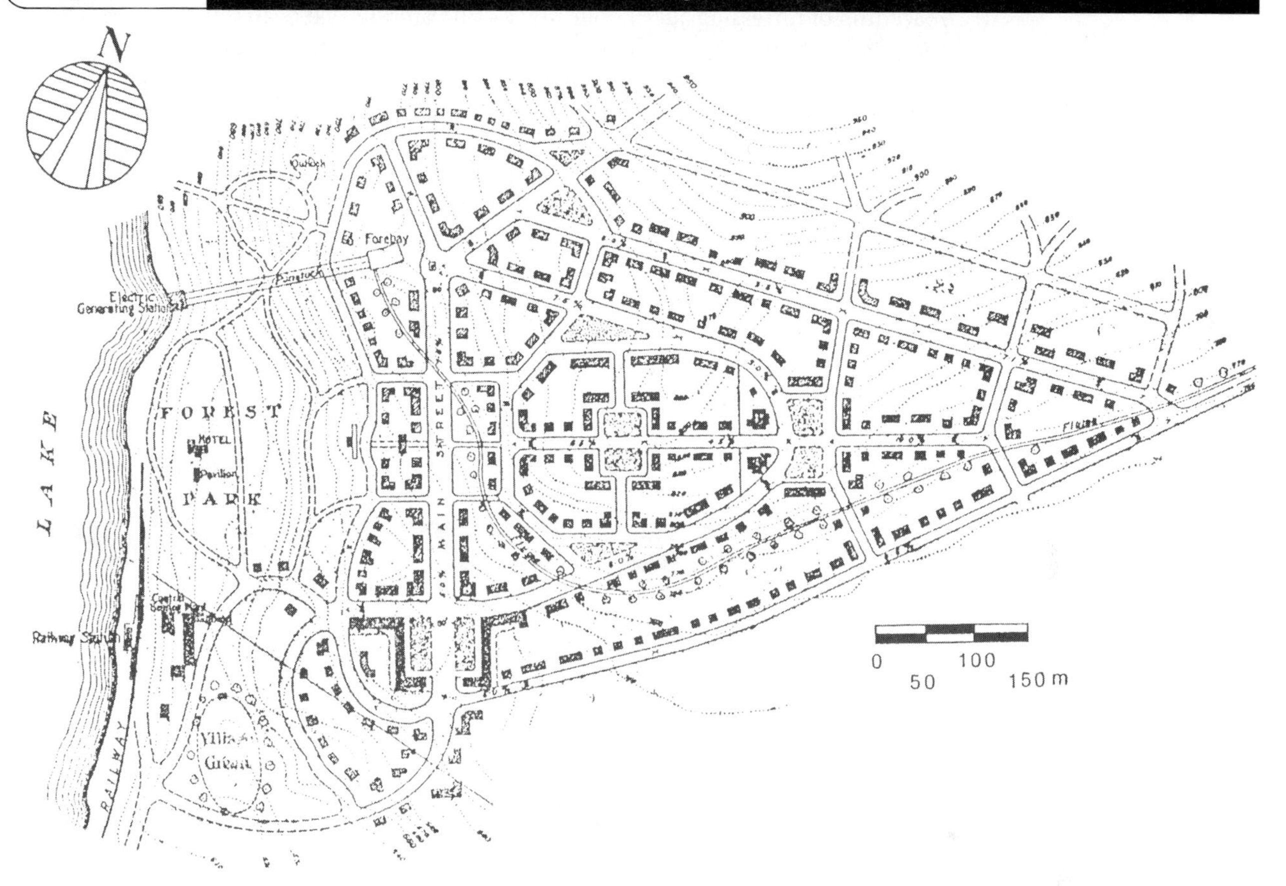

Resource producers in Canadian frontier areas have often sought to build livable communities to attract and hold their labour force. Thomas Adams was commissioned in 1917 to design this town where pulp and paper was (and still is) produced. He employed Garden City planning notions in the arrangement of streets, shops, public buildings, and parks.

Source: Commission of Conservation.

> The object of such plans should be to provide healthy conditions for the workers in the factories and the mills, together with convenience of arrangement to secure the most efficient methods of carrying on the industry, and not merely blind conformity to meaningless division of lines of a rectangular (gridiron) division.[27]

That Adams could prepare and have such plans accepted clearly indicates how readily planning ideas were accepted by the corporations for whom the towns were being built. Some companies actually sent representatives to Britain and other parts of Europe to study model towns that various industrialists had built in order to obtain for their employees housing more desirable than that available in the congested industrial revolution cities.[28] These intellectual connections have continued in the planning of many dozen resource development towns, mostly in the Canadian north, right up to the present day in such communities as Leaf Rapids, Manitoba, and Fermont, Quebec.

The Neighbourhood Unit

Around 1920, with better living conditions now their prime objective, community planners began to search for a workable unit of human scale around which housing and community services could be organized and designed. This search culminated in

1929 with the ideas of sociologist Clarence Perry for a neighbourhood unit.[29] Against a backdrop of increasing automobile usage—and auto-related deaths and injuries—Perry proposed a way both to insulate residential areas from traffic and to link the social needs of families to their environment. His concept was for residential areas to be organized in units of about 64 hectares, or sufficiently large to "house enough people to require one elementary school." The exact shape was not specified, but it was expected to provide an area within which young children had only about 400 metres walking distance to the neighbourhood school at the centre. Main streets would bound the area, not pass through it. Total population would be 5000 to 6000 people, or about 1500 families (see Figure 4.10).

Despite arguments that spatial units could not actually encompass, much less promote, a cohesive social environment, Perry's plan has been widely used. Community plans in Canada, the United States, and Europe have repeatedly used the neighbourhood unit notion in a variety of formats to structure the residential portion of the city. Suggested populations for units have ranged from 3000 to 12 000, but the essential characteristics of a school-oriented, traffic-insulated area have persisted. The

Figure 4.10 The Neighbourhood Unit Concept, 1929

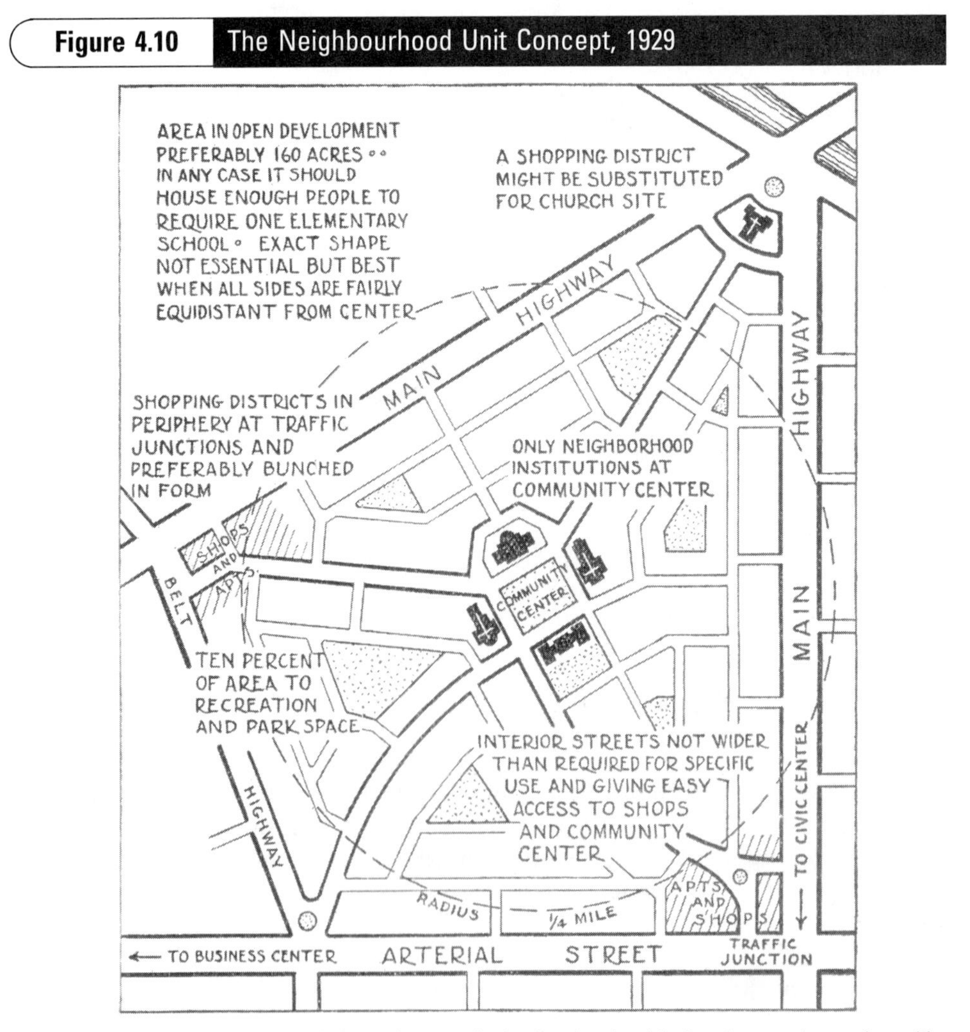

Clarence Perry's 1929 proposals, showing the central school and park, with sites for two places of worship. Shops and apartments are at the corners, and everything is within a radius of a 10-minute walk.

Source: Clarence Perry, "The Neighborhood Unit," in *Regional Survey of New York and Its Environs*, vol. 7 (New York, 1929).

neighbourhood unit became probably one of the strongest physical organizing principles in modern community plans. Its outcome is readily seen when one flies over almost any Canadian city (see Figure 1.1, page 10).

Radburn and the Greenbelt Towns

Another major planning concept during the period between the world wars was expressed in the "new town" plans of U.S. planners Henry Wright and Clarence Stein. Heavily influenced by the Garden City approach, they persuaded a private corporation to undertake such a venture on the undeveloped edge of New York City, in New Jersey. The first town was called Radburn and was, according to Stein, "not a Garden City as Howard saw it," but rather one planned for a society entering "the Motor Age."[30] Begun in 1928 and planned to grow to a population of 25 000, Radburn pioneered new design relations between houses, roads, paths, gardens, parks, blocks, and neighbourhoods (see Figure 4.11). It was one of the first practical applications of Perry's neighbourhood unit.

The main elements of the Radburn plan were:

1. the superblock, an area of 12 to 20 hectares with major roads on the perimeter so that through traffic would not intrude into housing groups;
2. specialized roads that would allow different traffic needs, from service vehicles for houses to through truck traffic, to proceed efficiently and with minimum impact on the community;
3. extensive use of culs-de-sac following Unwin and Parker's designs for Letchworth and Hampstead Garden Suburb;[31]
4. separation of pedestrians and automobiles by a system of walkways in different places and roads at different levels where they cross;
5. parks as the backbone of the neighbourhood, with open space left in the centre of superblocks and joined from one to the other in a continuous park; and
6. houses turned around facing gardens and parks instead of streets, with the latter becoming mainly service lanes for clusters of houses.[32]

In the Depression period of the 1930s, the U.S. government sponsored several more towns of the Radburn type. They were called Greenbelt Towns, and three of the four planned were built. These towns combined Garden City ideals, the neighbourhood unit, and Modern architecture advocated by Le Corbusier and other European designers. Although these federally sponsored towns were widely publicized, their impact on American suburban planning was not as influential as that of the 1932 President's Conference on Home Building. The research reports for this conference, prepared by Robert Whitten and Thomas Adams, featured the neighbourhood unit and Radburn-style cul-de-sac and crescents as the preferred alternatives to the inflexible gridiron pattern of subdivision used by most suburban builders. The model subdivision guides and mortgage insurance regulations that emerged from this era helped promote the neighbourhood unit and Stein and Wright's Radburn plan in the post-war expansion.[33]

Clarence Stein also left his mark in Canada in his 1951 plan for the aluminum smelting town of Kitimat, British Columbia (see Figure 1.1, page 10). And there is hardly a metropolitan suburb planned since the end of World War II, from Fraserview in Vancouver to Churchill Park in St. John's, that does not embody the principles of Wright and Stein to some degree.

Concern over the Environment

The turn of the century saw increased concern about public health in Canadian cities and the rise of the conservation movement and its concern over the natural

Figure 4.11 Radburn, N.J., Clarence Stein and Henry Wright, 1928

One neighbourhood unit was completed before the project collapsed in the Great Depression. All the culs-de-sac back onto a park and every child can walk to the central school without crossing a street, thanks to a pedestrian underpass below the central collector road.

Source: Regional Survey for New York and Environs, vol. 7, 1929.

environment. The 19th-century parks movement expanded from municipal parks to national parks in wilderness areas.

The Conservation Movement

The turn of the century gave rise to the recognition among the industrialized countries that industrial processes were consuming natural resources at an alarming rate, often leaving in their wake waste and pollution. The persistence of these problems in city and country alike led to strong national initiatives in the field of conservation of resources. This was the era in which the wilderness lands of the Rocky Mountains were secured for national parks (in both Canada and the U.S.).

In 1909, following the lead of the United States but going even further, Canada established a Commission for the Conservation of Natural Resources. Winnipeg lawyer and federal Cabinet minister Clifford Sifton was made chairman of the Commission of Conservation, as it came to be called. The commission concerned itself with many resource questions: lands, forests, minerals, fisheries, game and fur-bearing animals, waters and waterpower. Somewhat unique was the commission's commitment, from the outset, to human resources. Dr. Charles Hodgetts was one of the first to be appointed to the permanent staff as the Advisor on Public Health. In his years with the commission, he would often echo the maxim "Population is our most valuable national resource."[34]

Hodgetts, in his report to the commission in 1912, showed that he was fully aware of the scope of planning for an entire community. He set down the "essentials of town planning" in very sophisticated terms:

> The questions involved are more numerous and complicated than the mere building of a house. The various constituent parts of a modern town have to be considered and arranged in such a manner that they will form an harmonious whole.... [A] plan for town extension contemplates and provides for the development of the whole of every urban, suburban, and rural area that may be built on within from thirty to fifty years.[35]

By the beginning of World War I, the commission's involvement with community planning had become very extensive. First, the commission drafted a model "Town Planning Act for Canada," which it hoped each province might adopt in order to promote local planning. Second, it hosted the National City Planning Conference at its sixth convention in Toronto in 1914. These annual meetings were already well-established gatherings at which planners from North America and Europe exchanged ideas. Thomas Adams, a prominent British planner who had been associated with the Garden City movement, came to the attention of the commission at these conferences. Third, the commission hired Adams as its Town Planning Advisor in 1914.

By promoting community planning, the commission provided a national forum sponsored by the federal government, which, undoubtedly, meant a quicker, wider dissemination of planning ideas than would have otherwise occurred in such a large and sparsely settled country. By giving Thomas Adams the central role in this endeavour, it highlighted the role of the professional in the planning of communities, and by linking up community planning with the resources sector, it helped incorporate in planning the general economic values espoused by the commission. Clifford Sifton revealed these views at the 1914 convention:

> People must appreciate the idea that town planning is not born with the intent of spending money, it is simply not a new kind of extravagance, but is conceived with the idea of preventing extravagance and preventing waste and getting good value for the money which is expended.[36]

Parks Movement

The municipal parks movement continued to expand in the early 20th century. Canada's first resident landscape architect was Frederick Todd, who trained in the Olmsted office before opening his own firm in Montreal in 1900. Todd designed public open spaces across the country, including St. John's Bowring Park; the Battlefields Park in Quebec; Montreal's St. Helen's Island; Stratford's Avon River parks, and Winnipeg's Assiniboine Park. These park projects often represented the first major efforts of these communities to shape and give character to their physical environment and humanize their development patterns.[37]

Background **Thomas Adams**

The person who did the most to establish the substance and credibility of the professional side of planning in the first quarter of the 20th century was, of course, Thomas Adams. Born in 1871, in Edinburgh, Adams began by studying law but is referred to variously as a journalist, as a surveyor, and, in the initial roster of the Town Planning Institute of Canada (TPIC), as a landscape architect. This may well be a measure of his several talents and interests. In any case, he was acquainted with Patrick Geddes and the latter's work in Edinburgh, and also with Ebenezer Howard and the Garden City movement. For six years, Adams was secretary of the company that undertook to build the first Garden City at Letchworth. He was instrumental in promoting passage of the benchmark 1909 Housing and Town Planning Act in Britain, and was subsequently selected to organize the Local Government Board, which was to oversee municipal compliance with the new act. He was a founder and the first president of the British Town Planning Institute in 1914; half a dozen years later he became the first president of the Town Planning Institute of Canada. As one of his biographers, planner Alan Armstrong, notes, "on his arrival here he was already well-known as an eloquent author and speaker on the Garden City movement, on agricultural land use and on housing and town planning aspects of local government."[38]

Thomas Adams was enticed to Canada by Clifford Sifton to assume the high-profile post of Town Planning Advisor to the Conservation Commission of Canada. This post allowed Adams access to the highest government circles in Ottawa and in the provinces, since senior provincial ministers sat on the commission. It allowed him flexibility for travel, of which he never seemed to tire, in order to address groups in government and in business and the public-at-large in communities from coast to coast to persuade them of the virtues of planning. It also offered him a platform as a writer both in the commission's excellent journal, *Conservation of Life,* and in numerous other magazines. Thomas Adams stayed with the commission until it was summarily abolished by the government of Arthur Meighen in 1921. He continued as a consultant in Canada until 1923 and then was appointed to a new trend-setting planning venture as Director of the Regional Plan of New York and Environs.[39]

Todd's 1903 report to the Ottawa Improvement Commission is perhaps the most valuable contribution of the parks movement in this era.[40] This report recommends that the federal and local governments develop a system of parks that covers the entire national capital region. Todd proposed that natural reserves, suburban parks, and urban squares be linked by a system of riverside parkways to form an integrated network of open spaces, following the model of Boston's "Emerald Necklace." Although Todd only designed Macdonald Park in Ottawa, his regional open space system was the basis for the parks and parkways that grace Canada's capital today.[41]

Concern over City Efficiency: The City Scientific

The "City Efficient" began to replace the City Beautiful as the main focus of Canadian planning between 1910 and 1920.[42] While both notions contained compelling values, most of the pre–World War I beautification schemes sat on a shelf, unimplemented. Edward Bennett's 1915 comprehensive plan for Ottawa and Hull languished, despite the resources of the federal government. Only Confederation Square was built before 1945.[43]

The Public Health Movement

Although the early 20th-century cities in Canada did not suffer the alarming epidemics of half a century earlier, there were serious health problems. Many of these problems were attributable to the unsanitary housing conditions under which most new urban immigrants had to live. It was common, at the time, to attribute the cause to the customs of "foreigners," rather than to slum conditions. But since disease affected the wealthy as well as the poor, action was eventually taken. Public health laws were passed to try to ensure pure water supplies, to eliminate slum dwellings, and to provide for the purity of milk and other foods.

Public health advocates made slow progress in addressing disease and slum housing problems by direct municipal expenditures. By the late 19th century, they turned to regulatory structures rooted in statutes at the provincial and national levels. The first provincial Board of Health was set up in Ontario in 1883 and headed by Dr. Peter Bryce, a vigorous supporter of better health measures at the community level. He was also responsible for establishing better systems for recording vital statistics, which could be used to substantiate the needs in public health. Several other provinces also passed public health acts in this same period.

Dr. Charles Hodgetts, who succeeded Bryce at the Ontario Health Board in 1904, mounted a strenuous campaign for the improvement of housing. Toward this end, he advocated the use of town-planning techniques he had witnessed in Europe. Toronto's Medical Officer of Health (MOH), Dr. Charles Hastings, argued equally vigorously for better building standards and sanitary and water systems in this same period, as did Winnipeg's MOH:

> [We must] ensure that every place occupied as a dwelling unit within the City—no matter how humble it may be—is perfectly sanitary and a fit and proper place in which to bring up Winnipeg's most valuable asset—her children.[44]

This strong role of the local MOH in city development continues to the present time in most provinces.

The Medical Officer of Health wielded considerable power, owing not only to the public health legislation but also to the status accorded medical practitioners. Other bureaucratic structures were established to enforce building codes and conduct inspections of food and water. Building inspectors were appointed and building departments were established in many communities. Frequently, the new "experts" were paternalistic and authoritarian in their approach. The thrust of many of the new prohibitive measures was not social reform but enhancement of those with an interest in continued growth and expansion. The 1909 Manitoba Tenement Act may well have been framed with the fear of contagious disease in mind but, as Weaver notes, it was enforced as if its prime concern were the protection of property values.[45]

Public Works

The engineer–planners of the City Scientific era built impressive infrastructure systems while the City Beautiful's civic centres remained long-term dreams in most cities. City-wide systems for fresh water, sewerage, electricity, gas, paved streets, and electric streetcars had improved public health and the quality of life for most urban citizens. And the new discipline of traffic engineering produced impressive innovations to deal with the emerging technology of the automobile, such as parking garages, traffic signals, and limited-access highways.

The public works projects of the early 20th century often addressed a variety of objectives simultaneously. For example, the 1912 waterfront plan prepared by the Toronto Harbour Commissioners (THC) not only provided for a new port, but also for major new sewers, popular public parks along the western and eastern beaches, and a landscaped boulevard along the entire waterfront.[46] The project was under the direction of THC chief engineer E.L. Cousins and the public parks, beaches, and boulevards were built to the highest quality. They were designed by Boston's famous Olmsted landscape architecture firm.

By the mid-1920s, the functional approach to planning had become the official position of the Town Planning Institute of Canada. In a front-page editorial in the first volume of the *TPIC Journal*, the president, Thomas Adams, declared:

> Town Planning is a Science . . . knowledge and art—and action on these possessions—constitute the foundation of social progress. The most important first step in creating a sound town planning policy in Canada, therefore, is to develop the science of town planning.[47]

The leading example of the City Scientific planning techniques is Vancouver's 1929 comprehensive plan, prepared by American engineer-planner Harland Bartholomew, with Canadian engineer Horace Seymour as the resident planner (see Figure 4.12).[48] Although it provided plans for a park system and a City Beautiful–style civic centre, it is filled with City Scientific analyses—population growth, and traffic, streetcar, and harbour use. Bartholomew and Seymour's comprehensive plan includes city-wide systems for roads,

Figure 4.12 Vancouver, B.C., Major Street Plan, 1929, by Harland Bartholomew & Assoc.

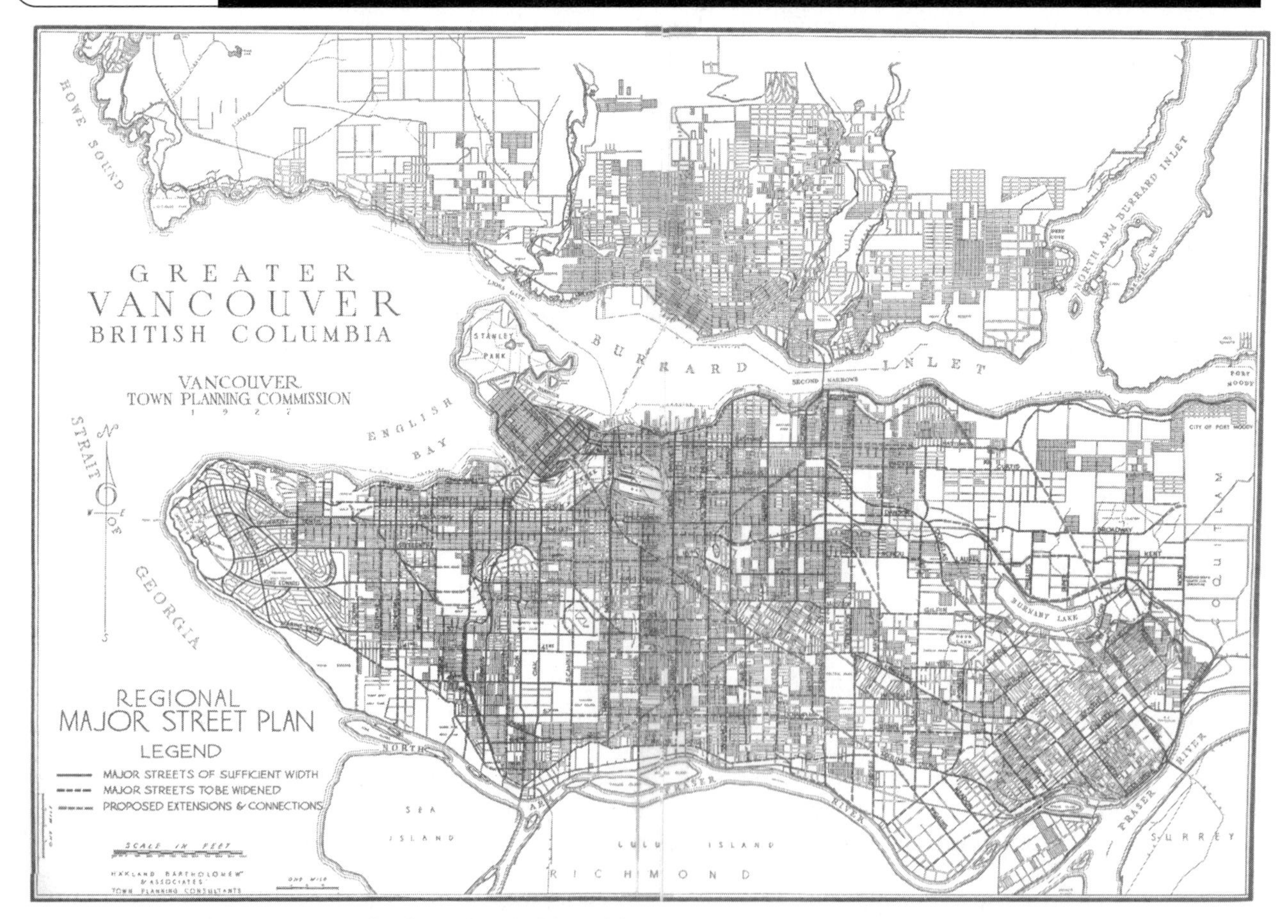

A regional roads system provided in the City Scientific style. Many of the street widenings were implemented, but the elaborate road network for the University Endowment lands at the extreme left was not.

Source: H. Bartholomew & Assoc. *A Plan for the City of Vancouver, BC.* (Vancouver: Vancouver Town Planning Commission 1929), plate 8.

bridges, streetcars, and utilities that were a guide for public investment for several more decades. Its land-use proposals were furthered by a sophisticated zoning bylaw that was adopted immediately by both the City of Vancouver and Point Grey municipality[49] and became a landmark in Canadian planning. Bartholomew's firm produced many similar plans for other American cities,[50] while Seymour went on to become the chief provincial planner for Alberta.[51]

The Civic Reform Movement

The final major influence on community planning was a widespread movement to improve the quality of local government. There were three fairly distinct areas of concern at various times: the social welfare of city dwellers, public ownership of basic utilities, and the efficiency of local government organizations. The issues ranged from procrastination of local councils to technical incompetence to municipal graft, and sometimes all three.

The earliest concerns had to do with such social problems as disease, poverty, crime, and poor housing. These efforts had gained considerable momentum in the 1880s, aided by new popular newspapers like *The Montreal Star*, *The Ottawa Journal*, and *The Vancouver News-Advertiser*, and by reform-minded clergymen such as Winnipeg's J.S. Woodsworth. Their approaches, although motivated by a mixture of humanitarianism and professional and business self-interest, were characterized by the use of statistics, the advocacy of government regulation, and the promotion of the use of experts and professionals.

The second stream of civic reform concerned the provision of municipal utilities. The waterworks, street railways, and electric power and telephone systems had been developed largely by private interests on franchises offered by municipal councils on "extremely generous terms to the entrepreneurs."[52] However, as growth pressures mounted, the services were often found wanting, contracts were not being met, there were instances of influence peddling, and private utilities were often reluctant to expand into new suburban areas. Moreover, civic leaders, who were mostly drawn from the business community, wished to appear progressive and willing to have industry in their communities. As the Mayor of Medicine Hat said at the time, "Municipal ownership (of utilities) and industrial progress go hand in hand."[53]

Guelph purchased its gas works and electric power system in 1893; Edmonton had a publicly owned power system in 1902; and similar efforts can be cited from coast to coast. But the progressive-looking municipal utilities were, not infrequently, a mixed blessing: sprawl was increased when premature land development was served, operating surpluses were not always forthcoming, and inept administration was not uncommon. Despite these difficulties, the move to municipal ownership established, from then on, the rights and the responsibilities of communities to provide community-wide services.

The third target of the urban reformers was urban government itself. The scale and rapidity of urbanization in this period was unprecedented; as well, much of the technology was new and local government was relatively inexperienced. Reformers sought to reduce the power of city councils and their committees and substitute special boards and commissions. These, it was contended, could administer the various functions that communities needed to handle their growth—waterworks, transportation, parks, police, schools, libraries—more efficiently because they would not be subject to political pressures. The attempts to separate politics from the administration of public services did not always result in increased efficiency, and it had two long-term effects on the planning and governing of communities. First, it led to a proliferation of agencies, each responsible for a separate function while not having to coordinate their activities with those of either the city government or other bodies. Second, because the special bodies needed to develop special skills, they fostered the use of experts in the solution of urban problems. The special boards and commissions tended to acquire considerable discretionary powers and vested these in professionals and administrators and thereby further increased their independence.

The various urban reform movements helped Canadian communities to cope more rationally with expansion, to ensure more stability in the provision of services, and to maintain a degree of humanitarian concern for the less fortunate of their citizens. But they also initiated much more complex means of local government, which necessitated the cooperation and coordination of many bodies. This situation remains a source of frustration for community plan-makers to this day in many urban areas.

Local Government: The Residue of Reform

Civic administration was transformed in the 1900–1920 period in both substance and structure. Local government was limited to a role of "providing services to property," while matters involving economic development and what we now call the "quality of life" were assigned to parallel local agencies. Municipal government was weakened by the reform epidemic. Not only was its competence questioned but it also failed to provide for broad public discussion on community issues. Municipal administration was favoured over municipal political debate; councillors often became preoccupied with administrative details, and administrators with policy-making. This rather introverted view of municipal government effectively excluded participation by the public while it gave special interests in business and the property industry, as well as administrators, the "right" to run the local government. The quality of municipal administration improved over the next 25 years (1920–1945) but, when the public came to demand a voice in the big new development decisions after World War II, local governments were generally unprepared.

Further, most provinces introduced various "overseer" methods for controlling municipal activities. These took two forms. On the one hand, provincial departments or ministries of municipal affairs were established to provide for consistency in the day-to-day operation of the municipalities the provinces had created. Guidelines and legislation covered financial affairs, administrative practices, and legal foundations, many of which required regular reports to a provincial ministry. On the other hand, many provinces also established quasi-judicial tribunals for the purpose of reviewing such municipal activities as borrowing capital. Examples are the Municipal Board in Manitoba, the Quebec Municipal Commission, the Local Authorities Board in Alberta, and the Ontario Municipal Board. These boards dealt with individual municipalities on an issue-by-issue basis, frequently as a form of appeal body against decisions made by municipalities. In recent decades, such boards have handled appeals in community planning matters. The various supervisory methods used by the provinces with respect to their municipalities thus made for a master–servant relationship between the two.

Private Group Involvement

Another response to urban problems at the beginning of this century was the direct involvement of elite groups in the planning of cities. This took two forms. "Civic Improvement Associations" sprang up in many places in the decade from 1900 to 1910. (Later, they were promoted by the Commission of Conservation in communities where they did not exist.) They often took upon themselves the task of preparing a city plan, as with the Toronto Guild of Civic Art, or of urging the city government to prepare one, as in the case of Kitchener.[54] The other form of involvement was by such varied interest groups as boards of trade, chambers of commerce, the Canadian Club, and social and professional associations of architects and engineers. These groups sponsored lectures by prominent persons in planning and municipal government—Thomas Adams was a favourite speaker—and actively lobbied for civic improvements, especially visual enhancement of their city.

There are two important facets of this private-sector interest in planning: (1) it clearly served the interests of property owners; and (2) it presumed a strong role for property and business interests in community planning and local government affairs. The advocacy by these groups of visual order, efficient transportation, and, later, municipal reform, was in the interest of protecting property values and promoting new opportunities for land development. What developed was a "proprietary interest," so to speak, in civic affairs that translated, over ensuing decades, into a seeming right to membership on planning boards, municipal councils, and other local boards and commissions that endures to this day.

Establishing Community Planning Institutions

Coincident with efforts to restructure local government was the pioneering of ideas about the aims of planning and the nature of its practice. This led to the emerging institutions of community planning being affected by the then-current reform outlook on local government and the constitutional setting established by the province. We deal below with each of these in turn.

The Social and Political Context

In many ways, community planning became a handmaiden to the interests promoting civic reform. Planning was seen to embody many of the same values reformers sought for local government—technical rationality, efficiency, and order. It became commonplace to champion planning either through existing organizations, such as boards of trade, or through new groups. Kitchener (then Berlin), Ontario, had its Civic Association, Calgary and Edmonton had the Alberta Housing and Town Planning Association, and Montreal had its City Improvement League. Some groups actually sponsored the preparation of a plan for the community and then urged its implementation on the municipal government. Another approach was to urge the city government to establish an independent planning committee for the purpose of preparing a plan. Notable instances of the latter approach were the Civic Improvement Committee, appointed by Toronto's city council in 1909, the Calgary City Planning Commission and the Winnipeg Town Planning Commission, both established in 1911, and the Regina Town Planning Committee of 1913.

The initial efforts to link action in community planning to the local government follow the model favoured by the civic reformers—that is, the establishment of a body outside the direct control of elected officials. This approach was propounded for U.S. communities in the same era of local planning. In contrast, British communities adopted the approach of appointing a planning committee from the membership of municipal councils. The rationale for the semi-independent planning committee in Canada was: (1) it would keep "politics" out of planning; and (2) community planning, being essentially a technical task, is better achieved by a small group devoted to it. Usually unstated was the reason that a special-purpose planning body would probably direct its efforts more narrowly to the development of private property.

The semi-independent planning bodies for community planning that prevail today in Canada came into being at this time. In the West they tended to be called by the U.S. name, "planning commissions," while in the East they were more likely called "planning boards" or "advisory planning committees." Where necessary, the provinces passed legislation permitting such bodies to be established. The primary function of the new agencies for community planning was to prepare a plan, or "town planning scheme" as they were then called, for the community. In larger cities, the agency might have its own staff, but more likely it employed consultants. The plan would then be passed on to the municipal council to be implemented through regulations and public works expenditures.

Despite its supposed advantages, as with every institution there were costs to bear. In the first place, unlike other special-purpose bodies favoured at this time, such as harbour commissions, park boards, and transit authorities, the planning commissions did not operate any facilities. They could only advise the municipality. The municipal council alone had the powers necessary to intervene in the development of private property or to create streets or to build needed public infrastructure. A second dilemma was that the recommendations of an advisory planning agency would have little, if any, direct bearing on the actions of other operating agencies. A third issue concerned the element of participation by the public. Although the semi-independent planning agency represented a form of populism, it fell outside the community-wide forum of municipal council. The views expressed often reflected only limited interests in the community, such as those of downtown merchants, the real estate industry, or the residents of a well-to-do neighbourhood.

In other words, the institution of the special-purpose planning agency created tensions within the local government structure. While attempting to keep community planning at arm's length from the political arena, it heightened the connection between the two. As Milner was to reflect later:

> Planning boards do not exist to take planning out of politics. In a democracy, planning is a political activity.... A planning board gives advice—I hope honest advice. It should not be too concerned whether the council takes the advice or not. This is council's business.[55]

The Constitutional Setting and the First Planning Acts

Many of the powers necessary for a municipality to prepare a plan did not exist in the early period of planning. For a plan to have statutory power over all properties in a community, the sanction of the province was required. If public works expenditures were envisioned to secure the plan's objectives, or if land-use regulations were considered desirable to achieve the plan, the legal authority would need to be obtained from the province. This is because there is a convention in Canada that a municipality may not do anything that its province has not empowered it to do.

A province must *enable* its municipal governments to carry out such functions. The need for provincial planning legislation was recognized early in the 20th century. In a period of less than 12 months, starting in the spring of 1912, four provinces passed planning legislation that would enable a municipality to prepare and/or carry out a plan. New Brunswick was first in April 1912, followed by Nova Scotia, Ontario, and Alberta. Within about a decade, four other provinces enacted similar legislation. No explanation has been offered for this profusion of provincial action in regard to community planning. It is especially interesting since, except for Ontario, most provinces had few cities, and these were not large in any case.

A number of factors probably influenced the provinces to enact planning legislation. For one, there was considerable subdivision of land in suburban areas and often chaotic development accompanying it. For another, the Commission of Conservation strongly promoted such legislation, especially through the efforts of, first, Charles Hodgetts and then, Thomas Adams; the premiers of both New Brunswick and Alberta sat on the commission. (Indeed, Premier Arthur Sifton of Alberta was the older brother of the commission's chairman.)[56] Other likely influences were the planning efforts in Great Britain and the United States, which received extensive publicity, in particular the Housing and Town Planning Act in Britain in 1909. And probably as important as any factor was the desire of Canadian governments to appear progressive.

The British Housing and Town Planning Act of 1909 provided the model for provincial planning legislation. The early Canadian acts shared three things in common with the British act:

- planning was confined to land in suburban or fringe areas that had the prospect for development (built-up portions of cities were initially excluded);
- landowners could claim compensation if public plans adversely affected private property; and
- local planning would be subject to close scrutiny by central government authorities.

The first and second provisions are evidence of the great reluctance at the time to intervene in existing property development. The third provision, notably, is still a fundamental part of Canadian planning institutions: most planning actions by communities usually cannot take effect until approval by the province or its agencies. More than anything else, this feature distinguished planning practice in Canada from that in the United States. The early Canadian planning acts were not, however, simply duplicates of the British act. Some variations were made to suit the Canadian approach, and several of these have endured to the present time. It will be helpful to elaborate on the five basic components of these 1912–13 planning acts.

1. The Scope of Planning

Three acts (in Nova Scotia, New Brunswick, and Alberta) permitted municipalities to prepare "town-planning schemes," and specified the aspects of the community that the plan should cover. To quote from the New Brunswick act,

> A town planning scheme may be prepared... with the general object of securing suitable provision for traffic, proper sanitary conditions, amenity and convenience in connection with the laying out of streets and the use of the land and of any neighbouring lands for building or other purposes.[57]

Plans were thus envisioned to encompass quite broad features of the built environment. However, they were to apply only to "land which is in the course of development, or is likely to be used for building purposes"—that is, to suburban extensions to communities. Provision was allowed for the province to authorize the inclusion of adjacent built-up land that might be affected by the plan. Nevertheless, this carefully limited the area covered by plans.

Municipalities were allowed considerable scope in carrying out plans. They could purchase, using powers of expropriation, properties needed to implement the plan, say for a park or a street extension. They could also make expenditures for such public works as water and sewer systems to put a plan into effect, but were required to specify in the plan how they planned to obtain these funds; we now call this a "capital-improvements budget." There was also provision for a municipality to remove any building that contravened a plan and to complete any work of a private developer that would delay the plan's coming into effect. The planning acts of New Brunswick, Nova Scotia, and Alberta also envisioned the province drawing up regulations for formulating and carrying out a plan. In particular, these regulations were expected to allow communities to control the density of development and the mixture of land uses—that is, to perform what came to be called **zoning**. However, no such land-use regulations were forthcoming until later years.

The planning legislation enacted in Ontario in 1912 allowed much less scope. The Cities and Suburbs Plans Act was enacted simply to control the subdivision of suburban land by private interests. It applied only to those cities with a population of at least 50 000 and covered the area within 8 kilometres of the city. All plans for subdivision of land in such locales had to be submitted to a provincial government agency for approval. Consideration of plans was limited to the number and width of streets, the location of streets, and the size and form of lots. Minimum street widths were specified at one **chain** (66 feet or 20 metres).

2. The Role of the Province

All four planning acts included a strong role for the province in the planning process, although the form differed in each case. To quote Alberta's planning act, "A town-planning scheme prepared or adopted by a local authority shall not have effect unless approved by written order of the Minister."[58] This power of ministerial approval also included the power to amend a local plan to suit provincial government standards. In Nova Scotia and New Brunswick, the provincial authority in planning lay with the Cabinet (or the Lieutenant-Governor-in-Council, as provincial legislation usually refers to it). In Ontario, the approving body was the Ontario Railway and Municipal Board (now called the Ontario Municipal Board). In Alberta, it was the Minister of Municipal Affairs who approved of local plans; Alberta was one of the first provinces to establish a provincial department to support and supervise its local governments. Such departments became common later in all provinces. Provincial planning legislation has only recently begun to be changed to reduce the central government's scrutiny of locally made plans.

3. The Local Planning Structure

In both the New Brunswick and Alberta planning acts, provision was made for municipalities to appoint a planning commission to undertake the local planning effort. To quote the New Brunswick act again, "For the purpose of preparing a town planning scheme and carrying the same into effect, a local authority ... may appoint a commission of not less than five or more than ten members."[59] Neither province's legislation required that the format of the semi-independent commission be used; it was one option. However, when the Nova Scotia act was revised in 1915, it required that a town-planning board be established in every municipality. It also specified the membership: the mayor, two other members of council, and at least two ratepayers.[60]

The creation of a separate board for making and carrying out a local plan was intensively discussed at the international meeting of planners held in Toronto in 1914. It seems clear, especially from the comments of Charles Hodgetts, that the reform mood permeating North America in regard to local government, as noted above, was the main reason for this feature in an otherwise British-style act. Hodgetts said:

> It is proposed to remove the important matters in connection with town planning out of the hands of our municipal councils.... I may say, after twenty-seven odd years of public experience, I am not impressed with the achievements or capabilities of "town councils."[61]

4. Effects on Private Property

In essence, provincial planning legislation involved delegating some provincial powers concerning property rights to a municipality. Given the strong belief in the sanctity of private property and individual enterprise, these were powers the provinces gave up reluctantly. The provisions that the province held for approval or rejection of local plans were one way of restraining municipalities from extending community rights over private interests. The acts also allowed a property owner to claim **compensation** "whose property is injuriously affected by the making of a town planning scheme," to quote the phrase in the Nova Scotia, New Brunswick, and Alberta acts.

There was also a counterpart to the compensation provision included in these early acts called **betterment**. It allowed a municipality to claim up to half of any increase in the value of a property beneficially affected by a plan. Planners had argued from the time of Ebenezer Howard and the first Garden Cities that proper planning could enhance the value of property. Thomas Adams was particularly forceful on this point, both before and after he came to Canada. He and others contended that this betterment of property was due to *public* efforts, and any rise in the value of a private property because of a public plan should thus be shared with the community. It is not clear whether any of the three provinces ever published the necessary regulations on the workings of compensation and betterment. In any case, these concepts proved difficult to define and do not appear in present-day planning acts.

The compensation/betterment notion cuts to the heart of the issue concerning the extent of public versus private property rights. While a private property owner may be able fairly easily to determine "injury" to his or her property, it is notoriously difficult to determine the extent to which a property's value may have been increased by a plan. A major difference is that the injury to a property is usually apparent when the plan is being implemented, whereas an increase in a property's value that is due to a plan may not materialize for many years. These early planning acts thus recognized some prior rights for municipalities by deeming that compensation could not be claimed if the provisions of the plan were for the purpose of "securing the amenity of the area." These included provisions that might "prescribe the space about buildings or limit the number of buildings to be erected, or prescribe the height or character of buildings."[62] This clause is the precursor for allowing zoning controls over height and lot coverage, although at the time (ca. 1912) the provisions could be applied only to specific areas for which plans were submitted.

5. Land for Parks

The Alberta Town Planning Act of 1913 took one further important step. It allowed the municipality to acquire up to 5 percent of a new subdivision area for park purposes at no cost to itself.[63] This is called **compulsory dedication** and is the same kind of requirement made of land developers to provide road access to the building lots they create and then deed the road allowance to the municipality. While it is clear that residences deserve road access as a matter of rights, how, in a new subdivision, is the simple amenity of a park to be provided except from the land that would normally be subdivided for houses? Further, there is the question of who should pay for the new park land. Alberta legislators took the approach that those benefiting from the park most directly should bear the cost—the developers who could offer a higher quality subdivision and the home buyers whose property values would more likely be sustained. They may also have been influenced by the converse of this principle: that is, the community as a whole should not have to bear the cost of acquiring land later, the value of which could be much higher. Thus, the Alberta planning act was especially perceptive and its requirements were modest—the equivalent of only one building lot in every twenty. It took over three decades, however, before this groundbreaking step was followed in other provincial planning acts.

The Development of Planning Tools

The first planning acts established the formal milieu for community planning, but they did not assure the preparation of plans or their implementation. In large part, this was because support for the idea of planning advanced faster than the development of the tools to carry it out. There were very few professional planners available at the time. There was also relatively little experience anywhere in the world with carrying out local planning under government sanction. Thus, Canada cannot be said to have lagged in supporting planning: 1909 is the date both of the first British planning act and of the first North American planning conference in Boston; by 1912, three provincial planning acts had been passed in Canada, and in 1914, the sixth planning conference was held in Toronto. The first comprehensive zoning bylaw was that of New York City in 1916. Thus, the tools that Canada's planners would need to use were still being "invented" as the formal machinery of planning was being assembled.

The planning of cities and towns by the communities themselves under the aegis of their local governments was virtually untried until the last decade of the 19th century. The prototypes that existed, whether Garden Cities, model industrial villages, socialist utopias, or Garden Suburbs, were all conceived as having a single or collective ownership of the land and facilities. Planning tools had not yet been devised for communities with multiple property owners whose interests must be reconciled through self-governing institutions. Planners thus struggled to devise acceptable tools that would provide:

1. **Planning policy for the entire community.** Some way to define the quality and direction of development for the entire community for some future period, for existing built-up areas as well as for vacant areas, is desirable. It should include both public and private development efforts.
2. **A framework for guiding private development.** Since the bulk of the land in the community is privately owned and will be developed by private interests, some ways are needed to achieve development that will be consistent with the aims of overall policy. The approach for development on vacant land will likely differ from that for already built-up areas.
3. **A framework for public capital investment.** Since the community must provide support services for private development as well as public facilities, some means are required to allocate public funds that will achieve the aims of the overall policy.

Today, we would think of the above tasks, respectively, as (1) preparing the community plan, (2) formulating the land-use and subdivision regulations, and (3) preparing the capital improvements program. But the meagre experience available to planners at the time made these formidable tasks. Let us examine the first two of these provisions briefly for the dilemmas they posed and their eventual resolution. (We deal with capital programming in Chapter 14.)

The Community Plan

The earliest planning acts empowered a municipality to prepare a town-planning scheme. However, such a scheme was not meant to be a plan for an entire town. It was more like what is now called a subdivision plan, showing the layout of streets, building lots, open spaces, and public utilities for a new residential suburb, similar to German town-extension schemes that were widely admired at the time.[64] These are typically submitted by a private developer or public agency desiring to open up a parcel of vacant land. But how would a community's plan be structured to cover a multiplicity of such subdivision plans where, moreover, much of the land would not be developed until far into the future? And how would it integrate the new development with already built-up areas, with needed transportation links between all parts of the community, and with the infrastructure of services and facilities?

The 1912 Ontario City and Suburbs Plans Act refers to communities having a "general plan into which a town-planning scheme might fit." This concept is repeated in the much-expanded Planning and Development Act, which Ontario passed in 1917. It gives the following definition:

> Such a plan shall show all existing highways and widening, extension or relocation of the same which may be deemed advisable, and also all proposed highways,

parkways, boulevards, parks, play grounds and other public grounds or public improvements, and shall be certified by an Ontario land surveyor.[65]

Although this definition still emphasizes the rationalization of road patterns, the act represents a major departure—it distinguishes the *general* plan from the *detailed* subdivision plan.

The most notable early Canadian plan of this general type appeared in 1915. It was prepared by Edward Bennett for the report of the Federal Plan Commission, which was charged with preparing a plan for Ottawa and Hull.[66] Indeed, the authors called it a general plan and presented proposals in map form for various land-use districts, for future population expansion, and for streets, parks, and waterways. Accompanying these plans are analyses of population density, land use, industrial employment, and public transit. This plan is very modern in its approach and not too dissimilar to those that would be common by the 1960s.[67] A second notable example is the 1923 plan prepared for the adjacent cities of Kitchener and Waterloo, Ontario, by Thomas Adams and Horace Seymour. It is also supported by extensive analyses of traffic, land use, and population, and includes a "skeleton plan" for the region.[68]

An examination of both the Ottawa and Kitchener plans, however, reveals a dilemma that affected the development of planning tools. Whereas nowadays the acknowledged approach is first to prepare an overall community plan and then to devise the regulations, bylaws, and other instruments to attain the goals of the plan, the reverse seems to be proposed in both of these historic documents: that is, the community plan's role was seen as providing support for land-use regulations. Two provincial planning acts of the 1920s, the 1925 British Columbia act and the 1929 Alberta act, made significant departures in regard to this situation. First, they clearly distinguished the roles of the community plan and the zoning bylaw. Second, they permitted them to be used independently.

However, because the early provincial legislation permitted municipalities to engage in the planning activities they preferred, and because local governments wanted to handle development problems directly, rather than simply provide guidelines for others, this led to the widespread adoption of zoning bylaws without corresponding community plans. Thus, in the 1930s in Alberta, 26 municipalities used the new planning act and passed zoning bylaws, and a further 31 followed suit in the 1940s, but most had never adopted a community plan.[69] In Ontario, as late as 1971, it was found that 544 municipalities had enacted a zoning bylaw, while only 356 had adopted a community plan. The requirement that a municipality adopt a community plan before adopting a zoning bylaw did not appear in provincial planning legislation until well after World War II.

Land-Use Regulation

It became clear to public authorities in this period that they could not ensure that private land, which made up the largest part of their cities and towns, would be developed to high standards. It was further realized that private land development fell into two general categories: that occurring on tracts of vacant land and that occurring on individual parcels of land within built-up areas. These two needs spawned their own land-use controls—subdivision control and zoning, respectively. In general, there is a sequence to these two controls: land is first subdivided and then it is zoned. They are discussed below in that order.

Subdivision Control

The subdivision of a tract of land transfers ownership and thus requires that all new owners be given the institutional protection of their land titles. The need for subdivision control became insistent in the decade preceding World War I. The relatively simple tasks in earlier times of transferring ownership and registering deeds for individual parcels of land were complicated by the land boom of the time. As well, the long-term consequences for the entire community of the individual new subdivisions was recognized and made obvious the need to assure compliance by the new landowners with various planning and development standards.

Further, the process of dividing up the land may not be connected with immediate building upon, or even sale of, the land. Frequently, subdivision precedes actual urban development by a considerable period of time. From its perspective, the community must know the obligations it might face to service the new parcels of land as well as the consistency of the new layout with adjacent tracts. The main concerns are for street layouts, open space provisions, drainage, and the size and shape of building lots. Also important is the time at which development is expected to occur.

The requirement that plans for new subdivisions be submitted for approval established the right of the community (and the province) to review plans for development on private land. In later land booms, especially during the late 1920s and the post–World War II periods, the administration of subdivisions became the central planning activity for many municipalities. The subdivision process is now usually accorded a separate section in provincial planning acts and elaborate systems of checking subdivision plans have been set up in provincial ministries. (See Chapter 131 for present-day practices of subdivision control.)

Zoning

The planning tool that deals with the use of land and the physical form of development on individual parcels of privately owned land is called districting or, more commonly, zoning. It deals, essentially, with (1) the use that may be made of a parcel of land, (2) the coverage of the parcel by structures, and (3) the height of buildings. Long before zoning, there were regulations dealing with these factors to assure public health, structural safety, and fire prevention.[70] However, what distinguishes zoning regulations is the application of use/coverage/height standards to districts or areas within a city. The outlook of early planners in Canada in regard to zoning is captured in the following extract from the recommendations for the 1915 plan for Ottawa and Hull:

> . . . that the authorities take steps to segregate industry into certain areas, to control the districts devoted to business and light industry, to control and protect the residential districts and to control the height of buildings.[71]

The first efforts at city-wide districting are credited to German cities in the 1890s.[72] They were based on the experience that similar land uses tended to congregate in common districts and also that dissimilar uses had a disturbing effect on one another when mixed together. The separation of nuisance uses was the prime aim of early zoning. Six different land-use districts were envisioned in the plan for Ottawa and Hull. Prior to this, attempts had been made simply to separate residential from non-residential uses within cities. The earliest Canadian bylaw was enacted in 1903 in London, Ontario. The City of Toronto's 1904 bylaw to control "the location, erection, and use of buildings for laundries, butcher shops, stores and manufactories" in effect excluded them from residential districts. Through Ontario legislation enacted in 1912, the exclusion of "apartment and tenement houses" from residential districts was also allowed in large cities.

In Toronto, as in other cities, the various health, safety, and occupancy regulations were usually enacted on a district-by-district basis. While this procedure commended itself for being flexible and responsive to the needs of individual districts, it was also complicated to administer and, not infrequently, subject to inequities between areas. Some areas were not regulated at all, or only partially, and these often turned out to be the areas where poor people lived as tenants. Planners began to urge comprehensive zoning bylaws that would provide not only greater uniformity of application of regulations but also a city-wide view of private development, consistent with the view of the community plan. As it turned out, this was no easy task.

The districting of a city according to specified land uses is a double-edged sword. Since zoning usually respects the kinds of uses and buildings already in a district, existing owners, or those with the desire to develop land in a like manner, will have their investments justified. But those with aspirations to build differently from existing uses, whether with real intentions or simply because they want to have free rein on the use of their property, will be restrained. Planners, seeking a position that would reconcile these two outlooks, chose the argument that zoning would stabilize and protect land values. That is, it would provide more certainty for landowners about what they could expect to be built in their areas and for municipal governments about what expenditures they would have to make for services, as well as the tax revenues they might receive. This was a strong argument, but it was usually grudgingly received, first, because there was little immediate experience in most places, and second, because it depended on the current prospects of the land market. If property prospects were not good or if they had been erratic, acceptance of zoning was usually facilitated. In buoyant situations, such as that of mid-1920s Vancouver, it could take several years for all the interests to agree on the final version of a zoning ordinance.[73] Opposition from real estate interests killed the Ottawa Town Planning Commission's proposed 1926 zoning bylaw, and the City of Ottawa did not adopt a comprehensive zoning bylaw until almost 40 years later.

Related to the reluctant response to early zoning is the criticism over the exclusionary approach that many zoning bylaws took. There were instances of zoning being used to exclude some classes of people from certain areas by excluding the commercial uses with which those classes were associated. Immigrant groups were often blatantly discriminated against in this way because, for example, they usually chose to live in the same building as their place of business.[74] The other, more common, form of exclusion was to separate types of dwelling units into different zones. Thus, two- and three-family dwellings would be excluded from single-family zones, and apartment buildings (or tenements) in yet another zone, because renters were considered less stable than homeowners. And further discrimination might also have been made between low- and high-income homeowners by requiring houses to be built on larger, and thus more costly, lots in areas where the well-to-do wished, as Milner noted, "to prevent intruders from spoiling an already established area."[75] In fairness, while zoning does involve separating incompatible land uses, its over-zealous application is not inherent in this planning tool. Indeed, approaches to zoning change as planning experience accumulates so that, today, the mixing of different residential types with retail and commercial uses is now considered desirable.

There are two fundamental problems with zoning. The first is the dilemma between protecting existing land uses and promoting the planning of future land uses. Zoning bylaws often seek, in Levin's terms, "to create those elements in the physical environment which the community finds desirable where they do not exist."[76] In other words, zoning seeks to protect present as well as future land uses and values. This gives zoning a dynamic role, but one that it cannot easily play because it is a legislative tool that must be uniformly applied and not readily amended. A second problem has to do with the administration of zoning bylaws. Although the intent is that all the properties in a zone be treated equally under the bylaw, physical conditions affecting the land in a zone may make this impossible. The topography of an area may render some parcels of land difficult to build upon and still meet zoning requirements, for example. Or an old street pattern may have left an awkward shape of building lot. In such situations, the bylaw is generally regarded as causing an "undue hardship" on a property, and mechanisms have been devised to allow a property owner to seek a **variance** from the provisions of the bylaw. These requests are heard today by a locally appointed appeal body called a Zoning Board of Appeal or Committee of Adjustment. Such appeals are meant only for minor changes to the bylaw and not, say, to changes in the use of land. The latter requires an amendment to the bylaw by the planning board and council and only after scrutiny of the implications for the community plan (see Planning Issue 4.1).

A Planning Profession Emerges

A community vests the responsibility of identifying and advising on many solutions to its important problems in its professionals, such as doctors, lawyers, architects, and engineers. Increasingly, communities in the first few decades of the 20th century sought the advice and skills of town planners. By the end of the World War I, over 100 people practised town planning in Canada and they formed, in January 1919, a Town Planning Club, preparatory to establishing a formal institute. In May of the same year, the Town Planning Institute of Canada (TPIC) was formed, with 117 members and branches in four cities: Ottawa, Toronto, Winnipeg, and Vancouver. The organization exists to this day under the name of the Canadian Institute of Planners and has several thousand members.

The TPIC received its charter in 1919. The efforts of Hodgetts, Adams, Cauchon, and a score of others had borne fruit. They could now promote the acceptance of planning ideas in Canada, and also identify Canadians possessing the necessary technical skills to carry out planning ideas. Indeed, by 1919 all but two of the provinces had passed substantial planning statutes, and Canadian planners had helped in drafting them. Cities across the country now sought the assistance of planners, from both Canada and abroad, in planning civic centres, suburban extensions, and parks. The skills of these hundred or so new "town planners" included those of several traditional professions. The constitution of the TPIC originally set out that membership was limited to architects, engineers, landscape architects, surveyors, sculptors, artists, and sociologists. Lawyers could seek associate membership. This array of professional skills reveals a good deal about how planners saw the task of planning: a concern for building design, physical layout, the natural environment, civic design, social factors, and legal and administrative processes.

The names of a few of these pioneer Canadian planners should be acknowledged because of their groundwork in the modern planning of many of our cities. Among them were architects Percy Nobbs of Montreal, J.M. Kitchen of Ottawa, and J.P. Hynes of Toronto; engineers James Ewing and R.S. Lea of Montreal, Noulan Cauchon and Horace Seymour of Ottawa, W.A. Webb of Regina, and A.G. Dalzell of Vancouver; landscape architects Frederick Todd of Montreal and Howard Dunnington-Grubb of Toronto; and E. Deville, the Surveyor-General of Canada.

Developing a Canadian Planning Outlook

Thomas Adams helped significantly to define an approach to planning Canadian communities. In philosophy, he espoused a utilitarian view similar to John Stuart Mill and Jeremy Bentham, who championed the notion that the aim of society should be *to produce the greatest good for the greatest number*. In planning terms, this might translate into a community not having to bear the burden of traffic congestion caused by faulty street layouts in private development projects. Or it might also justify a community providing adequate sanitary facilities for all inhabitants, so that the costs of disease and epidemics need not be borne by others.

Planning Issue 4.1 Council Fires Board of Variance

Vancouver Sun
July 1, 2006

Council Fires Board of Variance: Panel Had Blocked Plan for a Luxury Marina in False Creek
William Boei

VANCOUVER—City council has fired all five members of the city's board of variance, two days after the board blocked a Concord Pacific plan for a luxury marina in False Creek.

However, Councilor Peter Ladner said the firing had nothing to do with any of the board's decisions, and was the result of concerns about budget overruns and a disagreement over the board's jurisdiction.

The decision was made in a closed council meeting.

The board's legal expenses had ballooned from less than $1,200 in 2001 to more than $70,000 last year—48 per cent over budget—and were on track to be even higher this year, Ladner said.

The board of variance had also asked for extended powers to deal with legal challenges to its decisions, he said.

"They felt that their jurisdiction should be wider than it is. Following up on that conviction has resulted in these budget overruns."

The board is the only avenue of appeal to city zoning and permitting decisions, other than the courts. It has made several controversial decisions recently, including the False Creek marina ruling, which upheld an appeal by a residents' group of a city decision to approve the marina.

Earlier, the board ruled that a developer should sell two lots on Salsbury Drive in east Vancouver to the city parks board, rather than tear down two heritage cottages and a community garden. The developer has taken the case to court.

Councilor Raymond Louie said he thinks the decision to fire the board was a political one and called it "autocratic and inappropriate."

"If they say it's because of budgetary issues, why is it that all five needed to be replaced?" he asked. "Why not just the chair, who controls the organization and is responsible for it? That's why I say that it's politics."

Ladner said he "totally" disagrees with Louie.

"The board has been unwilling to cooperate with the city in trying to control those costs, and there are other costs too, staff costs," he said.

"And the board is not willing to acknowledge that it has an obligation to rein in those costs and to adhere to city policies on working within its budget."

Louie said the independence of the board of variance should be preserved. "We should try our best to maintain that level of separation between ourselves and that body, which is meant to be the safeguard for decisions made by the city," he said.

"When you get this outright high-handed, autocratic action by the NPA and [Mayor] Sam Sullivan, I say that is inappropriate."

In the False Creek marina appeal, Rider Cooey, a spokesman for the False Creek-Alert residents' group, said city bureaucrats appearing at variance hearings had questioned the board's jurisdiction, saying the board could not consider the project in the light of broader development issues.

The False Creek-Alert group had argued that people heading for the marina would cause serious congestion in a lane that would also be the only access to the parkades of two residential towers.

Cooey said he had been "surprised but delighted" by the board's decision to uphold the appeal, but was shocked by the firing of the board.

"What they've done, I think, is destroy the credibility of the board of variance for the city of Vancouver, which is extremely unwise because that's the only independent civic body that reviews neighbourhood developments."

Ladner said the city will now "put out a call for people who would like to be on the board of variance, and we will appoint a new board."

But he said the city won't be getting involved in board decisions.

"That's outside [its] jurisdiction."

The utilitarian ethic embodied a number of principles that affected planning outlooks profoundly:

1. The notion of ***social progress***, carried out with public consensus;
2. The *application of **reason*** to determine solutions to social problems that will lead to progress; and
3. The acceptance of ***government intervention*** to achieve the public good.

The utilitarian outlook thus provided a rationalization of planners' views about the need to clear slum areas and to ensure proper suburban extensions to cities. It also provided the logical underpinning for the planning objectives of promoting *order and efficiency* in the development of communities. A Canadian planning historian characterizes the early 20th-century planner's view as follows:

> By creating well-planned urban environments, happy and healthy homes would be made available to working families, constructive social intercourse would be facilitated, the economic and social efficiency of the nation would be enhanced, and the general happiness would be increased.[77]

One finds this view widely espoused by Adams and his planning colleagues in Canada in the 1920s, possibly most succinctly on the masthead of the *Journal of Town Planning Institute of Canada*:

> Town planning may be defined as the scientific and orderly disposition of land and buildings in use and development with a view to obviating congestion and securing economic and social efficiency, health and well-being in urban and rural communities.

There are three fundamental aspects to the outlook of the early Canadian planners that, although not much discussed today, are no less a part of the intellectual makeup of contemporary Canadian planners. Community planning is concerned with

1. the functioning of the city rather than with its beauty, and its approach is rational (i.e., "scientific").
2. the social well-being of the community as a whole, as exemplified in Adams's dictum that "town planning includes every aspect of civic life and civic growth."[78]
3. finding technical solutions to planning problems for which planners are equipped to supply.

The first two of these foundation stones distinguished Canadian planners from their U.S. and British counterparts. The broad social view of responsibility for community health and housing is derived from the British; the functional view of arranging streets, utilities, and the use of zoning is distinctively American. The Canadian planners shunned City Beautiful approaches, and possibly as a consequence did not develop concepts of overall community design. They did, however, succeed in building a legislative base for the planning of communities through provincial planning acts. Canadian planning professionals concentrated their approach on administrative and legislative processes, borrowing broad planning statutes from the British and local zoning laws from the United States.

Lastly, the acceptance of a planning profession in Canada meant that community planning was moving out of the phase of being a "cause" propounded by a "movement." It was beginning to be legitimized and institutionalized. It also put in planners' hands the opportunity to develop, evolve, and mould the social values of planning. Thereafter, the development of planning ideas became bound up with the outlook of the planning professionals. These people became important actors in the planning process, for they, in many ways, had invented it.

The Collapse of the Canadian Planning Movement

The gains of the planning movement in the 1920s were almost eliminated in the Great Depression of the 1930s and the war period that followed. With the onset of the Great Depression, community planning lost a great deal of momentum in Canada. Many promising efforts in planning were halted, such as the closing of Alberta's Provincial Planning Office. The *TPIC Journal* ceased publication in 1931, and the Institute itself collapsed in 1932 for lack of dues-paying members; national planning conferences ceased and local planning commissions often stopped meeting. (In the United States, by contrast, the concepts of public housing and regional planning became cornerstones in the Roosevelt New Deal programs for coping with the Depression, thereby nurturing planning ideas and the planning profession.) By the end of World War II, only one city in Canada—Toronto—had a formal planning department, and fewer than two dozen towns had zoning bylaws and plans (most of them outdated). The state of Canadian community planning in 1945 was essentially comatose, and it would take a mighty push to revive it.

Reflections

By the end of World War II, a distinctively Canadian planning framework was in place. The provincial planning acts, the various institutional arrangements through which planning is accomplished and adjudicated, the planning tools, and the professional staffs could truthfully be said to be "made in Canada." There had, of course, been much borrowing of concepts (and personnel) from our two main influences, Great Britain and the United States, but they were gradually moulded to suit Canadian conditions, social outlook, and governance.

The legacy of the formative decades from 1900 to 1945 should not be forgotten; nor should it be forgotten that the institutions that emerged to promote and facilitate community planning were the product of human effort. That is, the elaborate sets of local and provincial planning institutions that screen, evaluate, and reconcile land-use decisions, and shape our communities, grew out of and reflect the values and norms of the people who nurtured and built them. The planning problems of the past were responded to with solutions that were acceptable to professional and political interests and to citizens at the time. Planning arrangements, tools, and processes, being social constructs, change with new social currents, new technology, and new generations. For example, broad citizen participation, which plan-makers today cannot avoid, was almost entirely the purview of the elite and property and business interests until the 1950s.[79] This legacy raises the following question:

- ***What social situations do you anticipate will affect planning processes, tools, and institutions over the next two decades?***

Endnotes

1. Walter Van Nus, "The Fate of City Beautiful Thought in Canada, 1893–1930," in G. Stelter and A. Artibise, eds., *The Canadian City: Essays in Urban History* (Toronto: Macmillan, 1979), 162–185.
2. John C. Weaver, *Shaping the Canadian City: Essays on Urban Politics and Policy, 1890–1920*, Monograph No. 1 (The Institute of Public Administration of Canada, 1977), 40.
3. Godfrey L. Spragge, "Canadian Planners' Goals: Deep Roots and Fuzzy Thinking," *Canadian Public Administration* 18:2 (Summer 1975), 216–234.
4. Elizabeth Bloomfield, "Town Planning Efforts in Kitchener-Waterloo, 1912–1925," in A. Artibise and G. Stelter, eds., *Shaping the Urban Landscape* (Ottawa: Carleton University Press, 1982), 256–303; and David Gordon, "A City Beautiful Plan for

Canada's Capital: Edward Bennett and the 1915 Plan for Ottawa and Hull," *Planning Perspectives* 13 (1998), 275–300.

5. J. William Brennan, "Visions of a 'City Beautiful': The Origin and Impact of The Mawson Plans for Regina," *Saskatchewan History* 46:2 (Fall 1994), 19–33.
6. William H. Wilson, *The City Beautiful Movement* (Baltimore: Johns Hopkins University Press, 1989).
7. Thomas H. Mawson, *Calgary: A Preliminary Scheme for Controlling the Economic Growth of the City* (London: T. H. Mawson & sons, city planning experts, 1914).
8. Harland Bartholomew & Assoc., *A Plan for the City of Vancouver, BC* (Vancouver: Vancouver Town Planning Commission, 1929); and William T. Perks, "Idealism, Orchestration and Science in Early Canadian Planning: Calgary and Vancouver Re-Visited, 1914/1928," *Environments* 17:2 (1985), 1–28.
9. Elizabeth Bloomfield, "Ubiquitous Town Planning Missionary: The Careers of Horace Seymour, 1882–1940," *Environments* 17:2 (1985), 29–42.
10. Kenneth Greenberg, "Toronto: The Unknown Grand Tradition," *Trace*, 1:2 (1981), 37–46.
11. Paul-André Linteau, *The Promoter's City: Building the Industrial Town of Maisonneauve, 1883–1913* (Toronto: Lorimer, 1985).
12. Van Nus, "The Fate of City Beautiful."
13. Gilbert Stelter, "Rethinking the Significance of the City Beautiful Idea," in Robert Freestone, ed., *Urban Planning in a Changing World: The Twentieth Century Experience* (London: Spon, 2000), 98–117; and Harold Kalman, *A History of Canadian Architecture*, vol. 2 (Toronto: Oxford University Press, 1994).
14. Arthur B. Gallion, *The Urban Pattern* (New York: Van Nostrand, 1950), 376; and Le Corbusier, *The City of Tomorrow and Its Planning* (London: John Rooker, 1929).
15. Eric Mumford, *The CIAM Discourse on Urbanism, 1928–1960* (Cambridge, MA: MIT Press, 2000).
16. J.L. Sert, *Can Our Cities Survive?* (Cambridge, MA: Harvard University Press, 1943).
17. Cf. Thomas Roden, "The Housing of Workingmen," *Industrial Canada* 5:7 (1907).
18. Marc H. Choko, *Une cité-jardin à Montréal* (Montreal: Meridien, 1988).
19. G. Frank Beer, "A Plea for City Planning Organization," in *Report of the Fifth Annual Meeting* (Commission of Conservation, 1914), 108–116.
20. Mervyn Miller, *Letchworth: The First Garden City* (Chichester, UK: Phillimore, 2002); and Mervyn Miller and A.S. Gray, *Hampstead Garden Suburb* (Chichester, UK: Phillimore, 1992).
21. Susan Klaus, *A Modern Arcadia: Frederick Law Olmsted Jr. & the Plan for Forest Hills Gardens* (Amherst, MA: University of Massachusetts Press, 2002).
22. Larry McCann, "Suburbs of Desire: Shaping the Suburban Landscape of Canadian Cities, 1900–1950," in Richard Harris and Peter Larkham, eds., *Changing Suburbs—Foundation, Form and Function* (London: Routledge, 1999), 111–145.
23. Jill Delaney, "The Garden Suburb of Lindenlea, Ottawa: A Model Project for the First Federal Housing Policy, 1918–1924," *Urban History Review* 19:3 (February 1991), 151–165.
24. Larry McCann, "Planning and Building the Corporate Suburb of Mount Royal, 1910–1923," *Planning Perspectives* 11 (1996), 259–301.
25. John Weaver, "Reconstruction of the Richmond District in Halifax," *Plan Canada* 16:1 (March 1976), 36–47.
26. Nigel Richardson, "A Tale of Two Cities," in L.O. Gertler, ed., *Planning the Canadian Environment* (Montreal: Harvest House, 1968), 269–284.
27. Thomas Adams, *Rural Planning and Development* (Ottawa: Commission of Conservation, 1917), 66.
28. Institute of Local Government, Queen's University, *Single-Enterprise Communities in Canada*, A Report to Central Mortgage and Housing Corporation (Kingston: Queen's University, 1953), 24.
29. Clarence Perry, "The Neighborhood Unit," in *Regional Survey of New York and Its Environs*, vol. 7 (New York, 1929).
30. Clarence Stein, *Toward New Towns for America* (New York: Reinhold, 1957), 19.
31. K.C. Parsons, "British and American Community Design: Clarence Stein's Manhattan Transfer, 1924–74," *Planning Perspectives* 7 (1992), 191–210.
32. Clarence Stein, *Toward New Towns for America*, 39ff.
33. Robert Whitten and Thomas Adams, *Neighbourhoods of Small Homes: Economic Density of Low-cost Housing in America and England* (Cambridge, MA: Harvard University Press, 1931); and Eran Ben Joseph and David Gordon, "Hexagonal Planning In Theory and Practice," *Journal of Urban Design* 5:3 (December 2000), 237–265.
34. As quoted in Alan H. Armstrong, "Thomas Adams and the Commission of Conservation," in L.O. Gertler, ed., *Planning the Canadian Environment*, 17–35. See also Nigel H. Richardson, "Canada in the 20th Century: Planning for Conservation and the Environment," *Plan Canada*, Special Edition (July 1994), 52–69.
35. C.A. Hodgetts, "Housing and Town Planning," in *Report of the Third Annual Meeting* (Commission of Conservation, 1912), 136.
36. Clifford Sifton, "National Conference on City Planning," in *Report of the Sixth Annual Meeting* (Commission of Conservation, 1915), 243. See also David Lewis Stein, "The Commission of Conservation," *Plan Canada*, Special Edition (July 1994), 55; and Michel Girard, *L'ecologisme retrouve: Essor et declin de la Commission de la conservation du Canada* (Ottawa: Les Presses de l'université d'Ottawa, 1994).
37. Peter Jacobs, "Frederick G. Todd and the Creation of Canada's Urban Landscape," *Association for Preservation Technology (APT) Bulletin* 15:4 (1983), 27–34.
38. Armstrong, "Thomas Adams," 28. See also David Lewis Stein, "The Commission of Conservation," 14–15; and Michael Simpson, *Thomas Adams and the Modern Planning Movement: Britain, Canada and the United States, 1900–1940* (London: Alexandrine Press, 1985).
39. David Johnson, *Planning the Great Metropolis: The 1929 Regional Plan of New York and Its Environs* (New York: Spon, 1996).
40. Frederick G. Todd, *Preliminary Report to the Ottawa Improvement Commission* (Ottawa: OIC, 1903); and David Gordon, "Frederick G. Todd and the Origins of the Park System in Canada's Capital," *Journal of Planning History* 1:1 (March 2002), 29–57.
41. Gordon, "Frederick G. Todd," 43–50.
42. Walter Van Nus, "Toward the City Efficient: The Theory and Practice of Zoning, 1919–1939," in A. Artibise and G. Stelter, eds., *The Usable Urban Past* (Toronto: Macmillan, 1979), 226–246. Carleton Library No. 119.
43. David Gordon and Brian Osborne, "Constructing National Identity: Confederation Square and the National War Memorial in Canada's Capital, 1900–2000," *Journal of Historical Geography* 30:4 (October 2004), 618–642.
44. Canada, Commission of Conservation, *Report of the Third Annual Meeting* (Commission of Conservation, 1912), 141.
45. Weaver, *Shaping the Canadian City*.
46. James Lemon, *The Toronto Harbour Plan of 1912: Manufacturing Goals and Economic Realities*, Working Papers of the Canadian Waterfront Resource Centre, Royal Commission on the Future of the Toronto Waterfront (Toronto, 1990).
47. Thomas Adams, "Editorial: Town Planning is a Science," *Journal of the TPIC* 1:3 (April 1921), 1–3.
48. H. Bartholomew & Assoc., *A Plan for the City of Vancouver*.

49. William Perks, "Idealism, Orchestration and Science in Early Canadian Planning."
50. Norman Johnston, "Harland Bartholomew: Precedent for the Profession," in Donald Krueckeberg, ed., *The American Planner: Biographies & Reflections* (New Brunswick, NJ: Rutgers CUPR, 1994), 217–241.
51. Bloomfield, "Town Planning Efforts in Kitchener-Waterloo."
52. Paul Rutherford, "Tomorrow's Metropolis: The Urban Reform Movement in Canada, 1880–1920," in G. Stelter and A. Artibise, eds., *The Canadian City* (Toronto: Macmillan, 1979), 368–392.
53. As quoted in John C. Weaver, "Tomorrow's Metropolis Revisited: A Critical Assessment of Urban Reform in Canada, 1890–1920," in Stelter and Artibise, *Canadian City*, 393–418.
54. Elizabeth Bloomfield, "Town Planning Efforts in Kitchener-Waterloo."
55. J.B. Milner, "The Statutory Role of the Planning Board," *Community Planning Review* 12:3 (1962), 16–18.
56. P.J. Smith, "The Principle of Utility and the Origins of Planning Legislation in Alberta, 1912–1975," in A. Artibise and G. Stelter, eds., *The Usable Urban Past*, 196–225.
57. New Brunswick, *An Act Relating to Town Planning*, Chap. 19, 2 Geo. V, 1912, Sect. 1(I).
58. Alberta, *An Act Relating to Town Planning, Statutes of Alberta*, Chap. 18, 1913, Sect. 1(6).
59. New Brunswick, *An Act Relating to Town Planning*, 2(2).
60. Alberta, *An Act*, 1913.
61. Canada, Commission of Conservation, *Report of the Sixth Annual Meeting* (Commission of Conservation, 1915), 271.
62. Cf. New Brunswick, *An Act Relating to Town Planning, and Alberta, An Act Relating to Town Planning*, 6(2).
63. Alberta, *An Act Relating to Town Planning*, 6(2).
64. Stephen Ward, *Planning the Twentieth Century City: The Advanced Capitalist World* (New York: Wiley, 2002), 26–31, 52–56.
65. For a full discussion of this act, see J. David Hulchanski, "The Evolution of Ontario's Early Urban Land Use Regulations, 1900–1920," a paper presented to the Canadian–American Comparative Urban History Conference, Guelph, 1982.
66. Canada, Federal Plan Commission, *Report of the Federal Plan Commission on a General Plan for the Cities of Ottawa and Hull* (Ottawa, 1915).
67. David Gordon, "A City Beautiful Plan."
68. Bloomfield, "Town Planning Efforts in Kitchener-Waterloo."
69. J. David Hulchanski, *The Origins of Urban Land Use Planning in Alberta, 1900–1945*, Research Paper 119 (University of Toronto, Centre for Urban and Community Studies, 1981).
70. Raphael Fischler, "Markets, Politics and Social Science in Early Land-Use Regulation and Community Design," *Journal of Urban History* 24:6 (September 1998), 675–719.
71. Federal Plan Commission, *Report*, 46; see also the description of these zoning districts in Gordon, "A City Beautiful Plan."
72. Thomas H. Logan, "The Americanization of German Zoning," *Journal of the American Institute of Planners* 42:4 (October 1976), 377–385.
73. John C. Weaver, "The Property Industry and Land Use Controls: The Vancouver Experience, 1910–1945," *Plan Canada* 19:3,4 (September–December 1979), 211–225.
74. Ibid.
75. J.B. Milner, *Development Control*, (Toronto: Ontario Law Reform Commission, 1969), 12.
76. Earl Levin, "Zoning in Canada," *Community Planning Review* 7 (June 1957), 85–87.
77. P.J. Smith, "The Principle of Utility and the Origins of Planning Legislation in Alberta." See also David Sherwood, "Canadian Institute of Planners," *Plan Canada*, Special Edition (July 1994), 20–21.
78. Thomas Adams, "What Town Planning Really Means," *The Canadian Municipal Journal* 10 (July 1914).
79. For an excellent review of the evolving context of Canadian community planning, see Jeanne M. Wolfe, "Our Common Past: An Interpretation of Canadian Planning History," *Plan Canada*, Special Edition (July 1994), 12–34.

Internet Resources

Chapter-Relevant Sites

Planning Canadian Communities
www.planningcanadiancommunities.ca

Our Common Past: An Interpretation of Canadian Planning History—Part 1
www.cip-icu.ca/English/plancanada/wolfe.htm

Canadian Urban Policy Archive
www.urbancentre.utoronto.ca/policyarchive.html

Thomas Adams in Canada
www.cip-icu.ca/English/plancanada/stein.htm

Hampstead Garden Suburb
www.hgs.org.uk

Canberra—An Ideal City?
www.idealcity.org.au

Planning Canada's Capital: Six historic plans
www.PlanningCanadasCapital.ca

Canadian Council on Social Development
www.ccsd.ca

5

Chapter Five

The Growth of Canadian Community Planning, 1945–2000

[E]lbow room is a desirable characteristic...

Macklin Hancock, 1954

The second half of the 20th century posed as many challenges for community plan-makers as did the first half. It began with a pressing demand for housing and urban infrastructure that had been put aside because of the Great Depression and World War II, as well as a need to refurbish planning institutions. City populations that had swelled as people flocked in to fill wartime jobs continued to grow, and grow quickly. As the century continued, the baby boom, post-war economic expansion, and increased automobile use and home ownership further fuelled urban expansion into the suburbs and exurbs.

While the challenges were new, the same sets of concerns as had energized planning approaches before and after the beginning of the century came again to the fore. The visual appearance of cities, their living conditions, natural environments, and efficient functioning each vied for attention as the public's concerns about their communities waxed and waned. Communities became more diverse, with changing family structures, more women in the labour force, and massive immigration, especially to the largest cities. New concepts appeared, such as Smart Growth and New Urbanism, as well as new technologies, such as Geographic Information Systems (GIS), and new modes of working, such as telecommuting. Probably most noteworthy was the strong emphasis on protection of the natural environment and, more latterly, sustainability. In other words, the activity of community planning was called upon to modify its processes and its tools as it has always done with changing times. This chapter discusses the evolution of community planning in Canada from 1945–2000, the period that immediately underlies contemporary planning practice. With this perspective, keep the following question in mind:

- *In which ways do today's plans and plan-making processes reflect the concerns of planners in the latter half of the 20th century?*

Post-War Planning Challenges

Canada changed from a rural to an urban nation in the first half of the twentieth century. Demographers point to 1931 as the first year in which the majority of Canadians lived in urban areas.[1] However, the strong urban development of the 1900–1930 period stalled during the Great Depression of the 1930s, and the rapid expansion of Canada's industrial facilities for the war effort created a housing crisis in many cities. Community planning institutions were also left to languish during the same period. After the war, the federal government pushed to revive these institutions, which would soon be sorely needed to guide the growth and development of Canadian cities and towns. As they struggled with these problems, citizens and planners developed some new methods and models that were widely admired in the last decades of the century.

The term **exploding metropolis**[2] aptly captures the state of urban affairs of the 1950s and of the decades that followed. Cities grew in population, expanded their boundaries, and increased in both density (with taller buildings at the centre) and area (with more houses and factories on the outskirts). Three forces combined to produce the metropolitan environments we are familiar with today. First, there were demographic forces due to (a) the migration of people from rural areas to urban centres; (b) people emigrating to Canada from abroad, most of whom went to the larger cities; and (c) a dramatic increase in the natural growth of the population, known as the "baby boom."[3] Second, there were economic forces. The nation's economy continued to expand and produce jobs and rising incomes for the new and old populations of cities. There was also a pent-up demand for housing and other urban accoutrements created by its suppression during 15 years of depression and war.[4] The third force, and in many ways the most momentous in its effect on the form of cities, was the vast expansion in automobile use. These three forces combined to reshape the metropolis, and Canadian community planning had to face a number of challenges whose resolution would establish foundations for its future.

Advent of the Baby Boom, 1946–1965

In addition to the surging city populations of the wartime period, communities across the country were soon to feel the impact of an unanticipated boom in population that would last 20 years—the baby boom. Family formation and birthrates in the 1946–65 period increased dramatically. To appreciate just how much, consider that the annual birth group size for the previous two decades (1926–45) had been approximately 250 000; it then boomed to 425 000 per year, an increase of 70 percent. In other words, an additional 3.5 million births occurred in the post-war period up to 1965.[5]

Although the birth groups of this period are often referred to as the "baby boom generation," they do not constitute a true generation, but rather two 10-year age cohorts. Their significance for community planning is that these cohorts spurred sudden increases in demand for housing, schools, playgrounds, and health and other social services. And, over ensuing decades, this led to a demand for more colleges and universities, and more housing and health and social services. This demographic bulge still continues to add new dimensions to community planning as it contributes to a surging seniors' population that will expand dramatically within half a decade.[6]

Mass Automobile Use

Although we now seem accustomed to large, spreading cities, the tenor of the 1950s regarding this change is caught in the following quotation:

> Of all the forces reshaping the American metropolis, the most powerful and insistent are those rooted in changing modes of transportation. The changes are so big and obvious that it is easy to forget how remarkable they are. The streetcar has all but disappeared, the bus is proving an inept substitute, commuter rail service deteriorates, subways get dirtier, and new expressways pour more and more automobiles into the centre of town.[7]

These writers might well have added that more and more automobiles were pouring into the outskirts of cities because that is where the growing populations could be housed most easily, and it is also where the old public transportation systems did not usually reach, or service well if they did. Mass automobile ownership and expressways gave Canadians a transportation alternative that was private, convenient, flexible, and fast (See Figure 5.1). Public transit ridership, concomitantly, declined from 13 percent to 5 percent of all trips between 1950 and 2000.[8]

Two more auto-oriented facilities shaped the post-war city: the shopping centre and the industrial park. Both of these facilities signalled a reorientation of major community functions away from a single dominant centre. Major shopping facilities in pre-1950 cities were tied to downtown areas because most personal transportation focused on downtown, especially streetcars and buses. Industries were often tied to rail or harbour access. When the automobile's widespread availability favoured the development of suburbs beyond the reach of public transit, it meant that stores and businesses not only had a market, but could also take advantage of easy truck transport of goods on suburban roads. Shopping centres and industrial parks subsequently became common elements in the fabric of Canadian communities and injected a new set of relationships into planning because they both needed considerable land and easy automobile access to a large area. These changes also had an impact on downtown cores, often leaving the older central stores wanting for business, and industrial areas in decline.

Meeting Housing Needs

The change in housing options was driven by less visible but equally powerful preferences. A suburban single-detached home offered privacy, more space, personal control, green surroundings, and a long-term real estate investment that many families preferred over a rental unit in the city.[9] The federal government had tested the development of large-scale suburban housing projects at big wartime plants that needed worker accommodation. The federal agency Wartime Housing Ltd. built 32 000 rental homes for workers and returning veterans between 1941 and 1947.[10] These homes were sold, and mass home ownership for other middle-class and working-class families was made possible by changes in mortgage arrangements described below. The home ownership rate increased from 57 percent in 1941 to 65 percent in 1951 and has remained at approximately that level since.[11] In metropolitan areas, the housing stock increased between 10 percent and 20 percent between 1945 and 1950. In this period,

Figure 5.1 Impact of Expressways on the Metropolitan Landscape: Longueuil, Quebec

In an aerial photograph of any Canadian metropolitan area, the most prominent artificial feature will probably be an expressway. It was hoped that these massive modern facilities would relieve downtown congestion and facilitate movement among the new suburban housing and industrial areas. Building expressways significantly changed the physical structure of urban areas because they consumed lots of land and became barriers in the urban landscape.

Source: Courtesy of the National Film Board of Canada.

11 000 new dwellings were built in Winnipeg, 3000 in Halifax, 21 000 in Vancouver, and 31 000 in Montreal.[12]

Suburban development further accelerated in the 1950s and 1960s. The total effect was an extraordinary transformation: by 1996, most Canadians lived in suburbs built after 1945[13]—the country had changed from an urban to a suburban nation. Surprisingly, these momentous social and economic forces that reshaped Canadian communities were anticipated and somewhat encouraged by national planning efforts for the post-war period (as we shall see later in this chapter), as well as by early metropolitan planning in Winnipeg and Toronto, described in Chapter 9.

Thus, the three great post-war challenges—population growth, the housing boom, and mass automobile use—contained elements of the main themes of Canadian community planning from the late 19th century: city appearance, living conditions, the environment, and city efficiency (see Figure 5.2). The emphasis among the broader themes ebbed and flowed during the rapid metropolitan development over ensuing decades. City efficiency and appearance were major concerns in the construction boom of the 1950s and 1960s. Engineers and architect–planners were busy with vast infrastructure projects and major urban renewal schemes, and almost every Canadian community was engaged in comprehensive land-use planning. By the late 1960s, the social and environmental impacts of these programs were being questioned, and sustainable development became a dominant issue in Canadian planning in the late 1980s. At the end of the century, community design and infrastructure issues re-emerged in Smart Growth, New Urbanism, and transit-oriented development. More sophisticated and complex planning approaches characterize the last few decades and mark the growing maturity of the Canadian planning movement.

Figure 5.2 Evolution of Canadian Planning Ideas, 1945–2000

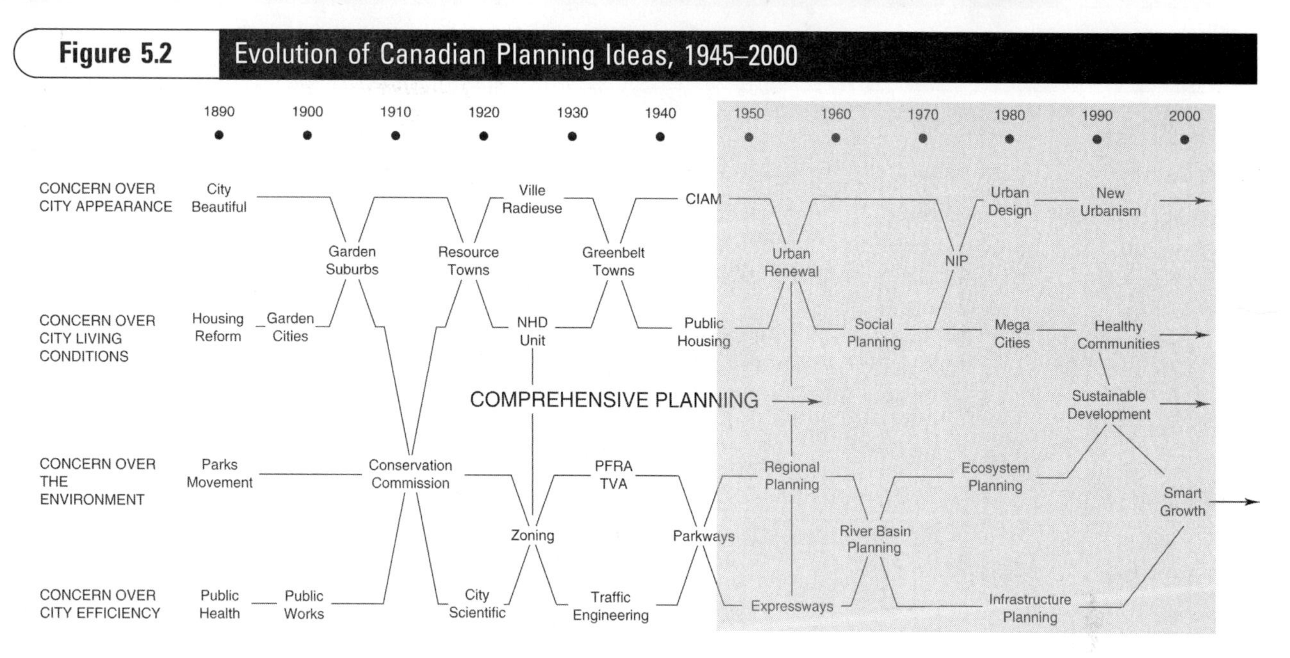

Today's planning themes of comprehensive planning, sustainable development, Smart Growth, New Urbanism, and Healthy Communities continue to echo the four basic planning concerns of more than a century ago.

Concern over City Appearance

The organizing and theoretical work regarding the form and functioning of cities done by the CIAM (Congrés Internationaux d' Architecture Moderne) in the 1930s paid off after World War II. Their message about the obsolescence of historic city centres and the need to rebuild them using Modern planning principles found a willing audience in the reconstruction of European and Japanese cities destroyed by war. The CIAM principles from the Athens Charter entered North America through architecture and planning schools, which were usually combined in that era.[14] Most North American schools switched from Beaux Arts to CIAM approaches during the 1950s.

Rebuilding the Inner City

One of the consequences of this shift to Modern planning concepts was to focus professional attention on large-scale rebuilding of the inner city. This manifested itself in four overlapping programs—urban renewal, neighbourhood redevelopment, public–private partnerships, and private megaprojects—that unfolded over the next several decades (see Figure 5.3).

Urban Renewal

Urban renewal emerged in recognition of the fact that a large portion of the buildings in Canadian communities were half a century or more old. The Central Mortgage and Housing Corporation (CMHC) reported in 1956 that almost 900 000 housing units were of this vintage and, furthermore, that one-tenth of all housing was in need of major repairs.[15] "Blighted" conditions, as they were then called, could also be found among old factory and warehouse buildings, many of which were being abandoned for sites in suburban industrial parks. The central parts of many cities, which were also the oldest sections, suffered most from physical deterioration. Urban renewal

Figure 5.3 Rebuilding the Inner City

	Urban Renewal	Neighbourhood Development	Public/Private Partnership	Private Megaprojects
Time period	1950–1970	1965–1990s	1975 +	1990 +
Project type	Slum Clearance	New Precinct	Waterfront/CBD	Railyards
Typical projects	Jeanne-Mance [Montreal] Regent Park [Toronto] West End [Boston]	False Creek S. [Vancouver] St. Lawrence [Toronto]	Harbourfront [Toronto] Quincy Market [Boston] Battery Park City [NYC] Docklands [London] Granville Island [Vancouver]	False Creek N [Vancouver] Railway Lands [Toronto] Mission Bay [San Francisco] Riverside South [NYC]
Land-use change	Residential to Residential & Institutional	Industrial/Transport to Residential/Mixed	Industrial/Transport/ Commercial to Mixed Use	Transport/Industrial to Mixed Use
Leadership	Housing Authority	Redevelopment Agency	Redevelopment Agency & Private Sector	Private Sector
Early capital	Public Sector	Public Sector	Public & Private	Private Sector
Change in ownership	Private to Public	Public to Public	Public to Private	Private to Private
Plan responsibility	Housing Authority & City	Redevelopment Agency	R. Agency & Developer	Developer & City
Planning mode	Master Plan	Site Plan & Urban Design Guidelines	Design Guidelines & Site Plan	Site Plan & Urban Design Guidelines
Key success factors	Public Grants Expropriation Power	Public Grants Expropriation Power	Private Investment Expropriation Power Public Borrowing	Land Ownership Private Investment Political Approval
Problems	Displacement No Market Response	Limited Public Funds	Uncoordinated Development Few Social Benefits	Loss of Public Control Few Social Benefits

Source: David L.A. Gordon, "Phases of Redevelopment in North American Inner Cities," in M.H. Yeates, *The North American City,* 5th ed. (Longman, 1998), 401–403.

was an effort to restore the commercial attractiveness of downtown areas and avoid the loss of investment they represented. Hoping to eliminate substandard housing, the federal government established a program for financing these "slum clearance" projects. Just as the name suggests, dilapidated housing would also be bulldozed. Residents of the area would be re-housed in public housing projects (a companion program to urban renewal), sometimes on the same site. Occasionally, the newly razed area was considered more valuable as a place for new commercial development, and the displaced residents were forced to move to public housing in another district, or to crowd in with friends or neighbours if they did not want to leave their neighbourhood.

Urban renewal was initially supported by the leaders of social reform, such as Catherine Bauer, Humphrey Carver, and Albert Rose. Planning for the first Canadian project, Toronto's Regent Park, had begun in the 1930s and had solid public support when it began development in 1948.[16] Many blocks of 19th-century housing in poor condition were demolished to create a master-planned project with 2100 public housing units. Similar projects were begun in most North American cities, but an avalanche of criticism from social scientists in the 1960s brought these programs to a halt.[17] Urban renewal became very unpopular because of the displacement of residents and destruction of close-knit communities such as that of Halifax's Africville, a community that had existed for over 100 years, which was razed in 1964.[18] Citizens were outraged by expropriation of their homes, and politicians were disappointed when some of the bulldozed sites remained vacant due to lack of market response. And the social and physical deterioration of high-density urban public housing projects magnified the perception that urban renewal was a planning disaster.[19]

The programs that replaced urban renewal were influenced by the ideas of noted urban critic Jane Jacobs.[20] Her advocacy for mixed uses, higher density, greater diversity, and retention of older buildings supported new programs for urban infill and rehabilitation. Although urban renewal was phased out in the late 1960s, it embodied some important features for community planning. In the first place, its focus on the need to rebuild cities complemented the planning concerns with new development on the fringes. In the second place, it established the need for the involvement of senior levels of government to facilitate such action, especially in its financing. Urban renewal programs instituted the idea of "shared cost" approaches in community planning where, for example, the cost of a site designated for urban renewal would be shared 50 percent by the federal government, 25 percent by the provincial government, and 25 percent by the municipality. A third, and possibly most important, feature was the requirement that the municipality have a community plan that took into account the future use and role of redevelopment areas. One finds these principles in use today in various new programs for renewing communities.

Neighbourhood Development

Citizen protests over urban renewal and expressway projects in major Canadian cities were the impetus for the election of reform-oriented municipal councillors in the early 1970s in cities across the country. In a burst of creativity, the reform councils sponsored inner-city development such as Vancouver's False Creek South and Toronto's St. Lawrence Neighbourhood.[21] The planning principles that guided the St. Lawrence Neighbourhood were strongly influenced by Jane Jacobs, who had moved to Toronto in 1968 (see Figure 5.4).

False Creek and St. Lawrence were postmodern urban projects that challenged the Modern planning principles embedded in the CIAM Athens Charter. Their mixed-use, mixed-income, mixed-tenure approach was judged a success and widely admired.[22] However, they relied on large federal and provincial grants for land expropriation, infrastructure, and housing subsidies. When these grants were cut back in the 1980s, it became difficult to start similar projects, and the cost of environmental remediation made redeveloping other former (brownfield) industrial sites too expensive.

Public–Private Partnerships for Urban Redevelopment

After the federal and provincial urban renewal funds disappeared, some North American cities turned to partnerships with private developers to implement urban redevelopment projects using the techniques discussed in Chapter 14. Many cities had underdeveloped land in the downtown and waterfront areas. Boston's Quincy Market, Baltimore's Inner Harbour, and New York's Battery Park City were early examples of public–private partnerships, where local governments attracted private developers to implement urban revitalization projects using well-located land, good urban design, and limited infrastructure funds to attract investment.[23] The Canadian response was to establish federal or provincial agencies for waterfront redevelopment in Vancouver, Winnipeg, Toronto, Montreal, Quebec, and Halifax. The record on these initiatives is varied. Vancouver's Granville Island Trust used an innovative plan by Hotson Bakker Architects to transform a declining industrial site into a vibrant, mixed-use waterfront with a distinctive sense of place.[24] The Winnipeg, Montreal, and Halifax waterfront projects have proceeded slowly, but with generally positive results. Toronto's Harbourfront started out with great promise and some fine re-use of industrial buildings, but dissolved into an intense controversy over parks and building heights.[25] In the meantime, other public authorities, such as hospitals, airports, universities, utilities, and housing agencies, began to use the public–private partnership tools pioneered on the waterfront to attract private investment to their sites.

Private Megaprojects

Another type of inner-city project emerged in the 1990s, when very little public funding was available for urban redevelopment. Private corporations with ownership of large, well-located sites close to the central business district, such as old railway yards, began to negotiate with local governments for redevelopment plans. Although the railway companies often started the planning process for sites such as San Francisco's Mission Bay, Manhattan's Riverside South, Vancouver's False Creek North, and Toronto's Railway Lands, large-scale private developers eventually acquired most of the sites. The planning techniques for these megaprojects used the approach of site plan/urban design guidelines pioneered in new urban neighbourhoods (see Figure 5.4).

Figure 5.4 Comparison of Planning and Urban Design for Urban Renewal Projects and New Urban Neighbourhoods

Urban Renewal Projects	New Urban Neighbourhoods
Planning Factors:	**Planning Factors:**
• Single income group (low/moderate)	• Mix of income groups
• Single tenure (rental)	• Mix of tenures (co-op, condo, rental)
• Single developer	• Variety of developers
• Single management and owner	• Diversified management and owner
• Separation of land uses	• Mixed-use development
• High density	• High density
• Site clearance and redevelopment	• Preservation of existing buildings
• Master plan	• Site plan and urban design guidelines
Urban Design Factors:	**Urban Design Factors:**
• New precinct/separate	• Extension of the city fabric
• Superblock/ring road	• Extension of the street grid
• Separation of vehicles and pedestrians	• Street as focus of activity
• Project address	• Address from a street
• Large parcels, coarse grain	• Small parcels, fine grain
• High-rise buildings	• Low-/medium-rise buildings
• Family housing on internal pathways	• Family housing on local streets
• Single landscape treatment	• Many landscape treatments
• Single architect	• Many architects

Source: David Gordon, ed., *Directions for New Urban Neighbourhoods: Learning from St. Lawrence* (Toronto: Ryerson SURP, 1990), Table 4-1.

The municipalities had little bargaining power to extract public benefits during the megaproject approval process since they were not contributing land or capital to any significant extent. The railways bargained hard, and if they didn't like the deal, they could just wait, as happened in Montreal's Outremont. As a result, projects such as Vancouver's Concorde Pacific Place and Toronto's City Place have substantially lower proportions of social housing and open space than the adjacent False Creek South or Harbourfront projects. The Vancouver planners had some success on urban design issues by negotiating a commendable water's edge promenade and pioneering a point-tower and townhouse block hybrid-building type that combines high density with good streetscapes.[26]

Evolving Planning Perspectives

Emergence of Urban Design

The urban renewal debate gave large-scale master planning a bad name in the 1960s, and concern over city appearance almost disappeared in Canadian planning over the next two decades.[27] Although Modern master-planning principles from Regent Park (see Figure 5.4) were discredited, it was eventually acknowledged that while physical design could not solve social problems on its own, it could help create a better built environment. Urban design emerged as a new interdisciplinary specialty addressing the craft of "designing cities without designing buildings," as Jonathan Barnett phrased it.[28] Urban design was influenced by postmodern planning ideas that avoided large-scale master planning by a single architect in favour of flexible site plans and guidelines to direct a variety of designers over the many years it takes to build a city.[29] Urban design guidelines became important parts of plans for downtown areas and redevelopment projects in the 1990s.

Historic Preservation

Heritage preservation was another area where the CIAM planning principles were rejected in the 1970s. A few of the more enlightened urban renewal plans, such as Philadelphia and Kingston, Ontario, incorporated some historic buildings into their new designs.[30] But the Modern *tabula rasa* approach had led to the destruction of many fine 19th-century structures. Citizen opposition increased when important public buildings such as New York's Penn Station and Toronto's post office were razed and replaced with mediocre Modern structures. Heritage advocates first rallied public support to preserve threatened city halls and railway stations. Then developers were encouraged to incorporate historic buildings into their redevelopment projects, as happened with Montreal's Maison Alcan. Finally, planning agencies began to protect and enhance whole districts of historic buildings, starting with the National Capital Commission's Sussex Drive in the 1960s and extending to Vancouver's Gastown, Vieux Montreal, Quebec City's Lower Town, and Halifax's Historic Properties. By the end of the century, many Canadian communities, from small towns to metropolitan centres, had planning policies in place to protect their built heritage.[31]

Redevelopment of Modern Projects

Many of the earliest Modern housing projects were reaching the end of their useful life by 2000, creating an interesting planning problem. The United States and Britain had begun large-scale programs to demolish their most distressed high-rise public housing, starting with St. Louis's Pruitt-Igoe project in 1972. Perhaps the most successful Canadian redevelopment project is Montreal's Benny Farm, where part of a 1940s housing project is being renovated.[32] Social housing, condominium apartments, and new community facilities were added by infill and selective demolition.

The latest "green building" techniques were included and the project has achieved the stature as a model of 21st-century urban redevelopment. After an extensive public consultation process, Toronto decided to gradually demolish Regent Park, replacing it with a mixed-use, mixed-income, medium-rise neighbourhood modelled on the St. Lawrence neighbourhood.[33] The Woodward's project (Vancouver) is another interesting case that mixes a heritage department store, and social and market housing with public spaces.[34]

Conventional Suburban Development

While the urban renewal battles raged downtown, another, more profound, change in planning principles was happening in the suburbs, where smaller projects and self-building were the norm before 1950.[35] Modern planning principles such as the superblock, separation of uses, and the neighbourhood unit had been demonstrated in Radburn, New Jersey, in the late 1920s and were well-known in the immediate postwar period.[36] Canadian projects followed these examples, such as Cité Jardin in Montreal (1947), designed by Samuel Gitterman,[37] and Wildwood in Winnipeg (1948), designed by Hubert Bird.[38]

A more important design precedent emerged in 1952 when industrialist E.P. Taylor assembled an 800-hectare parcel of rural land in North York. He financed, planned, and implemented Don Mills, Canada's first large-scale suburb produced by a private developer. Taylor produced a model suburban project, perhaps because he took a chance and hired a young landscape architect, Macklin Hancock, to plan Don Mills. Hancock had studied the Modern "new towns" near London and Stockholm and drew on these precedents and the neighbourhood unit to design a satellite community that was closer to a "new town" than a bedroom community.[39]

The Don Mills community (see Figure 5.5) would house 35 000 people and create jobs for 20 000. The core residential area comprised four neighbourhood units connected by a ring road. Each neighbourhood unit had an elementary school and places of worship at its centre, and an interconnected open space system that allowed most children to walk to school without crossing a major street. The centre of the community included a shopping centre, apartment blocks, high school, and other community facilities; all were planned using the best Modern architectural design. Traffic was sorted by a firm hierarchy of streets, with loops and crescents at the local level, rather than Radburn's culs-de-sac. An equal amount of land was set aside for industrial parks and soon filled with factories and offices designed by Canada's best architects. Finally, the area's ecological features were protected, with the original drainage system and Don River Valley preserved. Many other companies followed Taylor's innovations and established large development corporations to buy up huge tracts of farmland, build infrastructure, and sell building lots. But few developers provided the mix of housing types, community facilities, good landscape design, and jobs–housing balance of the original Don Mills model.

By the late 1980s, low-density, automobile-oriented suburbs had changed from an interesting alternative to urban living to the dominant model for Canadian family life. Shopping followed the residents, first in shopping centres, and then into isolated big box stores. Finally, jobs started flowing out to big office parks and manufacturing plants near freeway interchanges.[40] The result was a vast expanse of low-density separated land uses, which Melvin Webber christened the "non-place urban realm."[41] Some urban economists hailed suburban development as a pure expression of the people's preference for single-detached homes and automobile travel, citing Los Angeles as the new model for urban life.[42] Others called it **urban sprawl** and a new design movement arose to attack its worst features.

Figure 5.5 Don Mills Plan, 1954

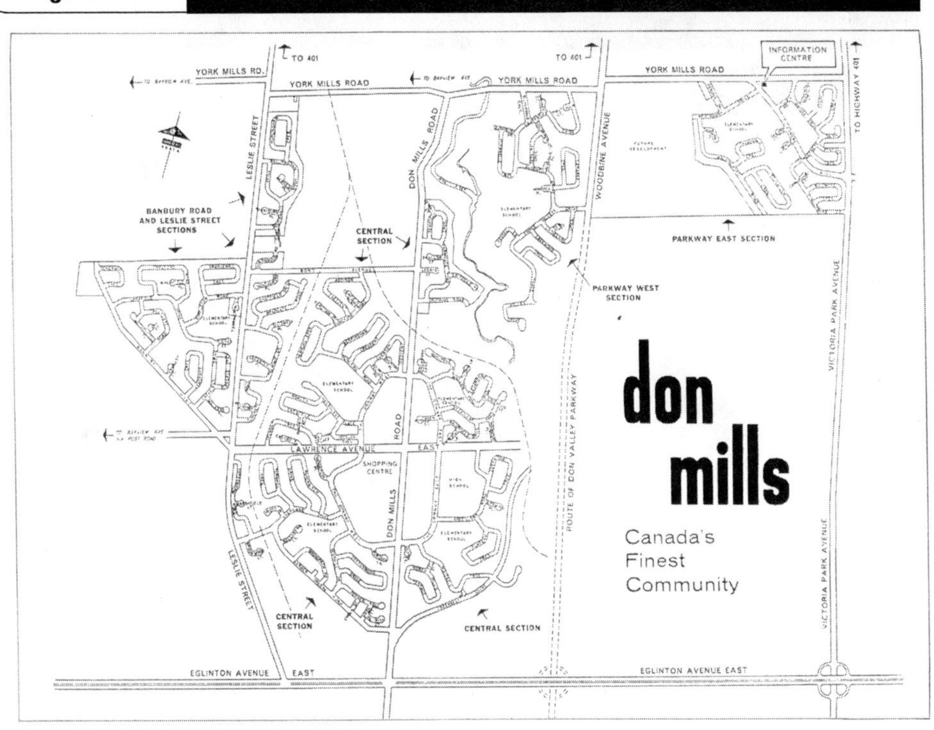

Macklin Hancock's 1954 plan for Don Mills, Toronto, started with four neighbourhood units, connected by the collector ring road in the centre of the diagram from this contemporary sales brochure. Employment areas ("industrial parks") were located to the north and south; the open space in the valley to the west was disrupted by construction of the Don Valley Parkway (dotted lines).

Source: John Sewell, "Don Mills: E.P. Taylor and Canada's First Corporate Suburb," in James Lorimer, ed., *The Second City Book* (Toronto: Lorimer, 1977), 25.

New Urbanism

By the late 1980s, the conventional suburban development model was degraded as amenities disappeared, lots grew narrower, and two- to three-car garages popped out the front of homes that were often referred to as "snout houses" (see Figure 5.6) Some suburban streets began to resemble a wall of garage doors. By the early 1990s, postmodern urban design ideas began to affect Canadian suburban planning, perhaps first at Bois Franc in Montreal.[43] Calgary's McKenzie Towne and Markham's Cornell were designed by Andres Duany and Elizabeth Plater Zyberk, both leaders of the Congress for the New Urbanism (CNU).[44] The CNU issued its own *Charter of the New Urbanism* to rival the CIAM's Athens Charter, but with quite different planning principles. New Urbanists call for a variety of planning approaches, depending on regional location (see Figure 6.7, page 159), and advocate mixed uses, mixed tenures, mixed building types, and a high standard of urban design for public places. The CNU charter recommends that street networks be an interconnected web that accommodates cars, but primarily designed to meet the needs of pedestrians, transit, and cyclists.[45]

These design principles recall the best pre-Radburn neighbourhood unit designs, such as New York's Forest Hills Gardens and London's Hampstead Garden Suburb. But, more controversially, the New Urbanists suggest that the design of buildings in the neighbourhoods take cues from local precedents. Seaside Florida, the first New Urbanist project, drew ideas from southern U.S. towns, while Cornell in Markham drew from the adjacent Victorian-era villages and Toronto neighbourhoods (see Figure 5.6). Many Modern architects decry this historicism, and some cheap knockoffs by developers have led some critics

Figure 5.6 Conventional Suburban Development ("snout house") and New Urbanist Streetscapes in Markham, Ontario

Source: Authors' collection.

to dismiss the whole movement as "sprawl with porches."[46] More serious criticism is that most New Urbanist projects, built in the post-1995 drought for social housing funding, have little affordable housing. This has contributed to a reduction in the social diversity they advocate; market forces and consumer preferences for socially homogenous housing also affected such projects.[47] However, when carefully planned, as in Markham, New Urbanist projects appear to support much higher gross densities than the conventional suburban development of the 1980s, while accommodating good ecological planning.[48]

Gated Communities

While social critics are divided on the New Urbanism, most are quite critical of gated communities, the latest design trend to make its way north from the United States, where they make up a substantial proportion of new suburban construction.[49] Recent Canadian research, somewhat surprisingly, has shown that gated suburbs were already appearing in Canada, particularly in B.C. seniors'/lifestyle communities. Although the numbers are small at present, it raises important planning issues about private roads, control of the public realm, visual impacts of walls, and increased separation of the affluent from the rest of society that have yet to be addressed by Canadian municipalities.[50] The advent of gated communities also turns attention back on long-standing concerns about city living conditions.

Concern over Housing and Living Conditions

After years of advocacy by social activists, Canada finally instituted a public housing program in 1949. Regent Park in Toronto was followed by the Jeanne Mance (Montreal) and Mulgrave Park (Halifax) projects, each in the slum-clearance tradition. Municipal housing authorities later used federal and provincial funds to build low-income housing on undeveloped suburban sites, and over 250 000 public housing units were built from 1948 to 1986.

Social Planning and the Attack on Urban Renewal

> This book is an attack on current city planning and rebuilding. It is also, and mostly, an attempt to introduce new principles of city planning and rebuilding, different and even opposite from those now taught in everything from schools of architecture and planning.

So begins one of the most influential books on city planning, Jane Jacobs's 1961 *Death and Life of Great American Cities*.[51] She used detailed observations of her own neighbourhood in Manhattan's Greenwich Village to critique urban renewal and develop new principles for mixed uses, density, and conserving older buildings. This critique focused attention on the social aspects of cities and away from aesthetic and functional planning issues.[52] A wave of public activism against urban renewal swept across Canada, from Strathcona in Vancouver, to Portage and Main in Winnipeg, to Milton Park in Montreal, and to Quinpool Road in Halifax. The opposition of citizens caused a series of policy reviews and a major change in Canadian urban renewal and housing policy in 1973 toward small-scale, more locally sensitive approaches.

The Neighbourhood Improvement Program (NIP) was one of the approaches that replaced slum clearance. Cities received small-scale grants to upgrade amenities and public infrastructure following Jane Jacobs's ideas. Public housing funding was phased out, and replaced by a range of programs for construction of cooperative and nonprofit housing. The Residential Rehabilitation Assistance Program (RRAP) upgraded existing housing. Municipalities combined these programs in innovative ways, leading to such projects as Vancouver's False Creek South and Winnipeg's Core Area Initiative.[53] These programs were driven by their social agenda for affordable housing, community control, and neighbourhood revitalization, and social planners were key members of the professional teams.

Many of the social housing and neighbourhood improvement programs were cancelled during the 1980s and early 1990s. A wave of conservative governments was elected at the federal and provincial levels, slashing housing, social services, and urban redevelopment programs to help reduce deficit spending. And what couldn't be cut was sometimes downloaded to other levels of government, until municipal budgets were near crisis. Eventually, the federal and some provincial governments began limited programs for housing in the most desperate situations after the numbers of homeless people sleeping in streets and public spaces of major cities became a national embarrassment.[54]

Social planners took up the safety issue, another of Jacobs's themes, starting in distressed public housing projects. While good aesthetics could not overcome poor social conditions, poor physical design could make a public housing project quite dangerous. Defensible space techniques were used to modify or redesign some projects, and identify others for demolition.[55] Feminist planners pushed safety higher up the planning agenda, and neighbourhoods, transit systems, and institutions were designed with Crime Prevention through Environmental Design (CPTED) techniques.[56] (See Chapter 15 for contemporary views on this topic.)

Healthy Communities

The last decade of the 20th century saw a revival of public health issues in community planning, starting with the Healthy Communities movement. The Canadian Institute of Planners, Canadian Public Health Association, and Federation of Canadian Municipalities vigorously adopted this program, which originated in the World Health Organization's (WHO) European office. The approach was different in each municipality, being guided by local priorities. Common themes included environmental clean-up, reduced pesticide use, emphasis on cycling and walking, and expanded community recreation programs.[57] But the local approach and social agenda of Healthy Communities were different from the public health crusades against epidemics led by doctors earlier in the century. The most recent connections between public health and community planning are more research-driven, as scholars investigate the effects of our "automobility" and suburban environments on air quality and obesity.[58] It is not yet clear whether our suburbs make us fat, but the connection between community design and public health may give additional traction to urban health issues. The effect of a small urban epidemic—the 2003 Toronto SARS outbreak—reinforced the public's understanding of the need for public health planning.[59]

Social activism on housing and slum-clearance issues changed the nature and techniques of community planning during the 1970s. The planning schools filled with social scientists and design studios almost disappeared. In the municipal planning offices, social planners were hired and community organizers often replaced the architects and engineers who ran the urban renewal projects. Citizen participation and advocacy planning techniques were developed to engage the community in the planning process (see Chapters 11 and 12). By the end of the century, social planning had begun to reconsider the importance of the community design through safety and public health issues. And social activists made strong connections to environmental advocates through the concept of sustainable development.

Concern over the Environment

The environment became a major Canadian planning focus again in the 1970s, more than a half-century after the demise of the Commission of Conservation. Although some river basin planning occurred in post-war Ontario and on the Saint John, South Saskatchewan, and Fraser Rivers, it took Rachel Carson's *Silent Spring* (in 1962), which warned of the dangers of pesticides and other chemical pollution of the natural environment, to revive the North American environmental conservation movement.[60]

Environmental Planning: Designing with Nature

Yet another path-breaking book, Ian McHarg's *Design with Nature*, established a method for environmental planning.[61] Taking a lead from Patrick Geddes' valley section (see Figure 9.1) McHarg drew transects through natural regions to analyze how ecological processes might interact with human activity. More importantly, he prepared maps of ecological, physiographic, and socioeconomic features that were overlaid to identify the areas of least environmental impact for urban expansion and highway projects. The evaluation factors were originally mapped on layers of clear plastic acetate; McHarg's method was later converted into computer use as an early version of Geographic Information Systems (GIS).

The federal and provincial governments, in the mid-1970s, all adopted environmental impact assessment (EIA) or environmental assessment (EA) processes. New provincially chartered regional planning agencies, such as the B.C. Agricultural Land Commission and the Niagara Escarpment Commission, protected large-scale agricultural and environmental features under stress from urbanization (see Chapter 9). A new cadre of environmental planners was trained to run the environmental agencies and manage the EIA and EA processes, often producing alternative development scenarios. The ecological basis for environmental planning improved steadily in the final two decades of the century. Landscape ecology developed more advanced techniques to design large regions for improved biodiversity[62] and these techniques were applied to large urban regions such as Ottawa and Hamilton, and smaller suburban areas such as Markham.[63]

Environmental planning came to entire metropolitan areas with the work of Michael Hough and his attention to the ecological values of urban open-space systems and incorporating natural processes into urban development. Good planning practice now protects environmentally sensitive areas such as wetlands, connects parks and river valleys with greenways, and incorporates storm-water ponds into neighbourhood open space. Degraded urban valleys such as Toronto's Don River are being rehabilitated and re-naturalized, and involve extensive citizen participation.[64] Hough also had a major influence on the Crombie Royal Commission on the Toronto waterfront, pushing it to expand its frame of reference to the entire Toronto bioregion (see Figure 9.3, page 236).[65]

Brownfield redevelopment is another theme in current environmental planning practice, focusing on the rehabilitation of former industrial sites that often contain contaminated lands. Most Canadian cities have brownfield sites, as a result of the migration of industries from old downtown and waterfront sites to new and larger sites served by expressways on the urban periphery. If the contamination issues can be addressed, brownfield sites often have the potential for new commercial and residential development that can take advantage of existing infrastructure and improve the context of adjacent downtown neighbourhoods.[66] Redevelopment of former industrial sites is not a new idea, but extensive and expensive remediation of contaminated lands is a characteristic of projects such as Victoria's Dockside Green, Ottawa's LeBreton Flats, and the Moncton Yards. Innovative financial tools are slowly emerging to make these brownfield projects more competitive with new greenfield development.[67]

Sustainable Development

Environmental planning, social planning, and economic development came together in the 1990s as sustainable development. The 1987 Report of the World Commission on Environment and Development (often called the Brundtland report) defined sustainable development as: "ensur[ing] that [development] meets the needs of the present without compromising the ability of future generations to meet their own needs."[68] This broad concept became a primary goal in many Canadian community plans at the end of the 20th century, pushed along by public concerns about global warming, climate change, and other future legacies. Individuals and communities were invited to reduce their "ecological footprint"—the size of the natural area required to sustain the impact of their lifestyle on the environment.[69] As the new century began, many cities were examining how their community plans might improve sustainability, and ecological planning became a standard component of Canadian planning practice. The Greater Vancouver Regional District has cast most of its activities, including planning, under the umbrella of its Sustainable Region Initiative.[70] Ottawa's greenbelt has been re-planned as an ecological feature rather than merely a urban separator,[71] and the Oak Ridges moraine is being protected as the north boundary of the Greater Toronto Area.[72]

Concerns over City Efficiency

Physical infrastructure construction literally paved the way for the metropolitan expansion in the post–World War II era. Like the CIAM architects, the civil engineers had earlier developed new prototypes for urban growth but had little opportunity to deploy the expressways and regional utility networks during the period of depression and war. A new generation of engineers was taught the techniques of municipal utilities, traffic engineering, and transportation planning, and was given an opportunity to build infrastructure on a scale unrivalled since the railway boom a half-century before.

Infrastructure Planning

The utility work to support metropolitan expansion, in the 1950s and beyond, started with 19th-century technology. Most Canadian cities were providing treated water for drinking and firefighting in pressurized pipes throughout most of the inner city.

These areas often had combined sanitary and storm sewers that drained into the nearest water body, hopefully downstream of the water-treatment plant. Meanwhile, adjacent rural townships usually could not afford piped services, so most of the scattered pre-war suburbs relied on wells, septic tanks, and privies. The Prairie cities had sewage treatment plants in place before the war, and other inland Canadian communities were forced to treat their sewage after 1945, when rapid metropolitan expansion put tremendous strain on adjacent water bodies. The Ottawa River was basically an open sewer downstream from the national capital in the 1950s, and the federal government contributed to intercepting sewers and a primary treatment plant as part of the National Capital Commission's program. Federal infrastructure assistance was extended nationally when CMHC made grants for water and sewer plants from 1960 to 1980. The provincial governments started to contribute, beginning with the Ontario Water Resources Commission, in 1956.[73]

Storm-water planning also saw a major change in the post-war period. Many of the new suburbs required separate storm sewers to collect rainfall and discharge it to local streams and rivers rather than mix it with the sanitary waste that was treated in the new plants. More sophisticated hydrological models allowed planners to predict the flow of rivers after a storm and identify areas prone to flooding. Municipal land-use plans then prohibited future development in flood zones, or rebuilding in the worst areas after a disaster. Major floods gave special emphasis to disaster planning, with Edmonton and Toronto protecting their valleys, while a major floodway was built to bypass Winnipeg, where much of the city was occasionally endangered by the Red River.

Large-scale suburban development required reforms to local government organization and financing for trunk sewer and water infrastructure. At Don Mills, E.P. Taylor demonstrated that a large-scale developer could provide the local water mains and sewers that previously had been built by a municipality. He even contributed to a sewage treatment plant on the Don River. But some form of large-scale capital financing was needed for regional water-treatment plants, trunk mains, and major roads; the tax base of the rural townships in the region could not provide it. The regional metropolitan governments described in Chapter 9 were, in many respects, innovations for infrastructure construction. The regional governments could issue bonds to finance the big pipes and roads out of future revenues; the developers built the local roads and pipes to city standards and handed them over as part of the land subdivision process (see Chapter 13).

Transportation Planning for the Automobile

The metropolitan boom of the 1950s and 1960s was facilitated by a thoroughly planned expansion of road networks and parking to support automobile use. The transportation planners prepared detailed specifications for all elements of the vehicular network. Zoning bylaws required houses to have garages, and all commercial and institutional uses to have parking and loading spaces. Community and regional plans reserved land to widen existing roads and protected corridors for future arterials and expressways. New industrial parks were planned near suburban expressway interchanges, for easy truck access, with zoning requirements for employee parking. Downtown redevelopment plans featured large new parking facilities built at public expense. And most municipalities adopted infrastructure standards that required wide roads, high-quality pavements, and generously sized parking spaces to make driving fast, cheap, and convenient, especially compared to public transit. These plans were quite popular and had broad political support because, at first, they worked. A high-speed drive on an uncongested expressway through a major city could be an exhilarating experience in the early 1960s.[74]

Toronto had a draft metropolitan expressway plan prepared in 1943 by Eugene Faludi, and other Canadian communities prepared transportation plans in the 1950s and 1960s using increasingly sophisticated computer models imported from the U.S. to analyze traffic flows associated with various land-use scenarios. The recommendations typically called for a regional network of expressways supported by a grid of arterial roads. The expressways allowed rapid travel throughout the urban area, and were built through suburban farmland, down valleys, and along corridors through under-utilized land along waterfronts and rail lines. But in the inner urban areas, the connecting links had to be built in trenches (Montreal's Decarie Expressway), on stilts (Toronto's Gardiner Expressway), or in corridors created by demolishing houses, using urban renewal.

This last technique caused a citizens' revolt that stopped expressway construction before it began in Vancouver, in mid-flight in Toronto, and deflected two highways into Ottawa's greenbelt. Toronto's experience was perhaps the most dramatic, drawing strength from successful freeway revolts in San Francisco and Boston. The Stop Spadina movement defeated a proposal to extend an expressway into downtown Toronto through a ravine and a stable residential neighbourhood,[75] and the elevated Scarborough Expressway was literally halted in mid-air just short of the Beaches neighbourhood (see Figure 5.7). Canadian cities were left with significantly smaller expressway networks than their American counterparts.[76]

Transit Planning

In the late 1970s, transportation planning in larger Canadian communities placed more emphasis on public transit. The transit systems had never been completely abandoned in the major cities. Toronto retained its signature streetcars after the war, and started construction of North America's first new subway system in 1954. It also adjusted its land-use policies to promote high-density re-redevelopment around

Figure 5.7 Scarborough Expressway Pillars

The end of the Scarborough Expressway. This elevated freeway was demolished in the late 1990s and replaced with a boulevard and bike path. Some of the expressway pillars were left as a monument to a previous era.

Source: "F.G. Gardiner dismantling project". Photo by du Toit Allsop Hillier. Engineer: URS ColeSherman. Artist: John McKinnon. Landscape Architect: OTH. Used courtesy of du Toit Allsep Hillier.

stations such as Yonge–Eglinton, and development of transit-supported city centres in suburban Scarborough and North York. These measures helped keep Toronto's per capita transit ridership higher than any North American city until the 1990s.[77] Montreal built its subway for the 1967 World's Fair, and continued to expand it over the next two decades. The central business districts of both cities developed extensive networks of underground passages lined with shops, interconnecting the office buildings and subway stations.[78]

Edmonton (1978), Calgary (1981), and Vancouver (1986) each developed Light Rail Transit (LRT) systems, and, in 1983, Ottawa developed North America's most effective busway rapid transit system.[79] But the pace of construction of transit systems slowed during the public financial crisis toward the end of the century, and transit's share of work trips continues to decline. Moreover, Canadian cities have had little success in attracting transit-oriented development (TOD) around the suburban LRT stations as had been advocated.[80]

Environmental Issues and Smart Growth

Infrastructure issues became more aligned with environmental issues in the last decades of the century. There was a brief burst of planning for energy efficiency after the 1973 oil crisis,[81] but it died away as prices stabilized. Declining air quality and increasing road congestion, however, dragged infrastructure issues back onto the public agenda. Increased automobile use is closely connected to urban air-quality issues; and as the 21st century began, urban Canadians began to hear about smog days and ozone alerts. Suburban road congestion proved to be a particularly intractable problem since there are usually few alternatives to driving for most trips in these areas.[82]

Infrastructure expenses, sustainable development, and New Urbanism came together in the Smart Growth Network established in the mid-1990s by the U.S. Environmental Protection Agency and Congress for the New Urbanism. Smart Growth has proved to be politically appealing, advocating infill and brownfield development in inner cities, and compact suburban development and preservation of ecological systems ("green plans") in suburban areas.[83] The American approach has been largely based on state and federal incentives, due to their weaker metropolitan and regional planning agencies. Smart Growth has also been a useful umbrella policy in British Columbia and Ontario for brownfield development, New Urbanism, transit-oriented development, and ecosystems planning, but with limited results.[84]

New Planning Tools

Several provinces refurbished their planning acts in the post-1945 period, the most notable being Ontario's Planning Act of 1946. This legislation did not so much invent new planning approaches as consolidate the concepts and experience gained previously, and establish a clear and workable framework for municipal land-use planning. The act's framework is still largely intact today and has been much copied by the other provinces. Its main features provide for:

1. the creation of planning units, usually one or more municipalities.
2. preparing, adopting, and approving "official plans," as well as specifying the legal effect of these statutory plans.
3. a system of subdivision control.
4. the delegation of powers to municipalities to enact zoning bylaws.
5. a quasi-judicial appeal procedure with respect to municipal planning decisions.

6. a plan-making body composed of citizens—the planning board—to advise municipal council.
7. involvement and education of the public through public meetings at various points in the planning process.

A comparison of provincial planning parameters is found in Figure 8.4 (pages 218–19).

The spate of planning activity that took place in the two decades after World War II is amply illustrated by Ontario's experience. In 1946, only 36 municipalities had established themselves as planning areas, only one municipality had an official plan, and only one had a comprehensive zoning bylaw. Within a decade, 200 more communities were in planning areas, 57 had official plans, and 48 had zoning bylaws. By 1965, over half of the municipalities were in planning areas and 75 percent of the province's population was covered by official plans in effect in municipalities. And in each of the decades from 1946 to 1955 and from 1956 to 1965, almost 10 000 subdivision plans were processed as nearly three million people were added to the Ontario population.[85]

An important addition to the planner's kit of tools during the 1950s was that of **development control.** Development control is an extension of the power that municipalities possess to undertake zoning and gives them the right to review proposed plans for new development in already built-up areas. Under normal zoning regulations, there is no stipulation that developers reveal their plans for a vacant or redeveloped site, as long as they adhere to the provisions for land use, height, and lot coverage (and all building regulations). Thus, whether a building will "fit" into an area in terms of its scale, its appearance, and its effect on traffic may often not be known until the building is completed. In rapidly growing downtown areas in the 1950s, this caused planners many problems. They therefore adopted the largely British method of requiring development plans to be submitted for scrutiny before construction could begin even, though they were in compliance with zoning provisions (see Chapter 13). In this way, not only might the community be saved trouble but the developer might also be alerted to more attractive alternatives.

A number of other new tools for plan implementation were developed over the past few decades that expanded the approach of development control. One might call them "hands-on" approaches, and they are now a distinguishing feature of Canadian planning practice. Transfer of Development Rights and bonus-zoning were introduced as incentives to developers in order to attain the kind of development desired in the location desired. Other municipalities used Comprehensive Development Zoning and Planned Unit Development techniques in order to be able to work with developers before a project is fully designed. This kind of planning has become increasingly common as planners seek ways to obtain better-quality physical surroundings.[86] (See also Chapter 13.)

Reviving Canadian Planning Institutions

Community planning did not quite disappear in Canada during the Great Depression and World War II. A number of infrastructure projects and some rural planning initiatives were undertaken during the Depression years (see Chapters 4 and 9). More importantly, the economic crisis inspired some innovative thinking about social planning, most notably by the **League for Social Reconstruction** (LSR). The LSR comprised some of the country's leading intellectuals, including Humphrey Carver, King Gordon, Leonard Marsh, Frank Scott, and Frank Underhill. Their 1935 report, *Social Planning for Canada,*[87] was a comprehensive review of Canadian conditions that proposed much of the post-war social safety net—unemployment insurance, pensions, and medicare. Community planning (although then called town planning) was also regarded as part of a progressive social planning agenda.[88]

Planning Issue 5.1 We Need Mass Transit, Not Wider Streets

Halifax Daily News
June 5, 2006

We Need Mass Transit, Not Wider Streets
Charles Moore

Twelve Chebucto Road homeowners have received notice that Halifax Regional Municipality (HRM) is planning to expropriate large chunks of their front yards in a street-widening project to ease congestion on the Armdale rotary. The city proposes to compensate the property-owners for the "fair market price" of their land, but an abstract assessment based on square footage doesn't take into account diminishment of resale value of the main properties, nor diminishment of quality of life imposed by having heavy traffic roaring by at closer quarters, felling of trees, and so forth.

Reportedly, a petition opposing the project has been signed by nearly 200 neighbourhood residents.

The rationale for this street-widening scheme is largely cost. It's a relatively cheaper way to address traffic congestion at the rotary, originally designed nearly 50 years ago to handle a volume of 5,000 vehicles daily, but now clogged with more than 10 times that number.

Easing the logjam by adding another lane to Chebucto Road below Mumford Road is a quick and dirty fix that will just download the problem, which is too many cars for the finite geographic reality of the Halifax peninsula to reasonably accommodate, a few blocks further north.

What Halifax really needs is the sort of light rail system that serves many American cities with large suburban commuter traffic. A subway would be prohibitively expensive for a small city like Halifax, but above-ground light rail wouldn't, especially since corridor infrastructure is already in place. Particularly promising would be a variant known as diesel light rail (DLR), such as Ottawa-Carleton's OC Transpo system, which can operate on existing tracks but has the potential to reduce operating costs by up to 40 percent compared with conventional trains thanks to one-person operation, use of standard bus components,

The need to deal with slum housing conditions was actively pursued throughout the 1930s by a cadre of socially minded architects, social workers, planners, and citizens, led by Humphrey Carver.[89] In 1939, this group encouraged people from across Canada to hold a national housing conference on the problems of slum housing. Among the participants were George Mooney of the Canadian Federation of Mayors and Municipalities, planner Horace Seymour, Vancouver city councillor Grace MacInnis, Nova Scotian S.H. Prince, and Leonard Marsh, who would later be the principal author of the 1944 federal committee's report on housing and planning. This conference clearly demonstrated the wide support for what Carver called "social housing" or, simply, the concept that government should be responsible for the provision of adequate housing for all Canadians regardless of income.

In 1944, with the end of World War II in sight, the federal government began to consider the future, hoping to avoid another post-war recession as had occurred after World War I. The government established a powerful Advisory Committee on Post-War Reconstruction, which was chaired by McGill president Cyril James. It prepared an economic and social plan for the future, including what the authors called "community planning" (see Chapter 1). Many important elements of the post-war planning agenda were set out in its 1944 subcommittee report *Housing and Community Planning*, prepared by Queen's University economist Clifford Curtis and social planner Leonard Marsh. The Curtis subcommittee noted extensive "congestion, deterioration, misuse and blight" in Canadian communities.[90] It recommended broad-scale housing programs to accommodate the backlog of housing demands caused by the Depression and the war, new housing to lower fuel consumption, and potential for direct service into downtown cores via streetcar-style track extensions.

No single alternative to high dependency on private vehicle transportation will likely ever be wholly adequate. Besides light rail, the high-speed catamaran ferry from Bedford to downtown that was tested recently has much to recommend it, as would possible expansion of the existing Dartmouth ferry service and more enhancements to Metro Transit like the MetroLink Portland Hills Link and Woodside Link express services. Perhaps a ferry from Herring Cove to downtown with park-and-ride lots at the Herring Cove end would be a sensible plan as well.

One idea with considerable merit would be to encourage commuter car-pooling by limiting traffic on certain lanes and/or times of the day to vehicles carrying two or more passengers. With respect to the Armdale rotary, the logical candidate for that strategy would be one of the existing two lanes inbound or outbound on Quinpool Road between the rotary and Robie Street, implementation of which would be a lot cheaper than expropriating private property and widening Chebucto Road, as well as quicker to get up and running.

Another slowdown identified as affecting traffic flow at the rotary is drivers stopping at pedestrian crosswalks. While that might be addressed to a degree with better traffic-light coordination, the ideal, at least where there is significant volume of pedestrian traffic, would be pedestrian bridges or tunnels.

There are more advantages to curbing use of automobiles than relieving traffic gridlock. Incentives for driving smaller vehicles and penalties for running larger ones should be part of the mix, too, as should considering the commuter implications when issuing permits for off-peninsula condominium development. A holistic planning approach with access to convenient, efficient mass transit as an integrated element is what's needed.

Source: Excerpted from Charles Moore, *Halifax Daily News*, June 5, 2006, 10. Reprinted with permission from the author.

meet projected population growth, and the renewal of over 100 000 dwellings in the older parts of cities. It also alerted governments at all levels to the need for comprehensive community planning, including the establishment of a federal Town Planning Bureau, a program of public education, and professional training programs for planners at universities. Out of the idea for a Town Planning Bureau came the **Central** (now **Canada**) **Mortgage and Housing Corporation** (CMHC), which subsequently fostered community and university planning education programs, as well as housing and planning research, and undertook the provision of housing for the poor and programs of urban renewal.

CMHC took an active role in reviving the planning profession in the post-war period. Graduate planning programs were funded at McGill (1947), Manitoba (1949), UBC (1950), and Toronto (1951), and student scholarships were provided for over 50 years. The agency helped revive the Town Planning Institute of Canada (TPIC) in 1952. CMHC also established its own planning branch, staffed it with many expatriate British planners, and produced many useful planning reports.[91] The agency also commissioned pure and applied research on community planning, a role that continues to this day.

Developing a Role for Citizens

The Curtis subcommittee also strongly recommended involving citizens in the planning of their own communities. The approach they suggested would, today, be considered paternalistic: that is, "to struggle against public inertia . . . people will accept and support what they can understand." It was, however, a new initiative in Canadian community planning. There had been various citizen movements prior to this time to lobby for civic needs in public health, housing, local government organization, and beautification, and even groups of property owners (ratepayers) to protest planning measures. But they did not play a continuing role within the planning process. What was envisioned were citizen groups that could be "educated" and energized to support the proposals of the planners. For example, the Toronto Citizens' Housing and Planning Association, led by Harold Clark, which had been organized in 1944, was instrumental in securing the Regent Park slum-clearance project.

The recommendations of the Curtis Committee were embodied in the 1944 version of the National Housing Act and called upon CMHC to promote public interest in planning. The outcome of this was the formation of the **Community Planning Association of Canada** (CPAC) in 1946. It was a national organization, with provincial and local branches, that was largely underwritten by CMHC. For 30 years, CPAC provided a forum in which citizens, planners, and politicians could discuss the needs of Canadian communities. Although, as Carver notes, "CPAC did not evolve into an instrument of local activism" as did later citizen efforts,[92] it did provide a springboard for many good ideas, and for individuals to become knowledgeable about planning and to participate vigorously inside and outside CPAC. No little credit for this is due to Alan Armstrong, the association's director for the first decade and the editor of their influential journal, *Community Planning Review*. As well, CPAC was a more inclusive organization than the professional planners' institute, TPIC.[93] Susan Hendler's research has shown that women made up about 40 percent of CPAC's membership and much of its national council during a period when there were few women members of the TPIC. This "dammed-up reservoir of talent" sometimes had considerable influence over the direction of post-war Canadian planning at the national level as well as in the communities in which they lived.

The Progress of Citizen Participation

For a variety of reasons, there was dissatisfaction over the outcome of much planning in the 1945–65 period. In coping with vast urban growth, the planning solutions were often large-scale and disruptive: expressways sliced through residential and park areas; old neighbourhoods were levelled for new office and apartment complexes or public-housing projects; and new shopping centres either displaced old commercial areas or dispersed the new populations, or both. Moreover, the solution to one planning problem not infrequently created other problems for which "more of the same" seemed to many citizens to be the planners' prescription. Proposals by planners began to be questioned, often vociferously, by ordinary citizens and neighbourhood groups. Terms such as "citizen activism," "participatory democracy," "advocacy planning," and "NIMBY" were coined.

Throughout Canada, planning proposals became issues for public debate. Some protests were protracted and rancorous. Many of the disputes became nationally known and celebrated in books.[94] Their names were synonymous with the victory or defeat of planning ideas, depending upon which side one was on: the Pacific Centre in Vancouver, Sunnyside in Calgary, Trefann Court in Toronto, Lower Town in Ottawa, Concordia Village in Montreal, and Quinpool Road in Halifax, to cite some of the well-known ones. As citizen participation became more pervasive, it stimulated a number of significant changes in the conduct of community planning. Some were formal, such as new provisions that were written into provincial planning acts to ensure avenues for public comment and consultation. Many municipalities instituted new channels of communication with citizens' groups and, in general, made their planning processes more open with public meetings, newsletters, open houses, design charrettes, and so forth; none more so than Vancouver.

These latter moves increasingly involved citizens in the preparation of plans, of articulating objectives and proposing alternatives. They also opened up for debate the question of how to move citizens farther up the "ladder of citizen participation" Arnstein had proposed in 1968;[95] that is, the role for citizens in getting plans carried out rather than just offering advice. This, in turn, led to extensive discussion of new methods of participation, such as consensus-building, and the related issue of the inclusiveness of planning processes to involve women, the physically impaired, ethnic populations, the elderly, and so on. In other words, the traditional participants—the politicians, the planners, other bureaucrats, and politically well-connected groups—began to shift their positions in order to accommodate citizens at the plan-making table. (See also Chapter 12.)

Growth of the Planning Profession

In 1951, there were fewer than 30 professional planners employed in public planning agencies in Canada, and it is likely that no more than twice that number comprised all the planners in the country, including private consultants.[96] By 1961, their numbers had grown to over 300, and to more than 600 by 1967. By the latter date, almost all municipalities with a population over 25 000 had their own professional planning staffs, nine provinces had planners on staff, and there were nearly 100 private firms of consulting planners.

This demand for trained planners was met in three ways. The first was to shift people who had been associated with planning and development as engineers, architects, surveyors, and landscape architects into full-time planning jobs. The second way was to recruit from other countries, primarily Great Britain; CMHC obtained many of its early staff members in this way, many of whom subsequently went on to private and other public agencies. The third way was to recruit from the new professional planning education programs in Canada. By 1972, there were 11 different programs, including undergraduate degrees from the University of Waterloo and Ryerson Polytechnic University. The Canadian university programs in community planning and closely associated fields were supplying almost all the staffing needs for professional planners by the beginning of the 1970s, as they do today.

Membership in the Canadian Institute of Planners in 2006 was reported at over 6000.[97] More than half of these planners work for governments, mainly at the municipal and other local levels, and over a third work as consultants, developers, or advisors in private business. The latter figures show not only the growth of the profession generally, but also its increasing presence in other public and semi-public agencies, bringing the latter the capacity to frame development proposals. Sometimes, as seen in Chapter 14, this new capacity competes with a city's own planning program. Thus, the institutional perspective for community planning is both an evolving situation and one that is becoming ever more complex.

The Values of Planning

Community planning gained acceptance during the 20th century because it provides reasonable solutions to the problems of cities and towns, and because those solutions promote the general aims of the community and society—that is, community planning comes to embody values that, while specific to its efforts, are consonant with the community's values. The values of planning also change over time with changing community concerns, as this chapter has shown. Usually the range of values broadens as the complex world of our modern communities reveals more concerns and more effects of our planning solutions. And, within the array of values, the emphasis, or importance, shifts as the community places different priorities on its various concerns. For example, slum conditions are not so great a concern today as environmental conditions. Thus, housing **equity** for the poor has not been discarded as a value, but rather it has been surpassed by **environmental conservation** as a priority in many communities. In Chapter 15, a host of new and evolving challenges are discussed that will, arguably, lead to a further revision of value priorities.

Such shifts in value priorities may take place in some segments of the community but not in others. Poor neighbourhoods may see heritage preservation as a restraint on the jobs that accompany new construction, while the more affluent may advocate the historic values. Some communities in a metropolitan area may wish to limit their own growth and thus fight highway extensions that other communities need for their growth. Value conflicts usually occur when there are transitions in community thinking about growth and change and insufficient political leadership to find a consensus on the values that should prevail. In these situations, there may not be an agreed-upon social agenda to inform planning decisions. This may be a strong signal either that a plan is needed or that an existing one needs revision.

The planning process requires that value positions be examined and debated. The result is usually a hierarchy of long-term value orientations for the community. In general, the set of value orientations that is adopted (as goals and objectives) to shape the social agenda will be drawn from an already well-established field. Listed below are those values that are part of the ethos of planning and that reflect long-standing concerns in community building, some centuries old. They are listed in alphabetical order so as not to convey any sense of priority.[98] It is up to the community to determine its value priorities when making its plans.

Beauty/Orderliness. This is probably the planning value with the deepest roots. The search for the "ideal" physical environment goes back to ancient times, but whenever the search is renewed, even in present-day suburbs, it involves a desire to order the environment so that it meets community needs and reflects community ideals. Planners must cope with two elements in rendering this value: (1) public beauty must satisfy public taste; and (2) the choices about the built environment are long lasting. As the current interest in heritage planning shows, a community may lose its delight in its previous environment only to have those tastes recur. But even long-lasting environments may change in the meantime.

Comprehensiveness. Planning is concerned with long-run consequences, both of the decisions made in building communities and of the side effects of decisions. Thus, the planner will attempt to comprehend, as much as the knowledge and skills of the time will allow, the outcomes or results of a proposed plan in terms of (1) how well the development will perform in the future, along with any problems it might generate; and (2) how others outside the immediate locale or group of participants might be affected. Applying the value of comprehensiveness does not imply the making of a completely comprehensive plan for a community, a nearly impossible task. (In recent years, there has been a tendency to trade the word "integrated" for comprehensiveness.)

Conservation of Resources. This value owes its recognition to Patrick Geddes, who was not only concerned about indiscriminate uses of such natural resources as water, land, and forests, but also of housing and open spaces within cities. The Commission of Conservation nurtured this concern among Canadians. A more modern version of this same value is in the concern shown over environmental degradation in, for example, the strip mining of coal in southeastern British Columbia, the pollution of the Great Lakes, and acid rain. Perhaps the most widely used variation of this value in 21st-century community planning is **sustainable development**.[99] The concept of **bioregionalism**, which the Crombie Commission used in its proposals for the Lake Ontario waterfront, conveys much the same value.[100]

Democratic Participation. The principle of working at planning decisions through the organs of democratic local government, and occasionally through plebiscites of the citizenry, is long-lived. More recent is the acceptance that all people in the community have the right to participate in public decisions. No other local government activity generates more issues of concern to citizens than does planning and, as a result, citizen participation in planning comes in many forms. Public meetings, opinion surveys, design charettes, and advisory committees are some of the formal ways to obtain participation, and the spontaneous reaction of the public against planning and development proposals must also be accommodated. Another side of this value is the democratic responsibility to consult the public regardless of potential conflict. Usually the more widespread the participation, the better the planning decision, and broad consensus-building techniques are sometimes used to mitigate some of the difficulties with comprehensiveness and rational decision making.[101]

Efficiency. The value of efficiency expresses itself in many ways in community planning. Efficiency leads planners to oppose premature subdivision of land because of the uneconomic demands for public utilities, and it also leads citizens to oppose projects that would depress surrounding land values. Efficiency is most often expressed in economic terms, but it is also of concern in functional arrangements in communities, such as in the best pattern of roads to distribute traffic and in the location of schools and parks. Concern over efficient use of public resources is a core theme of the Smart Growth movement.

Equity. Social equity did not receive easy admission to the values of planning in our highly individualistic society, which often wishes to believe that there are no impediments to human achievement. Public planning encounters such built-in inequities as differences in housing quality, neighbourhood services, and access to jobs. The increasing cultural diversity of cities has introduced further areas of concern about equity. Public planners must not only try to eliminate old inequities but must also avoid the creation of new ones. This value is most frequently in competition with efficiency.

Health/Safety. This combined value is one of the cornerstones of modern community planning. Concerns over clean water, clean air, and safe surroundings are often expressed now as environmental concerns rather than as fears over epidemics. But the support systems for large concentrations of population are still considered to be fragile with regard to disease, fire, crime, injury, and natural disasters. On the positive side are concerns over the provision of services and facilities that aid self-development and allow the less fortunate access to them as well. The Healthy Communities and Safe Cities movements and concerns about the effect of suburban planning on the obesity epidemic are further evidence of this value.[102]

Rational Decision Making. A direct outgrowth of professional planners' roots in utilitarian philosophy is the great faith placed in reason as a means to determine solutions to community planning problems. Planning also grew up at the time of civic reform movements that stressed the use of rational administrative and management approaches to local government. As planning has evolved, its methods have become focused on providing information and conducting logical analyses in order to evaluate planning proposals. With increased citizen participation, the planner's reasoning has often been shown to reflect a particular rationale that could be at odds with others' analyses.

Reflections

The planning practice we witness today is an amalgam of accumulated knowledge and experience, especially those of the second half of the 20th century. This most recent period showed both growth and maturity for community planning, its institutions, and its participants. What may have appeared half a century ago, perhaps naively, as a means of "curing the ills" of the city had, toward the end of the century, become its own set of social, economic, and environmental realities to contend with. In the process, planning became an accepted, institutionalized part of community-building. There would not likely be much support today to discontinue planning activities. Further, planners are now much better equipped to know what does and doesn't work.

The legacy of these formative decades just past should not be forgotten, for it is the continuing foundation for today's practice of community planning. Nor should we forget that planning arrangements, tools, and processes are *normative*. Assumptions about the best way to proceed and who needs to be involved are built into them. And arriving at those assumptions, in the 1945–2000 period, involved a learning process for all participants to recognize better both the substantive aspects of how communities develop and the social relationships among participants. Wholesale urban renewal—bulldozing the slums—gave way to more incremental approaches to improve the built environment. Planners also came, somewhat grudgingly at times, to realize that citizens, even in their opposition, could make useful contributions to planning situations (as well as recognize their rights as citizens). Not least, the vital importance of caring for the natural environments of communities was accepted into the planning milieu. Community planning practice is different and richer today as a result of all these changes.

This was also a period in which the larger social dynamic shifted from a Modern to a postmodern view of the world. Big was not always better, and eclecticism became a norm in community-building as elsewhere. There will continue to be changes in community plan-making, but perhaps not as momentous as those experienced since 1945. Community planning will doubtless be called upon to respond to new issues, because it deals with values and norms, with new ideas and social forces. So, as one views contemporary community-planning practice, as presented in the next several chapters, it will be useful to consider the question:

- ***Do community planning tools, processes, and institutions reflect appropriately the needs of today's communities and their populations?***

Endnotes

1. Leroy Stone, *Urban Development in Canada* (Ottawa: Dominion Bureau of Statistics, 1969), 39.
2. The Editors of Fortune, *The Exploding Metropolis* (New York: Doubleday, 1958).
3. David Foot and Daniel Stoffman, *Boom, Bust and Echo 2000* (Toronto: McFarlane, Walter & Ross, 2000).
4. Albert Rose, *Problems of Canadian City Growth* (Ottawa: CPAC, 1950).
5. Foot and Stoffman, *Boom, Bust and Echo.*
6. Gerald Hodge, *The Geography of Aging in Canada* (Vancouver: UBC Press, forthcoming 2007).
7. The Editors of Fortune, *Exploding Metropolis*, 53.
8. IBI Group and Richard Soberman, *National Vision for Urban Transit to 2020* (Ottawa: Transport Canada, 2001), Ex. 1.1.
9. Richard Harris, *Creeping Conformity: How Canada Became Suburban, 1900–1960* (Toronto: University of Toronto Press, 2004).
10. Jeanne Wolfe, "Our Common Past: An Interpretation of Canadian Planning History," *Plan Canada* 34:4 (July 1994), 23.
11. John Miron, *Housing in Postwar Canada: Demographic Change, Household Formation and Housing Demand* (Kingston & Montreal: McGill-Queen's University Press, 1988), Table 1 and p. 168.
12. Rose, *Problems.*
13. Over 62 percent of people in the top 15 Census Metropolitan Areas lived in post-1945 suburbs; see also Trudi Bunting, Pierre Filion, and H. Priston, "Density Gradients in Canadian Metropolitan Regions, 1971–1996," *Urban Studies* 39:13 (2002), 2531–2552, Tables 1 and 5.

14. Bauhaus directors Walther Gropius and Mies Van der Rohe relocated to Harvard and the Illinois Institute of Technology in the 1940s, and CIAM secretary J.L. Sert became dean of the Harvard Graduate School of Design in the 1950s; see Eric Mumford, *The CIAM Discourse on Urbanism, 1928–1960* (Cambridge, MA: MIT Press, 2000).
15. Canada, Central Mortgage and Housing Corporation, *Housing and Urban Growth in Canada* (Ottawa, 1956), 9.
16. Albert Rose, *Regent Park; A Study in Slum Clearance* (Toronto,University of Toronto Press, 1958); Peter Oberlander and Eva Newbrun, *Houser: The Life and Work of Catherine Bauer, 1905–1964* (Vancouver, BC: UBC Press, 1999).
17. Jane Jacobs, *The Death and Life of Great American Cities* (New York: Random House 1961); Herbert Gans, *The Urban Villagers: Group and Class in the Life of Italian-Americans* (New York: Free Press, 1962).
18. D.W. Magill, *Africville: The Life and Death of a Canadian Community*, 3rd ed. (Toronto: McClelland and Stewart, 1999).
19. Alice Coleman, *Utopia on Trial: Vision and Reality in Planned Housing* (London: H. Shipman, 1990).
20. Jacobs, *Death and Life.*
21. John Sewell, *The Shape of the City: Toronto Struggles with Modern Planning* (Toronto: University of Toronto Press, 1993); Claire Hellman, *The Milton Park Affair: Canada's Largest Citizen–Developer Confrontation* (Montreal: Véhicule Press, 1987).
22. David Hulchanski, *St. Lawrence & False Creek: A Review of the Planning and Development of Two New Inner City Neighbourhoods* (Vancouver: UBC Planning Papers, no. 10, 1984); and Stephen Ward, *Planning the Twentieth Century City* (New York: Wiley, 2002), 219–224 and 288–294.
23. Bernard Frieden and Lynne Sagalyn, *Downtown Inc.* (Cambridge, MA: MIT Press, 1989); and David Gordon, *Battery Park City* (New York: Routledge, 1997).
24. John Punter, *The Vancouver Achievement: Urban Planning and Design* (Vancouver: UBC Press, 2003). In 2004, the New York–based Project for Public Spaces rated Granville Island first in a list of best North American public places. See "The 20 Best North American Districts, Downtowns, and Neighborhoods," *Making Places* (November 2004), available online at www.pps.org/info/newsletter/november2004/november2004_neighborhoods.
25. Royal Commission on the Future of the Toronto Waterfront, *Regeneration: Toronto's Waterfront and the Sustainable City, Final Report* (Toronto: Queen's Printer 1992); and David Gordon, "Managing Change on the Urban Edge: Implementing Urban Waterfront Redevelopment in Toronto," in G. Halseth and H. Nicol, eds., *(Re) Development at the Urban Edge* (Waterloo, ON: University of Waterloo Press 2000), 175–226.
26. Punter, *Vancouver Achievement*, Ch. 6; and Elizabeth Macdonald, "Street-facing Dwelling Units and Livability: The Impacts of Emerging Building Types in Vancouver's New High-density Residential Neighbourhoods," *Journal of Urban Design* 10:1 (February 2005), 13–38.
27. Herbert Gans, "Planning for People, Not Buildings," *Environment and Planning* 1:1 (1969), 33–46.
28. Jonathan Barnett, *An Introduction to Urban Design* (New York: Harper and Row, 1982).
29. Kevin Lynch, *The Image of the City*, (Cambridge, MA: MIT Press 1961); and Christopher Alexander, *A Pattern Language: Towns, Buildings, Construction* (New York: Oxford University Press, 1977); Gordon Cullen, *The Concise Townscape* (London: Butterworth, 1971).
30. Edmund Bacon, *Design of Cities* (New York: Penguin, 1974); and Gordon Stephenson and George Muirhead, *Kingston: A Planning Study* (City of Kingston, ON, 1961).
31. Mark Fram and John Weiler, eds., *Continuity with Change: Planning for the Conservation of Man-made Heritage* (Toronto: Dundurn Press, 1984).
32. Saia Barbarese Topouzanov architectes, *Benny Farm Redevelopment Plan* (Montreal: Canada Lands Company, 2003).
33. Regent Park Collaborative Team, *Regent Park Revitalization Study* (Toronto: Toronto Community Housing Corporation, 2002).
34. Helena Grdadolnik, "Woodward's Takes Shape: Nothing Like it in North America," *The Tyee* (May 11, 2006), available online at www.Tyee.ca.
35. Richard Harris, *Unplanned Suburbs: Toronto's American Tragedy, 1900 to 1950* (Baltimore, MD: Johns Hopkins University Press, 1996).
36. Kun-Hyuck Ahn and Chang-Moo Lee, "Is Kentlands Better Than Radburn? The American Garden City and New Urbanist Paradigms," *Journal of the American Planning Association* 69:1 (Winter 2003), 50–71; and Eugenie Birch, "Radburn and the American Planning Movement: The Persistence of an Idea," *Journal of the American Planning Association* 46:4 (October 1980), 424–431.
37. Marc Choko, *Une cité-jardin à Montréal. La cité-jardin du Tricentenaire, 1940–1947* (Montreal: Méridien, 1988).
38. Don Gillmor, "Wildwood Childhood," *Canadian Geographic* (July/August 2005), 54–64; and Michael Martin, "The Landscapes of Winnipeg's Wildwood Park," *Urban History Review* 30:2 (2001), 22–39.
39. Macklin Hancock, "Don Mills: A Paradigm of Community Design," *Plan Canada* (July 1994), 87–90.
40. Joel Garreau, *Edge City: Life on the New Frontier* (New York: Doubleday, 1991); and Harris, *Creeping Conformity.*
41. Melvin Webber, "The Urban Place and the Nonplace Urban Realm," in M. Webber, ed., *Explorations into Urban Structure* (Philadelphia: University of Pennsylvania Press, 1964), 79–153.
42. Peter Gordon and Harry Richardson, "Are Compact Cities a Desirable Planning Goal?", *Journal of the American Planning Association* 63:1 (Winter 1997), 95–106; and Reid Ewing, "Is Los Angeles-Style Sprawl Desirable?", *Journal of the American Planning Association* 63:1(Winter 1997), 107–126; Larry Bourne, "Self-Fulfilling Prophecies? Decentralization, Inner City Decline, and the Quality of Urban Life," *Journal of the American Planning Association* 58:4 (1992), 509–513.
43. Louis Sauer, "Creating a 'Signature' Town: The Urban Design of Bois Franc," *Plan Canada* 34:5 (September 1994), 22–27.
44. Andres Duany and Elizabeth Plater-Zyberk, "The Second Coming of the American Small Town," *Plan Canada* 32:3 (May 1992), 6–13.
45. Congress of the New Urbanism, *Charter of the New Urbanism* (New York: McGraw Hill, 2000).
46. Hok-Lin Leung, "A New Kind of Sprawl (New Urbanism)," *Plan Canada* 35:5 (September 1995), 4–5; and Todd Bressi, *The Seaside Debates: A Critique of the New Urbanism* (New York: Rizzoli, 2002).
47. Jill Grant, *Planning the Good Community: New Urbanism in Theory and Practice* (New York: Routledge, 2006); Andrejs Skaburskis, "New Urbanism and Sprawl: A Toronto Case Study," *Journal of Planning Education and Research* 25:3 (2006), 233–248; Julia Markovich and Sue Hendler, "Beyond Soccer Moms: Feminist and New Urbanist Critical Approaches to Suburbs," *Journal of Planning Education and Research* 25:4 (2006), 410–427.
48. David Gordon and Shayne Vipond, "Gross Density and New Urbanism: Comparing Conventional and New Urbanist Suburbs in Markham, Ontario," *Journal of the American Planning Association* 71:2 (Winter 2005), 41–54; and David Gordon and Ken Tamminga, "Large-scale Traditional Neighbourhood Development and Pre-emptive

Ecosystem Planning: The Markham Experience, 1989–2001," *Journal of Urban Design* 7:2 (Winter 2002), 321–340.

49. Edward Blakely and Mary Gail Snyder, *Fortress America: Gated Communities in the United States* (Washington, DC: Brookings Institution, 1997).

50. Jill Grant, K. Greene, and K. Maxwell, "The Planning and Policy Implications of Gated Communities," *Canadian Journal of Urban Research* 13:1 (Summer 2004), 70–88.

51. Jane Jacobs, *Death and Life.*

52. Gans, *Urban Villagers*; and "Planning for People."

53. Jim August, "Partnership for Renewal: Winnipeg's Core Area Initiative," *Plan Canada* 34:4 (July 1994), 80–81.

54. Jeanne Wolfe, "Reinventing Planning: Canada," *Progress in Planning* 57 (2002), 207–235.

55. Oscar Newman, *Defensible Space: Crime Prevention Through Urban Design* (New York: Collier, 1973); Coleman, *Utopia;* and Sewell, *Shape of the City,* Ch. 8.

56. Gerda Werkele and Carolyn Whitzman, *Safe Cities: Guidelines for Planning, Design, and Management* (New York: Van Nostrand Reinhold, 1995).

57. Hugh Barton, C. Mitcham, and C. Tsourou, *Healthy Urban Planning in Practice: Experience of European Cities* (Copenhagen: World Health Organisation, 2003); and David Witty, "Healthy Communities: What Have We Learned?" *Plan Canada* 42:4 (2002), 9–10.

58. Lawrence Frank and H. Frumkin, *Urban Sprawl and Public Health: Designing, Planning, and Building for Healthy Communities* (Washington, DC: Island Press, 2004).

59. Richard A. Matthew and Bryan McDonald, "Cities Under Siege: Urban Planning and the Threat of Infectious Disease," *Journal of the American Planning Association* 72:1 (Winter 2006), 109–117.

60. Rachel Carson, *Silent Spring* (Boston: Houghton Mifflin, 1962).

61. Ian McHarg, *Design with Nature* (New York: Doubleday/Natural History Press, 1969).

62. Richard Forman, *Land Mosaics* (New York: Cambridge University Press, 1995); and W. Dramstad, J. Olson, and R. Forman, *Landscape Ecology Principles in Landscape Architecture and Land-Use Planning* (Washington, DC: Island Press 1996).

63. Charles Hostovsky, David Miller, and Cathy Keddy, "The Natural Environment Systems Strategy: Protecting Ottawa-Carleton's Ecological Areas," *Plan Canada* 35:6 (1995), 26–29; and Ken Tamminga, "Restoring Biodiversity in the Urbanizing Region: Towards Pre-emptive Ecosystems Planning," *Plan Canada* 36:4 (July 1996), 10–15.

64. Michael Hough, *Cities and Natural Process* (New York: Routledge, 2004); and Toronto, The Task Force to Bring Back the Don, *Bringing Back the Don* (Toronto: City of Toronto, 1992).

65. Royal Commission, *Regeneration.*

66. Chris De Sousa, "Brownfield Redevelopment in Toronto: An Examination of Past Trends and Future Prospects," *Land Use Policy* 19 (2002), 297–309.

67. Casey Brendon et al., "Urban Innovations: Financial Tools in Brownfield Revitalization," *Plan Canada* 44:4 (October–December 2004), 26–29.

68. World Commission on Environment and Development, *Our Common Future* (Oxford: Oxford University Press, 1987), 43; David Brown, "Back to Basics: The influence of sustainable development on urban planning, with special reference to Montreal," *Canadian Journal of Urban Research* 15:1 (2006), 99–117.

69. M. Wackernagel and W. Rees, *Our Ecological Footprint: Reducing Human Impact on the Earth* (Gabriola Island, BC: New Society Publishers, 1995).

70. A description is available online at www.gvrd.bc.ca/sustainability/about.asp.

71. Richard Scott, "Canada's Capital Greenbelt: Reinventing a 1950s Plan," *Plan Canada* 36:5 (September 1996), 19–21.

72. Robert Lehman, "The 50 Year, 10 Million People Plan," *The Ontario Planning Journal* 19:6 (November 2004), 22–25.

73. Jamie Benedickson, *Water Supply and Sewage Infrastructure in Ontario, 1880–1990s,* The Walkerton Inquiry Commissioned Paper 1 (Toronto: Queen's Printer, 2002). By the 1990s, 69 percent of the Ontario population had tertiary treatment vs. 39 percent nationwide (p. 76). Several large Canadian cities (Victoria, Vancouver, Montreal, Quebec, Saint John, Halifax, St. John's) continue to send partially treated sewage into local water bodies.

74. Reyner Banham, *Los Angeles: The Architecture of Four Ecologies* (London: Allen Lane, 1971).

75. David Nowlan and Nadine Nowlan, *The Bad Trip: The Untold Story of the Spadina Expressway* (Toronto: New Press, 1970); and John Sewell, *Shape of the City.*

76. Michael Goldberg and John Mercer, *The Myth of the North American City* (Vancouver, BC: UBC Press, 1986).

77. Peter Newman, "Public Transit: The Key to Better Cities," *Sustainable Cities White Papers* (New York: Earth Pledge, 2000).

78. John Zacharias, "Pedestrian Behaviour and Perception in Urban Walking Environments," *Journal of Planning Literature* 16:1 (2001), 3–18.

79. Robert Cervero, *The Transit Metropolis: A Global Inquiry* (Washington, DC: Island Press, 1998); Ottawa also began to experiment with LRT in 2001.

80. Peter Calthorpe, *The Next American Metropolis* (New York: Princeton Architectural Press, 1993); and Hank Dittmar and G. Ohland, eds., *The New Transit Town: Best Practices in Transit-Oriented Development* (Washington, DC: Island Press, 2004).

81. Peter Boothroyd, "The Energy Crisis and Future Urban Form in Alberta," *Plan Canada* 16:3 (September 1976), 137–146; and Hans Blumenfeld, "Some Simple Thoughts on the 'Energy Crisis,'" *Plan Canada* 20:3 (September 1980), 145–153.

82. Anthony Downs, *Still Stuck in Traffic: Coping with Peak-Hour Traffic Congestion* (Washington, DC: Brookings Institution, 2004).

83. Anthony Downs and F. Costa, "Smart Growth: An Ambitious Movement and Its Prospects for Success," *Journal of the American Planning Association* 71:4 (Autumn 2005), 367–381; and John Frece, "Preserving What's Best about Maryland," *Plan Canada* 41:4 (October 2001), 21–23.

84. Ray Tomalty and Don Alexander, *Smart Growth in Canada: A Report Card* (Ottawa: CMHC, December 2005); and Pierre Filion, "The Smart Growth and Creative Class Perspectives versus Enduring Urban Development Tendencies," *Plan Canada* 44:2 (April 2004), 28–32; Grant Moore, "Immigration: the Missing Issue in the Smart Growth Deliberations," *Plan Canada* 44:1 (January 2004), 32–35; Pierre Filion, "Towards Smart Growth: The Difficult Implementation of Alternatives to Urban Dispersion," *Canadian Journal of Urban Research* 12 (2003), 48–70.

85. Ontario Economic Council, *Subject to Approval* (Toronto, 1973), 50.

86. Hok Lin Leung, *Land-Use Planning Made Plain,* 2nd ed. (Toronto: University of Toronto Press, 2003), Ch. 7; and R. Fischler, "Linking Planning Theory and History: The Case of Development Control," *Journal of Planning Education and Research* 19:3 (Spring 2000), 233–241.

87. League for Social Reconstruction, *Social Planning for Canada* (Toronto: Nelson, 1935).

88. Ibid. Canadian Prime Minister William Lyon Mackenzie King also regarded planning and housing as part of a social agenda. See W.L.M. King, *Industry and Humanity* (New York: Houghton Mifflin, 1918); and David Gordon, "William Lyon Mackenzie King, Town Planning Advocate," *Planning Perspectives* 17:2 (2002), 97–122.

89. The group was established as "The Housing Centre" at the University of Toronto; the flavour of this period in Canadian housing and planning is superbly described in Humphrey Carver, *Compassionate Landscape* (Toronto: University of Toronto Press, 1975), especially 49–57.

90. Canada, Advisory Committee on Reconstruction, *IV. Housing and Community Planning* (Ottawa: King's Printer, 1944), Report of the Subcommittee, 161.

91. Cf. Harold Spence-Sales, *How to Subdivide for Housing Developments* (Ottawa: Community Planning Association of Canada, 1950).

92. Carver, *Compassionate Landscape*, 90.

93. Susan Hendler, "A Dammed-Up Reservoir of Ability: Women on the National Council of the Community Planning Association of Canada," *Plan Canada* 45:3 (September 2005), 15–17.

94. Illustrative of this Canadian "participation literature" are the following: Graham Fraser, *Fighting Back: Urban Renewal in Trefann Court* (Toronto: Hakkert, 1972); Jack Granatstein, *Marlborough Marathon* (Toronto: James, Lewis and Samuel, 1971); Donald Gutstein, *Vancouver Ltd.* (Toronto: James Lorimer, 1972); Donald Keating, *The Power to Make It Happen* (Toronto: Green Tree Publishing, 1975); John Sewell, *Up Against City Hall*, (Toronto: James, Lewis and Samuel, 1972); and, not least, *City Magazine*.

95. Sherry R. Arnstein, "A Ladder of Citizen Participation," *Journal of the American Institute of Planners* 35:3 (July 1969), 216–224.

96. Gerald Hodge, *The Supply and Demand for Planners in Canada, 1961–1981* (Ottawa: CMHC, 1972), 17.

97. See Canadian Institute of Planners website, www.cip-icu.ca.

98. This listing of planning values has drawn heavily on "The Values of the City Planner," in Frank S. So et al., eds., *The Practice of Local Government Planning* (Washington, DC: International City Management Association 1979), 7–21.

99. World Commission, *Our Common Future.*

100. Royal Commission on the Future of the Toronto Waterfront, *Watershed: Interim Report* (Ottawa, 1990).

101. Judith Innes, "Planning Through Consensus Building: A New View of the Comprehensive Planning Ideal," *Journal of American Planning Association* 62:4 (1996), 460–472.

102. Witty, "Healthy Communities"; Werkele and Whitzman, *Safe Cities*; and Frank and Frumkin, *Urban Sprawl.*

Internet Resources

Chapter-Relevant Sites

Planning Canadian Communities
www.planningcanadiancommunities.ca

Our Common Past: An Interpretation of Canadian Planning History—Part 2
www.cip-icu.ca/English/plancanada/wolfe2.htm

Jane Jacobs: Ideas that matter
www.ideasthatmatter.com

St. Lawrence Neighbourhood Toronto
www.toronto.ca/planning/stlawrence_west.htm

Congress for the New Urbanism
www.cnu.org/

Urban Design and Planning:
www.cip-icu.ca/English/aboutplan/ud_welc.htm

The Centre for Sustainable Community Development
www.sustainablecommunities.fcm.ca

Urban Renewal: Making the Case for Culture
www.creativecity.ca/resources/making-the-case/urban-renewal-3.html

Healthy Cities
www.euro.who.int/healthy-cities

Smart Growth BC
www.smartgrowth.bc.ca

PART TWO

2

Community Plan-Making in Canada

INTRODUCTION

Central to the activity of community planning is the preparation of plans, especially the overall community plan. It, and its sub-plans for specific functions or areas within the community, aims to provide coherence to the various activities that comprise the building and rebuilding of a community. The focus of the community plan is on the built environment and how its land-use outcomes shape components of the natural, social, and economic environments toward achieving a community's aspirations.

Plan-making of this scope is normative as well as technical; it is as much concerned about reconciling individual and community values regarding the use of land as it is about marshalling the necessary facts and information needed by decision-makers. Seen in this light, the community plan is more than a design for improvement of the built environment, more than a statement of what the community wants to become. The community plan plays a distinctive role in *governing* a community. It is like the keystone in an archway on which many other aspects of a community's governance depend. This is true whether the "community" is a city, small town, or region.

The image above is from the Montréal Master Plan *project, Montréal, Québec, which received the Canadian Institute of Planners' Award for Planning Excellence, Category of Community Planning, 2005.*

6

Chapter Six

Focus on the Built Environment

Planners . . . when driven to the wall to define their special field of competence tend to fall back on land use planning.

Hans Blumenfeld, 1962

What is it that community planners plan? The many demands made on planners in the last few decades give the impression of a diverse range of concerns, interests, and focus of their practice: policy planning, impact analysis, growth management, the natural environment, social planning, economic development, and more. The diffusion of planners' practice is, however, more illusory than real. The focus is, and always has been, on the built environment of cities and towns. As the Alberta ministry responsible for community planning stated in 1978, the concern is with "the forces [that] influence the physical shape of communities."[1]

At first glance, this seems straightforward enough. Then we realize that professional planners usually talk in terms of **land-use planning**. This is a type of planning for the built environment that provides the focus to the endeavour we call community planning. It is in the nature of a social institution (of which community planning has become) to distinguish its activities and the responsibilities of its practitioners from those of other institutions. Thus, planning is not architecture, or engineering, or the practice of law, even though the professionals in these fields also help shape the built environment in their practice. Moreover, a good deal of the defining of planning is done by the planning practitioners in the field as they carry forward the aims they and the community have for the activity. This chapter identifies the basic dimensions of the built environment with which community planning is concerned. This will allow us to see how community planning's view of the built environment is circumscribed and how other interests view and participate in it. The key question to focus on in order to grasp community-planning practice is:

- —*In which ways does the community planner view the built environment and how is this reflected in plans for a community?*

Land Use and the Built Environment

One can distinguish the built environment of communities from that of, for example, an agricultural area by the prevalence of structures and other forms of development of the ground space (or land) for activities of people. In everyday terms, it comprises the houses, parks, industrial plants, institutions, stores and offices, streets and highways, and other transportation facilities. All of these elements, directly or indirectly, involve the existing and prospective use of land by both public and private interests.

One of the main tasks of community planning is to develop an understanding of how a community's physical elements function, the amount of land they require, and their relations with one another. Planners thus seek to identify typical patterns and trends in the use of the land of the community. In order to do this, they must understand how to organize their observations of these elements. Community land use can best be understood by separating it into three basic—and interrelated—components: (1) ***physical facilities*** that require space; (2) ***activities*** of people that use space; and (3) the ***functions*** that the land serves. These, in turn, are mediated in terms of three main dimensions that determine a community's land-use patterns and the trends in patterns: **location**, **intensity**, and **amount** of land required. The understanding thus obtained makes it possible to predict the land-use patterns that may be expected when proposals are made for new development or redevelopment of the ground space of a community. But before discussing planning proposals, we must establish more precisely the nature of the three main components of the built environment of communities.

Components of Community Land Use

One of the important characteristics of a city, Jane Jacobs reminds us, is its diversity, and nowhere is this more apparent than on a city's streets.[2] Here we see that there are many uses of the land, even on a sedate residential street: houses of various kinds, possibly some with separate garages; sidewalks, maybe with children walking on their way to school; the road, probably with some cars parked at the side and other cars or bicycles passing along it; delivery trucks, street trees, and possibly even a neighbourhood park.

In this commonplace scene, we encounter instances of each of the three components of community land use. There are physical *facilities* and features (houses, garages, sidewalks, street trees, the road, a park). There are *activities* that are both in view (the children walking to school, the delivery of goods, people moving on bicycles and in cars) and those carried on inside structures (people residing in houses, cars being stored in garages). And the physical features we see have one or more purposes or *functions* (the houses function as residences, the sidewalks provide a path for walking, and the roads provide a path for wheeled vehicles and space to store vehicles).

Figure 6.1 pictures the six basic uses of land in communities: that used for residences, commercial establishments, industry, transportation, institutions, and open space. It must be noted that each of these categories exhibit many physically different forms (i.e., facilities) and perform various functions in the contemporary community. For example, land used for commercial activities may take the form of a shopping street, local plaza, shopping mall, big-box outlet, corner store, or office building. And while all are commercial in nature of their activities, each performs a different function. The same is true of the physical forms and functions in other types of land use.

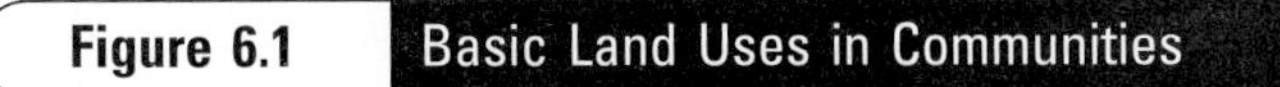

Figure 6.1 Basic Land Uses in Communities

(a)

(b)

(c)

(d)

(e)

(f)

Six land uses occupy most of the land in a typical community: (a) residences; (b) industry; (c) commerce or business; (d) institutions; (e) open space, and (f) the streets and roads that connect them all. As we see here, the land uses can take a variety of physical forms.

Source: Authors' collections.

Planners have devised ways, despite this diversity, of classifying the observations they make about land uses. The most highly regarded methods use three components—facilities, activities, and functions—and permit rigorous measurements to be made in each category.[3] These measurements would usually be done for each parcel of land in the community. The accumulation of such data allows the planner to determine, as Guttenberg points out, "how much space and what kind of facilities a community will need for activities in order to perform its functions at a certain level."[4] These components can be defined in more formal terms as follows:

1. **Facilities**: a description of the physical alterations made to parcels of land and public rights-of-way, especially buildings and other structural features. The type of building (e.g., detached house, office building) needs to be noted because this will indicate the form and quantity of indoor space available to users. Non-building constructions (e.g., pavement, power poles, and recreation equipment) may also need to be recorded.
2. **Activities**: a description of what actually takes place on parcels of land and in public spaces. This involves observing the various users and the form their use takes, usually focusing on the relationships of people obtaining goods and services and the mode of transportation involved. Thus, a house is normally for residential activities, a fire hall for protection activities, a parking lot for vehicle-storage activities.
3. **Functions**: a description of the basic purpose of an enterprise or establishment located on a parcel of land. Individuals, families, firms, and institutions use a specific location for places of residence, business, government, or assembly, and it is these latter purposes that need to be noted.

Planners have not always been assiduous in the classification of land uses in communities. It is not uncommon to find a set of categories such as the following:

Residential	Industrial	Roads	Commercial	Public
• one-family • multi-family	• light • heavy	• arterial • local	• retail • offices	• schools • recreation

There are several inconsistencies in such a list: the categories "residential," "commercial," and "industrial" specify the function of a piece of land while "public" denotes ownership and "roads" refers to a facility. And within categories there are references both to facilities (such as "school" and "offices") and to functions (such as "retail" and "recreation"). The industrial category, on the other hand, distinguishes types of activities between firms. Our ability to understand the built environment of communities is limited, if not flawed, by inadequate land-use classifications.

There are several other characteristics that enter into the description of land use in a community, although they are normally secondary to those above. Such physical characteristics as **slope, drainage, bearing capacity**, and **view** may play a part in determining the use for the land. In considering the stock of land in a community, we need also to distinguish that which is **developed** from that which is **undeveloped**. The latter category would provide an indication of space that could be available for future development. Also often included in land-use analyses are the **performance characteristics** of the activities on the land. Some activities generate a lot of traffic, some are carried on mostly in the daytime, others are nighttime activities or have peaks of activity at certain hours, and some may generate noise or odour. These characteristics are important because, although initiated by the land use on one parcel of land, they may spill over onto adjoining parcels or even affect a whole neighbourhood, causing *external effects* (as pointed out in Chapter 1). Activity characteristics often dominate the issues cited by both proponents and opponents when planning proposals are brought forth.

Performance characteristics of land uses are important to note because they enter into planning decisions regarding **mixed-use** development that Canadian planners are increasingly favouring.[5] Mixing of land uses at greater densities and diversity, often following the "traditional neighbourhood design" (TND) concept of the New Urbanism movement, aims at providing greater choice of housing and greater accessibility to other uses in contrast to the highly separated patterns of current suburbs.[6] The essential issue plan-makers face in mixed-use situations is the compatibility of different land uses (each with their particular activity characteristics) when located adjacent to or nearby one another.

Land Use at Different Scales

As suggested above, land use also occurs at different scales or spatial levels. These start from (1) **the individual parcel of land** and progress outwards to (2) **the district**, (3) **the community,** and (4) **the metropolitan area** or **region.** It can be seen that these are logically connected: parcels make up districts, which make up communities, which make up cities, and so on. The importance of making these distinctions is that the planner will emphasize different facets of land use at each different spatial level. For example, at the level of the single parcel of land, the land use is described in detail (e.g., building types), whereas at the district level, the dominant land use is usually the basis of the description (e.g., retail shopping). When we view an entire community, the land use is usually described by groupings of districts or groupings of uses (e.g., residential areas, local commercial areas). At the regional level, the description of land use will distinguish between urban areas, agricultural areas, open spaces, and their ecosystem components. Further, there is a progression of concern with the three land-use components (i.e., facilities, activities, functions) as one moves from the individual parcel level to the community-wide, or city, level, as the examples below show.

Parcel Level. When the subject is development or redevelopment of a single parcel, the planning issues tend to centre on the physical facility or improvements to be placed on the site: What type of building? How big will it be? What will it look like? There will also be concerns over the activities on the site and in the immediate vicinity of the parcel in question: Will there be a lot of traffic? Noise? Will the building block the sun?

District Level. As one moves to district-level planning, there is less concern with structures and physical appearance and more with patterns of activity. Of special interest are the patterns associated with pedestrian movements and auto traffic, as well as those associated with such district-wide facilities as schools, playgrounds, and shopping areas.

Community-wide Level. At the community-wide level, the land-use component dealing with the function or purpose of a facility becomes most important because this will affect the other facilities and areas in the community to which it is linked and its need for transportation connections.

Metropolitan/Region Level. At this scale the concern is usually with broad land-use purposes, such as those of a special district like the downtown, and with facilities that serve the entire area, such as an airport, regional park, or bridge.

An important parallel for community planners to the above land-use perspective exists also for the area of **ecosystems planning.**[7] It, too, is concerned with ecosystems at all scales from, for example, the neighbourhood stream to the regional watershed, and their interdependencies. Increasingly, planners are advocating not only that these natural environmental "land uses" be considered alongside built environment land uses, but that they be considered first.

Relations among Land Uses

The built environment of a community involves a complex set of relationships with many potential conflicts. Planners must be able to understand land-use relations, for example, to assess and plan for the impact of changes that may result from new development. The photographs in Figure 6.1 (page 143) show two of these relationships: first, how the land-use components (facilities, activities, functions) are related to one another at any single location and, second, how land uses at different locations are related and linked to each other, especially by streets and roadways. A third set of relations is associated with the distribution of particular land uses. Each is discussed below.

Locality Interaction. Each parcel of land in a community is linked directly or indirectly to every other parcel. Take, for example, the common situation of a parcel of residential land with a single-family house on it. The road (a facility) provides for the access to the lot and house (another facility), while the sidewalk (a facility) provides space for children to walk and to play (an activity). Analogous relationships can be seen in any shopping plaza between the store buildings, the parking lot, the walkways, shoppers, and drivers. Now suppose that the owner of a house on a large corner lot of a quiet residential street decides to demolish it and build a small shopping plaza The adjacent homeowners will likely be concerned about the impacts of traffic in their local street, a parking lot next to their front lawn, dumpsters and delivery trucks next to their back yards, and so forth. These latter land uses may not be acceptable without some form of buffer or separation.

Another important aspect that should be noted are the many activities and/or functions that physical facilities are required to accommodate. The road in front of a store may provide for customer access, for storing customers' and employees' cars, and for facilitating deliveries to the store. The store building may, at different times, house different establishments: a store, office, repair shop, even a residence. Thus, not only are the land-use components interrelated, but also the relationships may change. Through all this, the facilities may change little, if at all. Physical facilities have a high degree of permanence among the various land-use components, a fact that planners must keep in mind. Many planning issues centre on whether present facilities should be changed to accommodate new activities and functions: Should roads be widened? Old buildings demolished?

Community Interaction. The second important aspect involves how land uses in one part of a community are linked to and affect land uses in other parts. Households need to make purchases of food and other goods, and thus create the need for access to stores. If that access is in the form of automobiles, the stores, which depend upon customers, may provide a parking lot. The workers at factories, offices, and stores tend to come from households located in many different parts of the city or town. And the factories, offices, and stores will likely receive their supplies from warehouses and producers in yet other locations. The various facilities associated with transportation in our communities—roads, sidewalks, streetcar lines, rapid transit, bus stops—are the visible evidence of these necessary connections. Under the ground are utility lines, which also allow different facilities and activities to take place at a wide variety of locations and, in Toronto and Montreal, subways for transporting people among locations of activity. While these interactions are necessary for urban life, they may make impacts upon adjacent land. A shopping centre's owners may be delighted that a rapid transit station is proposed for the corner of their property, since it will bring more customers to their stores. But the residents of a collector street may not want a bus stop adjacent to their house, or buses travelling on their street.

Regional Distribution. A third set of relational considerations in community land use results from the fact that various facilities are distributed differently across an urban region. Houses, stores selling daily needs, and elementary schools are usually widely distributed, whereas factories, department stores, hospitals, airports, and auditoriums are not. The latter often tend to be concentrated in one or a few areas. Yet facilities (e.g., transit, roads, expressways) must be provided for their interconnection. Also, care must be taken over whether a change in one level of land use will affect changes in another at a different location, as, for example, in the case of shutting down a neighbourhood elementary school. If a community is experiencing growth, a major planning issue is where the new concentrations of stores, factories, or offices should go—or if they should simply be added to previous concentrations. Poor planning decisions for the regional distribution of land use may show up in wider inefficiencies rather than local conflicts. For example, local objections to shopping or employment uses may lead to a jobs–housing imbalance and possible single-use "bedroom communities" elsewhere in the region that, in turn, cause heavy traffic peaks on adjacent roads.

In conclusion, the built environment must accommodate a great variety of facilities, activities, and functions. Some have very localized impacts and others might affect an entire region. Moreover, the structures and surface improvements have a great deal of permanence. The latter tend to be either adapted or extended over time as new development needs to be accommodated, but are seldom removed. There are, therefore, many areas of potential conflict among the components of land use at any level of the community, as well as between levels. A good deal of the effort put into community planning involves trying to foresee conflicts between land uses, or incompatibilities, as they are often called.

How Planners View the Built Environment

It is the job of the professional planner to understand the workings of built environments of communities, for, as noted, a great deal of planning centres around two issues:

1. the **spatial impact** of proposals for new development on the built environment; and
2. the **spatial coordination** of the various functions and activities that comprise the built environment.

In facing these two issues, the planner must contend with the built environment's complexity. To do this the planner must find ways to "measure" the important features and relationships in that environment. The three basic components devised to classify land uses—facilities, activities, and functions—constitute one such measuring tool. To achieve a more complete picture of the community environment, five additional dimensions and concepts are required.

Dimensions of the Community Environment

The planner is primarily involved with planning for neighbourhoods, districts, or the entire community. This requires that the planner obtain measurements of the built environment larger than that of a single parcel. Since the larger scales are an aggregation of the land use on single parcels of land, the planner will need to combine the observations and also make comparisons between single land uses. Five dimensions provide the planner with most of the information needed about the community environment. These are type, amount, intensity, spatial distribution, and location.

1. Type

With this dimension, the planner considers the fundamental question of "What is it?" In ecological planning, the concern may be what sort of habitat is involved: Is the land occupied by a forest, a marsh, or a meadow? In land-use planning, the land may be occupied by residential, commercial, industrial, or institutional uses. In urban design, land could be occupied by a single-detached, semi-detached row, or tower type of building. Typological questions are usually analyzed by a classification scheme, which often allows for hybrids, or mixture of types—retail uses in the ground floor of an office building, or a row of townhouses attached to a point-block tower, like the new districts of downtown Vancouver.

2. Amount

With this dimension, the fundamental question of "How much?" may be broached: How much land is involved, how many dwelling units, how much traffic, how much commercial space, how much school population? It can be seen from these representative questions that the planner may be interested in land, dwelling units, traffic, people, spaces, or floor area. Whether one or several of these variables are involved will depend upon the size of the area and how comprehensive the planning is to be.

3. Intensity

With this dimension, the fundamental question of "How does this compare?" may be applied to a development proposal. The answer will determine differences between the proposal and either some development one is already familiar with or some planning standards. In general, the measures used are in the form of ratios: persons per hectare, persons per dwelling unit, dwelling units per hectare, floor space per lot area, employees per hectare, and autos per household. This dimension is associated with the common notion of "density," which is discussed later in this chapter.

4. Spatial Distribution

With this dimension, the distribution of facilities, people, and activities may be reckoned. On the one hand, the goal may be making facilities (parks, schools) and services (shopping plazas) available to people more or less equitably. On the other hand, the goal may be avoiding over-concentration of facilities (apartment buildings, stores, public buildings), which could lead to various public utilities and transportation modes being overtaxed. The planner often talks about accessibility and congestion in relation to this dimension.

5. Location

With this dimension, the planner is concerned primarily with the "relative" location of facilities, especially those of community-wide or district-wide interest; that is, how facilities relate to each other—homes to local schools and parks, major shopping areas to the road network, industrial plants to shipping facilities, and so on.

Patterns in the Community Environment

Planners have observed, measured, and analyzed the land uses in many different community situations and have thereby discerned certain patterns and relationships. Many of these, such as the amount of land commonly devoted to roads or the spatial distribution of elementary schools, provide ready guidelines in understanding the layout of the community. Observations about the intensity of various land uses, such as those related to different **building types**, provide the basis for the important concepts of **density and bulk**. Functional relations, or "linkages," between major land uses and districts have generated the concept of **accessibility** as well as the notion of **traffic conflicts**. Many of these patterns are so well established as to be used by planners as principles and standards in planning communities. These are part of the *lingua franca* of professional planners, and it is essential to have a grasp of them. The patterns are sometimes organized into hierarchies of use (e.g., elementary/junior/high schools) or of function (e.g., local/collector/arterial streets). New Urbanist planners have assembled these patterns into a broader synthesis called **transect planning**, which is described below.

Land-Use Requirements

When the land in urban communities is tabulated according to seven major urban functions, it is typically found to be distributed in the proportions shown in Figure 6.2. These land-use shares are derived from communities in the Vancouver Metropolitan Area, but are typical of other Canadian communities. When communities differ in land use, the difference usually appears in the increase of land required for institutions (e.g., Kingston, Ontario), or in the generous provision of parks (e.g., Edmonton), or in the large amount of industry (e.g., Trois-Rivières).

The type of terrain in a community can also affect the amount of space required by different types of land uses. For example, one that is hilly will probably have few industries and a greater share of residences, while a community that is a transportation terminal will probably attract a greater-than-average share of industry, commerce, and transport land uses. Communities may also vary in the quality of residences constituting the residential sector, depending upon the income bracket to which their housing market caters. And where two communities are adjacent, one may contain the bulk of the living areas while the other contains the working areas, because of old municipal boundaries.

Two further facets of the data in Figure 6.2 should be noted. The first is that the three land uses that are normally held in private ownership—residential, commercial, and industrial—constitute nearly two-thirds of the area of the community. The second is that one-fifth of the community's space is required for streets and roads. The latter, when combined with parks and some institutional uses of land, constitutes the public space of the community over which the local government has direct control; that is, only about 30 percent of the total community area.

Figure 6.2 Typical Land-Use Shares of Major Urban Functions

URBAN FUNCTION	PERCENTAGE OF URBAN AREA
Residential	51.0
Commercial	2.5
Industrial	8.0
Institutional	8.0
Transportation/Utilities	4.5
Recreation/Open space	5.5
Streets	20.5
Total urban uses	**100.0**

Functional Arrangements

In order to plan comprehensively for a community, whether for the entire community or only a portion of it, it is necessary to have an integrated view of it. This means moving beyond simple land-use composition and obtaining a picture of how the community functions. The knowledge that has accumulated about cities and towns allows the planner to utilize basic tendencies about how communities function in structuring the community's planning needs, as, for example, with the tendencies in community composition and the hierarchy of facilities in a community.

Functional Composition. Almost every community may be seen as comprising a number of major functional areas that reflect basic human activities. At its most fundamental level a community comprises **living areas, working areas**, and **community facilities**, all linked together by a **circulation system.** This simple concept of the functioning parts of the community contains the essential elements with which the planner is concerned—that is, people must have places to live and to work. The circulation system will provide for the necessary interconnections between living areas and working areas and also for access to community facilities. The latter are necessary adjuncts for the social, economic, and governing facets of a community. The category of working areas may be elaborated to distinguish between those providing goods and services to residents—commercial areas—and those involved in the processing and distribution of goods—industrial areas. These four functional elements become a focus of virtually every community plan.

Functional Reach of Facilities. It is easy to see in a community that some facilities are provided for the use of the community as a whole, such as a civic auditorium or a general hospital, while others serve only small parts, such as a local park or an elementary school. The same is true for commercial facilities and also for streets and highways. In other words, key public and private facilities each have a particular **functional reach** out into the community by virtue of the roles they play. Planners utilize this principle when proposing the distribution of various public and commercial facilities and in the designation of streets in a community plan. Further, there is ample justification for this from two points of view. First, it streamlines services and makes the best use of funds for public investments. Second, it allows districts—residential, commercial, or industrial—to serve their primary purposes as living, working, or shopping areas most effectively. There have been notable refinements in recent years in the locational arrangements of retail and educational facilities, for example, with big-box stores, cinemas, and specialty schools that modify traditional facility hierarchies.[8]

Density and Space Needs

One of the greatest concerns about city-building in the late 19th and early 20th centuries was congestion: too many people, too little space between buildings, too much traffic. The consequences for health and safety had been amply demonstrated (see Chapters 3 and 4), and much of the effort in community planning since then has been devoted to avoiding congestion in its various forms. Intensity of land use is, as we noted earlier, one of the main dimensions planners employ in measuring the adequacy of planning proposals. Since congestion is simply the too-intense use of a community's ground space, planners seek indicators that show when land uses would be approaching congested conditions, in order to avoid them. The most widely used indicators are those relating to density. **Density,** as planners define it, means the number of land uses or land users on a specified unit of ground space in the community. It is a ratio with a *numerator* (e.g., number of persons) and a *denominator* (e.g., an area of land, usually in hectares).[9] Most commonly, density indicators are used in regard to residential development: the number of persons per hectare or the number of dwelling units per hectare. There are analogous density measures for industrial and commercial areas, such as the number of employees per hectare. The appeal of the density measure is that it readily links the available land, structures, and activities that need space. Thus, by knowing the type of development that is proposed—for example, the construction of a certain number of single-family detached houses—the amount of land that will be required can be calculated by knowing the density at which such housing is normally built. Or, if a certain number of apartment units were proposed on a site, it could be ascertained whether this would result in an acceptable density. Equally important, it is a measure that is comparable to other development sites (when the same numerators and denominators are being used). Two variants of the density measure deserve attention:

- **Net density**, when used in regard to residential areas, refers to the number of dwelling units, households, or persons being accommodated on specified parcels of land. It does not include public roads, lanes or sidewalks, or other community land uses. It is sometimes called parcel density or net site density.
- **Gross density**, when used in regard to residential areas, also refers to the number of dwelling units, households, or persons but this time includes the specified parcels of land **and** the public roads, lanes or sidewalks, and other community land uses such as parks, schools, and churches considered *relevant* to the residential area. It is also referred to as neighbourhood density.

It can be seen in the case of gross density that the denominator can vary depending upon which features of the community are included in the land measure. Indeed, gross density ratios may be sought for a variety of areas, such as for one or several neighbourhoods or even for an entire community, using the municipal boundary as the land measure. Although a very useful measure, it is important when comparing gross densities that land areas be defined the same way.

Typically, planners measure residential density in terms of dwelling units per hectare because this has a direct connection to the built environment. When applied this way it carries with it the professional's knowledge of the type and intensity of such dwellings under average circumstances. It also carries with it, implicitly, some norm of acceptable and/or desirable density for the type of dwelling. Indeed, there is now widespread agreement among builders and planners about the density of dwellings that will contribute to an amenable community environment. The chart in Figure 6.3 describes the densities that typically result when different types of housing are built.

Comparing net densities using dwelling units per hectare assumes that the variation in the size of dwelling units is not important. This assumption may be adequate in suburban areas, but often is not appropriate in downtown areas, where the bulk of apartment buildings is a planning issue. The wide range in urban apartment unit sizes, from 50-square-metre bachelor units to 250-square-metre penthouses means that Floor Area Ratio (discussed on page 154) is a more appropriate net density measure.

Dwelling unit densities start, of course, from population densities: each dwelling unit is accommodation for a household. The number of people in households and the composition of household populations differ within a community. This generates the need for different kinds of accommodation. In Figure 6.3, household sizes that are typical of different types of housing have been applied to arrive at the net density ranges. For example, families with young children tend to prefer housing that has direct ground access and are most often found in low-density housing or row housing. The average household size of these families is commonly found to average 3.0–3.5 persons. By the same token, families with children tend not to seek high-density housing.[10] The latter housing usually accommodates an older population: young adults, couples,

Figure 6.3 Typical Densities (Net) of Different Forms of Housing

DENSITY	HOUSING TYPE	BUILDING HEIGHT IN STOREYS	DWELLINGS PER NET HECTARE	PERSONS PER NET HECTARE
Low	One-family detached	1–2	12–17	43–48
	Two-family	1–2	19–29	48–84
Medium	Row house, garden apartment	2–3	24–48	72–144
	Walk-up apartment	3–4	48–96	120–192
High	Multi-family	5–10	96–192	192–360
	Multi-family	10–16	192–240	360–480
	Multi-family	over 16	240–960	480–1680

families with teenagers, elderly couples, and single-person households, both young and old. In these accommodations, the household size tends to average 2.0–2.5 persons or less. The planner can, thus, estimate the various kinds of housing needs for a community by knowing the composition of the population and, following from there, the types of services and facilities that will be needed in the various housing areas. Hence, low-density areas with children need schools and playgrounds within easy access, while high-density areas that generate more traffic need easy access to major roads, and so on. Examples of the physical outcome of housing at different densities are shown in Figure 6.4.

As alluded to above, a measure of population density is often sought for an entire municipality, where the latter is the denominator. Among its uses is to estimate the amount of land that might be consumed by a new municipality or the amount of land needed when an increase in population is anticipated.[11] (Gross municipal area density figures are provided for municipalities and all other census reporting units by Statistics Canada in their "Community Profiles.")[12] Canadian municipalities tend to have gross densities of 1000 to 7000 persons per square kilometre, depending upon the age and size of community. Newer and smaller communities usually have lower densities, with the converse being true for older and larger communities. This raises a second issue about differences in density. There are not only differences because of the size and age of a community, but also because of the style of building that is acceptable in the community. For example, 125 dwelling units per hectare is considered high density in Kingston, Ontario, and is the maximum normally allowed, whereas densities several times higher are considered acceptable in other cities. The density indicator is thus used not only in determining housing and land requirements, but also in assessing the compatibility of proposed development with the existing physical form of the community.

Bulk of Buildings

Closely associated with the idea of density is the concept of **bulk.** The density on a site may increase by increasing the height of a building, covering a larger portion of the site with a building, or both. In general, the larger the area of a site that a building covers, the bulkier it will be, not only in actuality but also in our visual impression of the building. Bulk is thus a measure of the actual **volume** of a building (its height times its basic

Figure 6.4 Types for Housing Comprising Different Densities

The (a) **single-family detached** house dominates many low-density districts; (b) a **semi-detached** unit is divided by a wall; (c) **townhouses** arrange units in a row, with individual entrances; (d) **stacked townhouses** like these Montreal "plexes" provide 2–3 units with separate access to grade; (e) **garden apartments** provide housing at medium density and human scale; and taller apartment buildings can provide high-density living in either (f) **slab blocks,** (g) **perimeter blocks,** or (h) **point towers.** These point towers have townhouses defining the edges of their blocks, which has proved to be a successful inner-city urban design concept for Vancouver.

Source: Authors' collections.

footprint or area of ground covered). Building bulk needs to be considered in three different ways by the planner: its aesthetic aspects, its land-use implications, and its economic viability, as discussed below.

Aesthetic Considerations. The standards of taste and appearance by which members of a community assess the bulkiness of buildings have to be taken into account. A community that has no very tall buildings, such as Saint John, may want to limit the height of new buildings; a community that has lots of open space, such as Saskatoon, may want to limit the coverage of buildings on a site. In Vancouver, where the mountain vista is treasured by most citizens, the bulk of downtown buildings is a frequent subject of planning debate. Tall, thin towers may allow better views than short, wide slabs. Thus, zoning bylaws often contain the aesthetic concerns of citizens in their bulk regulations.

Land-Use Considerations. The bulk of a building also has land-use implications. The bulkier the building, in general, the greater the number of activities associated with it—that is, more dwelling units, commercial establishments, and so forth, will be located there. The planner is aware that more intense use of a site may put pressure on public services and facilities, such as sewer lines, roads, and schools. In addition, the bulk of the building will affect the amount of ground space that might be needed on the site for access, for parking, and simply for open space for the occupants of the building. Bulky buildings may affect the light and air available both to occupants and to people on adjacent properties. Tall buildings may shade lower buildings and nearby properties from the sun, an issue that has become important to the installation of solar-energy units.

Economic Considerations. The economics of building are directly associated with the matter of building bulk. The value of the land, the economical size of the establishment, and the costs of building all enter into the original proposal made by those wishing to develop a site. Take a simple case: apartment buildings more than 3–4 storeys tall usually require the installation of an elevator and extra fire exits, but it is normally not economical to build at less than 6–8 storeys in order to recover the extra costs. Thus, the taller building imposes the need for the land to accommodate twice the number of apartment units. How can this extra bulk be accommodated? For what else could the site be used? There are similar versions of this tradeoff argument for all types of structures built for commercial purposes in a city or town. An analogous issue pertaining to residential structures arose in the late 1980s with the advent of "mega-houses" (popularly called "McMansions") in the new-house market (see also Chapter 13).

Indexes of Bulk. In order to deal with the issue of building bulk, planners have devised an index of building bulk called the **Floor Area Ratio** (FAR) in order to respond to the various aesthetic, planning, and economic interests in these matters. Sometimes, it is called the Floor Space Ratio (FSR) or Floor Space Index (FSI). The FAR relates the floor area of a building to the area of the site. An FAR of 1.0 is the equivalent of a one-storey building covering the entire lot. A one-storey building may not suit the owner, who may wish, for example, to provide parking, so a 1.0 FAR might be translated into a two-storey building covering only half the site, or a three-storey building covering one-third of the site, and so on. At some point, the builder may find it economically unfeasible to build taller, or the community may find only certain heights of buildings acceptable. On the latter point, there appear to be community norms on downtown building bulk. For example, in Toronto an FAR of 12.0 is acceptable, but in Vancouver an FAR of 6.0 is considered high. Smaller cities seem to prefer an FAR of 3.0.

The Floor Area Ratio is used to determine more than just the acceptable levels of building bulk. There may be setback requirements from the front, back, and side lot lines, parking space requirements, and, in some cases, requirements to provide recreation space and landscaping. Each of these will reduce the amount of the lot that can be built upon at ground level, thereby forcing the building to be made taller. The FAR may also be used to obtain more open space at ground level than is normally required. In parts of cities where land costs are very high (as in many downtown areas and at key intersections) and where the economics of development demand the intense use of a site, a higher-than-normal FAR may be offered as a bonus in return for providing extra open space at ground level. The widespread appearance since the 1950s of plazas and sitting spaces in areas with tall office buildings and hotels is evidence of the workings of this incentive system.

Often overlooked is that the height of a building is not necessarily related to its density. Many citizen conflicts over increased density are actually a concern about changes in the type of building in their neighbourhood. Since it is possible to have many different types of buildings at the same density, it is often more important to ensure compatible building types. Most low-density, low-rise, urban neighbourhoods would find the 30-storey tower in Figure 6.5 to be an unacceptable change in scale. Yet the four-storey stacked townhouses have exactly the same net density, expressed in Floor Area Ratio, and are built with little complaint in many urban and suburban areas. It would be better to prohibit high-rise towers than to cut the density permitted.

Transfer of Development Rights. Another method that planners use to affect the bulk of buildings is called **Transfer of Development Rights** (TDR). Although every property owner has the right to a certain bulk or FAR, he/she may not wish to utilize it or the regulations may produce a building not suited for the site. In such instances, communities often permit these development rights to be used by the developer on another site. Developers themselves may offer to forego bulky development on one site (such as an historic building) in order to obtain greater bulk on a site they favour more. Development rights are often negotiated between developers and the community to allow unused bulk from one site at another site. Thus, through TDR, communities may be able to obtain open space, more amenable streetscapes, greater concentration of development, and so on.

Linkages and Accessibility

In the planner's view, the community environment is a set of living and working areas, each occupying a specific part of the ground space, and community facilities distributed at various locations. Implicit in this view is the notion that interaction must be facilitated among them. Essentially, this means there must be a means of circulation available, so that individuals, households, firms, and institutions that are separated from one another may be in contact. In this view of the built environment, the planner focuses on the patterns of interaction between people, firms, and institutions. Although there is a great deal of individuality and complexity in these interactions, important patterns can be discerned in the fairly repetitive routines that people or organizations follow when engaging in activities in the community.

Some activities are regularly patterned, such as the daily journey to work or regular food shopping; some are casual or infrequent, as in partaking of entertainment activities and visits for medical care. Commercial firms and public institutions have comparable patterns: they receive supplies, make deliveries, and are the destination for employees and clients. A commonplace example is the daily pattern of children, teachers, and other staff travelling to their local school. The planner thinks of these interactions in

Figure 6.5 Density and Built Form

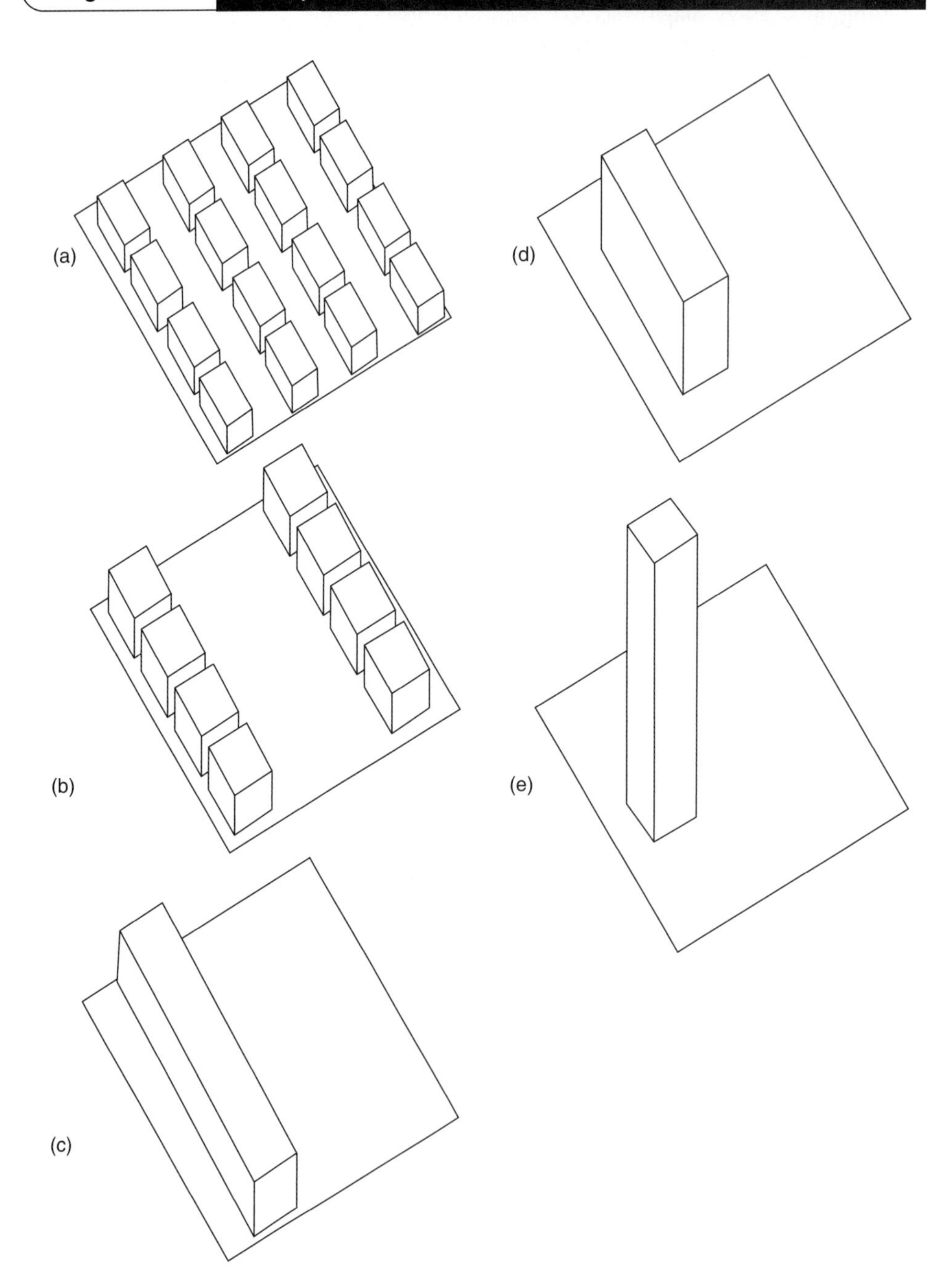

The density of development is only partially related to built form. The density of all five of the examples above is *exactly the same.* They all have a floor a Floor Area Ratio (FAR) of 1.0, which can be achieved by (a) two-storey townhouses covering 50 percent of the site; (b) four-storey townhouses covering 25 percent of the site; (c) six-storey deck access apartments covering 16 percent of the site; (d) a ten-storey slab apartment building covering 15 percent of the site; or (e) a thirty-storey point tower covering 3 percent of the site. Most arguments about intensification revolve around **building type,** not **density,** since the townhouses might be acceptable to many neighbours in single detached houses, but the point tower would not.

Source: Authors' collection.

Planning Issue 6.1 Development Near Stratford Town Hall Angers Citizens

The Charlottetown Guardian
May 30, 2006

Development Near Stratford Town Hall Angers Citizens
Nigel Armstrong

STRATFORD, PEI—Citizens are angry over plans to pack in new residents around the town hall.

Vacant land owned by real estate development company Kel-Mac Inc., which has Fay MacKinnon as president, wants to start developing the land bordered by Mason Road, Bunbury Road and the town hall.

The land now has three zones, single-family, two-family, and multiple-family residential.

There was to have been a public meeting on a request by Kel-Mac to have the whole area rezoned to Planned Unit Residential Development, or PURD.

On the day of the meeting last week, angry residents learned that Kel-Mac's application has been temporarily withdrawn.

Sandy McMillan, chair of the town's planning and heritage committee, was host. She allowed discussion of the Kel-Mac plans even though they were not officially on the agenda.

McMillan said the town is being guided by a public review of its official plan in 2003 that suggested a development goal is to "increase the density around our town hall."

There is a need generally in the real estate market for small, more affordable lots, said McMillan.

She said the town worked for six to nine months in negotiation with Kel-Mac to mutually agree on its development plan.

One suggestion was to create a new, modified zone, such as R1S with the "S" meaning lots that are smaller than standard.

Some residents were opposed to that; others just wanted council to prevent a large number of duplexes, town houses or apartment units from filling the area.

"You get apartments in there, people in rental (units) and it's going to change everything," said Lori Nelson. "The community will change and it will change fast. I have kids and I don't want them exposed to this."

Resident Karen Poley wanted to know why Stratford council was entertaining requests for rezoning or suggesting that high density is a worthy vision for the area when residents don't want that.

"We already have opinions on this," said Poley. "Why can't we stop this right now? We don't want apartment buildings."

The land in question is part of what Stratford is calling its "core area."

It held consultations last year on what should and should not be in the core area and now those ideas are being given formal structure by way of suggested bylaw and zoning changes.

Ekistics Planning and Design, a Halifax-based company is doing that work and will present its suggestions to the community within the next few months.

McMillan and Don Hickox, who spoke on behalf of Kel-Mac, said everything is on hold while Kel-Mac now consults with Ekistics in a further attempt to make its development proposal compatible with the Stratford core area planning project.

Source: Nigel Armstrong, "Development Near Stratford Town Hall Angers Citizens," *Charlottetown Guardian*, May 20, 2006, A2. Reprinted with permission.

terms of **linkages** and seeks to accommodate them, both by arranging a suitable pattern of land uses and by providing an appropriate transportation network.

An early step in community planning for land use is the consideration of which activities need to be linked and how close this access ought to be. Another consideration is the various means of transportation that are likely to be used. Accumulated experience in these matters indicates that closeness of access is more appropriately measured in terms of time and cost rather than pure distance, especially if the linkage must be made by automobile or public transit. Where access is usually achieved on foot, such as to local schools, parks, and shopping areas, distance is the limiting factor. The table in Figure 6.6 shows the time and distance standards that planners have found will provide a high level of convenience for most residents (probably 85 percent or more) in most urban communities.[13] Interestingly, these parameters, although derived in the 1960s, have remained relatively constant to the present day

Every linkage is, of course, accomplished by some form of transportation, and the planner must be cognizant of the modes of travel that might be used for different linkages. For some time now, perhaps since the planning of the model industrial villages over a century ago, planners have recognized the incompatibility of wheeled and foot traffic in many situations. The response has been to try and find ways to separate modes of travel: most simply, sidewalks are provided as well as roadways, or exclusive footpaths may be used, as in the designs for the greenbelt towns. Nowadays, bikeways and entire pedestrian shopping precincts may be set aside. Another facet of planning for traffic separation was recognized when automobiles came into widespread use in the 1920s. The car could not only travel much faster than all other modes, it could also cover greater distances. Yet the car holds no advantage unless it is allowed to move expeditiously. Areas not originally designed for such traffic may thus suffer through having to accommodate it. A perennial planning problem is to find ways of providing auto and truck access to living and working areas while also providing the means for large volumes of motorized traffic to move freely.

Transect Planning

Transect planning, a relatively new approach, is based on the notion of a continuum of environments, ranging from rural to urban, that each have their appropriate composition.[14] It derives from a cross-section that one might draw from the core of a city to the outer (rural) edge of the city's region as can be seen in Figure 6.7. (It is a variant of Patrick Geddes's "valley section," proposed a century ago, which underlies much of regional planning, as shown in Figure 9.1, page 227). These sets of environments vary in their degree of urban character. Further, transect planning draws heavily on ecological principles, dividing its continuum into six "ecozones," and seeks to find the "proper balance between natural and human-made environments."

Figure 6.6 Standards of Accessibility from Home to Selected Urban Land Uses and Facilities

DESTINATION	TIME/DISTANCE
Place of work	20 to 30 min
Central business district	30 to 45 min
Local shopping centre	0.8 km or 10 min
Elementary school	0.8 km
High school	1.6 km or 20 min
Playgrounds and local parks	0.8 km
Major park or conservation area	30 to 45 min
Commercial deliveries	30 to 60 min

Source: F. Stuart Chapin Jr., *Urban Land Use Planning,* 2nd ed. (Champaign-Urbana:University of Illinois Press, 1964), 376. Reprinted with permission from the author.

Figure 6.7 The Rural–Urban Transect

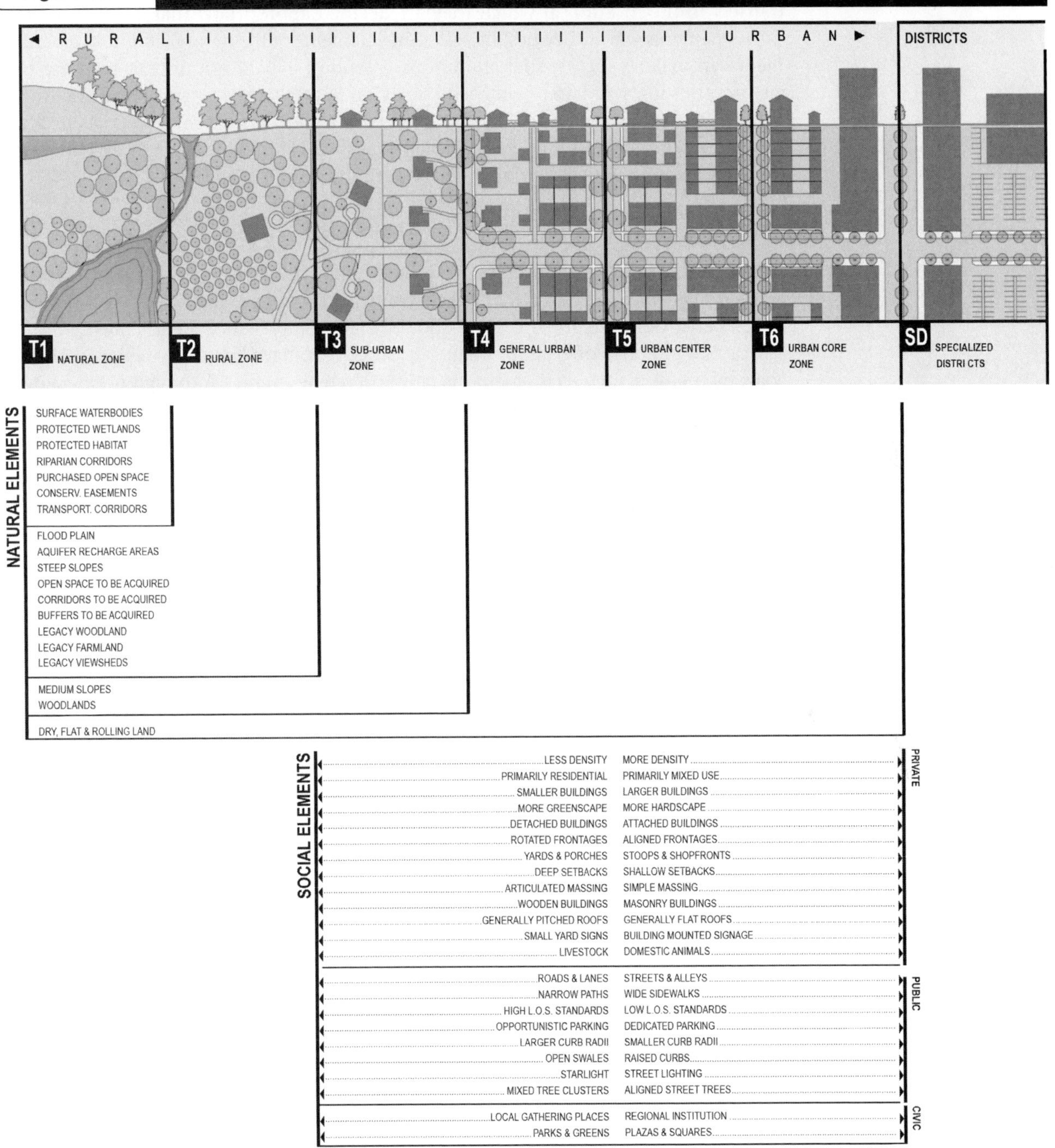

New Urbanist community planners recommend different types of urban design and built form for each zone in this rural–urban transect. Standards for housing types, streets, and parks should be different for suburban and urban districts. This idea may seem obvious, but many municipalities have one set of development standards for the mature urban area.

Source: Duany, Andres and Talen, Emily. "Transect Planning [Figure 1] Diagram of the transect system on page 248]. *Journal of the American Planning Association,* Summer, 2002, pp. 245–266. Reprinted with permission from the *Journal of the American Planning Association,* copyright Summer 2002 by the American Planning Association.

In practice, transect planning involves allocating urban elements spatially along the transect to the location at which they are deemed appropriate; that is, situating the urban elements where they best fit with other elements of the natural and built environments. In making such spatial allocations, attention is paid not only to land use but also to building types and frontages, streetscapes, roads, and open space. This leads, in turn, to a mixing of land uses in districts rather than single-use districts as with much conventional zoning. Transect planning thus complements the New Urbanism principles embodied in traditional neighbourhood design (TND).[15]

Planners' Principles of Community Land Use

Planners have developed a number of principles to use when arranging land uses in a community, which have grown out of a century of planning thought, analysis, and practical experience. These principles (or "should" statements), which are usually implicit in advice that planners offer rather than stated outright, have been accepted as desirable goals by builders, architects, engineers, and the general public. Although not heralded, their importance in the actions that give form to the community environment should not be underestimated. The basic ones are listed here:

1. Land uses should be located such that their activity characteristics do not conflict with one another and each is allowed to function effectively.
2. The pattern of land uses should provide for the integration of all functions and areas.
3. The circulation system should support the land-use pattern.
4. Social cohesion should be promoted by providing the opportunity for the proximity of home, employment centres, shopping opportunities, recreation areas, and schools.
5. Residential areas should be safe, attractive, and well drained, and have variety in their design.
6. Housing should be provided in a range of types to suit the income structure of the community and allow for a range of choice for residents.
7. Commercial and service areas should be located so as to be convenient and safe for clients and efficient for businesses.
8. Traffic with different movement, speed, and volume characteristics should be separated from one another.
9. The downtown area should be considered the social and business heart of the community.
10. The community's built environment and its natural ecosystem should be planned jointly so that they may function harmoniously.

This list of planning principles may be added to or modified as conditions and tastes change in a community. Two tendencies are often at work to modify these basic principles. The first is the realization that many of these principles emanate from earlier planning concepts that may now considered too rigid, such as the separation of uses or traffic.[16] The second tendency is the desire to establish new planning principles as new issues arise, such as for waterfront planning, environmental pollution, energy conservation, aging of the population, or affordable housing. Indeed, the 10th principle in the list above is a recent addition to Canadian planners' set of principles.

Types of Plans for the Built Environment

In planning for a community's built environment, a planner will almost certainly be called upon to prepare and/or assess several different types of plans. These will range from plans that focus on development either for already built-up areas or for undeveloped land, to plans for the community as a whole. In between will be plans for component districts of a community, such as a downtown, for specific community-wide function such as transportation, or for the natural environment of the community. The discussion in this section will focus first on the comprehensive community plan, because it is the cornerstone of local planning throughout Canada. Moreover, it provides the context for all other plans for the community whether they be for community ecosystems, specific functions, special areas, newly subdivided land, or the site plan for a housing project, each of which are subsequently discussed. It should also be noted that for larger communities a regional plan for the area may serve as the prime contextual plan for all constituent communities (see Chapter 9).

The Comprehensive Community Plan

A comprehensive plan for a city or town is a 20th-century planning concept. Its underlying premise is that of a long-range plan (possibly for 20 years) for the overall physical development of the community, which can be used to guide both public and private development efforts. Since developments in the built environment concern housing, the location and functioning of industry and commerce, the location and functioning of community services and amenities, and the health of the natural environment, the community plan also serves to help organize and direct social, economic, and political forces in the community in a rational and productive manner. This adds further to the notion of its comprehensiveness. It is usually the only such broad-based plan a community ever has at any one time. Although it has formal names that vary across the country, such as the Official Plan in Ontario and the General Municipal Plan in Alberta, the intent and scope is the same (see Figure 8.4, pages 218–19).

A community plan, if it is to be **comprehensive** and **long range**, will of necessity be a **general** plan. It will be a plan that deals with all the essential physical developments in the community environment but usually not in detail. It is not intended to be a blueprint document, but one that presents the major proposals for the community's future physical development in general terms. Thus, it uses a limited number of very broad land-use categories, shows only major transportation routes and public facilities, and provides only a general picture of the locations and sizes of major facilities and districts in the city or town. The aim is to provide a guide—both graphically and verbally—to all the major elements in the community built environment and the desired relationships between them. The general plan map for the City of Montreal illustrates the approach described here (Figure 6.8).

Thus, the focus of the community plan is on the **main issues** in physical development and the **major proposals** for future development. This is primarily to help direct discussion and debate in the community so as to arrive at agreement on policies and proposals that affect not only the overall functioning of the community but also the content of plans for specific areas and projects. Every plan is also a mixture of the general and the specific. Some elements in a community are so prominent—a civic centre, the main thoroughfares, a large institution, the downtown—that they cannot and should not be cloaked in generalities. What the plan will probably highlight is their relationship with other parts of the community. Some elements may be areas of special importance to the community's future form and character, such as a waterfront area, a district of historic buildings, special vistas, or a greenbelt. The community plan may contain distinct sections devoted to these special areas. Not least, the community plan will have been developed in concert with a plan for the natural environment.

The Natural Features Plan

Community planning, nowadays, starts with a comprehensive review of the features of the natural environment. The first step is to identify hazard lands, such as flood plains

Figure 6.8 Comprehensive Land-Use Plan Map: Montreal, 2004

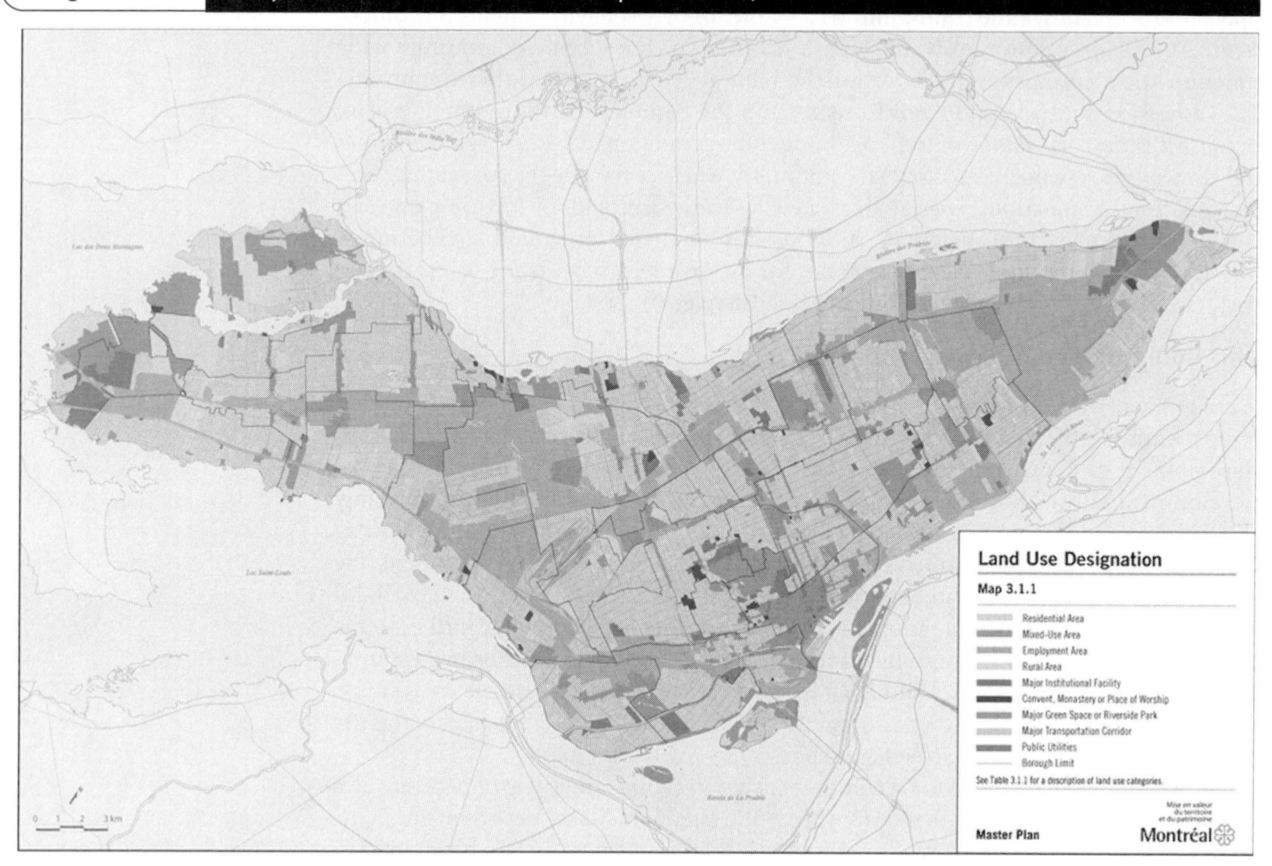

The main land uses for the entire island of Montreal are shown in this map from the 2004 Master Plan, which won a Canadian Institute of Planners' National Award of Distinction

Source: City of Montreal, *Master Plan* (2004), map 3.1.1.

and unstable slopes where natural features create a danger to human life.[17] Then, landscape ecology principles are used to identify important "patches" (such as forests and marshes) that need to be protected to preserve wildlife habitat. The "corridors" (river valleys, hedge rows) that connect the patches are also identified. Finally, the overall pattern, or "mosaic," of patches and corridors is examined to identify improvements that might be made in decades ahead to increase biodiversity, such as new connections, or enlarging the forest reserves.[18] With careful ecological design, the single-species agricultural lands that surround many cities might even see an increase in biodiversity, even as the cornfields are converted into new communities.[19]

Functional Plans

Some elements of a community's physical structure may be so important to its overall functioning as to require more elaboration than is possible within the comprehensive community plan. Separate functional plans are commonly made for the following four elements, especially in larger communities, as one sees in recent plans for Edmonton and Ottawa:

1. Transportation and circulation systems,
2. Parks and recreation facilities,
3. Economic development, and
4. Social services and facilities.

These functions either occupy large and strategic portions of a community's ground space, thereby demanding considerable public investment, and/or are of overarching importance to the community's economic and social well-being. Thus, separate plans are frequently prepared subsequent to the publication of the community plan, or sometimes they are drawn up in order to update the outlook of a comprehensive plan that is already in use.

A functional plan usually provides proposals at all levels of detail. A general concept plan is provided to encompass the main proposals and projects (such as a new bridge or conservation area), but there will also be detailed proposals for the location of facilities and, often, the acquisition of land by public authorities. Functional plans may be prepared by departments within the local government that are responsible for the particular function or, in some cases, by public agencies established outside of the local government to carry out the function.

Special-Area or District Plans

There are special areas in almost every community that will require, or perhaps deserve, more detailed planning. The central business district (CBD) is the area most commonly singled out for special planning consideration. This stems from the fact that it is usually the key to a community's vitality, both economically and culturally. The CBD often provides residents and visitors with the strongest visual image of a community; it represents the most dominant concentration of land values and investment in property, as well as providing a locus for social transactions. As such, the CBD needs to be continually maintained and renovated in order to function effectively. It is also a complex area requiring detailed planning analysis and design.

CBD plans tend to focus on three aspects of the downtown area. The first is the functional arrangement of land use, because most downtowns accommodate several functions: retail shopping, financial and business services, entertainment, and government institutions. In larger communities, these functions may occupy discrete, but often overlapping, parts of downtown. The second aspect is transportation and parking, for the downtown tends to contain those functions to which the entire community needs access. The third is the three-dimensional character—the urban design—of downtown. CBD plans often address the heights of buildings, prominent views, walkway systems, squares and open space, mall boulevards, and street furnishings. Concerns over aesthetics and architecture are usually greater in CBD plans than in any other physical plan.

Other districts may also merit special planning attention in a community, depending upon the historical and geographical circumstances. Seaport communities, often concerned over the demise of shipping and industry, prepare plans for revitalization of harbour areas, as did Halifax. In communities with clusters of historic buildings, such as the Old Town portions of Montreal and Victoria, there may be an impetus to plan for the conservation and upgrading of these areas. The future of inner-city neighbourhoods may also stimulate special-area planning, as the example from Calgary shows (Figure 6.9). Again, as with CBD plans, these special-area plans are normally framed within the context of the comprehensive plan and they, too, pay particular attention to detailed functional arrangements of land use, transportation, and visual design.

Land-Development Plans

Planners prepare or evaluate two types of plans dealing with new development of the ground space of a community. One is the **subdivision plan**, which involves dividing a large parcel of usually vacant land into numerous building lots. This is most commonly done for residential development, but may also be used in industrial and commercial development. The other is the **site plan**, which involves development on a single parcel of land that is usually either vacant or about to be made vacant by the razing of existing structures.

Figure 6.9 Special-Area Plan for a Residential District, Calgary

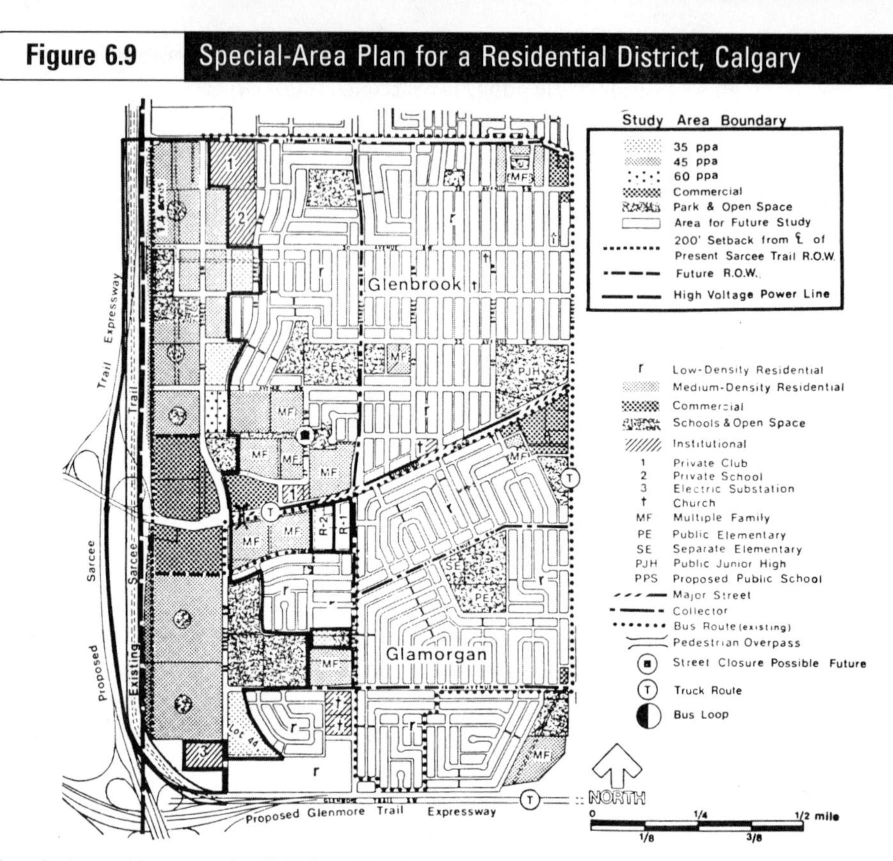

In order to provide a more detailed view of land uses, traffic, and facilities in different districts, special-area (or secondary) plans are prepared. This one is for several Calgary neighbourhoods.

Source: City of Calgary.

Subdivision Plan

A subdivision plan, in contrast to the general land-use plans referred to above, is a precise plan. It must show exactly the proposed property lines, street system, water and sewer lines, and topographic changes to the site. The proposed Ontario subdivision in Figure 6.10 illustrates such a plan. The reason for the detail is that, upon completion, the municipality assumes responsibility for the subdivision as an additional part of the community. Thus, careful scrutiny is given to subdivision plans so as to avoid such problems as those identified below by the Ontario government in alerting its municipalities:

> Water pipes, sewers, and roads might have to be run through vacant land to reach scattered subdivisions, thus increasing their length and the cost of services to the public. Subdivisions laid out on hilly ground could have street patterns that ignore the slopes. This also increases costs, makes them difficult to maintain, and more hazardous in cold weather. Others, laid out in poorly drained soil and provided with septic tanks and well water, become health hazards when the septic tanks pollute the wells or when the septic tank beds are subject to flooding.[20]

Subdivision plans must also indicate areas set aside for parks, schools, and any other public facilities, such as walkways, churches, and shopping areas. In a number of provinces, it is required that the subdivider deed 5 percent of the subdivision to the municipality for park use. However, before all these details are considered, the subdivision plan is examined for its conformity with the aims of the comprehensive plan, the requirements of the zoning regulations for the area in question, and street alignments and intersections with streets in adjacent subdivisions.

Figure 6.10 Draft Plan of a Subdivision, Ontario

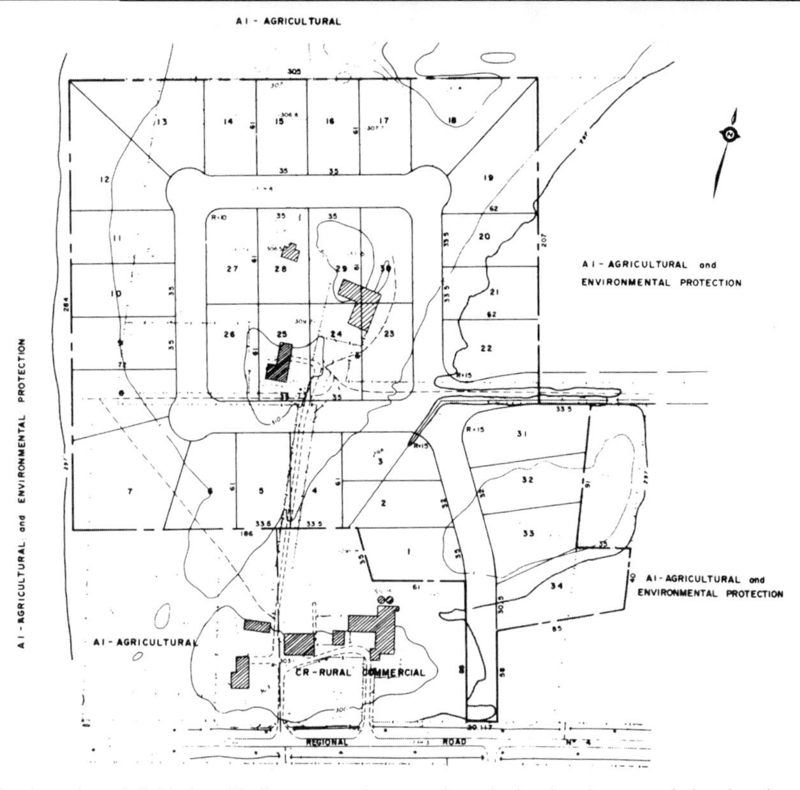

Before land can be subdivided and built upon, a plan must be submitted and approved showing the arrangement of lots, the alignment of streets, and the uses of land.

Source: Canada Mortgage and Housing Corporation, Residential Site Development Advisory Document (1981), 24.

Site Plan

A site plan refers to the proposed land-use arrangements, normally for a single parcel of land. It is usually prepared by the proponent of a development (not necessarily the owner of the property) for one or more new buildings or for making substantial changes to existing buildings. A site plan is a precise, blueprint-type plan that is concerned with the placement of buildings and the connections linking the site to the street system and to public services. All non-building types of facilities such as parking lots, recreation areas, and interior roads must also be included on a site plan, as must any topographic aspects that might affect drainage. Figure 6.11 illustrates a typical site plan.

Increasingly, site plans are required to be submitted so that they can be scrutinized for their consistency with community objectives. Aesthetic and functional considerations are also taken into account in this nearly final phase of land-use planning. In Ontario, this process is called site-plan control; elsewhere the term development control is often used to refer to the same local vetting of developers' site plans (see Chapter 13).

Other Views of the Built Environment

The land-use perspective that planners bring to the task of community plan-making comprises several dimensions and, as well, is subject to ongoing social and cultural refinement. In other words, land use and the built environment can be, and is, viewed in different ways at different times by other professionals who become involved in community

Figure 6.11 Site Plan for an Affordable Housing Project, Montreal

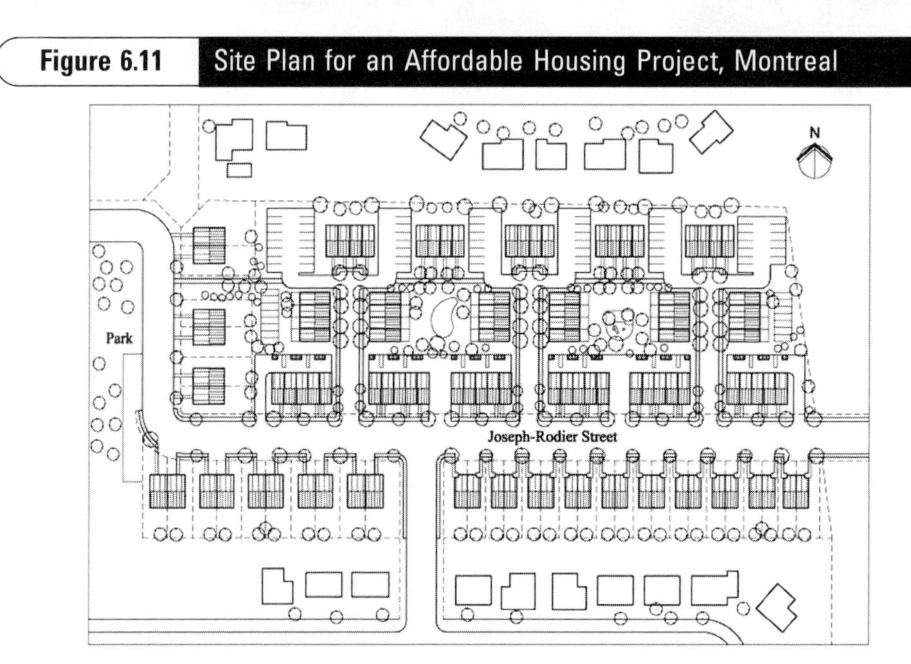

The precise location of buildings and arrangement of open space are shown on a site plan for a housing project, like this one for Cité Jardin Fontenaux, a "Grow Home" community in Montreal, designed by Groupe Cardinal Hardy. Such plans are often required when building permits and development-control agreements are being sought.

Source: Courtesy Avi Freedman.

planning. Other professions, for example, are involved in assisting a community shape its physical form, some as colleagues of planners within local government, some as consultants to developers, and some as advocates for the community. Further, how land is used in a community is a reflection of its citizens' values and, as we saw in preceding chapters, the aspects of land use considered important to plan for and regulate tend to change both over time and among communities. Thus, the planner's aim to strive for comprehensiveness in a community's plan must be able to accommodate diverse professional and cultural views of the meaning of land use that comprises the built environment.

As Seen by Other Professions

While the planner has the central professional role in planning the built environment of a community, views other than those of the planner also enter into such deliberations. Especially significant are the perspectives of several other professions. Lawyers, architects, engineers, landscape architects, environmental planners, and social workers all play substantial roles in the planning and development of communities. They bring their special concerns into community plan-making and, in so doing, add their distinctive views of the built environment to those of the planner. Below, we briefly examine the views of these professional areas regarding a community's built environment. Each of these affect the focus as well as the nature of the physical plans that are prepared but, at the same time, they help to better reflect the complex physical milieu of the built environment within which planners work.

1. The **lawyer** views the community environment as, essentially, a configuration of parcels of land, each of which has a legal description and carries a certain set of property rights. The lawyer thus seeks great precision in regard to plans affecting the parcels of land that constitute the community. Because planning proposals may have the effect of redefining property rights—limiting uses, heights, and placement of

buildings—lawyers have a natural concern that planning processes be conducted properly and that prior property rights not be arbitrarily taken away. The two types of physical plans that most closely reflect the built environment for the lawyer are the subdivision plan and the site plan because of the precision accorded to property lines. And, of course, the written land-use regulations and planning policies that have legal stature are of great importance to the lawyer involved with community-planning matters (see Chapter 13).

2. The **architect's, landscape architect's,** and **urban designer's** view of the community environment is mostly a three-dimensional one—that is, the architect is concerned with what can be built on the ground space of the community and what the resulting construction will look like. This concern extends from the design of individual buildings and other structures to groupings of buildings (e.g., an apartment complex), to streetscapes and the design of open spaces in the community. Architects and landscape architects may design subdivisions to achieve certain groupings of buildings and other features, and they usually prepare the site plan for a project. Except with respect to special district plans and the height regulations for buildings, many community plans have little to say explicitly about the visual outcome of land development. Comprehensive plans may state a policy of wishing "to preserve the character" of, say, an historic area, but it usually remains the prerogative of the designer to bring in the third dimension. As building and site aesthetics are often the subject of vigorous debate in a community, it is easy to see that visual qualities are very important in planning. They are, however, not easily resolved, impinging as they do on people's values about aesthetics, the extent of land-use control, and the nature of growth and change in the community. The landscape architect may also become involved in environmental issues regarding a site (see below).
3. The **engineer's** view of the community environment is primarily functional. Thus, the engineer is concerned with how well the various physical elements function in the community—the street system and other forms of transportation, the water supply and sewerage system, the electricity and communications systems—both individually and in conjunction with one another. The engineer's view is both comprehensive and detailed in regard to the community environment, but tends to be limited to providing a framework of services and streets within which land development can take place. A special concern of the engineer is the way in which natural drainage patterns may be affected by land development. It is the duty of the municipality to try to ensure that, when the surface of the land is modified in any land development, nearby properties are not affected adversely in regard to drainage. The engineer's concern is, indeed, among the traditional concerns of community planning that were addressed in the first planning acts in Canada: the alignment of streets, the efficient extension of public utilities, and drainage conditions. Subdivision plans and site plans that are required to address these matters directly often involve an engineer in their preparation and review.
4. The **environmental planner** and, not least, **environmental interest groups** have brought into sharp focus concerns over the effect of planning decisions on the natural environment of a community. Their view encompasses the land, water, and air of the community and the quality of each. The most pressing concerns are with the effluents generated by various land uses: sewage, other liquid effluents, smoke and fumes, noise, and solid wastes. Belatedly, it seems, we have come to know that these effluents, if they are not planned for and managed properly, can be dangerous to the health and safety not only of people on neighbouring properties but of the entire community. There are also related concerns with more passive elements of the environment, such as the disturbance of areas of natural vegetation (e.g., marshes and woods) and the preservation of views. Community plans

increasingly respect these factors, and planners scrutinize development proposals regarding their environmental impact, usually with the assistance of environmental planners and landscape architects. Nonetheless, environmental issues are still in many ways very technical, complex, incompletely understood, and often controversial.[21]

5. The **social worker's** view of the built environment is, primarily, from the perspective of housing. This involves issues of affordability and adequacy of housing, as well as the more recent issues of homelessness and housing for the burgeoning immigrant population.[22] Given that the largest use of community land is for housing, these are substantial issues that need to be addressed. A complementary role is that of the **social planner** in the few communities that provide such a governmental focal point.

Emergent Views

Just as we saw the physical planning perspective in previous generations evolving as new issues and ideals emerged (see especially Chapters 4 and 5) so, too, are views of the built environment being extended nowadays, sometimes quite dramatically. Here, briefly, are some of the more prominent new views being urged on planners and elected officials:

1. **Sustainable development** is perhaps the most widely mentioned view for the early 21st century. Economic, environmental, and social objectives (the "triple bottom line") are considered in combination so that current development does not compromise the needs of future generations. The **ecological footprint** is often used as a measure of sustainability.
2. **Bioregionalism** considers the built environment from the perspective of ecological principles and aims to counteract the environmental destruction caused by current city-building and other economic production. Communities and the environment are seen as integrated elements and, further, are bound up with their local regions and the "web of life," including the geology, soil, wildlife, water systems, and various human cultures.[23] The river basin or watershed is especially employed as an organizing principle, as was seen with the 1990s planning for the Toronto waterfront.[24]
3. **Smart Growth** is another recent approach being urged on planners, to deal with what proponents call "mindless sprawl" of suburban development. Smart Growth is advocated by a wide variety of interests—municipal managers, home-builders, transit advocates, conservation groups—and, consequently, is not always defined uniformly.[25] However, its basic elements are: limiting outward growth, promoting compact mixed-use development, reducing automobile dependency, preserving large amounts of open space, and creating a greater sense of community. A variety of financial incentives and approaches often accompany these proposals, as well as innovative neighbourhood designs. An integral part of the Smart Growth approach is the institution of **urban growth boundaries** within which to contain future urban growth, often for 25 years or more. Communities from Nanaimo, B.C., to Summerside, P.E.I., have implemented these boundaries.
4. The **energy-efficient community** was an idea that came into prominence during the energy crisis of the 1980s and prompted the examination of built environments for opportunities to conserve energy. It involves reducing the amount of travel by residents (especially by automobile), clustering land uses, and orienting buildings to take advantage of solar-heating possibilities. More recently, this concept is being enfolded within Smart Growth and New Urbanism initiatives.
5. The **postmodern suburb** is a reaction to the monotony and sameness of most post-World War II suburbs and has led to a variety of ideas: creating town centres, infill

schemes, and new social spaces for the enormous and diverse populations that now live there.[26] The **New Urbanism** is another reaction to prevailing modes of urban development, in which proponents seek to recreate "community values" such as neighbourliness and safety through a mixture of land uses, public space, and pedestrian access.[27] This is promoted through the concepts of **traditional neighbourhood design, transit-oriented development,** and **transect planning** (see Figure 6.8, page 162).

6. The **Healthy Community** concept has (re)entered the purview of planners, motivated by a much-expanded definition of health (compared to the public-health movement of the past) and broad support from federal and provincial health ministries. It grew out of the need to shift from traditional institutionalized health care to the more holistic approach of improving community environments, including such areas as food safety, housing, and economic development. The goal is to enable people to support each other more effectively and achieve a better overall quality of life, as examples from Alberta and Quebec demonstrate.[28] It is an approach embraced by Canada's professional planners[29] and echoes the concerns of some of the country's earliest planners (see Chapter 4).
7. The **multicultural city** is a fact for Canada's large cities, especially Toronto, Vancouver, and Montreal, where "ethnic enclaves" are already well formed. This has significant implications for land-use planning and regulation. Sets of community services (e.g., spiritual institutions), housing (e.g., family and household size), and neighbourhood arrangements (e.g., the mix of commercial firms) required by ethnic minorities may diverge from community norms. Conflicts may then arise over land use and confront a key premise that planning (and zoning) should be based on use and function and not on people.[30] To respond appropriately will necessitate understanding the basic factors at work in the formation of these new patterns of residence.[31]
8. The **Senior-Smart Community** is a call to recognize that Canadian communities have seen rapid growth in the number of seniors (65+) over the past three decades. With the aging of the baby boom generation, starting in 2011, most will see a doubling of their seniors' population.[32] This will further emphasize that this population group has special needs in three areas: housing, transportation, and support services. Responding to seniors' needs will require a broadening of the scope of community planning to be more "people oriented" and to engage with a much wider array of public and private agencies, as the City of Regina showed when implementing such plans.[33]

These emergent views make demands on planners to expand and refine the ways in which they view community land use. For example, bioregionalism demands inclusion of several levels of the natural environment, while the healthy-communities approach demands inclusion of social and economic environments, and so forth. But which to choose? Some, like New Urbanism and Smart Growth, seem to encompass the full array of physical, social, and economic elements in a community, but each in its own limited and distinctive way. Clearly, today's planners recognize in these viewpoints a more varied reality of what constitutes land use, as well as the more complex task of community planning that faces them.

Latent Views

By and large, most of the viewpoints identified above deal with the physical form of a city or town, that is, with its "external" appearance and arrangement. Consider, however, that the shape of a community's land use should be a response to its "users," as we see with the emerging multicultural and senior citizen dimensions. Yet, not until recently have we deliberately sought out the views on land use from those "inside" the community, so to speak—its *citizens*. In one innovative study, Hok Lin Leung conducted a survey to find out how Ottawa citizens perceive their city environment.[34] To them, the city's land use has many more dimensions, and needs to be expressed in different ways than planners usually do. Variables of age, gender, income, and workplace, for example, make for many of the differences between citizen and planner viewpoints. Probing further into each of these variables also seems necessary to obtain a more accurate internal view, as British planner Clara Greed (and others) points out regarding the view of women in a community. She notes that the "everyday life" of women involves short journeys, the need for localized facilities, access to a mixture of land uses, as well as concerns over safety and child care.[35] In a parallel way, planners will need to become more aware of the issue of safety in the city and of the accessibility needs of the ill and impaired.[36] Recognizing the diverse needs of these and other citizens will bring with it a clearer and more comprehensive view of land use and its planning. (This discussion is further elaborated on and extended in Chapter 15.)

Reflections

The set of land uses that comprise the built environment of a city, town, or rural community is the realm of the planner, and the practice of community planning consists in influencing the development of these land uses toward a desired physical shape for the community. To this end, planners devise means for systematically describing land uses regarding their space needs, functional relations, and linkages to one another. Through the experience of past practice, an array of plans has evolved to deal with different aggregates of land use. The outcome has been a nested set of land-use plans, with the comprehensive community plan encompassing all the land uses in the community and subsidiary plans covering districts, specific land-use functions, and individual sites, as well as developed or undeveloped land.

Understandably, this system for analyzing and planning land uses, which evolved in the practice of planning communities over, especially, the past half-century, reflected a particular perspective on land use—the professional planner's perspective. Although that perspective has been shaped and refined by the ideas and approaches to land use of associated professions, it is relatively recently that views of land use from outside the professional spectrum, from the citizenry, have insisted on being incorporated into community plans. The latter initiatives and others are challenging long-held assumptions that land-use planning is only about land and physical facilities and not about the people who use the land or the natural setting for land use.

Traditional planning practices and tools have begun to acknowledge these emerging needs and concerns and incorporate them into the comprehensive community plan. That these recent viewpoints are all contending for recognition and a role in the community plan indicates the importance of this planning tool, but it also means that it (and its subsidiary plans) will have to continue to change, possibly more profoundly than in the past. The thrust is for comprehensive community plans to become more truly comprehensive, to integrate the physical, social, and natural elements of a community. This is a long-touted, but seldom achieved, goal of planners. Thus, as the scope of community-planning practice is elaborated in ensuing chapters, it is cogent to ask:

- ***How and to what degree can community plans and the processes for achieving them better incorporate emerging physical, social, economic, and natural environmental land-use perspectives?***

Endnotes

1. Alberta, Department of Municipal Affairs, *Planning in Alberta: A Guide and Directory* (Edmonton, 1978), 1.
2. Jane Jacobs, *The Death and Life of Great American Cities* (New York: Random House, 1961).
3. Cf. Albert Z. Guttenberg, "A Multiple Land Use Classification System," *Journal of the American Institute of Planners* 25 (August 1959), 143–150; and Gerald Hodge and Robert McCabe, eds., "Land Use Classification and Coding in Canada: An Appraisal," *Plan Canada* 8:2 (June 1968), 1–28.
4. Guttenberg, "A Multiple Land Use Classification System."
5. Jill Grant, "Mixed Use in Theory and Practice: Canadian Experience with Implementing a Planning Principle," *Journal of the American Planning Association* 68:1 (Winter 2002), 71–84.
6. Congress for the New Urbanism, *Charter of the New Urbanism* (New York, NY: McGraw-Hill, 1999).
7. David L.A. Gordon and Ken Tamminga, "Large-scale Traditional Neighbourhood Development and Pre-Emptive Ecosystem Planning: The Markham Experience, 1989–2001," *Journal of Urban Design* 7:3 (2002), 321–340.
8. Cf. Ken Jones and M. Doucet, "Big-box Retailing and the Urban Retail Structure: The Case of the Toronto Area," *Journal of Retailing and Consumer Services* 7:4 (October 2000), 233–247; and Jim Simmons and Ken Jones, "Growth and Change in the Location of Commercial Activities in Canada: with Special Attention to Smaller Urban Places," *Progress in Planning* 60:1 (July 2003), 55–74.
9. John Hitchcock, *A Primer on the Use of Density in Land Use Planning*, Paper No. 41 (Toronto: University of Toronto, Program in Planning, 1994), thoroughly discusses this important measurement tool.
10. Martin Laplante, "A Simple Model of Urban Density," *Plan Canada* 45:1 (Spring 2005), 23–26.
11. Ibid., 8.
12. Available online at www.statcan.ca.
13. F. Stuart Chapin Jr., *Urban Land Use Planning*, 2nd ed. (Champaign-Urbana: University of Illinois Press, 1964), 376ff. More recent editions continue to confirm these parameters.
14. Andres Duany and Emily Talen, "Transect Planning," *Journal of the American Planning Association* 68:3 (Summer 2002), 245–266.
15. Congress for the New Urbanism, *Charter*.
16. Grant, "Mixed Used in Theory and Practice."
17. John Newton, "Exploring Linkages Between Community Planning and Natural Hazard Mitigation," *Plan Canada* 44:1 (Spring 2004), 22–24.
18. W.E. Dramstad, J. Olson, and R. Forman, *Landscape Ecology Principles in Landscape Architecture and Land Use Planning* (Washington, DC: Island Press, 1996); and Richard Forman, *Land Mosaics: The Ecology of Landscapes and Regions* (New York: Cambridge University Press, 1995).
19. Ken Tamminga, "Restoring Biodiversity in the Urbanizing Region: Towards Pre-emptive Ecosystems Planning," *Plan Canada* 36:4 (July 1996), 10–15; and Andre Daigle and Daniel Savard, "Designing for Conservation," *Plan Canada* 45:4 (Winter 2005), 27–30.
20. Ontario, Department of Municipal Affairs, *Three Steps to Tomorrow* (Toronto, 1972), 44; Pierre Filion "The Importance of Downtown," *Plan Canada* 46:1 (Spring 2006), 31–33; and Pierre Filion and Heidi Hoernig, "Downtown Past, Downtown Present, Downtown Yet to Come: Decline and Revival in Middle-Size Urban Areas," *Plan Canada* 43:1 (January 2003), 31–34.
21. Mary-Ellen Tyler, "Ecological Plumbing in the Twenty-First Century," *Plan Canada* 34:4 (July 1994), 69–176.
22. J. David Hulchanski, *Housing Policy for Tomorrow's Cities* (Ottawa: Canadian Policy Research Networks, 2002), Discussion Paper F27.
23. Kirkpatrick Sale, "Bioregionalism—A New Way to Treat the Land," *The Ecologist* 14:4 (1984), 167–173. Canadian planner Doug Aberley elaborates the bioregionalism perspective in Doug Aberley, ed., *Boundaries of Home: Mapping for Local Empowerment* (Gabriola Island, BC: New Society Publishers, 1993).
24. Ontario, Royal Commission on the Future of Toronto's Waterfront, *Watershed* (Toronto: Queen's Printer, 1990).
25. Anthony Downs, "Smart Growth: Why We Discuss It More than We Do It," *Journal of the American Planning Association* 71:4 (Autumn 2005), 367–378. See also Anthony Downs, "What Does 'Smart Growth' Really Mean?" *Planning* (April 2001), 20–25.
26. John Sewell, a former mayor of Toronto and royal commissioner on Ontario's Planning Act, explores this issue in his *The Shape of the City* (Toronto: University of Toronto Press, 1993). See also Joel Garreau, *Edge City: Life on the New Frontier* (New York: Doubleday, 1991).
27. Andrea Gabor and Frank Lewinberg, "New Urbanism," *Plan Canada* 37:4 (July 1997), 12–17.
28. Cf. Réal Lacombe, Julie Levesque, and Louis Poirier, "Villes et Villages en santé au Quebec: un idée qui a porté fruit," *Plan Canada* 42:4 (October–December 2002), 15–17; and Brett Hodson and Lori Anderson, "Okotoks: A Made-in-Alberta Healthy Community, *Plan Canada* 42:4 (October–December 2002), 18–19.
29. David Witty, "Healthy Communities: What Have We Learned?" *Plan Canada* 42:4 (October–December 2002), 9–10.
30. Mohammad Qadeer, "Dealing with Ethnic Enclaves Demands Sensitivity and Pragmatism," *Ontario Planning Journal* 20:1 (2005), 10–11.
31. Sandeep Kumar and Bonica Leung, "Formation of an Ethnic Enclave: Process and Motivations," *Plan Canada* 45:2 (2005), 43–45.
32. Gerald Hodge, *The Seniors' Surge: The Geography of Aging in Canada.* (Vancouver: UBC Press, forthcoming 2007).
33. Regina and District Seniors' Action Plan Steering Committee, *Seniors' Action Plan Report* (Regina: City of Regina, 2000).
34. Hok Lin Leung, *City Images: An Internal View* (Kingston, ON: Ronald P. Frye and Co., 1992), 271ff.
35. Clara Greed, "Promise or Progress: Women in Planning," *Built Environment* 22:1 (1996), 9–21.
36. Gerda Wekerle, "From Eyes on the Street to Safe Cities," *Places* 13:1 (Winter 2000), 44–49; and Rob Imrie, "Barriered and Bounded Places and the Spatialities of Disability," *Urban Studies* 38:2 (February 2001), 231–237.

Internet Sources

Chapter-Relevant Sites

Planning Canadian Communities
www.planningcanadiancommunities.ca

Visualizing Density
www.lincolninst.edu/subcenters/visualizing_density

The Land Centre
www.landcentre.ca/index.cfm

Urban Land Institute
www.uli.org

Smart Growth/Urban Sprawl—Ontario Nature
www.ontarionature.org/enviroandcons/smart_growth

International Initiative for a Sustainable Built Environment
www.greenbuilding.ca

Canada Green Building Council (CaGBC)
www.cagbc.org

Project for Public Spaces
www.pps.org

7

Chapter Seven

Steps in the Plan-Making Process

In order to be able to make a plan we must be able to predict; in order to be able to predict we must know; in order to know we must develop hypotheses or theories; in order to establish theories we must obtain and classify facts; we must observe.

John Dakin, 1960

When a community sets out to make a new plan, or to amend its existing plan, it is embarking on a process with a recognized set of steps and characteristics; it embarks on what is known as the **planning process.** The plan that the community adopts from this process marks the culmination of plan-making. There are then subsequent processes of implementing the plan (which is discussed in Chapters 13 and 14) and renewing the plan. In other words, the activity of community plan-making comprises both a **plan** and a **process**, each of which, in its own way, is an essential element. Unlike other kinds of decision making in a community, planning is not aimed at finding a solution to a specific problem. Community planning is better characterized as *preventative* than as *remedial* in its approach to community problems. As such, it requires time for deliberation, analysis, design, and evaluation, as well as for the involvement of diverse community interests—that is, the process needs time to be to be undertaken appropriately.

The community plan-making process comprises, essentially, two planning processes. One is the **normative process** that a community, usually through its municipal government, undertakes to determine its needs, objectives, acceptable courses of action, and whom to involve in the deliberations regarding its plan. The other is the **technical process,** primarily followed and guided by the professional planners, or their counterparts, of studying the community and designing the plan. Both of these processes derive from a long-standing paradigm (or conceptual view) of what constitutes a good plan-making process, which is grounded in the ideal of a *rational–comprehensive planning* process. In this chapter, as we describe each of these planning processes, a question worthy of consideration is:

- ***Where do the ideal and actual planning processes converge and diverge from one another and with what implications for planning effectiveness?***

The Concept of a Community-Planning Process

One analogy that is often used to describe the community-planning process to those new to the idea is that it involves the same sorts of decisions that go into choosing a daily wardrobe. There are, of course, certain similarities—assessing the occasion, considering the possibilities, and making a choice between alternatives—but the community-planning process has some distinctive differences. Indeed, the community-planning process is notably different from other planning tasks—personal, corporate, or institutional—in that it involves planning for and with a community.

The community with which community planning is concerned must be perceived both as a physical community of buildings, streets, and open spaces, and as a human community of individual people, groups, and social institutions. Each of these facets of community obliges special consideration in the planning process. On the one hand, the physical community—the built environment—is the substance of the community-planning process. The types of buildings that are built, their location, and their relation to one another and to streets and open space are the outcome of the process. The inherent permanence of physical structures means that the choices made by community planners are long-range in nature; the results of city-building, as noted in the opening chapters, persist long into the future. Furthermore, the built environment is never completely built: it grows, ages, deteriorates, and is transformed by successive development projects. It also comprises unique districts and neighbourhoods, each of which has its own integrity and which, in combination, constitute the character and image that belong to the particular community. In the terms used by planners, the planning for a community must comprehend both the parts and the whole of the city—that is, it must be **comprehensive** in its physical outlook.

On the other hand, the *human* community is the recipient of the outcome of the planning process, as well as the proponent in that process. People initiate, manage, deliberate upon, and decide on the results of the planning process. Thus, the diverse values, objectives, and interests that are inherent in any community become part of the process. This diversity obliges that community planning must deal with **multiple objectives**, and this alone differentiates the process from that of most other planning efforts. Where the concern is with the planning of a daily wardrobe, or the production schedule of a manufacturing plant, or the deployment of military weapons, or the fund-raising strategy of a charitable organization, the objectives are much more clear-cut and limited in number. Often only one main objective motivates such efforts, in contrast to the multiple objectives involved in shaping a community.

The diversity of objectives with which community planning contends brings with it a competition among various goals (and those who advocate them). This requires the process to select the courses of action reasonably, not capriciously. Thus, the community planner must not only take into account all of the important ends of the whole community, but must also maximize the attainment of those ends that are considered most important. What developed in community planning to meet these criteria is called the **rational–comprehensive planning process.** It is, at heart, a logical decision-making process aimed at meeting the needs of communities in their plan-making efforts.

The Rational–Comprehensive Planning Process

Until the 1950s, the process of community planning was much less systematic than it is today. In its earliest days, it was practised in traditional (architectural) design terms, with the aim of producing aesthetically pleasing and efficient layouts. In turn, the public-health and housing-reform movements provided planning with an extensive set of statutory controls. This led increasingly to intervention in land development and land use in order to minimize negative effects that the actions of one land user had upon the interests of others. As discussed in Chapter 4, planners then put more and more emphasis on community planning as a tool for achieving efficiency in city-building. The "efficient" use of land in a city carries with it the assumption that the planner can demonstrate attainment of this aim. Further, the idea of efficiency is a rational concept. Increasingly, traditional planning approaches were found to be deficient in their rationality; they were found often to be aesthetically, politically, or administratively arbitrary. The functioning of cities was not well understood, and little thought was given to the process of formulating land-use policies in a pluralistic society.

The groundbreaking attempt at systematizing (and critiquing) the community-planning approach was formulated by U.S. planners Meyerson and Banfield in the early 1950s. The rational–comprehensive approach, as it has since been called, contends that a planner would be acting rationally by following three general steps: (1) to consider all the possible alternative courses of action; (2) to identify and evaluate all of the consequences following from the adoption of each alternative; and (3) to select the alternative that would most likely achieve the community's most valued objectives.[1] It is, of course, greatly simplified in the three steps above and deserves to be more fully elaborated.

Canadian planner Ira Robinson melded the initial formulation of Meyerson and Banfield with the adaptations suggested by later planning thinkers and came up with the following five steps:[2]

1. Identify the problem or problems to be solved, the needs to be met, the opportunities to be seized upon, and the goals of the community to be pursued, and translate the broad goals into measurable operational criteria.
2. Design alternative solutions or courses of action (plans, policies, programs) to solve the problems and/or fulfill the needs, opportunities, or goals, and predict the consequences and effectiveness of each alternative.
3. Compare and evaluate the alternatives with each other and with the predicted consequences of unplanned development and choose, or help the decision-maker or decision-making body to choose, that alternative whose probable consequences would be preferable.
4. Develop a plan of action for effectuating or implementing the alternative selected, including budgets, project schedules, regulatory measures, and the like.
5. Maintain the plan on a current and up-to-date basis, based on feedback and review of information to adjust steps 1 through 4 above.

This concept of the planning process is sometimes called **synoptic**, in that it provides for all the principal parts of the community, physical and social, to be brought into the picture. Another term, more common today, that captures the same spirit of planning, is **holistic**. This view has venerable roots, drawing as it does upon the work of Patrick Geddes around the beginning of this century. Geddes put great stress on being able to see, to know, and to appreciate all the basic facets of any community before making plans for it. His view always encompassed the people, the geography, and the economy of the community, and he advised planners "that survey and diagnosis must precede treatment."[3] Just as important to Geddes was that the knowledge and appreciation of the community obtained by the planners should be shared with the community. This makes the community-planning process not only a logical process for decision making but also a participatory process. It obliges the planner and the policy-maker to keep all the items of a proposal for change—a new shopping centre, for example—and the issues that enter into their analyses and deliberations clearly in view. And the community, for its part, can and should become closely involved in vetting both the knowledge from the surveys and the issues raised in the decision making.

This ideal process of Geddes and others has not been without its critics. Its claims of being both *rational* and *comprehensive* have been challenged. How can planners be truly rational in achieving diverse and often competing goals? Whose rationality is used in choosing which goals to maximize? How can planners know what is in the public interest?[4] Similarly, how can planners fully comprehend land use and its many activities and purposes, which may vary by neighbourhood? And how can they, at the same time, comprehend the viewpoints of an increasingly diverse population? These criticisms are valid, although they do not mean the planning process cannot be rational. Rather, the plan-makers of communities (planners, politicians, citizens, developers) must accept that their rational approach is constrained and limited by their knowledge, culture, and values. Their approach can still be considered rational, but it is a **"bounded rationality."**[5] It is well to bear this constraint in mind when viewing community-planning outcomes.

The Flow of the Process

The rational–comprehensive planning process (its limitations notwithstanding) is a generally sequential process, with the community moving progressively from the identification of problems to the implementation of projects. Indeed, the community traverses all those steps, as shown in Figure 7.1, but not necessarily in strict sequence. In actual practice, many of the phases are linked to preceding phases by **feedback loops** (of which a few possibilities are indicated on the diagram). These provide for an essential review of the outcome of the planning process during the implementation phase, as well as to inform and adjust the initial phases of problem identification and goal articulation.

The secondary feedback loops permit conclusions reached at one stage in the process to be questioned, reexamined, and even for all or part of the process to be reiterated.

Figure 7.1 General Model of the Community-Planning Process

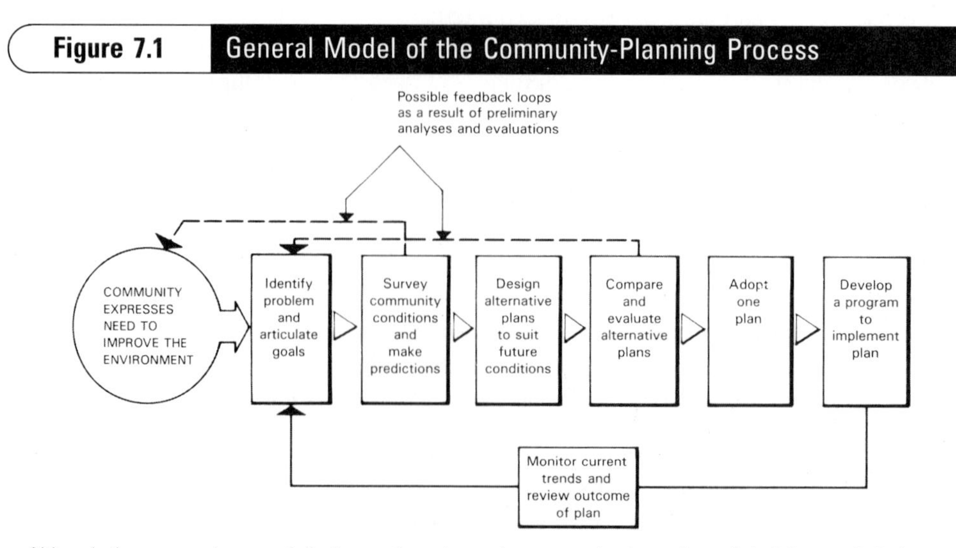

Although the process is essentially linear, there is ample opportunity for review of decisions and choices at the various steps and for reiteration of all or part of the process.

These feedback loops allow for any difficulties encountered in forecasting and in the normal debate over the outcomes of planning proposals to be taken into account. A community may find, for example, that the population forecasts indicate only limited growth; this would then limit the development alternatives for the community. Or it may find that alternative plans, while meeting overall community goals, are objectionable to certain neighbourhoods. This may cause a reappraisal of the goals that had been decided upon earlier. Formal community plan-making processes, prescribed in planning legislation, allow for some of these feedback loops in specified review phases, but not all. Most reappraisals, reviews, and new iterations take place aside from the formal process, and the need for them is readily recognized by those community members, groups, and agencies involved.

The Normative Process of Planning Community Land Use

The actual process a community goes through in planning its land use approximates the conceptual or rational process described above, but may deviate from the sequence or may accord special weight to certain steps. The main reason for this is that the land-use patterns of communities stem from the outcomes of various social, economic, and political behaviours. Myriad decisions made by individuals, groups, businesses, institutions, and governments are involved in the use to which the parcels of land in a community are put. Each of these "actors" behaves in the context of individual and shared values. Since there is never a perfect match of individual and community values in land use, one might say that the process of community plan-making is concerned to a large degree with determining **norms** by which the various value orientations may be reconciled. Thus, it is a normative process: it both recognizes and intervenes in the value system of community members.

It follows that the place given to the identification and articulation of goals is bound to be prominent in actual community planning. Indeed, any steps in the process that invoke the need to define goals for the community plan or to evaluate whether the plan will attain stated goals will assume more importance and visibility the more intensely is the planning process and the more extensive is the participation of the public in the plan-making. The technical aspects, the analyses, and the administrative requirements of the plan are less subject to community debate. Almost of necessity, then, as a community plans it may need to allow for many feedback loops in the process of refining goals and planning proposals. Let us examine briefly how the process is initiated and the arrangements that are made for its conduct in actual community settings.

The Determinants of Land Use

The planning process in a community is initiated by the actions of individuals and groups, in both the private and the public sectors, who desire to occupy and/or improve some part of the community's land base. These actions range over a host of possibilities, such as the redevelopment of an old residential area into an apartment complex, the acquisition of marshland for a conservation area, and the upgrading of a local road into an arterial highway. The outcomes of these actions may not necessarily result in a desired pattern of land use from the point of view of the community. This is true for the countless initiatives that occur regarding the location and pattern of land uses.

A community-planning process is put in place, so to speak, to harmonize the various factors and forces that determine land use. The determinants of land use may be categorized into three broad types, according to their value orientation—that is, those motivated primarily by economic values, those by social values, and those by public interest values. We shall examine each of these briefly; however, it should be noted that seldom are land-use decisions the result of only one of these determinants. Not only do the individuals, businesses, government agencies, and other groups who participate combine *all* the basic values to some degree, but also many land-use decisions that affect each determinant evolve concurrently.

Economic Determinants

The land use of a community is influenced by economic forces operating both outside the community and within its boundaries—that is, either by external determinants or internal determinants, or by both.

External economic forces are the trends and conditions in the larger (provincial and/or national) economy in which the community exists. They act mainly through the demand for the goods and services supplied by the community and may affect land use in several ways. They may account for the investment in key establishments for manufacturing, commercial, or institutional use and, therefore, in the buildings they use and the land they occupy. This is most obvious in a community dominated by a major manufacturing plant or public institution. These uses often require large amounts of land in strategic locations. Since the decisions regarding the future of these large establishments (e.g., whether to expand them or close them) are usually made outside the community, they may not always reflect the wishes of residents, and there may not be unanimity in the community about privileges that might be accorded these land users. Two other ways that external economic forces influence land use are in the amount and rate of land development. If the regional economy is expanding, for example, this may call for more land to be developed for houses, stores, parks, and schools. Moreover, the rate at which the larger economy is operating—whether in fast growth or stagnation—will influence the rate at which land undergoes development and buildings are built upon it.

Internal economic forces determine most of the land-use arrangements and the physical character of the community (in contrast to external forces, which affect its economic composition and vitality). This is done through the forces of supply and demand acting within the local land market. The land (or real estate) in a community may be looked upon as a commodity to be bought and sold and has a value because of its potential to produce income through some future sale or development. The actual market value of land varies between parcels in the same area and between different areas in the community. Each parcel of land is unique in its location, size, shape, form of building space, tenure, and other features and, as John Hitchcock notes, when it comes to housing, types of households differ and they make different demands for housing.[6] Housing is also strongly affected in its value and usability by the activities carried out on other parcels of land, especially those in the same vicinity. Thus, certain locations and districts come to be valued for one kind of use, such as those preferred for residential neighbourhoods; while for other uses, people find the attributes of other areas more appealing.

In any event, land is usually owned by a large number of different persons or organizations and developed by still others, all with their own personal aims, resources, and concerns. Take, for example, the process of creating housing in a community, the largest use of urban land. It involves landowners, the home-building industry, brokers and facilitators, and home buyers. Furthermore, all these actors are in the private sector. If the new housing is to be built on raw land, the interrelations among these actors could resemble those shown in Figure 7.2. Most of the same actors are also involved in converting already built-up land to other uses. Depending upon the planning proposals, some part of the array of values represented by these actors will enter into the planning process. It goes without saying that they seldom coincide with one another. Thus, a critical aspect of the normative planning process is to find a balance for the

Figure 7.2 Steps in the Conversion of Land to Residential Use

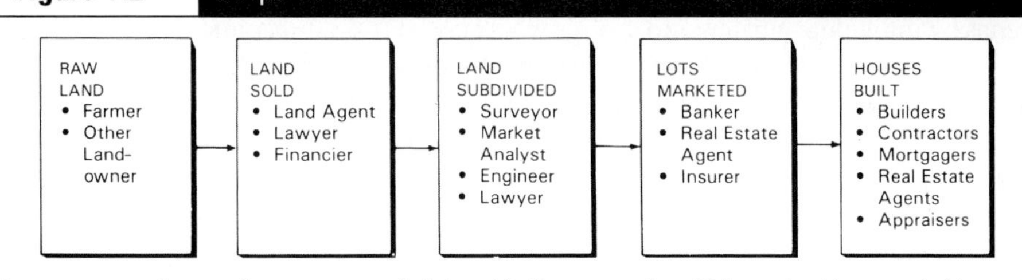

Many persons and groups have an economic interest in the process by which raw land is converted to use for housing.

diverse economic interests in community land development. Finding such a balance is often difficult when the economic stakes are high: for instance, householders defending their residential property values, which they see threatened by an unwanted development, or a big-box retailer seeking a rezoning against the wishes of local businesses. It has become a major task for planners in many communities to mediate such disputes, as discussed especially in Chapters 11 and 12.

Social Determinants

Although it often seems that all land-use decisions in a community are stated in dollar terms, many are influenced primarily by social factors. There are two main ways in which social factors influence land uses: one is the social "ecology" of the patterns of residence; the other is the social response to proposed changes in the community. People have strong feelings about the way their community or neighbourhood is arranged and functions. These feelings are rooted in values that, not infrequently, oppose values of economy and efficiency, as citizen protests against planning proposals frequently show.

Among the aggregation of people who constitute a community, one finds a pattern in which household units cluster together by virtue of ethnic, cultural, and/or income compatibility of populations to form residential districts. And they, in turn, may choose to live apart from commercial or manufacturing areas. Of course, the reverse may also happen. It is important to note that segregation of uses and districts is a continuous process of sorting the community into physically and geographically distinct parts according to values, attitudes, and social interests. Thus, some people like to live in mixed-use neighbourhoods while others prefer not to.

The prestige (and/or economic power) of some groups of people or firms, accordingly, leads to a hierarchy of areas in the community. Thus, some residential areas enjoy more prestige than others, and some commercial areas are more favourably located than others. The results of this sorting are not always beneficial for the living conditions and economic prospects of all residents. The tensions that this can create between areas of a community become part of the milieu of social values with which planning must frequently contend. Interest groups may be formed, such as those of ratepayers, businesses, and tenants. They may express concerns over the effects of a plan on land values and rents, but just as often, they voice concern over continued social cohesion of the areas in which they live or do business.

Over the past two decades, there has been a noticeable increase in the number of social values that community planners must take into account. This has happened, notably, with respect to four issues:

- The quality of the natural environment;
- The importance of the historical features of the community;
- The needs and concerns of women; and
- The increasing ethnic diversity of the population.

These issues are often advocated vigorously in various land-use decision-making arenas by individual citizens and groups who may have no direct interest in the affected lands through ownership, tenancy, or proximity of residence. Each of these issues areas thus introduces additional values about the substance and form of communities into plan-making. Not infrequently during such deliberations, flaws are revealed in the planning process itself. Canadian women planners have noted, for example, that women and other groups tend to be excluded from involvement in the planning process because of the time of day meetings are scheduled or the lack of transportation or the unavailability of child care and/or language constraints.[7] The community-planning process is being challenged to reconcile not only economic and social values but also differences in social values. With the increasing diversity of community populations, the latter challenge is certain to increase.

Public-Interest Determinants

Community planning not only mediates between private land interests, it is also an active determinant of land-use patterns on behalf of the public interest—that is, out of the values and institutions of community planning have arisen principles and standards for the location of both public and private facilities and the design of public spaces. The provision of water and sewer systems, the design and alignment of roads, and the allocation of park space are three ways in which public-interest factors control, or at least pre-condition, land-use planning. Whereas the private economic and social interests who participate in the community-planning process generally advocate on behalf of a single piece of land or a single issue (such as the environment), it is up to the community's planners to prepare plans that ensure sound, amenable development in the interest of the community *as a whole*.

The public interest, perhaps not surprisingly, is not based on a single entity or set of values. Although the local government of the community is usually the most important entity in this realm, other jurisdictions may also be involved. For example, the local or regional school board, the housing authority, the public utilities commission, and the provincial ministry of highways are some of the other public bodies one encounters. Each has its own mandate to interpret what is in the public interest in land-use matters within the community. Usually, the views of these other bodies are mediated through the local government and, nominally, it is the local governmental structure—the mayor, council, planning board, planning staff—that is the focal point for implementing public-interest matters in land use. Provincial statutes for planning and municipal affairs delegate this responsibility to local governments. It is largely up to the local government to decide the manner in which it will exercise this responsibility.

The Importance of Goals

The process of framing a community plan is, in large measure, a process concerned with sorting out the values and attitudes in the spheres of each of the three main determinants—economic, social, public interest—and putting in place what amounts to a commitment toward future land-use decisions that encourages these interests to act in harmony. That process, as we have indicated, is normative—it both recognizes and intervenes in the behaviour of community members. Moreover, community planning comprises more than sorting out differences in current viewpoints about land use. It is primarily concerned with establishing the context in which the future land-use pattern will evolve. The plan that emerges from the process is a statement of community goals and aspirations for its future physical form.

One of the central notions of modern community planning is that the identification of goals is an integral part of the planning process. Such other city-building activities as engineering and architecture, or such civic policy activities as transportation and recreation, tend to begin with a limited number of goals already given. In planning, the identification of goals and the design of courses of action to achieve them tend to proceed in a symbiotic fashion.[8]

Planning, as noted at the outset of this book, derives initially either from the need to solve problems and/or from the desire to achieve ambitions or aspirations for the built environment. The proposed solution to a community problem carries with it a goal to be achieved. For example, the shortage of parking space that promotes the idea of a proposed parking garage in a downtown area may be seen as consistent with the goal of maintaining the vitality of the downtown area. The desire of a community to provide adequate park space for each of its neighbourhoods, conversely, carries with it the need to designate specific areas that may be used for parks, and to design those parks. Thus, goals may justify proposed solutions to problems as well as stimulate a solution. In the actual planning process, regardless of whether it is initiated by a goal or a problem, the focus alternates from one to the other as goals, designs, or project ideas are refined.[9] Indeed, one planner sees the planning process as a progressive refinement of goals into projects.[10] Or, to put this another way, the schematic presentation of the planning process shown in Figure 7.1 has as its second step "goal articulation," and it is fair to say that the entire process is one in which goals are progressively being articulated.

Goals are basic to the planning process and thus become a cornerstone of the plan for the community. A plan is sometimes referred to as a statement of goals. The importance of goals stems from the fact that they provide the rationales by which the community, through its plan, may justify what needs are to be served and also whose needs are to be served by the proposed land-use arrangements.[11] Given the likelihood of competition among land users for some locations in the community and of conflicts between the values of different interests, goals are important because they represent the resolution that has been made in such controversies. They represent choices among the various value positions. Goals, in effect, say "when we encounter this situation we will probably act this way, for these reasons."[12] In addition, goals represent both a general commitment of the community to long-term planning and a willingness to communicate these aims to all members of the community so that they may participate effectively.

Goals, Objectives, and Policies

Before broaching the other important areas of the normative process, it will be helpful to sort out a few semantic difficulties. The term "goal" is used in a variety of ways in planning, along with a variety of such synonyms as aims, ends, purposes, objectives, and policies. Distinctions need to be made, in particular, between goals, objectives, and policies, because each reflects a different facet of the planning process. Hence,

- A **"goal"** refers to an ideal, a condition, or a quality to be sought in the community's built environment. It might be, for instance, to provide a maximum of access to the waterfront for all members of the community. Thus, a goal in community planning expresses a desire for the community's development to move in a particular direction.
- An **"objective"** is, by contrast, something that the community seeks to attain; it is more like a target that can be measured and reached (and thereby monitored). While the nature of a goal is general or abstract, that of an objective tends to be specific or concrete. To elaborate on the waterfront example above, a community may have the objective to provide a walkway for public use along its waterfront.
- A **"policy"** may be defined as the preferred course of action to be followed in achieving an objective or a goal. At its root, policy is about choices made by or on behalf of people; it is the course of action the local government thinks will be most acceptable to the diverse interests of the community. Thus, in regard to its waterfront,

a community's policy may be to purchase a strip of land along the waterfront for the purpose of constructing a public walkway. This example not only illustrates how goals, objectives, and policies form a sort of hierarchy of community intent, but also how they represent the progressive translation of general ideas into operational targets and then into actual, physical projects.

A good example of the use of planning goals and objectives is found in the 2005 Official Plan for the Regional Municipality of York. The plan has eight goals:[13]

1. Quality Communities for a Diverse Population
2. Enhanced Environment, Heritage and Culture
3. A Vibrant Economy
4. Responding to the Needs of Our Residents
5. Housing Choices for Our Residents
6. Managed and Balanced Growth
7. Infrastructure for a Growing Region
8. Engaged Communities and a Responsive Region

Each of these goals is, in turn, complemented by a number of "action areas" (i.e., objectives), examples of which are given here (and their links to the goals above):

- Creating Vibrant Urban and Rural Communities (Goal 1)
- Ensuring Clear Water and Air (Goal 2)
- Attracting and Supporting Businesses (Goal 3)
- Supporting Safe and Secure Communities (Goal 4)
- Providing Appropriate Housing Mix and Supply (Goal 5)
- Taking a Strategic Approach to Growth Management (Goal 6)
- Developing an Integrated Transportation Network (Goal 7)
- Being a Region That Involves Its Citizens (Goal 8)

The implementation of the plan is guided by policies that have been set by the region's council. And, in turn, the plan is to be monitored and evaluated for its success. The latter phase of plan-making is discussed more fully later in the chapter.

The Roles of Plan-Makers

The goals one finds enshrined in a plan for a community usually seem self-evident and uncontroversial. Yet a brief pondering of the diverse community interests involved in the outcome of land-use decisions reveals the maze of controversy that frequently arises in setting objectives. One of the main reasons for making a plan is to provide a basis for debate and discussion among competing interests in order to resolve present and future land-use conflicts.

Traditionally, plan-making was considered primarily a technical task to be carried out by professional planners and/or by a small group of knowledgeable citizens. The goals for achieving basic health, safety, and amenity seemed then to be self-evident. Today, the decision-making processes for planning are more open and democratic with, as a consequence, many more people involved in plan-making. As well, the issues have become more complex. Plan-makers thus face a dual problem: on the one hand, they must devise objectives and criteria that are relevant to a greater cross-section of the community; on the other hand, they must devise meaningful procedures for consultation about goals.[14]

In the Canadian community situation, it is the local or municipal government that is ultimately responsible for setting the objectives and making the plan. This means, in jurisdictional terms, the elected council of the municipality. But while there must be some ultimate plan-making authority, the issue is more complex than that. Community planning, as we have been stressing, is a process, as well as a function of local government. It is a process that might be likened to a cycle of behaviour involving the following four general phases: (1) experiencing needs and wants; (2) defining goals; (3) planning alternatives; and (4) deciding and acting.[15] Since this is a process involving many individuals and groups, it is probably more realistic to see it taking the form of several successive cycles, say the form of a spiral, as the concerns of various interests are progressively resolved.

As a process occurring under community auspices, it must have a visible, identifiable pattern of roles and responsibilities. Further, as a process that involves both normative, or political, judgments and technical judgments, it requires plan-makers who can cover each aspect as well as those who can link them together. There are three principal plan-makers in most Canadian communities and, as we shall see below, they and their roles correspond to three distinct aspects of the planning process: (1) to decide on the plan's goals and content; (2) to facilitate the public process; and (3) to provide technical assistance to plan-makers.

1. Deciding on the Plan. To become the official policy of the community, the plan will need to be validated by the local governing body. This role falls to the **municipal council** (or its counterpart). The council is the ultimate arbiter of what is acceptable to the diverse interests within the community, as well as what is in the best interest of the community. It almost goes without saying that the process of resolution at this level takes place within the "political climate" that characterizes the community. The planning policies that the council deems acceptable will reflect what is acceptable to elements within the community that wield influence. The council's role may be linked to presiding over the final cycle of community planning in which the plan is produced.

2. Facilitating the Public Process. Prior to deciding on the plan, the process demands that there be broad public deliberation on what goes into it. In the typical Canadian community, this phase is presided over by an **advisory planning body** established by the local council. These advisory bodies are called planning boards in some parts of Canada and planning commissions in other parts; such bodies may simply be a standing committee of the council. They play an essential role in examining private and public values, attitudes, and preferences regarding land use and its planning. It is their task to receive, consider, and refine planning proposals and then to advise the council on how to proceed.

In the normal course of plan-making, the advisory board provides advice to council only after it has deliberated over the proposals made by various interests in the community, the responses of other interests (e.g., the province, a regional government), and advice from the technical staff. In this sense the advisory board links the political (i.e., normative) and technical sides of planning.

3. Providing Technical Assistance. Planning proposals usually involve issues that require technical study and analysis. The technical sphere of planning is presided over by **professional planners**. For example, a proposal for a shopping centre requires an assessment of the effects of traffic on surrounding areas and road arteries, while the proposal for a subdivision near a water course requires an assessment of environmental effects. Such studies are normally carried out by the planning staff of the municipality, outside consultants, or by some combination of the two.

The issues and implications for the pattern of land use in a community have long since grown beyond those of health, safety, convenience, and amenity that stimulated the need for technical advisors in the planning process around the turn of the century. Professional planners must be able to grasp the way market forces tend to allocate and arrange land uses, the way cultural factors affect land-use patterns, and a variety of other concerns, including the natural environment, energy, housing, and heritage planning. The role of the planner, it has been said, is to provide a basis for a "balanced consideration of economic, socially rooted, and public interest factors throughout the land-use planning process."[16] This balance is provided through a systematic study of the issues raised by various interests, on the one hand, and through the design of alternative courses of action, on the other. In the latter endeavours, the planner tends to structure the planning process by directing the attention of the planning board and council to selected courses of action that respect both private and community interests. This, then, involves the planner at several stages throughout the planning process.

Tensions in the Planning Process

The actual planning process a community goes through (the normative process) varies in two major ways from the rational–comprehensive model. The first is the need to designate specific groups and individuals to preside over certain steps in the process. These appointments not only carry with them specific duties and powers—as, for example, the advisory planning board—but also are composed of people with their own personal and professional backgrounds and values. The second has to do with the substance of the social agenda that the plan-makers are asked to consider. For example, if planning proposals are contentious, such as a major citizen protest over a proposed big-box store, several iterations of the planning decision cycle may be required before goals are articulated clearly and discussion of alternatives can proceed. It is also common to find that results of analyses presented by the professional planner, such as the forecast of traffic generated by a new sports facility, may highlight other issues surrounding a proposed project.

The community-planning process has been criticized for being too optimistic, both about our analytical capabilities and about the altruism of community members.[17] A large part of the criticism centres on the difficulty, if not impossibility, of the process being truly comprehensive. In other words, can the plan-makers (including the planners, planning board, and council) really take into account *all* the parts of the community and foresee *all* the consequences attending upon *all* the alternative courses of action? There is, of course, validity in this criticism, and it contributes to one of the inherent tensions in the community-planning process.

Another tension concerns the importance given to rational analysis—identifying problems, translating goals into measurable criteria, predicting consequences, and evaluating alternatives, to draw upon Robinson's terms. These all are, or could be, the subject of rigorous analysis. Indeed, as our knowledge of the functioning and growth of communities has grown in recent decades, it has become possible to carry out more sophisticated surveys and diagnoses of community ills and planning proposals alike. But an almost inevitable condition appears in the wake of scientific applications in the community-planning process: more information usually exposes further gaps in our knowledge, as well as uncertainty over our predictive powers.

Community planning, as with other public policy-making, moves with difficulty from objectives to programs and projects. This means that, having identified *which* needs are to be served, it becomes necessary to translate those into *whose* needs are to be served. If, for example, a new expressway is deemed necessary to serve the entire community, its actual location will impinge on some neighbourhoods and some groups of people in the community more than on others (see Figure 7.3). Here we encounter one of the most fundamental tensions in community planning—between facts and values. Difficult value-based judgments must be made if the planning process is to progress toward community goals. Rational analyses, by providing relevant facts and information, bring planners to this threshold, but do not carry them over it. This is because

Figure 7.3 Citizens Protest New Hog Plant, Winnipeg

Inherent tensions in the planning process are exposed when neighbourhood residents dispute the decisions of planning bodies.

Source: Ken Gigliotti/Winnipeg Free Press, 2006, reprinted with permission.

beyond facts lie preferences, and, to this point, the scientific approaches to sorting out personal and group preferences are likely to be of little use to community planners. An Environics poll approach, for instance, will not work well within a community with diverse values.

Thus, the community-planning process involves several inherent tensions. They are a reality of the process and should not be denied or obscured. The most significant of these derive from the following dichotomies encountered in planning communities:

1. Neighbourhood/City. The final outcome of planning (the actual projects) occurs at the neighbourhood level of a community; thus, judgments arise regarding the status of the values of people in local areas against those of the entire community. The "not-in-my-backyard" (NIMBY) syndrome is evidence of this tension (see Chapters 11 and 12).

2. Natural/Built Environment. Protecting ecological sites may remove valuable land from development but may also enhance the ambience nearby development.

3. Long Range/Short Range. As the time frame increases, the degree of accuracy of predictions decreases and, further, the commitment of the community to projects well into the future will compete with desires to solve immediate problems.

4. Ameliorative/Developmental. The nature and pace of change may either be forced on a community by external circumstances or be sought by a community to achieve a certain environmental quality; either direction is a normal source of debate.

5. Fact/Value. Establishing the facts of trends, conditions, and impacts relevant to planning actions is different from establishing the social objectives of what might be done through community planning.

That the rational–comprehensive model of planning contains these "soft" areas for planners and decision-makers indicates the complexity of planning for communities. Moreover, the same tensions exist for any model of the planning process that might be invoked. The process of community planning is not simple, in any case. A major advantage of the rational–comprehensive approach is that it allows a holistic process to try to encompass these tensions. Importantly, the community-planning process should be open and accessible to community members, so that the judgments that have to be made in regard to each of the above tensions may be reviewed and debated if necessary.

Inherent in public decisions about land use is the notion of **intervention.** Both with regard to private profit making and social interests and to other agencies with public-interest mandates, the means of intervention must be developed to resolve the tensions on behalf of the community. Such a process of resolution takes place within the "political climate" that characterizes the community. In other words, the content of the plan and the means of intervention will reflect what is acceptable to those individuals, firms, and groups in the community who wield influence. The underlying tension in the normative planning process involves the need to bring together the various interests sharing a concern, including, not least, those that wield little or no economic or political influence. For a broader discussion of the process of resolving planning tensions, see Part Three, especially Chapter 12.

The community plan that results from the planning process thus becomes a record of the deliberations of the community in regard to the achievement of its goals. The judgments made about any of the inherent tensions should reflect a community consensus, something that may not be easily achieved if the plan invokes fundamental value cleavages.[18] However, once this consensus is achieved, the plan should become a matter of public record of what was agreed upon, as well as the basis for decisions. As one U.S. planner has put it, "if we could all remember what we did and why we did it, we could do very well without a plan."[19]

Technical Steps in the Plan-Making Process

Although the actual planning process may not follow precisely the steps of the rational–comprehensive model, it proceeds with an underlying sense of rationality. This is largely attributable to the studies and analyses that have become a standard underpinning of the various steps in the process. The professional planner has the obligation to bring all relevant facts to the plan-makers' attention and, thereby, establishes a parallel planning process. It complements the normative process and provides information for all those involved in plan-making—the planning board, the council, citizens, and groups—to enable them to understand the problems and the policy implications of the analysis.

The technical planning process has four main phases:

- Diagnosis
- Prediction
- Design
- Evaluation

These phases approximate the main decision-making phases of the rational model—problem identification, design of alternative plans, and evaluation of alternatives. This may give the impression of a sequential, step-by-step process, but as with the earlier descriptions, the steps in the technical process may overlap and be repetitive. For example, it is common for the staff planner or consultant to be asked to evaluate planning proposals submitted by developers or public agencies, thus initiating the evaluative phase. This may necessitate going back through the previous phases, or the process may be completed at this point. It should also be noted that the same phases and their studies and analyses are applicable regardless of whether the plan-making effort is for a single project or the entire community.

The material below is organized according to the sequence of basic phases in order to convey a sense of the coherence that does exist in the planning process, despite its variations. The technical planning process is based on objective studies. As planners have sought precision and clarity in their analyses, they have turned to quantitative methods and statistics. In this section, we do no more than outline these and place them within the context of plan-making as a process; for a full description, the reader should turn to texts devoted to planning analysis.[20]

Diagnostics Studies

The planning process is invoked most frequently by the desire to solve a problem in the development of a community. The goal may be to reduce traffic congestion in the downtown area or to increase the supply of housing for low-income families, for example; each community has its own distinctive set of problems. Normally, the planner would initiate efforts to identify and delineate the problem—to *diagnose* it. The term is used in the same sense that a medical doctor conducts a diagnosis of a patient's reported ailment—that is, he or she identifies the nature and scope of a medical problem. The planner similarly employs a number of diagnostic tools to describe both the extent of the planning problem and the community context into which it fits.

Planning studies at this stage are essentially descriptive. The most common methods the planner will employ fall into the realms of descriptive statistics and survey research. Using the two problem examples cited above—downtown traffic congestion and housing shortages—will help to demonstrate the diagnostic approach. In the case of traffic congestion, the planner would seek data on current conditions: the volume of traffic on affected streets, evidence of delays at certain points and times, the degree of use of parking facilities, and possible associated problems, such as businesses affected or the effect on pedestrian flows. In the case of housing shortages, the planner would seek data on the size and composition of the current housing stock, its condition, and vacancy rate, as well as data about the people reported to need housing, including family size, composition, age, and income. In both of these types of planning problems, the planner would also seek data on past conditions in order to examine trends over time, as well as seek data regarding comparable situations in other communities.

Pre-emptive Environmental Studies

Ecosystems planning must be done before planning for urban expansion, new subdivisions, and so forth. Experience has shown Canadian planners that the actions necessary to protect and restore a community's natural environment cannot be planned after planning has taken place for the built environment. Instead, a full understanding of the ecological features central to the effective functioning of local ecosystems must underpin all other planning in the community.[21] Further, local ecosystem plans should be based on a regional environmental inventory and strategy for protection, maintenance, and restoration. The community planner would, then, look first at the higher-level inventory and environmental plan for flood plains, environmentally sensitive area, ecological links, and so on before proceeding with plans for the built environment.[22] In this way the community (or neighbourhood) plan can respect ecologically sensitive sites that need restoration and maintain ecological paths and links. The recent regional plan for the (expanded) City of Ottawa provides such a regional inventory and environmental strategy for its constituent communities to follow.[23]

Built Environment Analytical Dimensions

There are two basic sets of dimensions of planning diagnoses for the built environment. The first set has to do with substantive information about the community, district, or neighbourhood that is being planned. This could be called the "what" of the diagnostic study. The second has to do with procedural perspective by which the planner is approaching the diagnosis, or what could be called the "how" of the study.

Substantive dimensions. These dimensions are of three main types: population, the physical environment, and the economy. All three recur throughout the various analytical phases of the planning process. In the diagnostic phase, the planner is mainly concerned with identifying who or what is involved and at what scale. The data are, thus, mostly numerical counts: the number of people, building types, dwelling units, automobiles, jobs. Alternatively, the average values of such data may be sought: average income, years of schooling, age of dwelling. At this level of analysis, interrelationships between characteristics may also be helpful, such as persons per dwelling unit, automobiles per employee, persons per unit of land (density). Figure 7.4 indicates typical diagnostic information for a neighbourhood planning situation. If the planning focused on a commercial area or an industrial area, the information would reflect the characteristics of such areas.

Figure 7.4 Types of Information Used in Neighbourhood-Planning Analyses

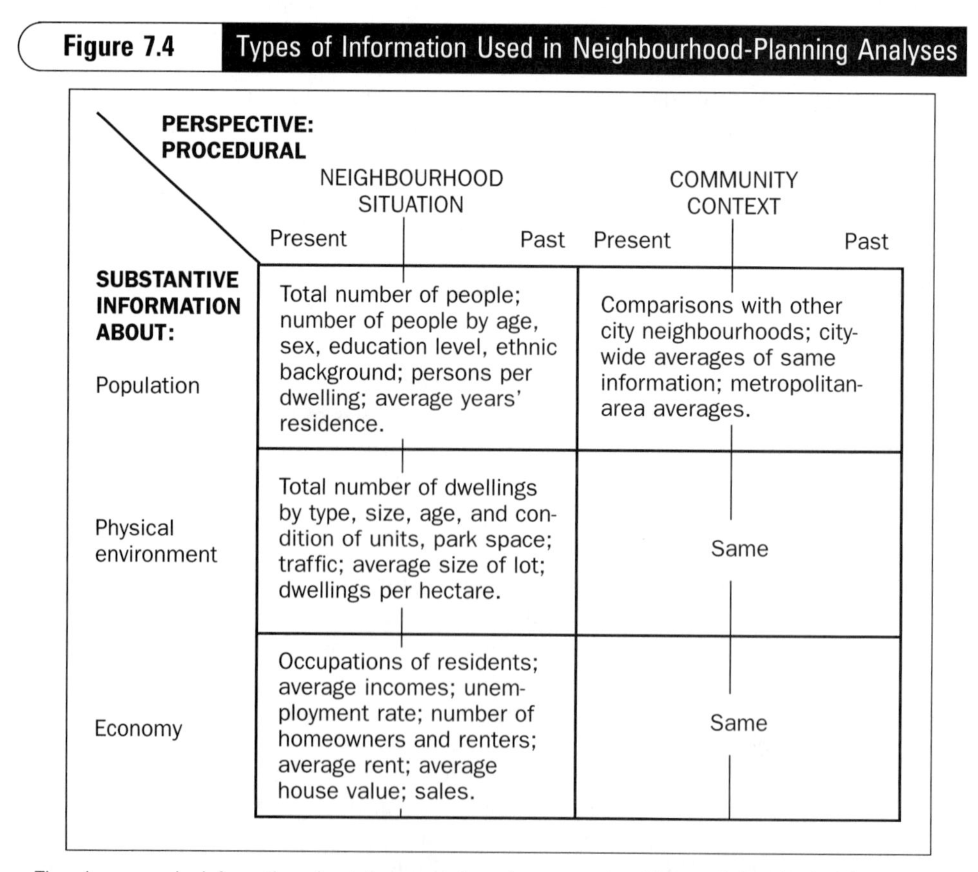

PERSPECTIVE: PROCEDURAL / SUBSTANTIVE INFORMATION ABOUT:	NEIGHBOURHOOD SITUATION (Present / Past)	COMMUNITY CONTEXT (Present / Past)
Population	Total number of people; number of people by age, sex, education level, ethnic background; persons per dwelling; average years' residence.	Comparisons with other city neighbourhoods; city-wide averages of same information; metropolitan-area averages.
Physical environment	Total number of dwellings by type, size, age, and condition of units, park space; traffic; average size of lot; dwellings per hectare.	Same
Economy	Occupations of residents; average incomes; unemployment rate; number of homeowners and renters; average rent; average house value; sales.	Same

The planner seeks information about the population, the economic milieu, and the physical features of a neighbourhood. Data showing present conditions and past tendencies allow forecasts to be made. Information about other neighbourhoods is used to compare present conditions and future prospects.

Procedural dimensions. These reflect how a planner thinks about a planning problem and its analysis. Specifically, integral to the planner's view is the twofold notion of the problem *and* its larger context, that any specific problem area is part of the surrounding community and may be affected by it, or vice versa. Even if there is no direct connection between the two, knowledge of comparable conditions in other parts of the community can aid in understanding the local situation. The planner also will want to know whether current conditions in a problem situation are stable or changing. For example, the housing in a neighbourhood may change little, but, as families age, the use of such facilities as schools and parks may change. Data on past conditions are often as important as those that portray the present in diagnosing whether conditions are improving or worsening.

Data Sources

Data for the inventory and analysis of the natural and built environment are usually obtained in the form of satellite images, aerial photography, or maps from other public agencies. Regional environmental planners must be skilled in interpreting data from all three sources, while urban planners may rely more upon local maps and aerial photographs. A municipal or regional planning agency in a fast-growing area might arrange to have aerial photography of their jurisdiction updated every two years, and some municipalities put their spatial information up on the World Wide Web for public access.[24]

A common source of data for many other diagnostic studies is the Census of Canada, especially the sections dealing with population, housing, and labour force. Every 10 years, the census provides a complete set of data for each incorporated city and town; every 5 years, it provides a limited set of population data. The value of the census lies in its accuracy, complete coverage, and consistency. This provides a community planner with an objective baseline of information at any time, and also with the data to trace trends in information over fairly long periods. It also allows easy comparison among communities throughout the country. One can obtain data for small areas (census tracts) within a city as well.

As useful as the census is to planners, it has relatively large time spans between publications of its findings. This can be a problem when a community is experiencing high population movements in and out or changes in its economy, which may happen between censuses. Moreover, since the census is geared toward providing nationally comparable data, it cannot cover distinctive local conditions. The planner, therefore, may need to seek other secondary data sources or to develop primary data sources within and for the community. Another source of data is the property assessment data that is collected in most parts of Canada by provincial agencies. The latter data are highly detailed, based as they are on individual properties. Also helpful to the planner are data gathered by other local agencies—for example, planning departments, public utility departments, and school boards.

However, there is hardly any planning problem for which diagnoses or other analyses can be completed on the basis of secondary sources. Each community has unique features, as does each neighbourhood and district. To get a complete picture of a place, it is necessary to gather first-hand information and even to gain first-hand experience. The planner obtains this through field visits, observing the continuous activities and functioning of an area, perhaps at different times of the day or week or year, and speaking informally with its users. Some planners may claim that this is not objective data; however, there is simply no substitute for the personal understanding, or local knowledge, obtained "on the spot."

The formal, objective approach to such knowledge employs questionnaires and other formal surveys that record observations about an area. These may take the form of personal interviews with users of an area or facility, such as householders who may be affected by a school closing or shoppers who may have to deal with reduced parking in a business area. Or they may take the form of inventories of, for example, an area's traffic, building conditions, or lot coverage. Survey research can yield high-quality information when the research instruments (e.g., questionnaires) are constructed thoughtfully and with scientific objectivity. The design of the research instrument is, therefore, important in enabling the planner to provide more penetrating observations, as through the use of statistical analyses. There are established methods for survey research that should be used in this regard.[25]

It has always been evident to planners that much useful data exist in maps or files in other departments within the same city, in adjacent cities, and in special-purpose agencies. Where these can be consolidated and combined, more broad-based planning-information systems can make diagnostic studies more complete. Metropolitan and regional planning agencies often play this role. There are still difficulties in combining data from some sources, owing to differences in format and dates of collection. Since most planning analyses are for small portions of the entire community, these difficulties are not insuperable and good case-by-case databases for analysis are also possible. These should, of course be conducted in a consistent manner so that they can become part of a larger data base.

The advent of **Geographic Information Systems** (GIS) and their continued refinement has enabled planners to make more accurate and complete analyses and maps.[26] Some of the typical planning applications include environmental impact assessment, land-use inventories, recording and enforcement of zoning maps and regulations, and mapping traffic congestion. The novelty and ease of use of GIS technology, however, carries with it problems of misuse and misinterpretation of data inputs and outputs, which planners must keep in mind.[27]

Predictive Studies

In the second phase of the technical-planning process, the planner moves from description to prediction; that is, the planner becomes concerned both with predicting what is possible within the context of the problems and with predicting (designing) likely solutions for problems. The first of these facets is largely grounded in such analytic approaches as the prediction of future population levels or economic growth. The second, the design of solutions, goes beyond normal analytic skills and calls for the planner to synthesize the various elements of the problem situation and, possibly, combine them with new elements. One often hears planning described as both an art and a science. In this phase, that is quite evident, for the planner needs to link analysis and design. Because of the subjectivity of the latter, the emphasis here is on objective analytical methods.

Analytical Approaches

In this phase, the planner draws upon methods of analysis capable of distinguishing the factors involved in the planning problem. On the substantive side, the population, economy, and land uses need to be taken into account to predict their outcome should conditions change in the future. The complexity associated with any one of the major factors is obvious; even more complex are the relationships among them.

The methods developed to cope with this complexity draw upon the concept of systems analysis. In this view, the various parts of a problem are perceived to be linked, forming a functioning whole, or **system.** A system's overall character affects the way in which the separate parts work, so that the parts cannot be adequately understood without understanding the whole. Thus, for example, a neighbourhood may be thought of as a system (more properly, a sub-system of the city system) in which the age composition of the population is linked to the need for housing, the use of schools and parks, and traffic considerations. A particular neighbourhood will also have a distinctive character, in terms of tradition, location, status of residents, and so on, which affects the functioning of the various elements within it. Other parts of the city may be similarly perceived. While this approach is persuasive, it must be acknowledged that analytical methods cannot offer us a complete view of the complexity of a city system. However, partial views are available through analytical **models** (the analyst's way of replicating a system) of the economy, population growth, and the housing market, to name the main ones. The planner's tendency to think in terms of interrelationships and interdependencies allows the gaps between partial models to be identified, even if not fully understood.

Population Forecasting. Because of the importance of population growth and change in a community, there is hardly a planning study anywhere that does not begin with this factor. Of a variety of methods for forecasting population, the simplest extrapolates data on past population levels into the future using graphical and/or mathematical means. This often provides the planner with a satisfactory overview of population tendencies. However, the planner may need to understand the role of births, deaths, migration, birth rates, age, income, or ethnicity in regard to a population change to plan for a particular client group, such as the elderly or visible minorities, both increasingly common concerns. For these needs, the planner will use more extensive data and more elaborate methods.[28] Among the most useful of these tools is the **cohort-survival method,** which allows each age cohort for either sex to be forecast separately.

Economic Forecasting. The future of the local economy is important because of its connection with the need for housing, public utilities, and transportation. Economic studies at the city or town level attempt to determine employment opportunities, rather than volume of business. Through the prediction of employment, the planner has a way of linking the economy of a place to the size and needs of the population. Probably the most common type of question raised in regard to a local economy would be one such as: "If we were to get that new factory with its 500 jobs, what effect would it have on the local economy?" The answer to such a question is usually approached through a **community economic base model,** or one of its variations. The foundation of these methods is to estimate the impact of the additional income brought to the community by firms that export their products to other communities and regions (which is the case with most factories). The community economic base model uses the notion of an economic multiplier to estimate the portion of the exporting factory's income that will accrue to the community and thus generate other jobs in the local economy.[29]

Of course, it is not just factories that may be considered "export industries." All forms of activities catering to tourists are almost wholly export-oriented, because they serve people from other communities. The same is true for firms in the business districts and shopping centres of cities and towns: a substantial part of their business volume is due to purchases made by people from the countryside and other communities. There are other more elaborate economic models that may be used in understanding and predicting economic impacts on communities (especially for larger cities and metropolitan areas), such as **input–output** and **industrial-complex** analyses. Indeed, the tools available for economic analysis are the best developed among those the planner has available. A general note of caution is appropriate at this point, for one frequently hears dramatic multiplier effects claimed by community boosters, such as "every job in the tourism industry generates five other jobs in the community." It is extremely rare for new export industries to generate more than one additional job each in the community, and the ratio is often much less.

Land-Use Forecasting. Methods for predicting **land-use changes** are less well developed. This is especially so at the detailed level of individual properties because here the reasons for land-use decisions and the responses to the decisions of others are affected by the personal views of those involved. Analytical approaches to land use thus favour a broader view of the community, such as predicting the amount of residential land that would be needed in light of population growth or economic expansion or the opening up of a new highway. There are also useful models in existence for predicting the impact of a new shopping centre on established businesses.[30] But, all in all, this area of analysis remains underdeveloped.

Before leaving this discussion of predictive studies, we should note again that the planner's view of a community is of a *linked set of factors*. Therefore, by knowing about the changes in one factor, the planner can estimate the changes in another. For example, when the future population of a community is known, the proportion of the population that will be in the labour force or the number of households that compose that population can be deduced. In this way, sometimes known as a **step-down analysis,** estimates can be made of future job levels in the economy and land-use needs for business and industry. Or, stepping down from population to households, estimates can be made of residential land use. Since these relations tend to be transitive, it is also possible to employ a **step-up analysis** to estimate population from a knowledge of the number of jobs expected in the community (see Figure 7.5). The results of such analyses are, of course, limited by the assumptions one makes about the connections between the factors.

Design Approaches

Analytical models and statistical studies will take the planner only so far in predicting possible alternatives for the future arrangement of a community's built environment. The numerical analyses do not provide a way of combining the various elements into a physical, visual community of houses, stores, streets, parks, and so on. That task falls into the realm of land-use design and it involves aesthetics. As Canadian planner Hok Lin Leung has stated, "Modern city planning is rather blind to aesthetics."[31] The concern that citizens often evince about new high-rise buildings or the removal of old trees is as much about aesthetics as anything else in planning. It often seems that today's planners are oblivious to visual and three-dimensional elements of city form and texture, with rare exceptions.[32]

The Design Task

It devolves on the planner to draw upon knowledge gained from experience with urban development and growth to conceive and present possible future physical designs for the community or for some part of it. The design task in planning a community's built environment is one of combining the right physical elements into a unified whole. And what is "right" for the design is informed by (a) the goals and objectives that have been agreed upon, and (b) the results of analyses that have been undertaken. It is a process of *synthesis* that is very important to the successful acceptance of the plan. European planner Andreas Faludi notes that for a plan to be viable it must incorporate a graphic organizing principle that conveys the way in which the planning area is to be arranged.[33] A striking Canadian example is the "greenbelt" in Gréber's 1950 plan for the Ottawa region; its recent counterpart for the Toronto region may have the same effect.

Further, the design task demands capacities in the planner to "see problems in new ways, to break out of conventional boundaries of thought, to use analogy and metaphor."[34] Proponents of the New Urbanism are, for example, injecting design metaphors

Figure 7.5 Chains of Reasoning Employed in Predictive Studies

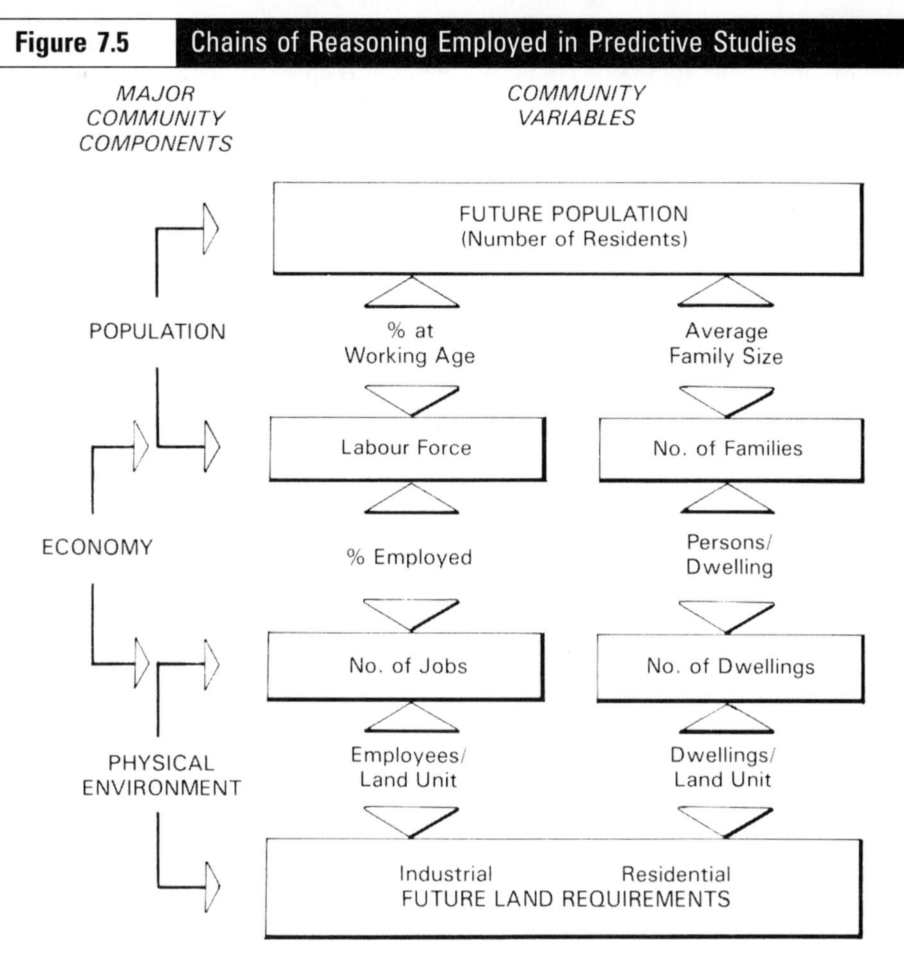

Planners use the functional relationships between people, jobs, and housing to make predictions about the future population, labour force, number of dwellings, and the land required for various community uses.

drawn from small towns into city designs.[35] Their aim is to improve community living situations through design solutions that recast the patterns characterized by typical suburban development (or "mindless sprawl," as it is often called) with "mixed-use, walkable, transit-served districts and neighbourhoods."[36]

Although this discussion focuses on the spatial design of the environment, it is important to mention that the design of policies, programs, and regulations for managing community-building involves the planner in much the same process of synthesis. A program for future housing needs, for example, also involves imagining future possibilities.

Design Development

Design is an elusive process to describe, whether it be for planning, architecture, engineering, or the arts. Prominent U.S. planner Kevin Lynch provides some helpful touchstones for our purposes. As Lynch notes, most environmental designs are adaptations of previously used solutions.[37] Many urban forms are now applied customarily in community design: for example, the neighbourhood concept for residential areas, or the layouts for local shopping plazas. Most are mere imitations of these stereotypes, but more imaginative solutions may also emerge by improving upon earlier designs. The converse of this adaptive approach is to seek optimum, or "ideal," design solutions that best satisfy the objectives of the community or the project. The design of entire new towns, such as

Kitimat, B.C., or the design of innovative housing, such as the Habitat '67 project in Montreal, are usually approached in this spirit. Another approach to design may be generated out of a SWOC analysis of strengths, weaknesses, opportunities, and constraints for each location. This approach focuses on the present reality and is appropriate where the planned changes will "intrude" on an existing situation, as, for example, in the location of new arterial roads or expressways in already built-up areas. There are a number of variations on these approaches, and the planner will likely gravitate to one or another for the planning problem at hand on the basis of both professional experience and personal preference, for all designing entails a high degree of personal involvement.

In the design phase of planning, there remains the question of how to introduce possible designs into the planning process. This question arises mainly for two reasons: first, there are always numerous possible design solutions to planning problems; and second, choices among designs will be made by both planning boards and citizens. The planner may employ two general strategies: (1) either generate a limited set of alternatives that reflect the likely range of possibilities; or (2) develop one reasonable possibility and refine it on the basis of verbal deliberations. The first of these strategies may be time-consuming and expensive. Sometimes the planner may have one preferred design; this may save money in the design phase if the selection is good, or incur waste if it is not. In actual design, some compromise is often made between these two strategies by first identifying the range of basic alternatives and then choosing one for full development.

Collaborative Design

The community's planners play a key role in structuring the design process for other plan-makers, either through providing alternative designs for the community environment or through encouraging collaborative processes involving a wide array of community members in actually making designs. The latter mode has been happening more and more in a planning-design process known as community **visioning** in large cities and small.[38] The notion is a simple one of developing a description of what the community should "look like" in the future through citizen-produced maps, drawings, and text. One of the most ambitious, and successful, uses of visioning is that employed by planners in the City of Vancouver, which started in 1997 to link neighbourhood planning to its citywide CityPlan.[39] The Community Vision process operates in the following way for each of the city's neighbourhoods:

> The Community Vision process is a 15-month, four-step process that includes extensive outreach; the identification of community needs, ideas, issues, and opportunities on all the CityPlan topics; the creation of Vision options and directions; broad community voting on preferred options and directions; and Council endorsation of the final Vision. Subsequently, the community works on setting priorities for Vision implementation.
>
> Each step provides a variety of ways for people in the community to be involved in creating, reviewing, and deciding on their Vision—including meetings, workshops and discussion groups, community events and festivals, brochures and surveys. The process also provides for an on-going Community Liaison Group [CLG] made up of people from the community.
>
> The Community Vision Program asks communities to look 30 years ahead. Communities will be given information to help them consider both the short-term and the 30-year future. But it is not necessary to plan now in detail for a 30-year end point for each CityPlan topic. The Vision is a framework, it will need to be revisited and revised over time as the communities and their needs, conditions, and ideas change.

This process, it has been found, not only produces design ideas for the city's plan-makers but also aids in building consensus regarding the plan's outcome.

A variation on visioning is the design charrette that brings together people with diverse expertise to create design solutions for planning projects.[40] The idea is also "to give visual form to ideas and policies." However, the charrettes are designed to be used by smaller groups of people rather than a whole community (often a variety of specialists) and are to be concluded in a short period of time, say from three to four days. Further, they tend to work best when focused on a given site, often one that is contentious, to develop options that would integrate with broader community planning. An extension of the community-design charrette to an entire urban region, using a 100-year planning horizon, was conducted for the Greater Vancouver Regional District with considerable success in an international competition.[41]

Evaluation Studies

Plan evaluation is a crucial step in plan-making. Not only is it an opportunity to review a plan before its implementation, it is also the phase where an appreciation of the "whole," of the comprehensiveness of the plan, is best obtained. Plan evaluation may occur at three different stages in the process of plan-making.[42] The first occurs before the plan is formulated to assess it for consistency. The second occurs after the plan (and alternative plans) have been formulated to test for unintended consequences. And the third stage occurs after the plan has been adopted, the so-called post hoc stage. The planner assists the plan-makers by providing means by which the relative merits of plans can be evaluated. When a community plans, it anticipates and prepares for change. Thus, the plan that is chosen will have consequences, and the planner's evaluative studies aim to determine the nature of those consequences beforehand. Much needs to be done both to clarify the need for plan evaluation and to develop the methods for accomplishing it.[43] Nonetheless, some tools are available to point out the direction that plan-makers need to pursue.

Plan Assessment

Community plan-making needs to be considered as a process of learning about future possibilities. For this there are few certainties; it requires a process of embracing errors when they are revealed. Thus, opportunities for evaluation should be sought from the beginning to the end of the process. In the early stages, the various components of the plan need to be assessed for consistency of goals and objectives, background data and projections, and so forth. As shown in Figure 7.1 (page 176), the findings of such preliminary evaluation may evoke the need for a feedback loop to obtain, say, a better diagnosis of problems. Two important areas of preliminary evaluation are (1) achieving a clearer expression of goals, and (2) assessing the impact on the environment.

Identifying Goals. One of the main starting points in plan-making is to identify the goals and objectives community members wish to achieve with the plan. A community's goals will tend to reflect a number of different facets for which change and/or improvement are sought (e.g., the economy, rate of population growth, appearance of the community, the natural environment). The example of York Region in Ontario cited earlier showed eight goals. Since goals are the foundation of a community's plan, it is important that they each contribute to its outcome. It is, therefore, valuable to assess whether a plan's goals are consistent.

Not surprisingly, any community's (or region's) goals are related, as they are each a reflection of the same community. The issue for the planner, almost at the outset of plan-making, is whether all of the goals can be achieved and to what degree. Behind this issue is the realization that each goal derives from the same social and economic milieu and, if achieved, will draw from the same pool of community resources. In essence, the plan's goals are implicitly in competition with one another for, among other things, capital and land resources, and attention. The desire for a new airport may vie with the desire to conserve land for agriculture, for instance. It is also true that some goals may complement other goals, as when an improved rapid transit system aids both economic development and residents' needs.

A method is available to identify the effect of achieving each goal on each of the other goals. It is called **cross-impact analysis** and involves the use of a square matrix with rows and columns for all goals.[44] This method, used by the Greater Vancouver Regional District in its 1976 *Livable Regional Plan* activities, employed a simple rating approach in which each cell in the matrix is filled in by asking: does achievement of the specified goal enhance (+), reduce (−), or have no effect (0) on each other goal?[45] In this way, plan-makers can check the efficacy of goals prior to planning and thereby improve both their research and implementation strategies.

Assessing Environmental Impact. Impact studies have become familiar, especially regarding the effect of new development on the natural environment. Environmental impact-assessment procedures are now in widespread use, although they are not required as a matter of course in community planning in Canada as they are in the United States. Such provincial government agencies as departments of highways undertake them; many large projects, such as energy projects, require them, both in southern Canada and in the Far North. The scope of environmental concerns in local planning is increasing. For example, the British Columbia government now has legislation requiring local plans to take into account impacts on fish stocks and fish-spawning streams, and the Ontario government has adopted an Environmental Bill of Rights.

The **environmental-impact statement** (as it is often called) attempts to forecast the consequences of a project for its surrounding natural environment, including plant life, wildlife, soils, water, and air conditions. The natural interrelations and interdependencies of phenomena in the environment require that impact assessment capture the ecological interactions. The actual complexity of natural systems as understood by biologists, botanists, and zoologists has, however, defied the development of precise means of environmental-impact assessment. The most common approach uses a checklist of potential effects in order to ensure that impacts are not overlooked and that those identified may be pursued in depth. More elaborate are matrix techniques that identify the interactions that occur when a project disturbs one part of the environment. The magnitude of each effect may be included in the matrix, either in absolute terms or on the basis of a rating of the expected impact.[46]

The concerns of environmentalists have also led to efforts to extend the same ecological concepts into other realms of planning practice. **Social-impact assessment** methods help determine the effects on people's lives, community functioning, and social and cultural traditions of possible changed conditions resulting from large projects. Such assessments are required prior to the initiation of large projects in the Far North; as well, they are becoming a more common part of the planning scene in other parts of Canada.[47] Indeed, the Environmental Assessment Act in Ontario requires impacts on social, cultural, and historical, as well as economic, "environments" to be assessed, and the federal government has a specific process for projects coming under federal jurisdiction to study and report on their environmental impacts that is administered by the Federal Environmental Review Office.

Plan Testing

As plan-making progresses, various alternatives emerge often with different consequences for resource use and support for basic goals. An important evaluation step is to understand the consequences of each alternative plan that may be put forward as a possible final plan. This phase has been called "pre-adoption evaluation."[48] Two evaluation methods are available for use in this phase: one for testing economic impacts, and the other for achieving goals. It needs to be noted that the responsibility for this testing falls on the community's planners.

Testing for Costs and Benefits. A form of economic impact analysis with strong roots in planning is **cost–benefit analysis.** Developed originally in the 1930s for use in river-basin planning, it is now widely used in other planning endeavours. The term "cost–benefit" has become something of a generic phrase in planning to show an awareness that proposed projects carry with them costs as well as benefits. Further, there is recognition that often neither all the costs nor all the benefits of a project can be quantified in dollar terms. In water-resource projects, for example, it was recognized that the costs of displacing people from their homes included many "intangible" costs, just as there might also be intangible benefits from the improvements in recreation. It is not difficult to imagine analogous costs and benefits from expressway-building, urban renewal, and airport development projects.

It is also not difficult to imagine that as the multitude of factors that the planner tries to take into a cost–benefit reckoning expands to include intangible (but still very real and pertinent) items, the more difficult the summation of costs and benefits becomes. Several variants of the approach have arisen, for example, limiting the reckoning to costs and revenues that can be rendered in dollar terms or broadening it to include an evaluation of effectiveness in achieving specified goals. Planning-programming-budgeting systems (PPBS) are a form of the latter, but the most pertinent for planners is the Planning Balance Sheet developed by Lichfield.[49] It mixes both "hard" and "soft" data about the plan in an effort to include measurements of all its effects. Weights are not assigned to the various impacts; rather, that type of judgment is left where it properly belongs—with the plan-makers.

Testing for Achievement of Goals. Economic efficiency is the essential criterion of cost–benefit approaches, but planners have never regarded this as the only or main measure of a plan's worth. The goals and objectives decided on by the community are considered the paramount criteria for evaluation by many planners. In other words, if a plan is to help achieve a community's goals, then the worth of the plan must be looked at in terms of whether it represents progress toward those goals. In response, planners have developed "goals-achievement" evaluation and other similar methods. These methods allow plan-makers to compare complete plans without the necessity of disaggregating them, as in the Planning Balance Sheet described above.

The best known of these techniques, the **goals-achievement matrix,** was framed by planner Morris Hill[50] and emulated by many others. The goals-achievement matrix is a way of summarizing overall performance with respect to each goal of each alternative plan. In its simplest form, the columns of the matrix represent the alternative plans or policies being considered; the rows represent the goals and objectives the community has set. Since not all goals are usually considered of equal importance, this method allows each to be weighted to reflect its importance to the community. Further, if some goals have several facets, as is often the case, then each row may be elaborated into several to capture the range of aims. The most thorough approach is to define quantitative measures, or scores, that reflect degrees of success in achieving each goal that could, in turn, be "summed." These may turn out to be only ranked scores, but this still provides a strong base of comparison.[51]

Post Hoc Evaluation

"Does planning work?" is a cogent question for planners and other plan-makers and one not easily answered.[52] It makes sense that, after adopting and implementing plans, plan-makers want to know if their efforts were successful in bringing about the desired changes. However, such post hoc evaluation poses a number of challenges, including the following that are key:

- The kinds of planning outcomes that are to be evaluated;
- The appropriate period of time to measure outcomes; and
- The methods by which to measure them.[53]

Planning outcomes may be looked at in terms of either the links between plans and actual development, called **conformance-based evaluation**, or whether the plan is used in decision-making processes, which is known as **performance-based evaluation.**[54] Evaluating the conformity of actual development to planned development is one of the most common approaches, especially in the area of environmental concerns. For example, evaluations have been made by assessing whether development permits for subdivisions achieved desired the protection of wetlands.[55] Ontario's regional municipalities tend to use quantitative measures such as numbers of housing units built, growth of population, level of water quality, and numbers of jobs created.[56] Still others combine quantitative and qualitative measures in their evaluations.

Timing, or frequency, of evaluations also vary, and the choice is likely to depend upon the amount of time required for results to show; environmental improvement may take longer to become evident than improvements in transportation, for example. The Official Plan of Ontario's Regional Municipality of York, which was referred to above, calls for its planners to assess success every three years of the next fifteen using a set of performance-type indicators (see Figure 7.6). A U.S. project to maintain riparian health utilizes citizens who are given training to monitor its watersheds every year, which proves both cost-effective and succeeds in raising public awareness.[57] The Canada Lands Company has adopted "triple bottom line" reporting to address the environmental, social, and economic objectives included in sustainable development.[58] These few examples indicate the increasing interest in monitoring outcomes of plans and evaluating them. More and more planning agencies are coming to realize the importance of not only displaying success to their councils and citizens but also of

Figure 7.6 Planning Success Indicators for York Region, Ontario

INDICATORS	2021 TARGET
Regional forest cover	25%
Waste diversion	75%
People both living and working in the Region	75%
Housing mix (% of housing stock that is single detached)	62%
Households on full municipal services (water/sewers)	80%
Urban transit modal split (peak period)	33%
Urban households within 500 metres (5–10 minute walk) of public transit	90%
Households within 800 metres (1/2 mile) of an elementary school	80%
Households within 1600 metres (1 mile) of a major green space	80%

Indicators of desired conditions in a community's future are used as a means of determining whether a plan is succeeding in meeting its goals.

Source: York Region Official Plan Office Consolidation, Nov. 30, 2005, reprinted with permission.

being more aware of planning processes, not least when shortfalls are revealed. However, much remains to be done to strengthen this part of plan-making, such as in better linking of goals and outcomes, choosing indicators, and giving the task due priority in the planning agency.[59]

The Planning Process and Community Learning

Not infrequently, situations arise in community plan-making (and implementation) that illuminate the differences between, for example, plan-makers and neighbourhood residents as to how each group views the plan's aims and outcomes. To better understand such differences, it is appropriate to characterize the community-planning process as a **learning process,** both for plan-makers and for neighbourhood residents; that is, one in which plan-makers and citizens learn from each other. In this way it can be seen that the difficulties over plan proposals arise out of differences in how each learns—not only differences in the amount of time each has to spend learning, but differences in perception of the substance of a proposal, in prevailing values, and in the source of knowledge. The professional planners, for example, have technical expertise grounding their views, while citizens bring their personal knowledge of an area into their viewpoint. Further, not everyone, including official plan-makers, enters the process either at the same time, or with the same background and the same propensity to learn what is needed to make a decision.

This realization has led to a call that a "more active relationship should be developed between planners and their clients."[60] Increased interaction, particularly through verbal communication, is aimed at mutual learning for both planners and clients. Mutual learning generally leads both to broader participation and to a less bureaucratic style of planning. It helps all participants to appreciate better the respective positions of others in the plan-making process. This includes the interactions of planners with developers, as well as with private citizens. Hok Lin Leung describes an excellent example of mutual learning in practice in a Canadian suburb. Scarborough, Ontario, undertook negotiations with developers over its planning criteria for office and commercial development. Municipal planners learned that they could "soften" their criteria for such stipulations as Floor Area Ratios and density and still attain their planning objectives. The developers learned about the "logic of land use designation . . . [and] came to appreciate the need to go beyond market demand and jobs and argue their case on planning grounds as well."[61] Not least, mutual learning in this case helped to streamline the negotiation process.

Community-planning learning involves plan-makers and professional planners in a process that accepts that it is natural for citizens and developers to want to be involved in planning their community. Further, it requires that citizens and others outside the formal plan-making process also be accorded the status of "clients" of the plan-makers, along with those who have plan-approval powers. As John Friedmann has noted, much of a community's planning is a "sequence of interpersonal relations"—that is, of planners and clients "communicating valid meanings to each other," or at least trying to. He saw the planning process as a set of "transactions" that lead to mutual learning.[62] Several instances of how this can occur in actual planning processes have been cited above, including design charrettes, citizen monitoring of watersheds, and Vancouver's Visioning Process. These are among the recent efforts where the emphasis in the planning process is on interaction and communication, on "collaborative planning," (or sometimes "communicative planning") as it is now called.[63] These emergent approaches to the process of community plan-making are also part of the discussion of succeeding chapters, especially Chapters 11 and 12.

Planning Issue 7.1 **"Park" Tag Irks Landowners**

The Now Newspaper, Surrey
March 9, 2005

"Park" Tag Irks Landowners
Ted Colley

A Surrey municipal councillor says the city goofed when it identified specific parcels of land in Grandview Heights as potential park sites.

Surrey is working on a general land use plan for Grandview Heights and as part of that process, six options were identified by planners as possible sites for an athletic park that won't be built for at least 10 years.

Because the options identify specific properties, Councillor Bob Bose believes the city has harmed the landowners' ability to sell to anyone but Surrey.

"Who are they going to be able to sell to if their land is identified as future parks?" Bose said yesterday.

"People may want to have several options in developing their land. When you start identifying certain properties, their options are reduced to one."

When the six options were presented to area residents at two recent public meetings, the plan ran into serious opposition from landowners who feared their properties would lose value if designated as park. Faced with that opposition, city staff decided to abandon its parkland identification plan.

Instead, staff recommended waiting until the various neighbourhood concept plans are drawn up for the various sections of Grandview Heights before designating park sites.

Council voted in favour of the new approach, though Councillor Marvin Hunt said Monday residents needn't worry about land values if their property is designated for a park. "We pay fair market value for land," Hunt said.

Source: Ted Colley, "'Park' Tag Irks Landowners," *The Now Newspaper, Surrey*, March 9, 2005. Reprinted with permission.

Reflections

Describing the community plan-making process only in words, as this chapter attempts to do, has a number of limitations that are due to the nature of the process. The process is promoted as rational; that is, it is promoted as a sensible, logical way to avoid chaotic decision making, and there are a number of understandable, acceptable steps toward this end. The process is also complex, as it involves many decision-makers and requires an understanding of elaborate human and physical systems. Implicitly, planning processes effect change, or intervene, among community interests on the basis of shared values and attitudes. However, one of the most confounding situations for many plan-makers, including professional planners, occurs when they put forward a planning proposal on which they have spent considerable time deliberating (including having media coverage) only to have, for example, a neighbourhood group object just before it is to be adopted. This not-infrequent occurrence results when the neighbourhood perceives the outcome of the planning intervention, but does not accept the rational basis on which plan-makers justify the scheme. Yet both perceptions are valid, which testifies to the complexity of the process of effecting change in the community environment. It is thus worth considering the following question regarding the plan-making process described in this chapter:

- ***How can we best balance the need for objective planning processes with the need to be inclusive of participants with diverse perspectives on achieving planning outcomes?***

Endnotes

1. Martin Meyerson and Edward C. Banfield, *Politics, Planning and the Public Interest* (Glencoe IL.: Free Press, 1955), esp. 312–322.
2. Ira M. Robinson, ed., *Decision-Making in Urban Planning* (Beverly Hills: Sage Publications, 1972), 27–28.
3. Patrick Geddes, *Cities in Evolution,* 3rd ed. (London:Ernest Benn, 1968), 286.
4. This is still an issue of debate in community planning; for example, see Jill Grant, "Rethinking the Public Interest as a Planning Concept," *Plan Canada* 45:2 (2005), 48–50.
5. As discussed in Herbert Simon, *Administrative Behavior* (New York: Free Press, 1965).
6. M.A. Qadeer, "The Nature of Urban Land," *The American Journal of Economics and Sociology* 40 (April 1981), 165–182. See also John R. Hitchcock, "The Management of Urban Canada" *Plan Canada* 25:4 (December 1985), 129–136, for an excellent discussion of supply and demand for housing.
7. Penelope Gurstein, "Gender Sensitive Community Planning: A Case Study of the Planning Ourselves In Project," *Canadian Journal of Urban Research* 5:2 (December 1996), 199–219.
8. One of the earliest and best discussions of planning goals is Robert C. Young, "Goals and Goal-Setting," *Journal of the American Institute of Planners* 32 (March 1966), 76–85; also helpful in this regard is Robinson, *Decision-Making,* 33–41.
9. This idea is expressed well in Ian Bracken, *Urban Planning Methods* (London: Methuen, 1981), 11–35.
10. John Friedmann, "Planning as a Vocation," *Plan Canada* 6 (April 1966), 99–124.
11. Bracken, *Urban Planning Methods,* 30.
12. Frank Beal and Elizabeth Hollander, "City Development Plans," in Frank S. So et al., eds., *The Practice of Local Government Planning* (Washington: International City Management Association, 1979), 153–182.
13. Regional Municipality of York, *Vision 2026,* November 2002, 13ff. Reprinted with permission.
14. William L.C. Wheaton and Margaret F. Wheaton, "Identifying the Public Interest: Values and Goals," in Robinson, *Decision-Making,* 49–59.
15. Cf. F. Stuart Chapin, Jr., *Urban Land Use Planning,* 2nd ed. (Urbana, IL: The University of Illinois Press, 1965), 29ff.
16. Ibid., 67.
17. A helpful review of the issues surrounding the rational–comprehensive model is found in Bracken, *Urban Planning Methods,* 11–36.
18. Judith E. Innes, "Planning through Consensus Building," *Journal of the American Planning Association* 62:4 (Autumn 1996), 460–472.
19. W.G. Roeseler, *Successful American Urban Plans* (Lexington, MA: D.C. Heath, 1982), xvii.
20. Useful books on planning analysis include Edward Kaiser, David Godschalk, and F. Stuart Chapin, Jr., *Urban Land Use Planning* (Urbana, IL.: The University of Illinois Press, 1995); and Richard E. Klosterman, *Community Analysis and Planning Techniques* (Savage, MD.: Rowman and Littlefield, 1990).
21. Ken Tamminga, "Restoring Biodiversity in the Urbanizing Region: Towards Pre-emptive Ecosystems Planning," *Plan Canada* 36:4 (July 1996), 10–15.
22. David L.A. Gordon and Ken Tamminga, "Large-scale Traditional Neighbourhood Development and Pre-emptive Ecosystems Planning: The Markham Experience, 1989–2001," *Journal of Urban Design* 7:3 (2002), 321–340.
23. City of Ottawa, *Ottawa 2020: Environmental Strategy* (2003); for background on this strategy see Chuck Hostovsky, David Miller, and Cathy Keddy, "The Natural Environment Systems Strategy: Protecting Ottawa-Carleton's Ecological Areas," *Plan Canada* 35 (November 1995), 26–29.
24. For example, see Vanmap, available online at www.city.vancouver.bc.ca/vanmap/setup/index.htm.
25. Nancy Nishikawa, "Survey Methods for Planners," in Hemalata Dandekar, ed., *The Planner's Use of Information* (Chicago: APA Planners Press, 2003), 49–78.
26. Britton Harris and Michael Batty, "Locational Models, Geographic Information and Planning Support Systems," *Journal of Planning Education and Research* 12:3 (1993), 184–198.
27. Robert E. Kent and Richard. E. Klosterman, "GIS and Mapping: Pitfalls for Planning," *Journal of the American Planning Association* 66:2 (Spring 2000), 189–198.
28. H. Craig Davis, *Demographic Projection Techniques for Regions and Smaller Areas* (Vancouver: UBC Press, 1995).
29. The foundation for this approach is found in Charles Tiebout, *The Community Economic Base* (New York: Committee for Economic Development, 1962); for a Canadian example, see Craig Davis, "Assessing the Impact of a Firm on a Small-Scale Regional Economy," *Plan Canada* 16 (1976), 171–176.
30. Robert W. McCabe, *Planning Applications of Retail Models* (Toronto: Ontario Ministry of Treasury, Economics and Intergovernmental Affairs, 1974).
31. Hok Lin Leung, *Land-Use Planning Made Plain,* 2nd ed. (Toronto: University of Toronto Press, 2003), 129.
32. Cf. Allan B. Jacobs, *Looking at Cities* (Cambridge, MA: Harvard University Press, 1985).
33. Andreas Faludi, "European Planning Doctrine: A Bridge Too Far?" *Journal of Planning Education and Research* 16:1 (1996), 41–50.
34. Michael Teitz, "Urban Planning Analysis: Methods and Models," *Journal of the American Institute of Planners* 43 (July 1977), 314–317.
35. James Howard Kunstler, *Home from Nowhere* (New York: Simon and Schuster, 1996).
36. Ellen Dunham-Jones, "New Urbanism as a Counter-Project to Post-Industrialism," *Places* 13:2 (Spring 2000), 26–31.
37. Kevin Lynch, *Site Planning,* 2nd ed. (Cambridge, MA: MIT Press, 1971), 270–288, provides all the references used in the accompanying discussion.
38. Amy Helling, "Collaborative Visioning: Proceed With Caution," *Journal of the American Planning Association* 64:3 (Summer 1998), 335–349; and Norman Walzer, ed., *Community Strategic Visioning Programs* (Westport, CT: Praeger, 1996).
39. The visioning process employed in Vancouver is described and available online at www.city.vancouver.bc.ca/commsvcs/planning/cityplan/Visions/. Reprinted with permission from the City of Vancouver Planning Department.
40. Canada Mortgage and Housing Corporation, *Sustainable Community Planning and Development: Design Charrette Planning Guide* (Ottawa: CMHC, 2002).
41. Canada Mortgage and Housing Corporation, *Tools for Planning Long-Term Urban Sustainability: The Cities^PLUS^ Design Charrettes* (Ottawa: CMHC, 2002).
42. William C. Baer, "General Plan Evaluation Criteria: An Approach to Making Better Plans," *Journal of the American Planning Association* 63:3 (Summer 1997), 329–344.
43. Cf. Emily Talen, "Do Plans Get Implemented? A Review of Evaluation in Planning," *Journal of Planning Literature* 10:3 (1997), 248–259.
44. This type of analysis derives from the work of the Tavistock Institute in London and other futurists in the U.S. Cf. Allen Hickling, *Aids to Strategic Choice* (Vancouver: University of British Columbia Centre for Continuing Education, 1976).
45. Cross-impact analysis has much more sophisticated mathematical antecedents that employ statistical probabilities in the cells. Cf. Norman C. Dalkey, "An Elementary Cross-Impact Model," in Murray Turoff and Harold A. Linstone, eds., *The Delphi Technique: Techniques and Applications* (Newark, NJ: New Jersey Institute of Technology, 2002), 317–329.
46. Robert W. Burchell et al., *Development Impact Assessment Handbook* (Washington: Urban Land Institute, 1994).
47. Peter Boothroyd, "Issues in Social Impact Assessment," *Plan Canada* 18 (June 1978), 118–134.

48. Kaiser, Godschalk, and Chapin, *Urban Land Use Planning.*

49. Nathaniel Lichfield, *Community Impact Evaluation* (London: UCL Press, 1996). An earlier book is Nathaniel Lichfield et al., *Evaluation in the Planning Process,* (Oxford: Pergamon Press, 1975); on cost-benefit analysis, a valuable text is E.J. Mishan, *Cost-Benefit Analysis,* 2nd ed. (London:G. Allen, 1971).

50. Morris Hill, "A Goals-Achievement Matrix for Evaluating Alternative Plans," *Journal of the American Institute of Planners,* 34 (1968), 19–29; this is also reproduced in Robinson, *Decision-Making.*

51. Cf. John C. Holmes, "An Ordinal Method of Evaluation," *Urban Studies* 9 (1972), 179–191.

52. Samuel D. Brody and Wesley E. Highfield, "Does Planning Work? Testing the Implementation of Local Environmental Planning in Florida," *Journal of the American Planning Association* 71:2 (Spring 2005), 159–175.

53. Mark Seasons, "Monitoring and Evaluation in Municipal Planning: Considering the Realities," *Journal of the American Planning Association* 69:4 (Autumn 2003), 430–440.

54. Lucie Laurian, Maxine Day, Philip Berke, Neil Ericksen, and Jan Crawford, "Evaluating Plan Implementation: A Conformance-Based Methodology," *Journal of the American Planning Association* 70:4 (Autumn 2004), 471–480.

55. Brody and Highfield, "Does Planning Work?"

56. Seasons, "Monitoring and Evaluation."

57. Bill Fleming and David Henkel, "Community-Based Ecological Monitoring: A Rapid Appraisal Technique, "*Journal of the American Planning Association* 67:4 (Autumn 2001), 456–465.

58. Online at www.clc.ca.

59. Seasons, "Monitoring and Evaluation."

60. Anne Westhues, "Toward a Positive Theory of Planning," *Plan Canada* 25:3 (September 1985), 97–103.

61. Hok Lin Leung, "Mutual Learning in Development Control," *Plan Canada* 27:2 (April 1987), 44–55.

62. John Friedmann, *Retracking America* (New York: Anchor/Doubleday, 1973), 171–193.

63. Patsy Healey, *Collaborative Planning: Shaping Places in Fragmented Society* (Vancouver: UBC Press, 1997); and Judith Innes, "Consensus Building and Complex Adaptive Systems: A Framework for Evaluating Collaborative Planning," *Journal of the American Planning Association* 65:4 (Autumn 1999), 412–423.

Internet Resources

Chapter-Relevant Sites

Planning Canadian Communities
www.planningcanadiancommunities.ca

Federation of Canadian Municipalities
www.fcm.ca

The Community Planning Website—helping people shape their cities, towns and villages in any part of the world
www.communityplanning.net

Service Nova Scotia Local Government Resource Handbook
www.gov.ns.ca/snsmr/muns/manuals/lgrh.asp

Internet Impact on Plan-Making
www.casa.ucl.ac.uk/planning/olp.htm

Centre for Advanced Spatial Analysis—University College London
www.casa.ucl.ac.uk

Geographic Information Systems Web Portal
www.gis.com

Resources for Urban Design
www.rudi.net

8

Chapter Eight

The Urban Community Plan: Its Characteristics and Role

The Master Plan is not an end but a directive. It cannot be definitive and inflexible, but must be constantly adapted to changing conditions. Such adaptations . . . are to be made only with the whole scheme in mind.

Jean-Claude LaHaye, 1961

The activity of community planning has many facets. It comprises several types of plans and a variety of processes, but it is not a random set of plans and processes. It has a coherence that is provided by the comprehensive community plan. Like the keystone in an archway, the community plan (master plan, general plan, municipal plan, official community plan) is the fundamental component in planning for communities large and small (see Figure 8.2, pages 218–19). It is the component that provides both the context and the raison d'être for detailed plans and regulations. It is the criterion for judging private development proposals and public investment decisions, and for making regulations with regard to land use. Indeed, the latter, which include zoning bylaws and capital budgets, are often referred to as "tools" for implementing the community plan, thus indicating their dependence on the overall plan.

Seen in this light, the community plan is more than a design for improvement of the built environment, more than a statement of what the community wants to become. The community plan plays a distinctive role in *governing* a community, whether it is an urban area, small town, or region. In this chapter, the focus is on the plan for urban communities, on its scope and content, on whom it serves and its relation to other planning tools. Since the scope and content of plans for larger (regions) and smaller (towns) entities have their own rationale, discussion of them is left to the following two chapters. Nonetheless, a question that is pertinent regardless of size of community is:

- *What is the role of the community-wide plan in the overall activity of community planning?*

The Scope of the Urban Community Plan

To describe the scope of a community plan is not a simple task, because it is not a simple device. A community plan spans several important dichotomies in the life and development of a community. First, there are *future* aims versus *immediate* needs; then, there is one concerned with the *ideal* view versus the *pragmatic* view; and, of course, there is the *citywide* view versus the *local* view. Implicit in each of these is the basic dichotomy that a community plan tries to address: the *planning* of land use versus the *control* of the physical development. This brings the community plan to a consideration of the basic values and objectives that inform the decisions of the governing body. In order to sharpen our perspective on the inherent complexity of the community plan, we should look first at the general concerns that a plan addresses.

Concerns of the Community Plan

The concerns addressed by a community plan were described in the preceding two chapters. The plan focuses on the built and natural environments, it frames a planning viewpoint, and ultimately it acts as a policy instrument. These concerns are reviewed briefly here.

1. Importance of the Built Environment

From the time of the first planning acts in Canada, the focus of community planning has clearly been the built environment. These acts very often carried a statement of general aims—namely, to allow a community "to plan and regulate the use and development of land for all building purposes." In other words, the focus has been on the "built" environment, both in terms of what already existed in the way of houses, stores, factories, parks, schools, institutions, roads, and so forth, and in terms of the prospects for built development on vacant land. Concern for the health and maintenance of the natural environment in which the community exists was added to this focus in the late 20th century.

Coinciding with the growth and refinement of professional planning practice, it became apparent that planning for the built environment required taking into account the natural environment and the social, economic, and financial aspects of the community. For instance, the type of people in the community, their level of affluence, and their values, as well as the kind of economic development, all affect the kind of built community that exists or will exist. More recently, the impact on energy use of different forms and patterns of urban development is being included in the focus of community planning. This expanded perspective asked of plan-makers does not mean that the community plan is a plan *for* the social, economic, natural, and energy dimensions of a community as well as for the built environment. Most of these other dimensions are outside the direct control of local councils, so it is vital to grasp the constancy and importance of the built environment in promoting community well-being. Two pragmatic issues are central here: first, almost everything that gets built has a long lifespan, and second, the public investments needed for support and service involve large capital outlays and must be financed over a long period of time. The primacy of the built environment in preparing a community plan has been reiterated in new planning legislation in several provinces. This sharp focus does not, however, preclude consideration of other dimensions of the community. Indeed, since the community's physical plan is usually the only overall plan it has, it can (and has come to) serve as the focal point for planning social, economic, and natural environmental factors in a coherent manner. This is especially evident in recent planning endeavours in larger urban areas such as Vancouver, Edmonton, and Ottawa.

2. Patterns and Processes of Land Development

A community's built environment develops on a land base that is mostly privately owned, some of which is built on and some of which is vacant. Furthermore, the built environment is ever-changing, either through the natural aging of buildings and facilities or in response to change through population and economic growth, new technology, and new values and goals. A good deal of the thrust of any community plan is to promote ways in which the land base should be developed in order to respond to anticipated change in the community. The solution is usually not self-evident because the land base is owned by a large number of different persons, groups, and organizations, each with their own vision of the future of their own land as well as their own aims for the community as a whole. A progression of typical questions that plan-makers must ask will help illustrate the range of concerns:

- Will change (e.g., growth) require additional land?
- Is new development best on the fringe of the community or located within already-developed sections?
- Where is vacant land available and for sale?
- Are there built-up areas that might be appropriate for redevelopment instead?
- Will redevelopment increase densities and place pressure on adjacent stable areas?
- What will be the cost to the community to provide public utilities and roads to vacant undeveloped land?
- Will growth in one section of the community lead to decline in other sections (as with a new shopping centre and the old downtown)?
- In sections of the community where stability and continuity of land use are sought, how can these sections be protected and also encouraged to renew themselves?
- Which kind of new development, internally or on the fringe, will have the least negative impact on the natural environment?
- Which should have precedence: maintenance of the health of local ecosystems or new physical development?

The above types of concerns mean that a community plan must be able to take into account the dynamics of land and building development. Further, it must provide an interface between the ideals and goals of the community and the need to manage land development activity. Land development is a vital and dynamic process often in need of control. While the community plan does not directly regulate land use, it does provide the criteria—the terms of reference—for regulatory efforts. The community plan thus acts to mediate such difficult questions as: Which land is to be developed? Where would development best be located? When? And whose land is to be developed?

3. Establishing Good Planning Principles

Experience in planning and building cities has shown that certain ways of structuring the built environment work best. Out of this have developed planning principles (such as we referred to in Chapter 6) that guide or motivate plan-makers. They tend to reflect, in turn, the basic values of planning: health, safety, welfare, efficiency, and amenity. Planning principles are thus like obligations for plan-makers. The community plan "ought to" promote their achievement because this will help ensure a "good" community environment.

Four important planning principles have come to direct planning behaviour in a consistent way in modern communities:

1. **Environmental integrity.** The land, water, and air of the community comprise a vital base upon which, and within which, land uses and development activities take place, and the impacts on this natural environment should be mitigated or minimized.
2. **Appropriate land-use assignment.** Each land use usually has distinctive locational and activity characteristics. Land uses should be spatially located where they will function most effectively and not conflict with other uses (e.g., separating heavy industry from residences; locating shops next to transit routes).
3. **Integration of activities.** Since the activities associated with various land uses need to be linked to one another, systems for moving people and goods should be provided that are convenient, economical, and safe (e.g., between homes and jobs; between industry and transportation terminals).
4. **Neighbourhood integrity.** Residential areas should be clearly defined, large enough to maintain their own character and values, fully protected from the hazards of major traffic routes, and have parks, schools, and stores within easy reach by walking.

It will be readily noticed that the fourth principle follows closely the "neighbourhood unit" concept propounded by Perry in the 1920s. Further, it parallels contemporary ideas about the neighbourhood planning propounded by the advocates of the New Urbanism.[1] This illustrates how deeply such physical planning design notions have penetrated community-planning thinking. Also, the idea of respecting the integrity of residential districts is carried over into the design principles for other districts—shopping precincts, industrial parks, historic districts, waterfront areas—so that their distinctiveness and effectiveness can be promoted. And respecting the integrity of the natural environment is rooted deeply in the conservation ethic that guided much early planning.

Planning principles for the built environment are, in many ways, at the heart of a community plan. They determine the actual physical conditions under which people will live, work, shop, and play in the community. They may or may not be made explicit in a community plan, but they are, nevertheless, embedded in it and in the implementing tools.

4. Coordination

There is probably no closer synonym in most people's minds for the idea of "planning" than "coordination." A good deal of the original justification for community planning was the chaos of traffic and slums in the centre of cities, and the helter-skelter subdivision of land on the fringes of cities. The wisdom of developing cities in an orderly fashion—aligning the street patterns of adjacent subdivisions, providing utility and transport lines at the time they are needed, providing schools and parks in residential areas—did not escape the notice of citizens. The coordination of city-building activities gave people confidence in the government's use of tax resources, added to their sense of physical well-being, and contributed to their aesthetic sense of a pleasant and smoothly functioning community.

Given the diversity of bodies that make decisions, in both the public and private sector, regarding the future built environment, it is clear that there must be a means of coordination. Further, the coordination must be intentional and provide a focus of responsibility. Those plan-makers primarily involved in establishing an overview of community needs—the local council, its planning board, and planners—are not themselves the agencies that actually develop and redevelop the community. Their overall plan provides a focus: for example, the parks department or transit authority might use the plan as a guide in designing their services; or builders, land developers, and business firms can know of the community's intentions and be guided by the plan.

5. The Need for Policy

Physical development (planning) matters occupy perhaps half the agenda of the average municipal council at its regular meetings. This attests to the important position that planning matters occupy in community government. But it is also important to realize that by the time planning matters arrive on the city hall agenda, councillors are caught up in the press of other day-to-day issues and must make decisions. The existence of agreed-upon policies for the physical development of the community enables councillors to judge development problems and proposals in light of ideas about the kind of community they and their citizens want, rather than on grounds of expediency.

In order to deal with the array of planning concerns discussed above, it is necessary to do more than just identify them: yes, it is sensible to have a certain arrangement of land uses; yes, we should apply the best planning principles; yes, it is wise to have coordination of investment actions. But in order to provide some assurance that these concerns can be met, a commitment must be made to the community's objectives for its built and natural environments—that is, there must be a **public policy.** The community plan is an expression of this policy, in effect stating, "in these kinds of situations, we will act in this way for these reasons." It states the community's position in advance and requires a persuasive argument before any deviation can be considered. Serious community planning demands that we go beyond speculation and idealization and make a commitment to strive for goals. Effective community planning requires a plan that embodies firm commitments.

Key Features of the Community Plan

In general, then, a community plan's main characteristics may be derived from the above set of concerns. In one sentence: *the community plan is a long-range, comprehensive, general policy guide for future physical development.* These are the four essential features of the plans prepared for urban communities, regardless of the name conferred on them by their respective provincial planning acts (official plans, general municipal plans, etc.). There are, in addition, several other features that provide a linking function to such aspects as the background analyses, staging of the plan, and capital investment needs.

The first planning acts offered planners the opportunity to prepare detailed "town-planning schemes," but experience seemed to show that a broad, policy-oriented plan must precede and give direction beyond the immediate development problem. In recent years, there have been suggestions for financial plans, social plans, energy plans, and environmental plans, but all these turn out to depend upon the general community plan. To reiterate, the four key features of a community plan are:

1. Focused upon the built and natural environments. The plan should encompass the entire land (and water) base of the community and take into account both man-made and natural features of the environment. The plan should deal explicitly with five basic physical elements of the community:

1. Natural environment—the local and regional ecosystem and its biodiversity;
2. Living areas—the areas comprising the residences of citizens;
3. Working areas—the areas comprising industries, places of commerce, and other forms of economic development;
4. Community facilities—the location and character of public and private facilities that provide community services for both the neighbourhood and the overall area; and

Planning Issue 8.1 Development Proposal Generates Positive Feedback

The Victoria Times Colonist
June 8, 2006

Development Proposal Generates Positive Feedback
Sandra McCulloch

CUMBERLAND, B.C.—The residents of Cumberland got their first look Wednesday night at a "master plan" of a new development on the doorstep to their historic community.

Trilogy Properties Corporation wants to develop 308 hectares at the crossroads of the Island Highway and the Comox Valley Parkway into retail and office space, housing and recreational sites.

"We want to show the full vision based on our work over the last eight months," said Trilogy CEO John Evans earlier Wednesday.

"We want to share with them, so there's no sense that we're not being completely transparent."

Cumberland Mayor Fred Bates said Wednesday afternoon the overall mood in the community is "pretty positive. Obviously, the devil's in the details but I think the community is feeling very good about Trilogy."

The developer needs an amendment to the Official Community Plan, followed by rezoning and development permits for specific parcels of land.

"We see the first applications happening probably within 30 days," said Evans.

"If we're able to stay on the timeline, and it is aggressive, we're hoping to have rezoning and development plans by mid-October."

He wouldn't give a dollar figure for the development, but said "clearly it's very significant, the overall investment."

The development would straddle the Inland Island Highway at the southern edge to the Comox Valley. It's prime land for development, said Evans.

"There's an opportunity for accommodation for tourism and overall growth of the community. What it's going to do for the village of Cumberland is provide a tax base that today is not there.

"The 2,500 people who live in the village have no commercial tax base . . . to support the development of amenities and infrastructure."

It's also another way for Comox Valley business to keep shoppers in the area, he said.

"When you look at the economics, the Comox Valley needs to develop the goods and services that keeps those expenditures at home, within their own communities."

Yet the village of Cumberland will be insulated from the development, he said.

"These lands are largely separated from the historic village," Evans said.

"It becomes an attraction and enhancement and becomes a way to get people off the Island Highway to get people to come visit the historic village."

But the details are yet to come, he added, saying the plans are illustrative.

"What the actual form will take—how many units, how many people—really will arise when we come back to the next stage. At this point we're talking land use."

Source: Sandra McCulloch, "Development Proposal Generates Positive Feedback," *Victoria Times Colonist*, June 8, 2006, B2. Reprinted with permission.

5. Circulation—the systems and facilities needed to enable people and goods to move between living areas, working areas, and community facilities, as well as between the community and its region.

2. Long-range and forward-looking direction. The time scale of the plan is determined by factors relevant to the particular community, such as population and economic growth, the condition of structures, and the need for utilities and amenities. Modifying the existing government, building new facilities, and paying for the public infrastructure all take considerable time. A common time horizon for plans is 20 years, as in Ottawa's 20/20 plan; York Region, on the other hand, uses a 15-year horizon (see Figure 7.6, page 198). Intermediate targets may be set for specific projects included in the plan, such as a new expressway or urban renewal project.

3. Comprehensive in viewpoint. The plan should comprehend, or take into account, "all significant factors physical and non-physical, local and regional, that affect the physical growth and development of the community"[2] That is, the plan should deal with the basic physical elements, as indicated above, as well as any other significant physical areas or features that are distinctive to the community. The viewpoint must also be broad enough to take into account conditions and trends in the larger geographical setting of the community.

4. General and broad-based in perspective. To be effective as a comprehensive instrument and as a policy guide, the plan should focus on the main concerns and issues of the community and the broad design components for its physical development. The plan is not a blueprint and should not include any details that detract from overall physical-design proposals and policies. Its primary purpose is to define the general location, character, and extent of desirable future development, and to be a guide against which detailed proposals may be evaluated. It may be necessary to provide some specific details in order to clarify the intent of policies or to provide physical images to which the community can relate.

Other Important Features of the Plan

In addition to the four key features of a community plan described above, there are several other important features that are included in many plans. Some plan-makers assign them an individual place within the plan report because of their importance in linking the general physical design with the policy role of the plan, as well as with related policy areas. These additional features are:

1. Linked to social and economic objectives. Even though a community plan focuses on the built environment, it is in many respects a vehicle for achieving social and economic objectives, at least in part. Indeed, a physical development plan that is not harmonized with these objectives may be less than successful. The two most obvious subject areas where we find strong interactions between physical and socioeconomic objectives are in housing[3] and in the provision of space for industry and commerce. It is vital that the process of community planning take into account social and economic factors. In addition, the plan must clearly state social and economic objectives that can be furthered by the physical development proposals. In this way, the community plan serves to focus attention on non-physical planning goals and to point out the need for coordination in their attainment.

2. Based on planning analyses. The analyses of current conditions and the forecasts of future conditions in the community combine to form one of the two cornerstones of the plan, the other being the community objectives. The analyses of population, economic base, land use, circulation, and the natural environment define the range of possibilities for the plan-maker. It is important that at least the main findings and the rationale behind the analyses be reported within the plan. Moreover, many of the analyses pertain to non-physical factors (e.g., population age, income, culture, employment), and this will assist in clarifying the relationships between the physical plan and social and economic factors in the community.

3. Implemented by stages. In contrast to older community plans, which presented a single long-range concept, it is now common for the progression of development to be stated. This is important because not all new development areas are likely to be opened up at the same time; if development is left to disperse, it could result in costly extensions to roads and utilities in the municipality; sprawling development can be inefficient for commuters as well as for those providing commercial services. The inclusion of a staging plan clearly signals the community's intentions to land developers, homeowners, business firms, and institutions. Additionally, the staging plan is often backed up nowadays with fixed limits to the areal extent of new building. They are called Urban Containment Boundaries by some and Urban Growth Boundaries by others.[4]

4. A guide for capital improvement. An overall plan is intended to help predict and anticipate the demands for public works and other capital investments. Most plans provide only a very general indication about needed investments, as, for example, when a new area is opened up for residential development and roads, utilities, parks, and schools are planned. A greater degree of specificity is, however, more helpful to the local council for scheduling capital projects. To this end, many communities prepare a Capital Improvements Program (CIP) in conjunction with their overall plan (see Chapter 14). The CIP is often structured to relate to the annual budgeting process of the community and also to project several years into the future as anticipated by the community plan. Thus, it makes the stages for development proposed in the plan more realistic.

5. A basis for community design. An ever-present value of planners and an expectation of citizens is that of visual beauty for the community. The community plan's proposed pattern of land uses is the base upon which the three-dimensional environment of buildings and open spaces is designed and constructed. Although community plans result in a particular community form, in recent years the design element has been notably absent from many community plans.[5] A community-design component of an overall place can be vital in communicating to citizens and developers the key built and natural features and special districts that have a particular significance for the community. It "lets them see" what the plan proposes.[6] (This issue is discussed further later in this chapter.)

Most provincial planning acts provide only a broad hint of the plan's scope, much less its content and characteristics. The above points thus attempt a synthesis of concepts and experience that reflects planning practice in Canada. It is perhaps unwise to seek a more precise definition, for every community plan will need to respond to the aspirations and problems of its particular community setting. The community plan must ultimately persuade citizens that it reflects their best interests as well as those of future citizens. In this task of human communication, community plans take a variety of forms. There are those that convey their message in a colourful booklet, those that use a newspaper tabloid format, those that are bound in ponderous tomes, and those that are laid out as a one-page poster But, regardless of format, the most effective community plans have the scope indicated above (see Figure 8.1).

The Policy Roles of the Community Plan

A community plan is foremost a **policy statement** about the future built environment. The content and scope that we have defined provide the necessary direction for its use in planning the community. But it is a document—an instrument—to be used in the realm of community governance. Thus, the community plan plays a number of roles, according to its uses in policy determination and implementation by the local council, planning board, and planning staff, as well as by citizens and developers. These roles call for the plan to have other more functional characteristics so that it may effectively achieve what the community wants for its future built and natural environment. The

Figure 8.1 Community Plans for All Sizes of Communities

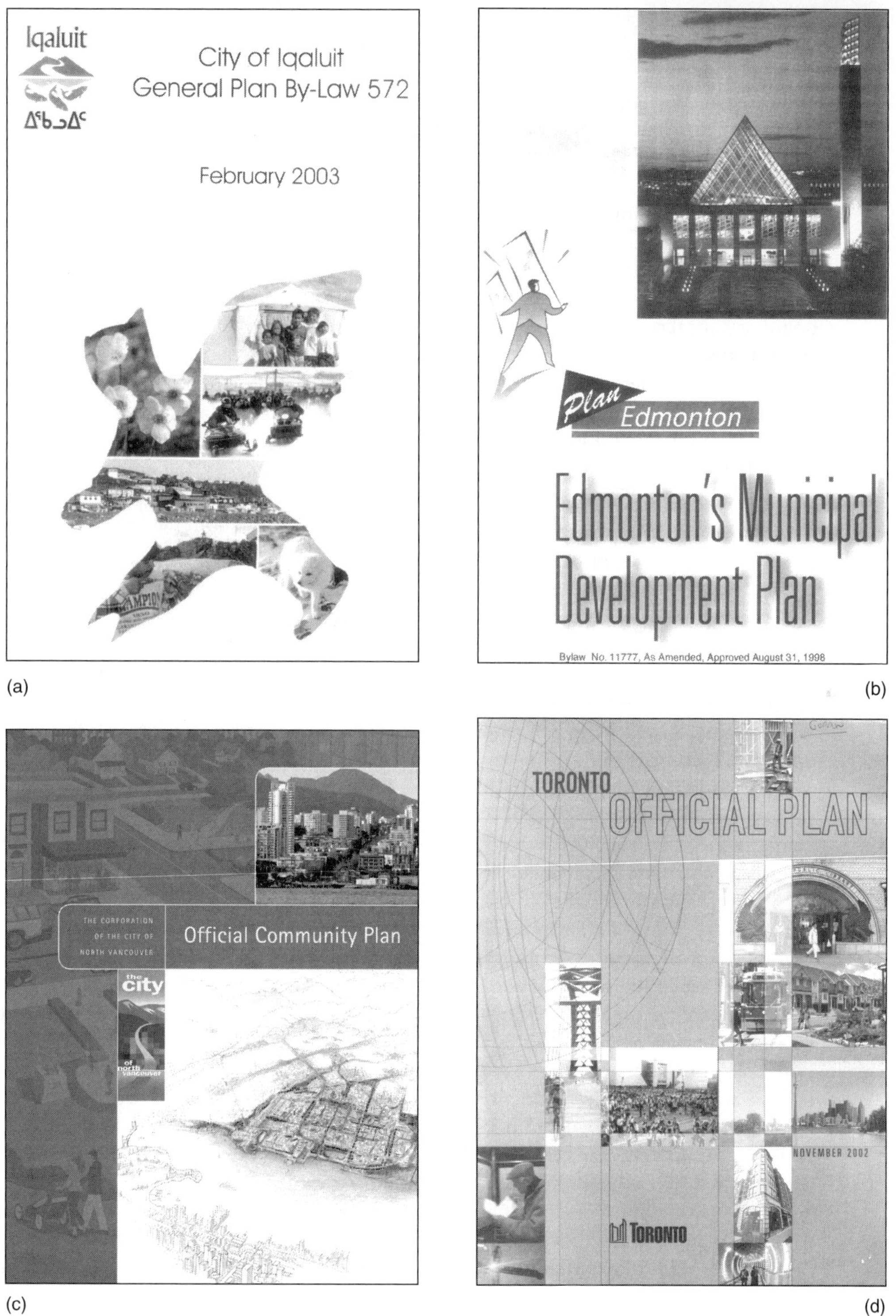

(a) (b) (c) (d)

Source: (a) Courtesy City of Iqaluit; (b) Courtesy City of Edmonton; (c) Courtesy City of North Vancouver; (d) Courtesy City of Toronto.

community plan must be a vehicle for both policy determination and policy implementation.

The Plan in Policy Determination

The community plan's role in policy determination is a progressive one. It begins with the preparation and initial adoption of the plan and continues through its regular review and evaluation as day-to-day physical development matters are considered. The community plan is an ultimate base of policy in its own right. It is also the basis for formulating secondary policies for, say, transportation, housing, and parks.

The efforts that go into the initial adoption of the plan are the most important, for when the plan is adopted it will represent the culmination of thorough deliberation about major alternatives by many sectors of the community, as well as by the planning board and local council. Ultimately, it is the council that adopts the plan, but the deliberations that lead up to that point are vital. Usually a draft plan is made available to community members to communicate the basic ideas and alternatives. The reactions, both positive and negative, are part of a process of educating the community and the council about the major issues in physical development. If this period of public learning and debate is effective, it will normally mean that the council is not surprised later by unresolved issues. There are three important policy characteristics of a community plan that can assist in the initial deliberations, as well as throughout the life of the plan:[7]

1. The plan should be in a form suitable for public debate.
2. The plan should be available to the community.
3. The plan should clearly communicate its proposals to the public.

The council's adoption of the plan is in effect a declaration of the policies that it intends to apply to future physical development proposals. This allows private interests (citizens as well as developers) to anticipate the probable reaction to proposals for development. The impact of the plan as a statement of policy is as important inside the local government framework as outside of it. Coordination of the actions of public officials in various city departments as well as in semi-independent commissions and boards is facilitated. Lastly, the plan's policies are an important guide for judicial bodies in appeals against land-use regulations.

While it is important that community plan policies be firm and be applicable over a long period of time, a plan should not be considered immutable. Conditions in a community do not remain static and, not infrequently, new problems arise that were not anticipated when the plan was adopted. For example, trends may change in population growth or job creation, or new information may become available that affects land-use decisions, as with new knowledge about ecological systems or the aging of the population. Moreover, some policies may prove unworkable or unrealistic. The council should thus be willing and able to amend the community plan if the situation or conditions warrant it. Proposals for large new projects—a new airport, sports arena, or big-box store, for instance—may trigger the need to review the compatibility of the project with the plan. Or a new transportation study may show how the road network could be refined. It is important that the plan be able to accommodate review and renewed debate. In short, plan-makers can never know all the things that will happen during the course of the plan's life. Thus, we should add a fourth characteristic to facilitate policy roles of the plan:

4. The plan should be amendable.

Although the plan should be amendable, it should not be subject to trivial challenges that would threaten its role as a continuous statement of policy. Moreover, frequent amendment probably indicates a lack of agreement and commitment on basic policies. If not abused, amendability of the plan is important, for it means that basic development policy can be debated and refined and the extent of community agreement broadened in the process. This also means that the plan is being consulted in regard to decisions that need to be taken. For all these reasons, there is in most provincial planning acts provision for the community plan to be reviewed on a regular basis. The Planning Act of Ontario requires a maximum of five years between regular reviews of the plan by the council. Some planners try to have their planning boards review the plan annually.

There are also a host of current topics brought before council that impinge on plan policies. These cover such matters as parking regulations, use of public parks for non-recreational activities, or the use of advertising signs in public thoroughfares. Usually, specific concerns such as these are not part of the community-plan policies. However, cumulatively, as part of the day-to-day actions of a council, they may affect the community environment. The plan's effectiveness may be impaired if the decisions are not consistent with its policies and/or the plan is amended too frequently. Nonetheless, council's actions may reveal its perception of changed conditions or outlooks, and thus the need to reconsider policies set forth in the plan. It is in the best interests of plan-makers to create a community plan that becomes an explicit part of the backdrop against which the local government and its officials make their decisions—that is, a community plan that becomes a "working plan." In this way, its policies are being used and frequently tested. This day-to-day use of the plan means its policy determination role overlaps with its policy-implementation role.

The Plan in Policy Implementation

The means for bringing a community plan into effect are increasingly specified in provincial planning acts. Local councils are thus required to take a number of formal steps that give legislative effect to the policies of the community plan. If land use is to be regulated by zoning, provincial planning acts require that the local council pass a zoning bylaw. It is now normal to require that the community have an overall plan in place before enacting zoning regulations and, moreover, that the zoning bylaw be consistent with the community plan. Provincial planning acts tend to include, in addition to zoning, such other important measures for implementing the plan as subdivision control, urban redevelopment, and site-plan control, and to specify council's role in enacting, approving, or amending proposals in these areas.

In general, there are two levels of action that councils take in regard to plan implementation. At the first level, the matters are legislative in nature; that is, the passing of local bylaws based on the principles and policies of the plan. The two most common are for the control of land use on already developed or developable land—**zoning**—and for the arrangement of new properties for development on vacant land—**subdivision control.** Some provinces provide for councils to pass **development control** bylaws that allow the community to review proposals for development on a property-by-property basis rather than on a district basis, as in zoning, and to take aesthetic considerations into account in their decisions. And in some places, the public **program for capital improvements** in the community is required to conform to the community plan. There are, in addition, detailed plans for development that a council is occasionally called upon to approve. These may include plans for downtown revitalization, street and highway improvements, and parkland acquisition and development. (Chapters 13 and 14 elaborate on the role of these tools and others in implementing the community plan.)

At the second level, there are many routine council decisions that arise in conjunction with implementing the bylaws that council has passed. Councils are usually required to approve proposed plans of subdivision, development-control agreements, and applications for rezoning, as well as public-works expenditures, under the capital improvement program. Other routine council decisions that may affect the plan's policies are requests for street closings, traffic regulation, transit routes, and the locations of fire halls, libraries, and schools.

Policy Perspective for the Community Plan

The perspective that has been described above for the use of the community plan is, essentially, hierarchical. The plan may be considered the keystone in an arch of planning activities, as in Figure 8.2. One can see the hierarchy of control measures, and the distinction between those that operate in the public and the private sectors, as well as the roles of various participants.

On the left side of the diagram are the initiatives that are taken by city officials and boards in support of the overall plan. These may be required under the planning act in some provinces. It can also be seen that, as one moves away from the locus of the community plan, they deal progressively with *specific* parts of the built environment. The zoning bylaw usually deals with the entire community; subdivision control is applied to specific areas within the community; and the capital-improvements program specifies individual projects. Further, as the general concerns shift toward the specific, the involvement in decision making shifts to a wider array of participants. The planning board and staff have roles in formulating the zoning bylaws and subdivision controls, and additional members of the municipal staff become involved in formulating the program for capital improvements. Council is thus dependent upon the advice and actions of several levels of participants within local government circles in preparing the plan and in providing the supporting legislative framework. Note also that the initial stages of

Figure 8.2 Key Role of the Community Plan in Land-Use Control

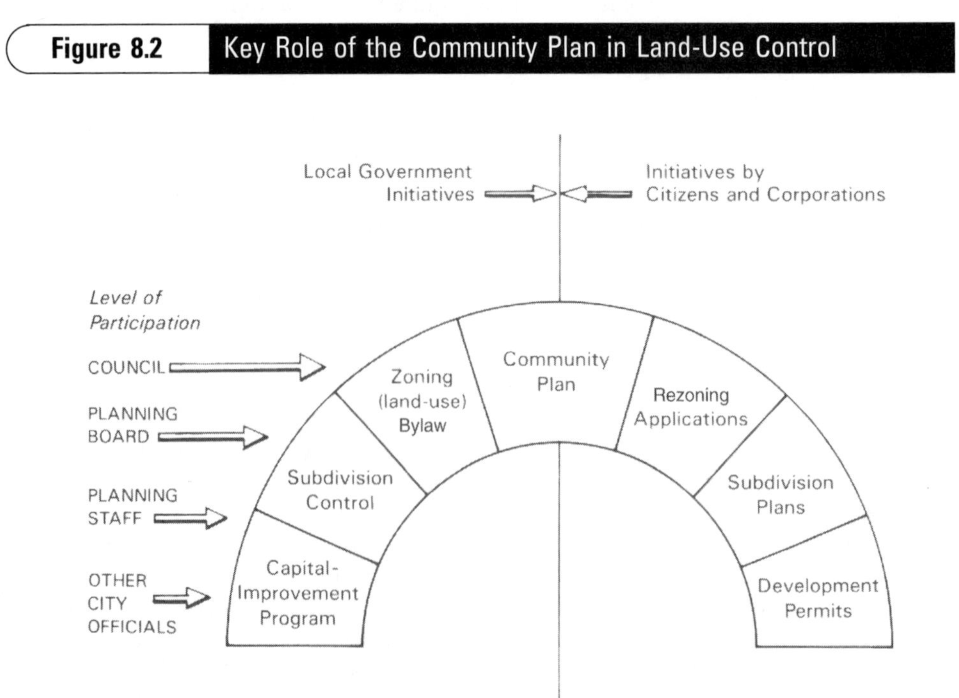

The community plan has the same position as that of a keystone in an arch. Initiatives of both the local government and the citizens and corporations depend upon its integrity.

plan preparation—basic forecasts, background studies, identifying alternatives, initial designs—are usually allocated to an advisory committee or board and to professional staff. Although the board and staff are advisors to council, theirs is not a passive role. Indeed, they function more as joint participants and can influence council on the progress of planning more than any other participants, if they choose.

Referring again to the diagram, the right-hand side comprises those initiatives normally taken by individual property owners and other development interests and agencies. In other words, any person or corporation that wishes to develop or redevelop land in the community must apply to the municipality. Each application to, for example, rezone a property to allow a different use or to construct new buildings on vacant land represents a potential step in shaping the built environment. The review and approval of such applications are important steps in carrying out the policies of the plan. In general, these applications ultimately have to be approved by the council, which can determine their consistency with the plan's policies. However, planning staff and other officials (such as building inspectors) play crucial roles in receiving the applications and judging their acceptability at the outset. The relationship between the municipal plan and development regulations and other policy instruments in Edmonton, shown in Figure 8.3, demonstrates the vital role of the community plan in this set of linkages.

In summary, once a community plan is adopted, it sets in motion efforts by various participants to implement its policies and proposals. There are, basically, two policy streams. One is used by the council to secure *compliance* (through bylaws, budgets, and regulations) with the policies it has approved in adopting the plan. The other is used to *communicate* to others (officials, citizens, and developers) the aims and policies, and the hopes and expectations, of council regarding the future built environment. Both streams frequently converge, as, for example, when experience shows the nature of applications from the private sector differing from forecast trends, or when budget constraints change capital spending. These may then show the need for amending the plan. An effective community plan is one that is put to use in the continuing development of the community. A plan that is used to evaluate proposals for development may need its own policies re-evaluated and, if necessary, its sights reset.

The Plan in the Context of Public Planning

Community planning, as we observed right from the opening chapters, is a widely accepted public activity in Canada. In this country, community planning is accomplished in a setting with its own statutory foundations, formal processes of plan-making, and structures for appeals. It is important to grasp this institutional setting, for Canadian community planning is characterized, as perhaps is the case in few other countries, by a highly structured legalistic/bureaucratic format that operates in conjunction with a rather open land market of multiple owners. It offers many opportunities for public initiatives to shape the outcome of community planning efforts that do not exist in the U.S., for example. Or, looked at from the point of view of constraints, the formal setting establishes the boundaries of planning action by both public and private actors.

Statutory Foundations

Since the 1920s, nine of the ten Canadian provinces have had substantial planning legislation in effect. (All the provinces and territories now have such statutes.) These

Figure 8.3 Relationship between Plans and Development Measures, Edmonton

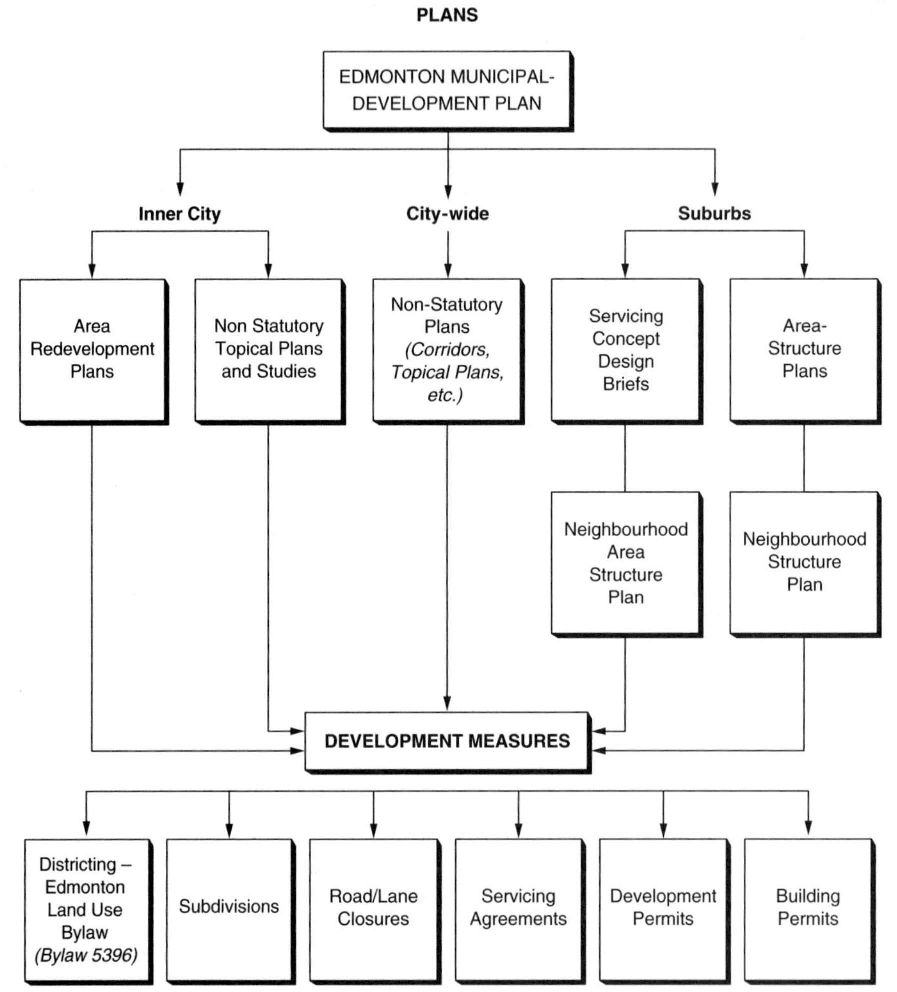

A community plan, once adopted, articulates and coordinates a host of planning, regulatory, and development efforts by public and private participants, as we see here with Plan Edmonton, the City's Municipal Development Plan.

Source: City of Edmonton Planning and Development Department, *Planning and Development Handbook*, March 2000, Figure 2. Used by permission.

planning acts, as we call them generically, are the foundations for land-use planning and implementation at the local government level. Planning acts are the type of statute often referred to as **enabling legislation** because they enable, or allow, a municipality to carry on a specified governing activity. Such legislation is necessary for local planning so that the powers over private property and land use residing with the province are available to the municipality; that is, are delegated to the municipality. Planning acts in most provinces are ***permissive*** in their provisions; they do not require municipalities to make plans and land-use regulations. (Some provinces are not fully permissive in their planning legislation in allowing municipalities to decide to make a plan. Alberta requires that municipalities over 3500 in population prepare a Municipal Development Plan and encourages those with fewer people to do so.[8]) In any case, when a community prepares a plan the act prescribes the planning content and the procedures to be followed. Then, any local plans and planning bylaws take on much of the force of the planning act; they too are statutes. This is the basis for describing Canadian community planning as ***statutory:*** it is empowered by statute, and its output (which falls within the terms of the act) has the power of a statute.

This statutory feature, which has characterized Canadian community planning for nearly a century, generates a high degree of dependence on the province by local planners. The province, until recently, has been the final arbiter of community plans in all provinces and of local land-use regulations in most. Provincial planning establishments oversee local planning efforts and refine the process of planning for land development by formulating guidelines for special situations and applying the planning act. Starting around the mid-1970s, eight provinces undertook extensive revisions of their planning acts. These efforts sought to refine (or "streamline") the elaborate planning legislation and its supporting bureaucratic structures. Thus, local planning in Canada operates within a provincially generated framework: seeming, not infrequently, as if the province were the ultimate client and the planning act an incontestable authority.

The planning act (or the applicable legislation in the province) specifies *who* may plan, *what* they may plan, and *how* they may plan for community planning to be statutorily correct. There are differences among provinces in the methods and the styles that are used, but the general thrust remains the same. Planning acts normally deal with the following five matters:

1. The creation of planning units;
2. The establishment of organizational machinery for planning;
3. The content, preparation, and adoption of statutory plans;
4. The format for enacting zoning, building, and housing bylaws; and
5. The system for subdividing land.

The pervasiveness of community planning in Canada has, in many respects, led to a new level of maturity among communities regarding planning. The provinces are, again, revising their planning acts to devolve their responsibility for approving plans and bylaws onto the municipality, and the province will limit itself to a review to make sure that provincial interests are protected.[9] Not a little of the initiative for this devolution comes, of course, from budget cutting and consequent reductions in the size of provincial planning staff. Nevertheless, it points the way to a much less paternalistic process.

To give a better sense of the statutory structures for planning that exist across Canada, Figure 8.4 lists the various legislative tools available in each province and territory. It will be noted that there are few instances where planning tools are not provided at all levels. The names may differ, but the pervasiveness of community planning potential is clear.

Before leaving this discussion of statutory foundations for planning, it must be noted that all provinces have other pieces of major legislation that affect the substance and organization for planning. Almost all provinces now have some form of legislation for environmental assessment and protection. Nearly as common are statutes covering the establishment of condominiums, the designation of historic buildings, the location of pits and quarries, transportation, water quality, and the use of natural resources. In British Columbia, for example, municipalities may have to deal with the provincial Agricultural Land Commission in regard to the zoning and subdivision of land. And, recently, the province passed legislation requiring that the care and quality of fish-spawning streams be taken into account in local official community plans. The administration of these statutes is most often in the hands of officials from ministries other

Figure 8.4 Comparative Provincial Planning Terminology

Level of Detail	*British Columbia*	*Alberta*	*Saskatchewan*	*Manitoba*	*Ontario*	*Quebec*
Provincial planning legislation	*Municipal Act*	*Municipal Government Act*	*Planning and Development Act*	*Planning Act*	*Planning Act*	*Land Use Planning & Development Act (Loi sur d'amenagement et d'urbanisme)*
Plan/Zoning Appeal body		Subdivision & Development Appeal Boards (local)	Development Appeals Board (local)/Saskatchewan Municipal Board		Ontario Municipal Board	Commission municipale du Québec
Regional plan	Official Regional Plan	Intermunicipal Development Plan	District Development Plan	Joint Planning Scheme	Upper Tier Plan	Land Use Planning and Development Plan (*Schéma d'aménagement et de développement*)
Municipal land use plan	Official Community Plan	Municipal Development Plan	Basic Planning Statement; Development Plan	Development Plan	Official Plan	Planning Programme (*Plan d'urbanisme*)
District plan	Area Development Plan/Comprehensive Development District (Vancouver)	Area Structure Plan/Area Redevelopment Plan	Local Area Plan	Secondary Plan/ Redevelopment Plan (Winnipeg)	Secondary Plan	Special Planning Programme (*Programme particulier d'urbanisme*)
Street & block layout	Neighbourhood Plan	Outline Plan	Plans of Survey	Local Planning Area Map	Tertiary Plan	Special Planning Programme (*Programme particulier d'urbanisme*)
Land subdivision	Preliminary Proposal/Plan of Subdivision	Proposed Plan/ Plan of Subdivision	Proposed Plan/ Plan of Subdivision/ Plans of Survey	Short-form/ Registered Plan of Subdivision	Draft/Registered Plan of Subdivision	Subdivision By-Law (*Réglement de lotissément*)
Zoning	Zoning By-Law or Registered Land Use Control	Land Use By-Law	Zoning By-Law	Zoning By-Law/ Zoning district map	Zoning By-Law	Zoning By-Law (*Réglement de zonage*)
Site plan review	Development Permit	Development Permit	Development Standards Site Plan Control Architectural controls	Development Permit	Site Plan Control Development Agreement	Site Planning and Architectural Integration Programme (*Plan d'implantation et d'intégration architecturale*)

than that containing the provincial planners. As with most provincial legislation, these statutes can provide opportunities for planning as well as constraints. Therefore, it is wise to be aware of the range of the provincial statutory foundations that could affect the outcome of a particular planning effort.

Formal Steps in Plan-Making

The last chapter discussed the process of plan-making in a normative context—that is, the refinement of community goals through a more or less rational set of steps comprising both technical studies and community participation. But how does this general planning process translate in terms of the statutory planning framework described in the preceding section?

Planning acts, as noted, not only specify *who* may plan but also prescribe *how* they may plan. A minimum series of steps that must satisfy the statutory requirements of the act is set forth. These formal steps parallel many of the steps in the normative process. The differences lie in various technical and participatory steps undertaken in the real community setting to accommodate the social, economic, environmental, and physical

New Brunswick	*Prince Edward Island*	*Nova Scotia*	*Newfoundland & Labrador*	*Yukon*	*Northwest Territories*	*Nunavut*
Community Planning Act	*Planning Act*	*Municipal Government Act*	*Urban and Rural Planning Act*	*Municipal Act, Area Development Act*	*Planning Act*	Planning Act
	Island Regulatory and Appeals Commission	Nova Scotta Utility and Review Board	Four (District) Appeal Boards	Yukon Municipal Board	Development Appeal Boards (local)	Development Appeal Boards (local)
Regional Development Plan	Regional Plan	Regional Municipal Planning Strategy	Regional Plan or Joint Municipal Plan	Regional Land Use Plan	Regional Land Use Plan	Regional Land Use Plan
Municipal Plans	Official Plan	Municipal Planning Strategy	Municipal Plan	Official Community Plan/Local Area Plans	General Plan	Community Plan (General Plan)
Development Scheme or Area Plan		Secondary Planning Strategy	Development Scheme or Comprehensive Area Plan	Area Development Scheme	Development Scheme	Development Scheme
Tentative Subdivision Plan	Proposed Subdivision	Subdivision Regulations		Proposed Subdivision	Proposed Subdivision	Plan of Subdivision
Tentative/Final Plan of Subdivision	Preliminary/Final Subdivision Plan	Tentative/Final Subdivision Plan	Subdivision Regulations	Preliminary Plan/Proposed Subdivision	Proposed Subdivision/Plan of Subdivision	Plan of Subdivision/Plan of Survey
Zoning By-Law	Zoning & Development By-Law	Land Use/ Zoning By-Law	Land Use Zoning Regulations	Zoning By-Law, Area Development Regulations	Zoning By-Law	Zoning By-Law
Development Permit	Approved Subdivision	Site Plan Approval/ Development Permit	Development Permit	Development Permit	Development Permit	Development Permit
Final Subdivision Plan						

conditions unique to that community. Thus, when a Canadian community plans, its normative process is articulated by the requirements of the provincial act that permits planning; in short, the community's planning actions are punctuated by several required steps. Figure 8.5 shows the general steps required to bring a community plan into effect. These steps are much the same as those required for making amendments to a plan and for enacting and amending a zoning bylaw. But, as we noted above, some provinces are eliminating the final step, that of provincial approval, thereby shortening the process and leaving responsibility with the community.

One feature of the normal process of plan-making is that planning decisions are the product of several bodies, not just the local council and its advisory committees. In some provinces, such as Saskatchewan, the outcome of a community plan or zoning bylaw or subdivision plan may be the result of modifications made at three levels: the community, the ministry, and the appeal board. Within the first two levels, a large number of other, non-planning agencies may also be involved in the outcome. Within the municipality, provision is usually made to have planning proposals reviewed by all technical departments and public utility companies. At the provincial level, there is also widespread distribution of municipal planning proposals to various ministries and Crown corporations. Somewhat in contrast is the rather limited role for citizens—the general public—in the formal process. It seems, therefore, that in light of the efforts taken to preserve the statutory basis of the plan, and of the inherent paternalism of the provinces, one would be justified in speculating on *whose plan* is represented in the final output. Among the provinces, Alberta, Ontario, B.C., and Newfoundland have evolved the formal plan-making system that most respects the integrity of locally made planning decisions.

Structures for Appeals

The statutory nature of Canadian community planning imparts a legalistic bias to many of its activities, but especially to the structures for appeals of planning decisions. Most provinces have established some form of quasi-judicial body to hear appeals from people or agencies that object to an official plan, zoning bylaw, or subdivision plan. The Ontario Municipal Board (OMB) has been the model for some planning-appeal bodies.[10] Its proceedings are conducted much like those of a courtroom (through the adversarial process), and its decisions have the force of law. Objectors (including citizens and citizen groups) and proponents are expected to be represented by legal counsel, although other representations will usually be heard.

In order for an appeal to proceed, a number of specific steps must be taken. First, no appeal can be made until the bylaw approving the plan, zoning change, and so on, has been passed by the local government body. Second, there is only a limited period of time after approval of the bylaw—30 days is common—in which an appeal can be lodged. Third, the appeal body or the minister in question, depending upon the province, will decide whether the objection is trivial, in which case a hearing would prove dilatory to the planning process and the appeal is not heard.

In many respects, hearings of provincial appeal boards tend to become hearings *de novo* (as lawyers call them), or brand-new considerations, of the planning pros and cons. This is not only time-consuming and costly to all participants, but also may bring in planning evidence and arguments not heretofore considered by the community and its council and planning bodies. In recent years, limitations have been placed on some aspects of appeal-body powers. Still, in the provinces where they are used, provincial appeal boards create a significant and additional avenue through which planning decisions may be made, one that is likely to be developed with its own rationale and criteria for its decisions. This means it has the potential of becoming planning *for* a community rather than planning *by* a community. This is one of the major dilemmas of the legalistic style of planning, for it must honour judicial concepts, including the right to appeal to a third party. Legislation was introduced in Ontario in 2005 to deal with such issues by confirming the ultimate role of a local council to determine land-use matters and to restrict the admission of new information into OMB appeals.

Not all provinces use a provincial appeal body with broad-based powers of intervention. Saskatchewan, for example, has two levels of appeal on planning matters. Each municipality with a zoning bylaw is required to appoint its own Development Appeal Board to deal with appeals on plan amendments, rezoning applications, minor variances, development permits, and subdivision plans (where the local council has approval authority). Appeals of the decisions of the local appeal board may be made in many instances to the Planning Appeals Committee of the Saskatchewan Municipal Board. The latter committee must hold a public hearing on appeals made to it. Similar arrangements also exist for subdivision and development permit appeals in Alberta municipalities. The exception to local appeals in Alberta is when it is determined that the application has a provincial interest (e.g., within specified distances to provincial highways and water bodies). In the latter instance, the appeal goes to the Municipal

Figure 8.5 The Process of Plan Preparation by Municipalities

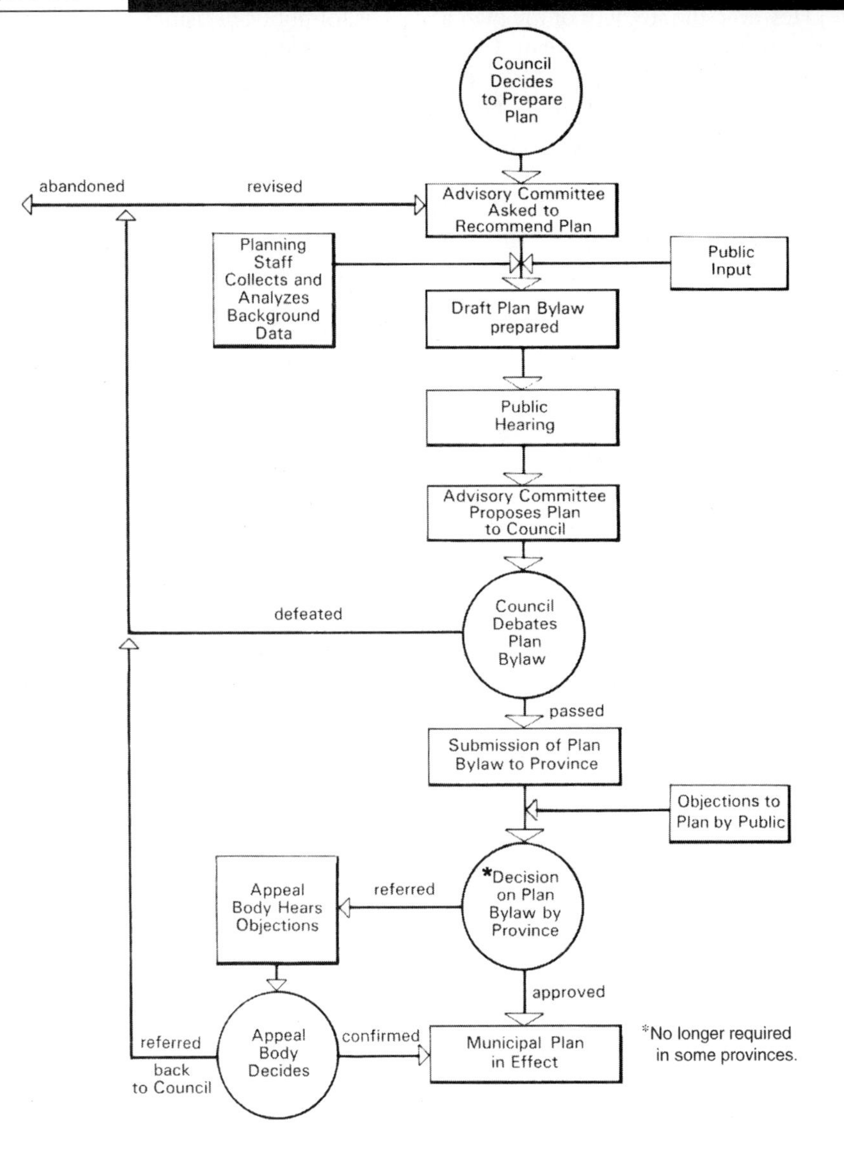

These steps apply generally to all municipalities. Some differences exist in the requirements for public input, as well as in appeal procedures from one province to another.

Government Board.[11] Ontario is proposing such an approach in order to reduce appeals to the Ontario Municipal Board. The right of appeal to the courts on any planning matter also exists, as is the route in B.C., but usually only if there have been procedural irregularities or disputes on points of law in previous hearings.

The Community Plan as a Legal Document

Plans for Canadian communities prepared under provincial legislation are legal documents. They require formal council endorsement and usually approval by the province. (Contrast this to the system in the United States, where community plans were until recently considered local policy initiatives and their existence a matter of local prerogative.) However, the strength of the plan as a basis for public decisions, especially for regulating private land development, was not much tested until four or five decades ago. Provincial planning legislation has tended to be permissive, e.g., "... a municipality *may* prepare a plan for...." But as land-use controls became more prevalent, the courts have been resorted to in order to challenge the community's right to place constraints on the development of private land. Since land-use controls are almost inevitably enacted to achieve some intended form for the built environment, it has become necessary to provide the courts and others with the policy rationale—the plan—on which these are based. The plan performs a function similar to that of a national constitution; indeed, some observers have referred to the community plan as having a "constitutive" function.[12]

Another important aspect of this shift in approach is the introduction of flexible land-use control practices and instruments. Early zoning practice, largely justified because it could provide stability to neighbourhoods and land values, did not envision the dynamics of land development occasioned by the urban growth of the past several decades. In response to these pressures and needs, planners have looked for ways to accommodate, for example, new styles in building subdivisions, industrial parks, retailing, and office buildings, as well as managing the ubiquitous automobile. The aim has been to devise means by which individual development proposals could be appraised by the planners. One avenue was to institute **development controls** requiring, under a local bylaw, that a permit be sought for the siting of buildings, traffic access, and architectural qualities. Another avenue was **planned unit development**, which encouraged developers to plan their projects in conjunction with local planners and thereby find ways of mutual compromise regarding development plans. Yet another was the use of negotiated agreements with individual developers, commonly called **development agreements.** Although zoning remained as the general framework, it was no longer the only prescription for what might be done with properties (see also Chapter 13). The constant in this increasingly diverse planning milieu is, of course, the overall community plan.

Plans as Community Images

Lastly, plans are more than words. Plans have the ability to convey images in graphic terms of visions of a community's future that other instruments of public policy lack.[13] Plans help connect people to places, and when a plan's concepts can be conveyed in graphic terms, it can enhance the plan's capacity to promote consensus. The influential European planner Andreas Faludi argues strongly for a plan to have a "spatial organization principle," a graphic image that will "stick" in people's minds if it is to be a politically viable and physically attainable plan.[14] A striking Canadian example of such a spatial principle is the "greenbelt" contained in Jacques Gréber's 1950 plan for the Ottawa region; that concept has become a firmly entrenched reality 50 years later.[15] The designation of a greenbelt around Toronto and Greater Vancouver's Green Zones could play a similar role, as could urban growth boundaries. This graphic aspect of a community plan, though somewhat sidetracked in the planning literature in recent years, has venerable roots. The two-dimensional diagrams and three-dimensional models made by those such as Ebenezer Howard and Le Corbusier, which were examined in earlier chapters, are its forebears. The negative urban image of "mindless sprawl" that spurs many of the New Urbanism and Smart Growth efforts shows how evocative a planning image can be. We have come to expect in the process of plan-making to find out the "what," "when," and "how" of a community's future; the plan should also tell us "where." Community plans, after all, deal with real geographical spaces.

Reflections

Recent practices in plan-making encourage diversity and innovation in development from New Urbanism to Smart Growth, mixed land use, brownfield development, and environmental restoration. While there is much to commend in each of these initiatives and in others, it is important to recognize that the more diverse our approaches, the more the need that we proceed with them in an integral way. Without a common plan-making perspective, there are bound to be problems associated with coordinating decisions, not to mention the risk of arbitrariness in such decisions. This perspective is provided by the community plan, and it comes to assume a more vital role as a backdrop against which diverse development concepts and decisions can be assessed. The question raised in one of the references in this chapter, "Does planning need the plan?" deserves a resounding yes.[16]

How community plans are initiated is yet another question to be debated. The issue is whether a plan imposed by a higher level of government can be as meaningful as one generated by local needs and experience. Certainly, community plans produced to comply with provincial legislation are often complex and ponderous in language in an attempt to forestall legal challenges. There are, however, examples of plans for communities large and small that fulfill the legal requirements yet offer a vigorous rendering of community aspirations. The crafting of a community plan demands skill and sensitivity on the part of plan-makers.

Up to this point, the primacy of the community plan in the planning activity of an urban community has been emphasized. Now, it is necessary to put the community plan into a larger perspective that is both deeper and broader. On the one hand, a plan exists to be implemented and, thus, requires an array of regulatory tool and policy instruments to realize its goals and objectives. The plan's perspective deepens when it is seen as encompassing these means for plan effectuation (discussed in Chapters 13 and 14). On the other hand, a plan's perspective must also be broadened to encompass the region within which it exists. For each community fits within its region and is subject to the growth dynamics of that region, as well as to the plans made for that region. In the next chapter, as the regional planning perspective is discussed, it will be helpful to consider this question:

- ***How can community plans and regional plans be mutually supportive?***

Endnotes

1. Cf. Andreas Duany and Elizabeth Plater-Zyberk, "Neighbourhoods and Suburbs," *Design Quarterly* 164, 10–23.
2. T.J. Kent, Jr., *The Urban General Plan* (San Francisco: Chandler, 1964), 99. Much of the discussion in this section is based on the concept of a community plan posited by Kent. This influential text was reprinted in 1991 (Chicago: Planners Press, American Planning Association).
3. Provincial planning legislation in British Columbia requires Official Community Plans to include a statement of housing policy, especially regarding affordable housing. B.C. Municipal Act, Sec. 945, 2.1.
4. Canada West Foundation, "Lines in the Sand: Are Urban Growth Boundaries Effective?" *Western Cities Project Update*, June 2004; see also Smart Growth BC, available online at www.smartgrowth.bc.ca.
5. Jeremy Black, *Maps and History: Constructing Images of the Past* (New Haven: Yale University Press, 1997).
6. Michael Neuman, "Does Planning Need the Plan?" *Journal of the American Planning Association* 64:2 (Spring 1998), 208–220.
7. Kent, *Urban General Plan,* especially 119–123.
8. Alberta Municipal Affairs, *The Legislative Framework for Municipal Planning, Subdivision and Development Control* (March 2002), 2.
9. Ontario, Commission on Development Reform and Planning in Ontario, *New Planning in Ontario, Final Report* (Toronto, 1993); and Stan Clinton, "Changing Times: Newfoundland's Municipal Planning and Implementation Systems," *Plan Canada* 37:2 (March 1997), 18–20.
10. Cf.Bruce Krushelnicki, *A Practical Guide to the Ontario Municipal Board* (Toronto: Lexis Nexis Butterworths, 2003).
11. Cf.Frederick A. Laux, *Planning Law and Practice in Alberta,* 3rd ed. (Edmonton: Juriliber, 2002).
12. Stephen Griffin, "Constitutionalism in the United States: From Theory to Politics," in Sanford Levinson, ed., *Responding to Imperfection* (Princeton, NJ: Princeton University Press, 1995).
13. Neuman, "Does Planning Need the Plan?"
14. Andreas Faludi, "European Planning Doctrine: A Bridge Too Far?" *Journal of Planning Education and Research* 16:1 (Spring 1996), 41–50.
15. Richard Scott, "Canada's Capital Greenbelt: Reinventing a 1950s Plan," *Plan Canada* 36:5 (November 1996), 19–21; and David Gordon, "Weaving a Modern Plan for Canada's Capital: Jacques Greber and the 1950 Plan for the National Capital Region," *Urban History Review* 29:2 (March 2001), 43–61.
16. Neuman, "Does Planning Need the Plan?"

Internet Resources

Chapter-Relevant Sites

Planning Canadian Communities
  www.planningcanadiancommunities.ca

Vancouver City Plans
  www.city.vancouver.bc.ca/commsvcs/cityplans

Coquitlam BC (search for Corporate Strategic Plan)
  www.coquitlam.ca

City of Winnipeg
  www.winnipeg.ca/ppd/planning.stm

York Region Official Plan
  www.region.york.on.ca/About+Us/York+Region+Official+Plan

Greater Sudbury
  www.planningsudbury.com

Ottawa 2020 Plan
  www.ottawa2020.com

City of Montreal
  www.ville.montreal.qc.ca/plan-urbanisme

City of Fredericton
  www.fredericton.ca/ed2.asp?607

9

Chapter Nine

Planning Regional and Metropolitan Communities

Planning has to be at a scale which is large enough to work out a strategy for growth within a setting as broad as the entire urban-centered region.

Leonard Gertler, 1966

Many planning problems have effects on areas well beyond their source. The drainage of storm water from a subdivision into a watercourse has the potential of causing pollution to areas downstream, for example; or a major sports facility in one community may generate large amounts of traffic flowing through normally quiet residential areas in another. Probably the classic example is **suburbanization,** when those who work in an older central city choose to reside in a newer, lower-density community on the edge of the city and shop in yet another. Indeed, planning problems and/or their solutions are seldom confined within an individual community. The reasons stem from the interrelationships of various economic and social activities in a community, and the ways in which the effects of development are transmitted through space by transportation, water flow, air currents, and economic transactions, among other factors. Thus, it is often necessary to undertake planning for areas larger than one community, areas large enough to encompass the major effects of development.

Planners have traditionally proposed the regional approach for two kinds of planning problems: one, involving the growth and spread of cities into the countryside; the other, the development and conservation of natural resources in rural and non-urban settings. Both concerns have generated distinctive forms of regional planning, and there are many examples of each in Canada. This chapter reviews the nature of planning for large areas and describes the Canadian experiences with this facet of planning called regional planning. A key question to keep at the forefront when exploring this material is:

- *How does the scope of regional planning differ from that of community planning in terms of both the space being planned for and the content of plans?*

The Regional Planning Perspective

Planning for the physical environment is commonly talked about on two levels—city or urban (community) planning and regional planning. Although the two are obviously complementary, each has developed in its own distinctive way. This fact stems partly from basic differences in the planning issues and objectives addressed by each, and partly from differences in jurisdictional arrangements. Regional planning is concerned with many facets of the built environment, the natural environment, and the social and economic activities occurring in large areas that may include city and towns. As well, such planning takes place in spatial situations that, in most instances, do not have well-defined political boundaries. Indeed, starting from its roots, regional planning has developed a different outlook from that of community planning, which it will be instructive to review.

The Roots of Regional Planning

Regional planning is a product of the intellectual and social ferment of Europe and North America in the mid- and late-19th century. It parallels in many ways the responses of the proponents of utopian communities of that period, which were discussed in Chapter 2. Many observers of the growth and spread of cities of that time believed that the natural environment and agrarian activities were being threatened by the demands of modern, capitalist, urban society. Among the most prominent were Frederick LePlay in France and, later, Patrick Geddes in Britain. They turned their attention to promoting ways to make the earth more habitable and, especially, to achieving a balance between human and natural factors.

Patrick Geddes was the first among latter-day planners to sense the need for larger area planning. He observed the spread of urban development in 19th-century England and coined the term "conurbation" to capture the interdependent quality of these linked cities. He also propounded the need to plan together all the features of a river basin (e.g., the land and natural features, as they are affected by agriculture or industry or water control) and the needs of a population for land for recreation and for residences (see Figure 9.1).What Geddes noted was how areas are unified by either the problems of their development or by their resource base, and sometimes by both. He called for planning for regions as well as for their constituent communities. Geddes also propounded his famous trinity of factors to be taken into account in spatial planning: *Folk* (the people of the region); *Work* (the economy of the region); and *Place* (the geographical and natural environmental dimensions of the region).[1] The interrelations among these factors not only signify the integrity of any region to Geddes, they also provided the foundation for undertaking regional planning.

Geddes tried many ways to demonstrate his notions; his hypothetical geographical unit, "The Valley Section" (Figure 9.1) is one example. The Valley Section presented the spatial relations of a set of settlements, each with distinctive geographical and human values in the form of a transect from wilderness to urban use. Within it are embodied two basic principles of regional planning: (1) the need to take a synoptic approach to regional problems in order to encompass the interrelationships of areas; and (2) the planning of each area in coordination with adjoining areas. Thus, the planner sees in a given region that the different factors interact such that change in one leads to changes in others. Of course, the use of a river basin to illustrate

Figure 9.1 The Valley Section of Patrick Geddes

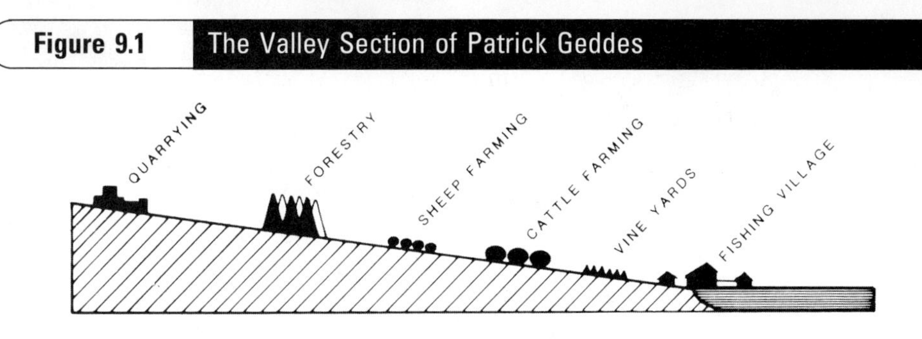

Geddes used the example of activities at different levels in a valley to emphasize the need to plan for them together. This integrated view of planning, dating from 1892, is the forerunner of modern ecosystems planning.

these principles made very good sense, for much of human settlement has taken place in such regions, and the parts of river basins are clearly linked to each other. The concept of the river basin as a natural region of high order proved to be a powerful one in planning, as discussed below.

Contemporary connections to the Valley Section are found in the New Urbanism concept of transect planning.[2] Instead of a cross-section through a river valley, the modern concept features a cross-section through an urbanized region from urban core to rural fringe (see Figure 6.7, page 159) Like its Geddesian forbear, the urban–rural transect sees the urban and rural environments (and those in between) as being deeply connected and displaying basic ecological principles such as a sequence of habitats, interrelatedness of habitats, and internal diversity.

Some other enduring ideas and ideals in planning are also associated with regional planning. For example, the Garden City movement, initiated by Ebenezer Howard and advanced by Thomas Adams, was not concerned exclusively with planning single new towns. It was based on a concept of how best to organize the territory around large cities—that is, to concentrate populations and provide open space between towns rather than let cities sprawl. Figure 9.2 illustrates Howard's regional ideal for the new towns and how they would be connected to the large city by railways. Howard's 1898 diagrammatic concept for satellite Garden Cities has been far-reaching. A greenbelt of farms and forests would separate all communities, but major transport routes would interconnect them. Metropolitan plans for Ottawa, London, and Washington owe much to this model. As well, some present-day planners tout Howard's model as a prototype for sustainable development of metropolitan regions.[3]

The River-Basin Planning Region

There have been many efforts in river-basin planning throughout the world: in the United States, Colombia, Russia, India, and even in Canada. The most dramatic and successful of these began in the 1930s for the Tennessee River, a major tributary of the Mississippi River, in the southeastern United States. This region of several thousand square kilometres was for many decades subject to major floods, owing to inordinate cutting of its forests and debilitating agricultural practices that rendered its soils incapable of holding water. Since the basin spread over several states, the U.S. federal government established a special agency, the **Tennessee Valley Authority** (TVA), with the power to plan and economically develop the region. The mandate of the TVA was to rehabilitate the region for the benefit of its inhabitants through the construction of dams, generation of electric power, reforestation, promotion of improved agricultural methods, irrigation, and the building of new towns. This broad mandate was based on the same principles of the interdependence of the

Figure 9.2 Ebenezer Howard's Concept of the Metropolitan Region

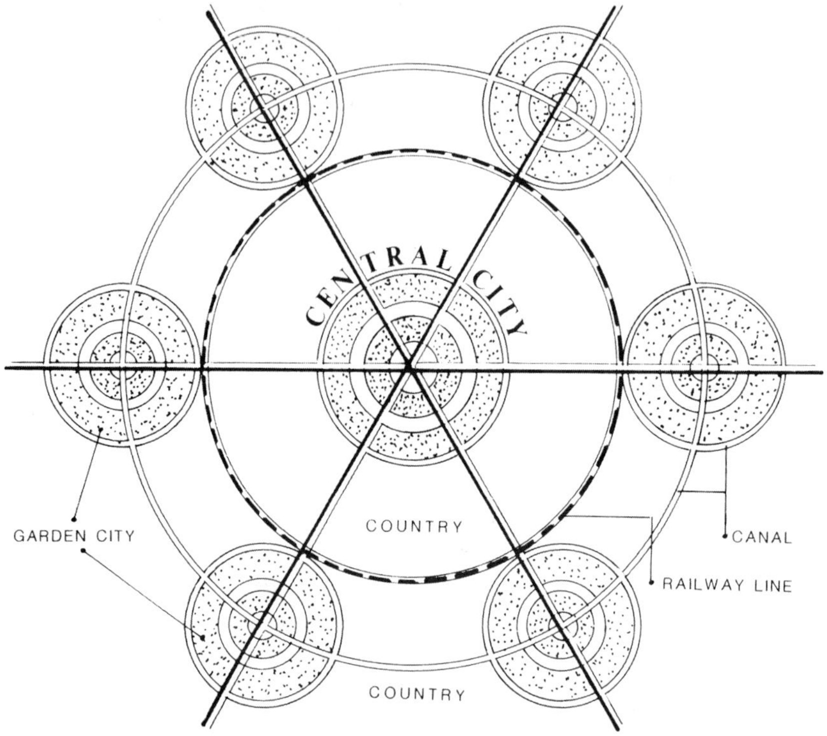

Howard's 1898 diagrammatic concept for satellite Garden Cities has been far-reaching. A greenbelt of farms and forests would separate all communities, but they would be interconnected by major transport routes. Metropolitan plans for Vancouver, Washington, and London, Ontario, owe much to this model.

parts of the region that Geddes espoused. Urban historian Lewis Mumford commented as follows about the TVA:

> The Tennessee Valley project, with its fundamental policy of conservation of power resources, land, forest, soil, and stream, in the public interest, is an indication of a new approach to the problems of regional development.... The river valley has the advantage of bringing into a common regional frame a diversified unit: this is essential to an effective civic and social life.[4]

There is more than a hint in Mumford's remarks of a strongly held philosophy about planning for human communities. For him and others who witnessed the growth and spread of large cities in the 1920s and 1930s, and the consequent debilitation of resources and the environment, the region was a special place created by people interacting with their environment. The natural region could be a bulwark against massive urbanization and the standardization of culture. This regionalism—a belief in the profound connection between humankind and the territory it inhabits—has re-emerged in Canada and elsewhere in recent years under the name **bioregionalism**.[5] Many of the same sentiments have always pervaded regional planning and may be detected in, for example, the 1969 plan for the Mactaquac River Valley in New Brunswick.[6] There are also echoes of regionalism in New Urbanist planning.[7] This humanistic view, it must be noted, has not always been easy to reconcile with the rigidities of political boundaries, bureaucratic jurisdictions, and economic determinism. Such tensions are characteristic of the history of regional planning in Canada.

Characteristics of Regional Planning

In general, regional planning is rooted in the importance of using natural resources wisely. From this resource perspective emerged three venues for regional planning in Canada:[8]

- **Planning for watersheds.** In the tradition of the TVA, the planning is for the control and use of water resources to prevent floods, to provide irrigation to agricultural lands, to generate electricity, and to create recreational opportunities, or some combination of these. The conservation authorities that were established for a score of river basins in southern Ontario in the late 1940s epitomize this type of regional planning in Canada.
- **Planning for rural land resources.** There are two venues for rural region planning. One is in the vicinity of large, expanding cities with efforts to achieve a harmonious balance between urban and rural land needs. The other is in planning for "completely rural" regions—those regions with broad areas of resource use and land occupancy by resource producers and with many small settlements.[9]
- **Planning economic development in resource regions.** These regions depend heavily on the economic performance of their natural-resource sectors. In many of these regions, the resources have become obsolete (as with coal) or depleted (as with minerals and timber), or the technology for exploiting them has become outdated (as with farming and fisheries).

There is now a fourth important venue for regional planning that must be noted:

- **Planning for large urban regions.** Canada is a leader in the planning of growing and expanding metropolitan areas. (This form of regional planning is examined in the final section of this chapter.)

Even though the substantive concerns of each of these approaches of regional planning differ, all the approaches have certain characteristics in common.

1. Regional Planning Deals with Large Areas. "Supra-urban space" is the criterion used by John Friedmann[10] to describe the spatial scope of regional planning. Regional planning efforts encompass areas larger than a single city, from metropolitan regions of a few thousand square kilometres to large resource regions of several hundred thousand.

2. Regional Planning Is Concerned with the Location of Activities and Resource Development. Since large areas are being planned, it is vital to know where activities and/or resource development occur (or might occur) in a region, so that their spatial relations (i.e., transportation, communications) with each other and with other regions can be planned for effectiveness and efficiency, both socially and economically.

3. The Scope of Regional Planning Includes Social, Economic, and Environmental Factors. Regional planning is intimately tied up with social and economic questions, more so than city planning, and with environmental issues, through its connection to natural resources. Indeed, it has a special concern with the relationship between human use and the natural landscape and environment; the need to achieve "man/land harmony," or balance, as we call it today, is often cited in the earliest regional-planning literature.

4. Planning Regions Have No Constitutional Basis. Regions for regional planning are not part of Canada's governing structure. Thus, planning regions literally have to be invented each time we want to conduct regional-planning activities. The power to define a planning region lies with the province and is usually superimposed on existing local governing units (municipalities, special functional districts, etc.). British Columbia is a major exception to this situation by having established a system of two dozen regional districts covering almost all the province, each of which may undertake regional planning. In almost all cases, regional planning is involved with more than one governmental jurisdiction, and very often with more than one governmental level.

Although regional planning is a public, governmental activity, it does not always take the form of direct government action. Many times, regional planning is only an *advisory* activity. This reflects the need to blend the governing powers of all the public units that are involved with the planning needs of the region. Thus, regional planning in Canada has been undertaken under a variety of formats. In some cases, an agency has the authority to make and carry out plans; in others, the agency only has advisory powers on the implementation of plans. The former category includes the James Bay Development Corporation, which was established by the province of Quebec to plan and develop the hydroelectric resources of an immense area in the northern part of the province. The latter category includes regional-planning commissions such as those in New Brunswick, which have the authority to make plans for their regions but wield almost no powers of implementation. In between are a variety of other arrangements for regional planning as, for example, the use of Regional Growth Strategies in regional districts in British Columbia, which require constituent municipalities to commit to achieving regional growth objectives.[11] Other devices include establishing development corporations with limited powers, interdepartmental committees, and federal–provincial planning agreements.

The Problem of Regional Boundaries

Often, "regional planning" seems a nebulous term. This arises from the fact that what constitutes a region from one point of view may not constitute a region from another. Regional planning boundaries cannot be drawn with precision because of the variety of concerns involved. Even the boundaries of a watershed, which define the extent of water resources of a region, may not encompass the human interactions of commuting to work or the shipment of goods in and out of the same region. Moreover, since regional planning is a public effort to bring improvement to a region, that public effort will need to be expressed within the appropriate governmental jurisdictions: that is, regional-planning boundaries must reflect the boundaries of the governments involved. This recognition often involves compromise between government units to determine an acceptable planning region. Moreover, governmental boundaries seldom follow the patterns of natural regions. As an example, when a river is used as a boundary between government units, the natural region is split.

The issue of regional boundaries for planning purposes can never be finally resolved. There may be a unique set of river-basin regions, but human activities cannot be defined so neatly and, significantly, their spatial configurations change over time. Political boundaries, meanwhile, tend to remain fixed over long periods of time. This ambiguity must be tolerated in regional planning. Rather than pursue the elusive ideal boundary for a planning region, it is more productive to plan explicitly for the boundary areas in conjunction with the adjacent region. It is wise to remember that wherever the region's boundaries are drawn, there is another region on the other side.

Canadian Experience in Regional Planning

Regional planning began in Canada, as it did in several other countries, just prior to World War II. The practice that developed has seen regional planning address two distinct regional situations:[12]

- **Rural and non-metropolitan regions** of the country, where regional planning aims to maintain settlement systems, rejuvenate economies, and conserve resources.
- **Large urban and metropolitan regions**, where regional planning is employed to direct urban growth in viable, efficient, and harmonious patterns.

These two approaches have often coexisted in adjacent regions and have sometimes overlapped, as with planning for the countryside surrounding a large city. And while their paths and practice developed distinctively, the planning for each situation derives from the same roots and possesses the same general characteristics noted above.

Planning for rural and non-metropolitan regions is diverse, in response to the sheer size of Canada's territory (about 10 million square kilometres) and the variety of regional situations within it. Three separate streams of regional planning have emerged to allow planners more readily to grasp this diversity: (1) planning aimed at rural resource development, conservation, and the environment; (2) planning aimed at maintaining and protecting rural regions, especially their communities; and (3) planning for rejuvenating rural economies. Below is a brief description of these basic forms that shows the kinds of problems regional planning tackles, the way in which planning efforts are organized, and the effects this may have on local community planning. The Canadian experience in metropolitan planning follows in the next section.

Regional Planning for Resources, Conservation, and the Environment

Throughout the history of regional planning in Canada there have been many instances of concern over the quality and extent of natural resource development. It was one of the main stimuli for the creation of the federal Commission on Conservation in 1909 (see Chapter 4).

During the 1930s, the influence of the unified approach to planning river basins, such as the TVA, could be seen in this country. Although there was nothing as grandiose as the TVA in Canada, its imprint can be seen in the establishment, also during the Depression, of the Prairie Farm Rehabilitation Administration (PFRA) and, later, the Maritime Marshland Rehabilitation Administration (MMRA). The PFRA sought to develop various measures for soil and water conservation in the then drought-stricken Prairie region, and MMRA's mandate was to reverse incipient saltwater intrusion into coastal agricultural lands in the Maritimes.

In 1946, something close to the "classic" form of river-basin planning appeared in Canada. Ontario proclaimed a new Conservation Authorities Act. In it were provisions to establish regional conservation authorities to conduct multi-purpose planning for a dozen or more river basins in the province. The fruits of this planning are still enjoyed today in terms of flood-control measures, wetlands conservation, and water-based recreation. The exemplary work of the Crombie Royal Commission on the Future of the Toronto Waterfront was set within conservation authority boundaries.[13]

The doctrine of the connection between humankind and the territory it inhabits is also found to a greater or lesser degree in other regional planning efforts of the 1960s and 1970s. We see it clearly in the planning for the Mactaquac River Valley of New Brunswick,[14] in the planning for the impacts of the Diefenbaker Dam in Saskatchewan, in the plan for the Cumberland Sound region of the Northwest Territories, and in the Niagara Escarpment plan in Ontario. Contemporary efforts in the same vein can be seen in the federal-provincial-local-First Nations initiative to plan for the Fraser River Basin in British Columbia.[15]

Rural Region Planning and Community Maintenance

One of the most successful and long-lived regional planning experiences in Canada is found in the physical planning for rural regions and small towns. Regional planning commissions in Alberta and New Brunswick, county planning in Nova Scotia and Ontario, and regional–district planning in British Columbia are among the forms that provinces have devised over the years to respond to the planning needs of low-density settled regions.

The oldest of these agencies are five non-metropolitan regional-planning commissions in Alberta, dating from the 1950s.[16] The reasons for their establishment in Alberta, and later elsewhere, are obscure, but it was probably a response to urban-type problems in rural areas, including spillover from growing towns and the consequences of subdivisions and ribbon development along countryside roads. Curiously, they seem not to have been formed to deal with basic rural problems, such as the lack of growth in towns, loss of land used for crop production, and population decline. Nevertheless, the planners recognized that their regions did not need conventional urban physical planning.[17] Each town or township needed solutions to its specific problems, such as the provision of basic utilities, disposing of trash, refurbishing Main Street, or reducing residential scatter (see Chapter 10 for a full description of small-town planning problems and approaches). Several provinces also established agencies to provide planning services to small communities, such as Prince Edward Island's Land Use Services Centre and Manitoba's and Newfoundland's field planning offices.

Worth special mention are the efforts of two provinces, British Columbia and Quebec, to conserve agricultural lands. In 1973, B.C. set up the Agricultural Land Commission, which, in turn, established Agricultural Land Reserves (ALRs) to restrict urban development on areas with soil types that could support agricultural production (usually soil types 1, 2, and 3). The Land Commission continues to this day and has control over the subdivision of land in zones designated as ALRs inside and outside municipal boundaries. Its worth has been proven in protecting agricultural land in the rapidly growing Vancouver metropolitan area.[18] A similar step was taken in Quebec in 1978 with Bill 90, An Act to Protect Agricultural Land, which established a commission to protect agricultural land in "protected zones."[19] In both provinces there has been considerable success in preventing non-agricultural uses in protected areas.

Regional Planning for Rejuvenating Rural Economies

Toward the end of the 1950s, Canada's economy entered a new spatial phase. Metropolitan regions were replacing rural-resource regions in economic importance. And in the process, many of the country's agricultural, fishing, mining, and forestry regions exhibited signs of economic underdevelopment. Poverty, illiteracy, poor housing and infrastructure, inefficient technology, obsolete resources, and out-migration were in evidence. As these rural regions began to demand a share of the national prosperity, a period of extensive regional planning was embarked on that would last until the early 1980s.

The awareness of unequal development between regions of the country captured the attention of people and politicians at both the provincial and federal levels. "Regional disparities" became a familiar phrase, and the concerns led to considerable amounts of mostly federal funds (some have estimated $15 billion) being devoted to mitigating regional economic differences. This, it should be noted, was not a new phenomenon in Canada, as the Rowell-Sirois Royal Commission had pointed out in 1940: "[T]he income of the country is concentrated in a few specially favoured areas."[20] They were referring, in particular, to the urban–industrial corridor from Windsor to Quebec City that Canadian geographer Maurice Yeates would later call "Main Street."[21]

The first major effort at countering regional disparities came in 1961, with the passage of the federal Agricultural Rehabilitation and Development Act (ARDA). Under ARDA, low-income agricultural regions were targeted for programs of enlarging farms, establishing community pastures, and improving farm market roads. Over the next few years, 40 regional planning efforts were undertaken. Three of them deserve special mention, to show both the scope and difficulties surrounding ARDA regional planning. One is the truly grassroots redevelopment program in the Gaspé, the Bureau d'Amanagement de l'est du Québec (BAEQ), involving all sectors of the community in what nowadays would be called "self-management" of the region's future. A second is the "top-down" Newfoundland Outport Resettlement Program, which, while it succeeded in relocating 300 outports (of the 600 that were targeted), caused numerous splits in families and communities that still evident today. A third is the multifaceted Interlake Region program in Manitoba; it might be considered ARDA's success story.[22]

A host of other alphabetic agencies joined ARDA during the 1960s: ADA, the Area Development Agency; ADB, the Atlantic Development Board; and FRED, the Fund for Rural Economic Development. They were all gathered under the umbrella of DREE, the Department for Regional Economic Expansion, in 1969.[23] DREE's approach was generally comprehensive in regard to a region's needs and included housing, municipal infrastructure, transportation, and education, along with job creation. Throughout this period, the regional-planning efforts involved both federal and provincial levels of government, for although the necessary funds were available at the federal level, the responsibility for regional resources lay with the provinces. This was, as Gertler notes, "a period of experimentation" in regional planning, owing, as much as anything, to the need to evolve an approach to joint planning between two levels of government.[24] This vigorous 25-year period in Canadian regional planning is hard to characterize. Some would ask: Was it regional planning at all? There were seldom permanent planning staffs in DREE-designated regions. There were federal–provincial task forces, development corporations, and general funding agreements. But there was seldom a published regional plan. Not least, it was regional planning from *outside* the region. And this could sometimes be extremely upsetting to local councils and planners, as witness B.C. Hydro's plans to flood Arrow Lakes' communities[25] and the mega-industrial projects along the Canso Strait in Nova Scotia.[26] In any case, the "experiment" Gertler spoke of is now far behind us and unlikely to arise again in an era when both bioregionalism and globalism vie for attention.

The Bioregion and Other Approaches

Brief mention should be made of a number of other approaches to regional planning either for or affecting land use. One concerns the planning for regional transportation systems, especially expressway planning. While these are similar to the "functional" plans referred to in Chapter 6, they cover a much wider territory and involve many communities. Almost all the large metropolitan areas have had such plans prepared. Quebec took a major step regarding regional planning in 1980 when its Regional and City Planning Act established a new regional-planning framework.[27] It provided for the creation of 94 Regional County Municipalities (Municipalités regionales de comte, or MRCs), and one of their main tasks was to prepare a regional plan. The act defined very specifically the contents of such a plan. Briefly, the plan had to include proposals for land use; delimitation of areas to become urban; identification of environmentally sensitive, historical, and cultural areas; and the location of inter-municipal facilities and public utilities.[28] With this legislation, Quebec took the most comprehensive step in province-wide regional planning of any province because, significantly, it linked regional plan preparation to governmental powers of implementation.

More recently, **bioregional** approaches have begun to appear in Canadian regional planning. One of the first is found in the planning for the Kitimat-Stikine Regional District in northwest British Columbia in the late 1980s.[29] This and other initiatives draw upon the emerging field of bioregionalism, which is not a specific planning paradigm so much as an "action-oriented movement based on ecological principles."[30] It arises from

Planning Issue 9.1 Farmers Fear Loss of Land

The Globe and Mail
June 13, 2006

Farmers Fear Loss of Land: Push to Open Agricultural Land Reserve May Be Beginning of the End for Way of Life
Mark Hume

MAPLE RIDGE, B.C.—Bill Hampton, with the reins to his workhorses, Doc and Beauty, slack in his hands, sat on an old beat-up manure spreader yesterday and contemplated the future.

The Maple Ridge farmer, whose family first settled the lands along the south Alouette River in 1879, didn't like what he saw.

"I don't know for sure what's coming, but it don't look too good," said Mr. Hampton, 73, who raises 165 head of dairy cows on pastureland that has long been coveted by developers.

Not far from where he was sitting, authorities have marked out a highway right-of-way that will cut across his land and that of neighbouring farms.

It could be, he fears, the beginning of the end for the farmland in the area. "Right where those mailboxes are, the road will go through," Mr. Hampton said, pushing back the blue-and-white striped railway engineer's hat that shielded his tanned face from the sun.

"It will angle up through those fields and it's pretty well going to split my place in half. It's just sort of unfortunate, that's all, and it's going to make getting to my cattle difficult."

But that highway, Mr. Hampton said, is the least of his worries.

What he's really concerned about is the drive in Maple Ridge to open up the Agricultural Land Reserve to development. The highway, he said, will fragment farms, making them easier targets for developers.

Maple Ridge, a bedroom community in the Fraser Valley about an hour's drive from downtown Vancouver, is growing rapidly. There has long been talk about removing land from the provincial ALR for housing developments, and that talk has intensified in recent years, with a push by city council to establish an industrial park.

Mayor Gordon Robson, whose council is in the process of drafting an official community plan, said there is a need to create "employment zones."

He said the community does need to develop an industrial park somewhere, however, because the current mix of land use—92 percent residential, 6 percent commercial and 1 percent industrial—is not sustainable.

Matthew Laity, 31, whose Brookfield Farm is next to Mr. Hampton's, said he first started hearing talk about development 20 years ago.

Now he and his wife, Deanna, wonder what the area will be like when their children, nine-year-old Jeremiah and three-year-old Nadine are old enough to take over Brookfield Farm, which produces milk, composted manure, free-range eggs and corn.

Mr. Laity said he's well aware that he could make a lot more money by selling to developers than by working hard every day to milk his cows. But he has no desire to sell.

"It's about more than just making money," he said, looking around at the lush, green fields where his herd of black and white cows browsed.

"It's about environmental stewardship, it's about family traditions, and it's a good life. It's hard work, but it's clean work, and it's a great place to raise a family."

Yesterday, Mr. Laity, Mr. Hampton and about 50 others stood along a narrow country road and waived signs at passing drivers, most of whom honked in support. "No Industrial Park Here!" and "A.L.C.—No more exclusions," they read.

Among those standing at the roadside was Jeremiah Laity, who said that one day he hopes to run the farm his parents now operate.

"I just think it's crazy," the boy said of the proposal to develop the land. "It should always be protected. It's good land."

Source: Excerpted from Mark Hume, "Farmers Fear Loss of Land," *The Globe and Mail,* June 13, 2006, S1. Reprinted with permission from *The Globe and Mail.*

a particular set of values regarding the quality of a region's living space. It is holistic in its outlook, in that a bioregion is a system comprising three sub-systems: a "biophysical" sub-system (the natural environment); an "inhabiting" sub-system (communities, agriculture, transportation); and a "network" sub-system (the economic and political systems). Bioregionalism rejects existing political units in favour of "contiguous, mappable geographic regions" based on similarities of topography, plant and animal life, culture, and economy, such as watersheds.[31] In many ways, this is a similar perspective to the "regionalism" of the 1930s alluded to earlier in this chapter.

What tends to appear in Canadian regional planning is not full-blown bioregionalism but rather the application of some of its principles. The Crombie Commission on the Toronto waterfront utilized the watershed bioregion notion as its organizing principle (see Figure 9.3). The large municipality of Markham outside Toronto has conducted a comprehensive environmental-planning study emphasizing not only environmental protection but also restoration of ecological diversity and the integration of these with cultural patterns in the area.[32] A similar approach is taken in the updated plan for the greenbelt surrounding Ottawa, which propounds a "connected ecological system" and employs nodes, buffers, and links, as the bioregionalists call the landscape components.[33]

Bioregional concepts and values are increasingly influencing both regional and local planning with important implications for planning practice. For example, the bioregional approach links ecology and community and puts them at the centre of planning, rather than, say, land use and economic development.[34] And not the least will be the need to negotiate new regional boundaries and implementation powers. In the meantime, one of the key components of life—water, and its regional situation—is receiving considerable attention by planners. In the Calgary region, the stewardship of a number of watersheds is sparking interest in inter-municipal planning[35] while "integrated water management" is receiving attention by Quebec's planners to avoid over-exploitation of water and harmful impacts of conflicting water uses, and to plan for natural regeneration of regional water resources.[36]

Planning for Metropolitan and City-Regions in Canada

Canada is a leader in metropolitan planning in the Western Hemisphere. Almost all of the country's 27 metropolitan areas—the regions of our largest cities—have active planning agencies. This experience is now six decades old, having started with Winnipeg in 1943 and Toronto a few years later. In this section are described some of the highlights of Canadian metropolitan planning, with the aim of identifying the forms it has taken, the problems it has tackled, its accomplishments, its relationship to local community planning, and its new challenges as metropolitan areas have begun to expand into city regions. But before doing this, it will be helpful to examine the backdrop for metropolitan planning.

Nature and Origins

Metropolitan planning is a special form of regional planning. It not only deals with a large area, but it also deals explicitly with the growth and expansion of a major city on which the region is usually focused. Further, the planning normally is conducted in an institutional setting involving several municipalities. Broadly speaking, metropolitan planning is concerned with the allocation of land uses and the location of major public works throughout the region of a metropolis and, in the past decade or so, also concerned with protecting the natural environment.[37] In many ways, metropolitan planning is community planning "writ large."

Three factors must be remembered, however, about metropolitan planning. The first is the scale of the area involved. Metropolitan planning areas are usually defined to

Figure 9.3 Greater Toronto Bioregion

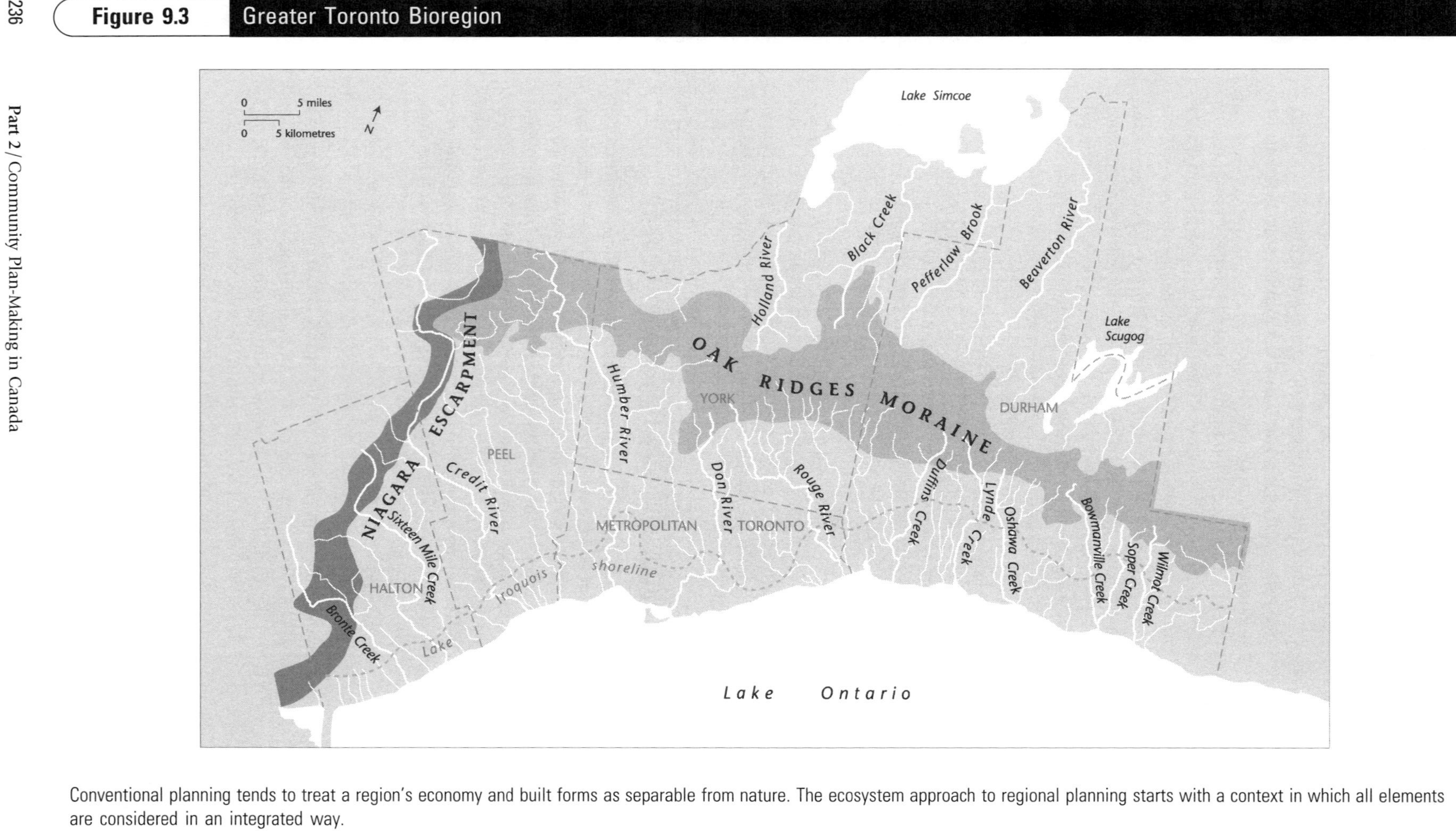

Conventional planning tends to treat a region's economy and built forms as separable from nature. The ecosystem approach to regional planning starts with a context in which all elements are considered in an integrated way.

Source: Gerald Hodge and Ira M. Robinson, *Planning Canadian Regions* (Vancouver: UBC Press, 2001), Figure 15, 223. Reprinted with the permission of the publisher. © UBC Press. Cartography by Eric Leinberger.

encompass the potential spread of urban development, and this often means 1000 square kilometres or more. Moreover, the area will be developed at different densities and interspersed with agricultural areas and open spaces for recreation. The second also has to do with scale, but this time with the size of population. Planning for large aggregates of population requires recognition of the increasing division of the metropolitan community into separate areas for work, residence, shopping, and leisure. Metropolitan planning must try to accommodate, as Hans Blumenfeld says, a number of "contradictory requirements," such as providing a "minimum need for commuting but maximum possibility for commuting."[38] The third factor is the inter-municipal setting for planning. Given the autonomy accorded to local governments in Canada, the planning for several of them simultaneously must seek ways to blend competing aims for development in the interest of all the citizens of the metropolitan community. Achieving this planning blend requires special organizational arrangements, and these have taken a number of forms, as will be seen later.

The emergence of metropolitan planning in Canada coincides with the end of World War II, when cities all over the world began to have dramatic surges of population growth and commercial and manufacturing development. For example, in 1941 Canada had 15 metropolitan areas, which comprised about 40 percent of the total population. By 1961, five more cities had been added to the metropolitan class, and metropolitan populations had become 51 percent of the total. Even more dramatic is the fact that in this 20-year period, 5 million more people crowded into Canada's largest cities, more than doubling the population living there. In 2001, the population of the current 27 metropolitan areas comprised nearly 19.3 million people, or 64 percent of Canada's total.[39]

This vast growth created a host of problems for both the central city and the surrounding region in which the new suburbs were being built. In the central city, the problems were a combination of an aging physical environment and a lack of vacant land for new development. There was deterioration of physical infrastructure and of some older residential and industrial areas as populations and factories moved to the outskirts, thus leading to the need for rehabilitation and redevelopment. (In modern planning parlance, the latter are known as "brownfields.") The new suburbs, at the same time, faced problems in providing services such as water, sewerage, garbage disposal, police and fire protection, roads, and schools. Their problems stemmed from the size of the new suburban growth and the meagre financial resources available to previously small municipalities. In addition, with the rapidly growing population and the vast areas being settled came the need for new facilities to serve the entire metropolitan area—among them hospitals, expressways, parks, airports, new sources for water supplies, and sewage-treatment plants.[40]

Metropolitan planning originated in this complex, large-scale urban development. It grew out of the realization that no single municipality in a metropolitan area could deal with an array of problems as intertwined as these, with the need to balance growth and to provide metropolitan-wide facilities. For example, the location of a large subdivision or a shopping centre in one part of the area may generate the need for improved highway access, new schools, or new trunk sewers. Conversely, local desires for development may depend upon the availability of metropolitan facilities and services. Achieving coordination in the land-use planning of diverse communities thus becomes a major aim of metropolitan planning. Manitoba, British Columbia, and Alberta were the first three provinces to establish such coordinating planning agencies, followed closely by Ontario.

Organizational Approaches

The first formal metropolitan-planning agency established in Canada (indeed in North America) was that created by the Manitoba government in 1943—the Metropolitan Planning Commission of Greater Winnipeg. The dozen or so municipalities that constituted the metropolitan area were members of the commission and contributed financially to its operation, as did the provincial government. It had a small and energetic staff, headed by Eric Thrift, and produced (by itself and with the help of consultants) an impressive series of reports on traffic, parking, the central business district, and parks. Although much was achieved through this inter-municipal effort, municipalities could choose whether to participate, and the plans that were made by the commission were not binding on any municipality. This same cooperative, advisory form of metropolitan planning was adopted soon after in Vancouver (1949), Edmonton (1950), and Calgary (1950).

It gradually became evident that metropolitan planning without a commensurate level of government authorized to implement planning policies could achieve only limited results. This was especially true in regard to decisions about the location and financing of facilities to serve the entire metropolitan area. It was also true for local land-use regulations and capital investments that could thwart the intent of the metropolitan plan. This division of authority among local units of government was, of course, the product of the primacy accorded municipalities in handling their own affairs a half century or more earlier by the provinces. The provinces, with the exception of Ontario, were slow to intervene in this tradition and form metropolitan governments.

Metropolitan Toronto

Origins and Context. In 1953, the Ontario government established the Municipality of Metropolitan Toronto, the first metropolitan government in either Canada or the United States. It followed the form of a **federation** of the 13 local municipalities that made up the metropolitan area at the time. (The alternative form, an **amalgamation** of municipalities into a single unit for the entire metropolitan area, was debated and rejected at the time.) Within the federation, each municipality retained responsibility for its own local planning and land-use regulation, and its own local public works. The metropolitan government, "Metro Toronto" as it came to be called, assumed responsibility for major regional services and such facilities as public transportation, water supply, and expressways. Metro Toronto also had the power to raise funds for capital-works projects.

When it was established, the metropolitan municipality was provided with an Advisory Planning Board like those of other municipalities in the province. There was one major difference, however, because the Metro Toronto Planning Board had jurisdiction in planning over a surrounding area that was twice as large again as Metro. The planning area covered a total of 1970 square kilometres, of which 620 square kilometres comprised the 13 metropolitan municipalities, while the remaining area comprised 13 fringe-area local governments. The reason for establishing such a jurisdiction was to give metropolitan planners some measure of control over the development occurring beyond their boundaries.

The Metropolitan Toronto Planning Board was charged with preparing an Official Plan to which municipalities both inside and outside of Metro were required to conform in their planning and public works projects. Within Metro, the zoning of local municipalities had to conform to Metro's Official Plan and Metro had to approve local subdivision plans. It should be noted that the Official Plan of Metro was only advisory and could not bind the local municipalities, but it was a strong persuasive force. An important reason for this, as Blumenfeld notes (he was Deputy *Commissioner* of Planning for Metro for many years), is that each municipality sent representatives to Metro Council, where they had a voice in adopting the Official Plan.[41] As well, the debentures for public works of each municipality were underwritten by the metropolitan corporation. Further, from the metropolitan-area side, Metro had the power to fund and construct area-wide facilities and, thus, see its own plan become implemented.

Later Changes. There have been a series of changes to the metropolitan government of Toronto since its origin. The first was 15 years after its founding, in 1968, when the continued pressure on the City of Toronto to consolidate all 13 municipalities into one single city resulted in a review of Metro's organization and functioning. The government structure was altered, the 13 local governments were reduced to 6, and a number of other refinements were made. In 1971, the province began instituting its program to establish regional governments, somewhat modelled on Metro, for all major urban areas in Ontario as well as for the urbanizing area surrounding Metro. The Regional Municipality of York was established to the north, Peel Region was instituted on the west, and Durham Region, on the east. The significant change was the elimination of the fringe portion of Metro's planning area, thereby reducing it to the size of the metropolitan area. The evolution of the structure of Metropolitan Toronto region continued again in 1997 with the amalgamation of the six constituent municipalities into one metropolitan unit, the new City of Toronto. The debate was at times rancorous, often centring on the loss of democratic rights in the way the decision was made and the loss of presumably more responsive local government entities.[42] Curiously, the province's proposal barely touched on the impact that amalgamation would have on the content and process of planning for what had become an even larger "city-region."[43] Similar amalgamations were also instituted at about the same time for other urban areas including Ottawa, Hamilton, London, and Kingston. Ottawa and Hamilton became single-tier governing arrangements rather than the two-tier federated models previously used, although two-tier arrangements remained in the regional municipalities surrounding Toronto.

Metropolitan Winnipeg

Metropolitan planning in Winnipeg was given governmental backing in 1960 when the province created the Metropolitan Corporation of Greater Winnipeg, the second metropolitan government in Canada. The corporation was given responsibility for water supply, sewage disposal, transportation, and planning. Winnipeg chose a two-tier system of government, with the local municipalities remaining along with the new metropolitan corporation, yet it was not a federation as in Toronto. Metropolitan councillors were elected directly rather than drawn from local governments. This created tension between the two levels of government. Further, the metropolitan corporation had no control over the capital spending of the local-level governments. These two conditions frequently resulted in a failure to implement metropolitan plans.[44] In 1971, Manitoba restructured the metropolitan government, this time in the form of a single-tier arrangement, the City of Winnipeg, or Unicity, and all levels of government and planning were integrated.

Other Metropolitan Areas

The approach of cooperative, advisory, inter-municipal planning remained in practice in the metropolitan areas of Vancouver and Victoria until 1966. Regional governments were then formed and assumed the planning functions previously carried out by regional planning boards. Both the Capital Region District (Victoria) and the Greater Vancouver Regional District have since prepared metropolitan-area plans. The Vancouver plan, The Greater Vancouver Livable Region Strategic Plan (see Figure 9.5, page 242) adopted in 1996, conceptually owes much to the official plan prepared in 1969 by the long-lived Lower Mainland Regional Planning Board. The metropolitan governmental setup in Vancouver is a federation, like the two-tier arrangement of Toronto before 1997.

Two other early metropolitan planning efforts, in Edmonton and Calgary, did not evolve into metropolitan governments. In both cases, the regional planning commissions were continued, but with increased powers. For example, not only were these commissions empowered to prepare a regional plan, but they were also the designated authority for approving subdivision plans for all parts of their respective metropolitan areas outside the central cities. As well, all cities in Alberta had planning authority over an additional 5-kilometre zone outside their boundaries. Calgary was a special case: the city annexed large areas on its fringes in the 1960s to bring the potential urban area under a single jurisdiction. The province withdrew its regional planning legislation in the early 1990s.

Planning the Form of the Metropolis

As much as metropolitan-area planning seems like a peculiarly contemporary challenge, European, especially German, planners and architects were broaching it more than a century ago.[45] Otto Wagner presented his modular *Grossstadt* plan for Vienna in 1893, not long before Ebenezer Howard put forth his plan for satellite Garden Cities (Figure 9.2, page 228). Other satellite models appeared in Germany in 1909 and 1911. A decade later, Corbusier contemplated the "city of three million" (see Chapter 4) and Thomas Adams soon after would present his Plan for New York and Environs. The advent of the widespread use of the automobile, and its impact on land use, is one of two defining characteristics of the latter-day metropolis that Canadian planners have had to confront. Canada's metropolitan areas did not grow as large or as early as had those in Europe and the United States, and the first challenge was the auto-oriented "exploding metropolis."[46] The other defining characteristic, now confronting planners, is the protection of the natural ecosystems of the metropolitan region. Each is discussed below.

Taming the Exploding Metropolis

The vast metropolitan growth that characterized the 1950s and 1960s (and since) did not bring forth compact cities. Rather, it dispersed new and old populations, businesses, and factories over large areas. Modern means of transportation—automobiles, trucks, and highways—and better communications systems came into their own in this period, enabling people and firms to seek an ever-widening array of locations. It seems that many people wishing to live close to the countryside sought the new suburbs, with country on one side and city on the other. This creates a dilemma, as Blumenfeld notes, for "as more and more people move out into ever widening rings of suburbs, they move farther and farther away from the city and country moves farther and farther away from them."[47] This perception led planners of that time to propose patterns of urban development that would be consonant with the new, large scale of cities, the need to blend the amenities of both city and country, and the necessity to provide a maximum of accessibility among all parts of the metropolitan area. In short, there was much seeking after urban forms that would give coherence and cohesion—a sense of community—to the "exploding" metropolis. Planning models were brought back from the past, adapted and reinvigorated, and/or borrowed from other jurisdictions.

Four patterns gained early prominence among planners: the "concentric" city; the "central city with satellites"; the "star-shaped" or "finger-plan" city; and the "linear" or "ribbon" city (see Figure 9.4). They are, of course, ideal types, and in practice they must be modified to fit the geography of the area and the past history of development. Nevertheless, each has had an influence at one time in metropolitan plans that have been drawn up in Canada.

The **concentric plan** is based mainly on sustaining the primary business centre by ordering new residential development and other activities more or less at equal distance around the centre. The aim is to keep travel distances to the centre for work or business at a minimum for all sectors of the community. The main means for accomplishing this

Figure 9.4 Four Forms of Metropolitan Development

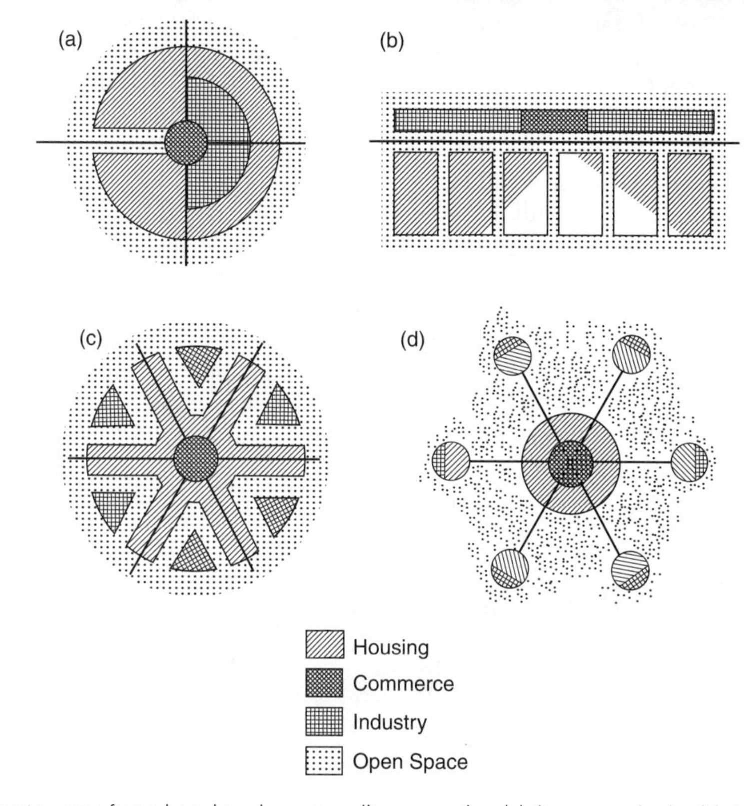

Four concepts were often adapted to plan metropolitan expansion: (a) the concentric city; (b) the linear city; (c) the star-shaped city or finger plan; and (d) the city with satellites. Each has been used as a model in planning Canadian metropolitan regions, suitably transformed to fit the prevailing topography. Each diagram is drawn to the same scale, illustrating the typical transportation axes, open-space pattern, and area required for similar land uses.

is by favouring transportation investments that concentrate travel movements in the centre, as with the subway investments in Toronto and Montreal and the light rapid transit (LRT) systems in Edmonton, Calgary, and Vancouver. In such large cities, the growth of suburbs gradually reaches limits where travel to the centre for all major activities becomes inefficient. At this stage, planners call for major sub-centres of business, or "new towns in-town" as they are sometimes called. The 1976 plan for Metropolitan Toronto[48] followed this latter concept, creating three new "town centres"—Scarborough, North York, and Etobicoke—each roughly 16 kilometres from downtown Toronto.

The **central city with satellites** derives from Ebenezer Howard's idea of Garden Cities surrounding a major city, described earlier in this chapter (Figure 9.2). This metropolitan concept became a reality with Patrick Abercrombie's plan to guide the growth of London, England, after the World War II.[49] A greenbelt of parks, agriculture, and exurban development was established to limit the physical expansion of the central city about 16 kilometres from the centre. Beyond the greenbelt, outwards about 30 to 40 kilometres, a dozen or more new satellite towns were planned to surround London and absorb its new growth, as well as to decant some industrial development from the central city. The new towns were planned to connect with the centre by high-speed transportation, but were also expected to be relatively self-sufficient. The towns would not be large, ranging from 30 000 to 60 000 in population, and would afford residents quick access to the countryside.

The 1969 plan for the Vancouver metropolitan area proposed a satellite scheme with four "regional town centres" east and southeast of Vancouver, all connected to the central city by a rapid-transit system. This plan was largely realized and the current Livable Region Strategic Plan, implemented in 1996, both expands and refines the concept. There are eight such centres in the new plan articulated by a more extensive transit system. Residential development will be focused on town centres with the aim of creating "complete communities" with a "better balance" in the availability of jobs, housing, public services, and transportation.[50] There is no formal greenbelt in the Vancouver plan, although an extensive network of "Green Zones" are used to establish a "long-term boundary for urban growth" (see Figure 9.5).

The 1947 Copenhagen plan is the best-known example of the **star-shaped** or **finger plan**. The essential feature of this concept is that development is confined to radial corridors emanating from the business centre, with green areas between each corridor. Major highway and rail transit routes serving the central city follow the corridors. This interpenetration of green space and urban development increases the distance to the city centre over that of the concentric plan, but also maximizes the access of city

Figure 9.5 Map of Vancouver's Livable Region Strategic Plan

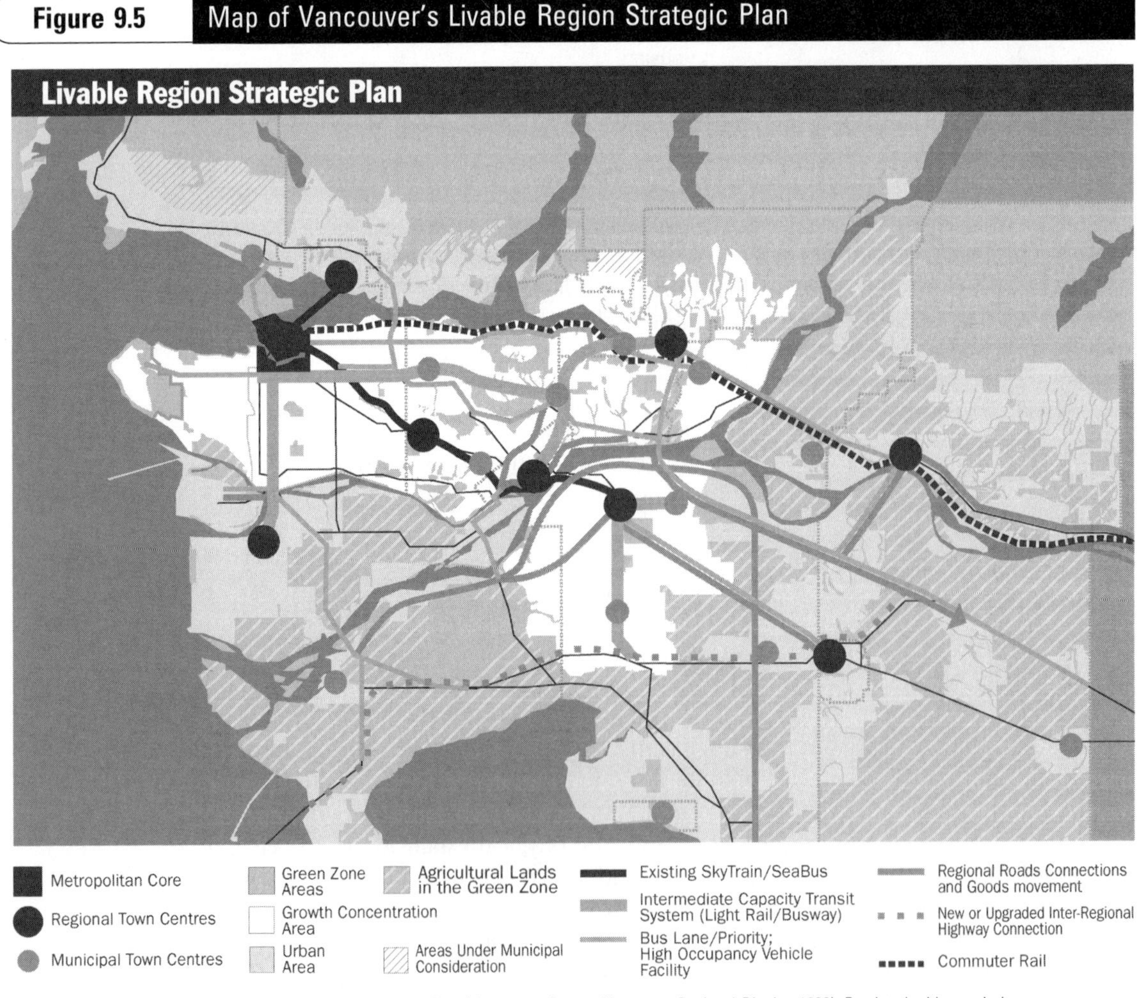

Source: Livable Region Strategic Plan (Vancouver: Greater Vancouver Regional District, 1996). Reprinted with permission.

dwellers to the countryside. The unimplemented 1974 plan for the National Capital Region of Ottawa-Hull proposed the finger-plan concept, along with a greenbelt. The plan called for a continuous open space system with "the penetration of rural wedges" into the urban area.[51]

The **linear** or **ribbon city plan** proposes urban growth in modules along major transportation routes. A greenbelt would separate the major uses of industry, residences, and transport from one another in this concept. A variation on this type was used in one alternative plan considered for the Toronto metropolitan region in a major transportation planning report in 1967.[52] The concept was for a spine of transportation running east and west from the city, along which would be arrayed, to the north and south of the spine, a series of communities separated from each other by greenbelts.

The Green Metropolis

In the closing decade or so of the 20th century, Canadian planners of communities large and small began to embrace the notion that the natural environment, and its protection, must become an integral part of decisions regarding the future physical form of the place. Nowadays, one is unlikely to find a metropolitan area plan that does not put its natural environment to the forefront. For example, the 2005 (Draft) Halifax Regional Municipal Planning Strategy[53] has "Environment" as its first substantive chapter and it begins: "Protection of water, land and air is a cornerstone of the Regional MPS." Another example is found in one of the major objectives in the 2004 Montreal Master Plan:[54] "Preserve and enhance the natural heritage." Canadian metropolitan planners have manifested the environmental prerogative in their plans in various ways and a few deserve discussion.

Ottawa and Its Greenbelt

In 1915, the Federal Plan Commission prepared a plan for Ottawa and its environs which proposed many of the projects that have since come to pass—for example, the relocation of railways out of the downtown area and the creation of the Gatineau Park greensward to the north.[55] Following World War II, French planner Jacques Gréber was commissioned to prepare a plan that encompassed a region on both sides of the Ottawa River, including the City of Hull in Quebec. This plan, submitted in 1950, proved to be very influential in creating the present physical environment of the capital region.[56] Among its main features was the development of a greenbelt around Ottawa, the expansion of Gatineau Park north of Hull, and various urban parks and parkways. The greenbelt covers about 20 000 hectares, about eight kilometres from Parliament Hill, and varies in width from two to eight kilometres (see Figure 9.6).

The federal government first attempted to implement the greenbelt using local land-use regulations, without success, because the rural townships preferred to permit low-density suburban development. In 1959, the federal government created the National Capital Commission (NCC), and hired Winnipeg's metropolitan planner, Eric Thrift, to manage the agency. As the NCC said in its 1974 plan for the region, "in implementing the Gréber plan... the NCC used as its basic planning and development tool the ownership of land."[57] Expropriating the greenbelt lands was controversial, but a crucial factor in securing it in perpetuity. In 1996, the NCC adopted a new plan for the greenbelt, which added to its extent and made it much more ecologically oriented. A primary component became its continuous natural environment, which included several core natural areas (i.e., large, sensitive, natural environments), natural-area buffers that surround the core areas, and natural-area links for maintaining continuity of plant life and facilitating animal movement within and beyond the greenbelt.[58] This new orientation adds an important dimension to what was initially just a way of shaping the urban form of

Figure 9.6 The Greenbelt in the National Capital Region

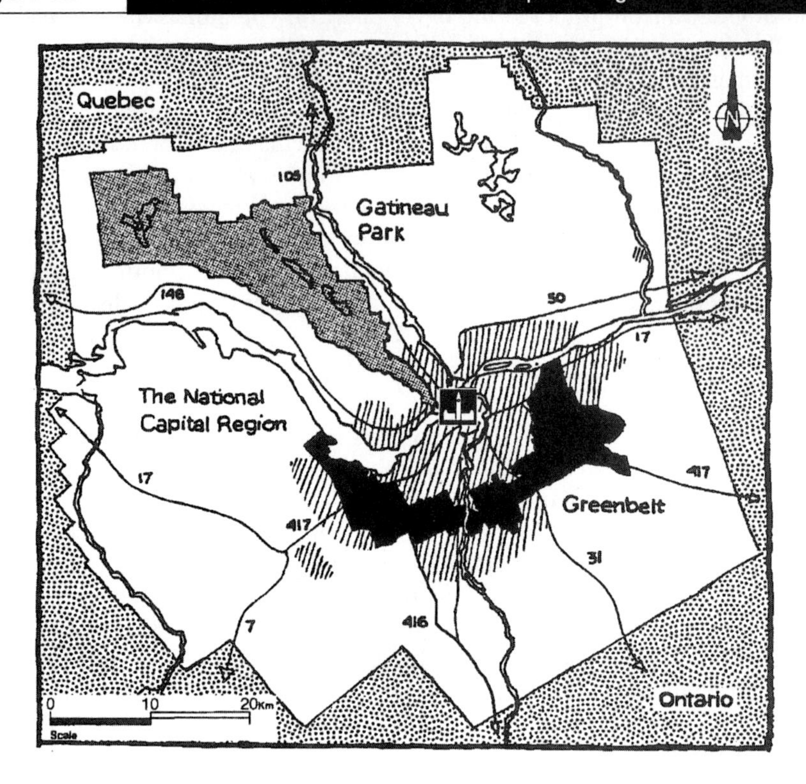

Source: Greenbelt Master Plan Summary (Ottawa: National Capital Commission, 1996), 7. Reprinted with permission.

Canada's capital (see Figure 9.6).In the most recent planning for the enlarged City of Ottawa, the greenbelt is linked to an even broader environmental strategy for the entire urban region within eastern Ontario.[59]

Toronto's Greenbelt

A major initiative regarding metropolitan Toronto took place in 2005 with the establishment of an extensive greenbelt surrounding the total urbanized region. The province instituted a greenbelt of nearly 7300 square kilometres north and west of the area, from Hamilton to Oshawa, a region in which over 6 million people reside (see Figure 9.7). The purposes of the greenbelt are stated as follows:[60]

- Protects against the loss and fragmentation of the agricultural land base and supports agriculture as the predominant land use;
- Gives permanent protection to the natural heritage and water-resource systems that sustain ecological and human health and that form the environmental framework around which major urbanization in south-central Ontario will be organized; and
- Provides for a diverse range of economic and social activities associated with rural communities, agriculture, tourism, recreation, and resource uses.

The new greenbelt enhances the protective measures of the Oak Ridges Moraine of 2001 and the Niagara Escarpment of 1990 (see Figure 9.3, page 236). Further, the greenbelt more than doubles the protected area.

Overall, the Toronto greenbelt has two goals: to contain sprawl in the urbanized area, and to permanently protect greenspace and rural areas, and activities that surround the urban area. Thus, while it is effectively an "urban growth boundary," it is not simply a demarcation line as is often advocated in conjunction with Smart Growth initiatives.

Figure 9.7 South-Central Ontario's Greenbelt Plan

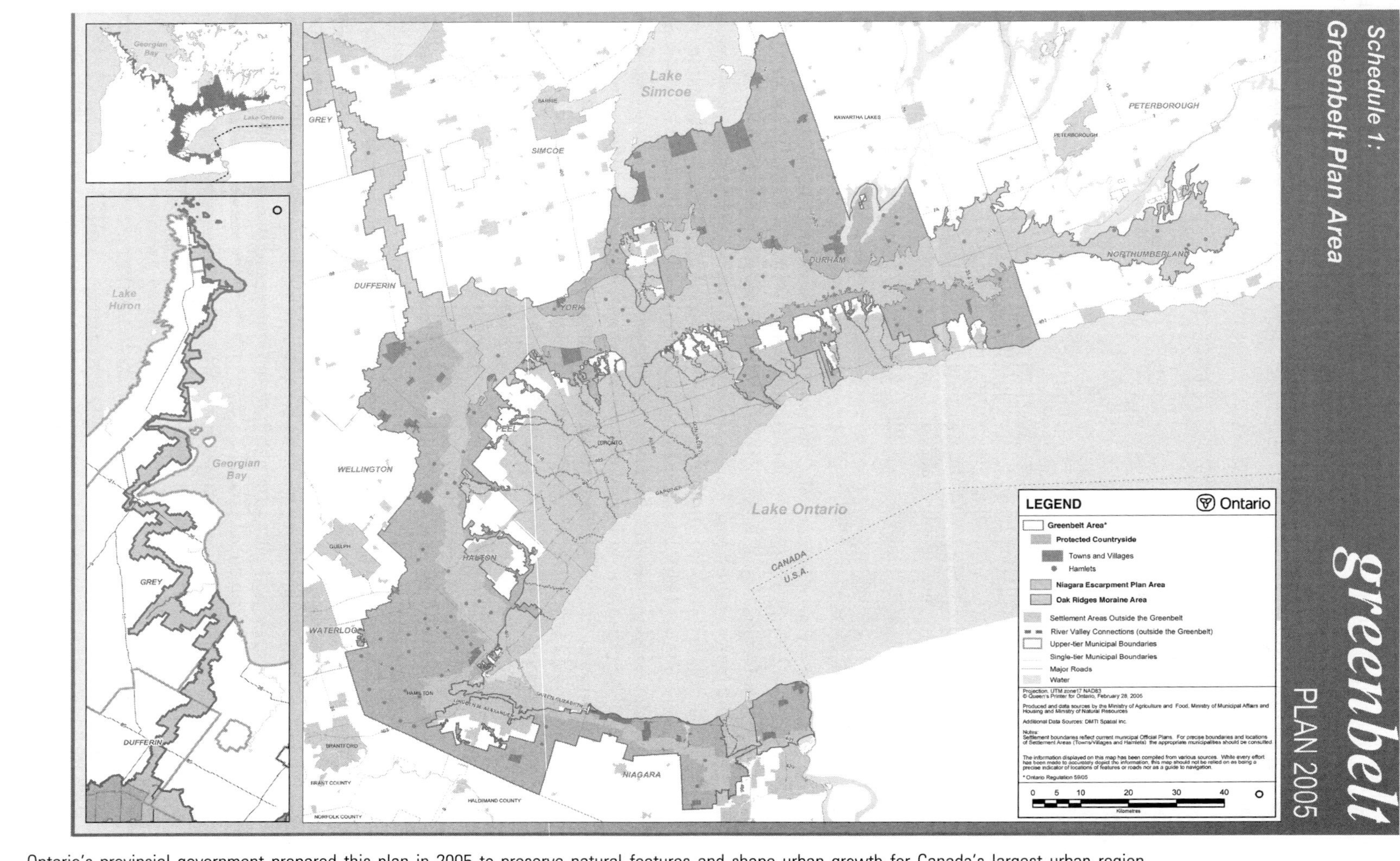

Ontario's provincial government prepared this plan in 2005 to preserve natural features and shape urban growth for Canada's largest urban region.

Source: © Queen's Printer for Ontario, 2005. Reproduced with permission.

The greenbelt's boundaries are partly based upon ecological features that are more amorphous. Further, the greenbelt area is, essentially, an extensive non-urban zone with environmental and rural imperatives to achieve rather clear boundaries with which development must comply. Although its status is not guaranteed by public ownership, as in the Ottawa situation, Toronto's greenbelt is backed by substantial provincial legislation and complementary regulations, which can override local planning policies.

Vancouver's Green Zones

Rather than a greenbelt, the approach taken in the Vancouver metropolitan region to emphasize the importance of the natural environment employs a set of Green Zones.[61] These comprise the natural mountain buffer to the north; protected agricultural lands in provincial Agricultural Land Reserves to the south and east; other ecologically important lands, major parks, and recreation areas; and, in combination, serve to define the limits to urban expansion. Protecting the Green Zones is the first of the four listed "fundamental strategies" of the metropolitan plan. In 2002, the Greater Vancouver Regional District (GVRD) adopted a Sustainable Region Initiative (SRI) to reinforce its commitment to protecting the environment.[62] It applies, first, to all programs and activities of the GVRD, and then to other governmental and public agencies with which it works. Sustainability measures include upgrading water reservoirs, transforming old gravel pits into parks, and integrating greenways and utility corridors.

The Emerging City-Region

The explosive development of Canadian metropolitan areas in the second half of the 20th century was a foreshadowing of an emerging new spatial form—the **city-region.** The new form is larger, more complex, more diverse, and more extensive than the metropolitan areas of the past. This form is already evident in the country's three largest urban regions: Toronto, Montreal, and Vancouver.[63] A distinguishing feature of city-regions, here and abroad, is that they have a spatial impact on people and jobs well beyond metropolitan-area boundaries, into outlying suburban, exurban, and rural areas. The urbanized portion of the Greater Toronto Region expanded from 482 square kilometres in 1967 to 1737 square kilometres in 1999, for example. These developments often induce severe planning problems in the form of congestion, pollution, and housing shortages for which there are few effective planning and governance responses.

The Toronto city-region is the prime example of this emerging challenge for planners and legislators. A provincial task force, reporting in 1996, felt the need for metropolitan planning to deal with what they called the Greater Toronto Area (GTA). This region is unique among Canadian metropolitan areas, because it comprises more than a single central city, its suburbs, and rural fringe. The GTA Task Force called it "a mixture of mature cities, growing suburbs, newer edge cities, and adjacent rural communities."[64] Further, this spatial mélange sprawls over an area that is currently 25 percent larger in area than the designated Toronto census metropolitan area. This presents a recurring issue of how to maintain appropriate boundaries for planning purposes and to plan the increasingly complex land-use, transportation, and urban-infrastructure needs of the city-region. Provincial governments, which have the responsibility of creating new planning and governing arrangements, will have to address these issues but none have yet done so effectively.

Reflections

Regional planning confronts planners and others with a dilemma: What is the appropriate area for which we should be planning? It's a dilemma because of the interconnectedness of social and economic activities, not to mention the connections we have with the natural environment and it with us. We know that whatever spatial extent we choose for planning—municipality, county, census metropolitan area, river basin, soil region—its boundaries will never be sufficient to encompass all its relationships. Yet the regional planning that has taken place in Canada, at its best, has taken this dilemma in its stride—it has recognized that human activities and natural phenomena may have correlates on the other side of the region's boundary, and has recognized the need to deal with the planning problems within the region.

Out of this dilemma have come two important realizations about planning in general: the first is how human and natural facets of our world are interconnected in spatial terms; the second is that, despite the need to plan with predetermined spatial units, it is often necessary to consider the *interface* with other spatial units, adjacent and otherwise, to deal fully with the problems planners are confronting. Thus, we are always going to need to do regional planning; there is an "irrepressible imperative" to use the region as a platform for public-policy efforts, as British planner Urlan Wannop has said.[65] At the same time, both our planning regions and the content of their planning will vary as the need arises. We should not expect uniformity, or even seek it, in our regional planning; it is one of its strengths that it can respond in different forms. There is, however, a crucial element in this perspective, and that is the dependence upon the province to recognize and act to create, support, and, where necessary, revise both the boundaries of planning regions and the resources given to regional planners to plan with, for no regional planning can take place without the approval of the province.

While some planning problems often seem to demand a larger area for their comprehension, just as often other planning problems do not; they are limited to a specific area. In the next chapter, the spatial focus shifts to that of the planning needs of individual small towns that occur in most regions. Thus, in consideration of this scale change, a guiding question should be:

- ***How do the size and locational characteristics of small towns and rural areas affect planning problems and procedures?***

Endnotes

1. See Artur Glikson, *Regional Planning and Development* (Leiden: A.W. Sijthoff, 1955), 70–85, for an incisive description of Geddes's concepts.
2. Andres Duany and Emily Talen, "Transect Planning," *Journal of the American Planning Association* 68:3 (Summer 2002), 245–266.
3. Stephen M. Wheeler, "Planning for Metropolitan Sustainability," *Journal of Planning Education and Research* 20:1 (2000), 133–145.
4. Lewis Mumford, *The Culture of Cities* (New York: Harcourt, Brace, 1938).
5. For a Canadian statement see Mike Carr, *Bioregionalism and Civil Society* (Vancouver: UBC Press, 2004); and Doug Aberley, *Boundaries of Home: Mapping for Local Empowerment* (Gabriola Island, B.C.: New Society Publishers, 1993). The seminal work in bioregionalism is Kirkpatrick Sale, *Dwellers in the Land: The Bioregional Vision* (Philadelphia: New Society Publishers, 1985).
6. Cf. L.O. Gertler, "Regional Planning in Canada," *Plan Canada* 45:3 (Autumn 2005), 22–24.
7. Duany and Talen, "Transect Planning"; Peter Calthorpe and W. Fulton, *The Regional City* (Washington: Island Press, 2001); Michael Leccese and Kathleen McCormick, eds., *Charter of the New Urbanism* (New York: McGraw-Hill, 1999), especially Section 1, "The Region: Metropolis, City and Town."
8. Gerald Hodge and Ira M. Robinson, *Planning Canadian Regions* (Vancouver: UBC Press, 2001).
9. Ibid., 141.
10. John Friedmann, "Regional Planning as a Field of Study," *Journal of the American Institute of Planners* 29:3 (August 1963), 168–178.
11. Sharon Fletcher and Christina Thomas, "Coping with Growth in the Nanaimo Regional District of BC," *Plan Canada* 41:4 (Winter 2001), 16–18.
12. Hodge and Robinson, *Planning Canadian Regions*, 12–15, provides the basis for this section.
13. Canada, Royal Commission on the Future of the Toronto Waterfront, *Watershed*, 2nd Interim Report (Ottawa: Ministry of Supply and Services, 1990).
14. L.O. Gertler, *Regional Planning in Canada* (Montreal: Harvest House, 1972), 86–96.
15. Fraser Basin Management Program, *2nd Anniversary Report* (Vancouver, 1994).
16. Alberta, Department of Municipal Affairs, *Planning in Alberta* (Edmonton, 1978), 77.
17. A good review of the history of Canadian rural planning is found in Wayne J. Caldwell, "Rural Planning in Canada," *Plan Canada* 45:3 (Autumn 2005), 25–28.

18. Barry E. Smith and Susan Hald, "The Rural-Urban Connection: Growing Together in Greater Vancouver," *Plan Canada* 44:1 (Spring 2004), 36–39.
19. Evelyne Power Reid and Maurice Yeates, "Bill 90—An Act to Protect Agricultural Land: An Assessment of Its Success in Laprairie County, Quebec," *Urban Geography* 12:4 (1991), 295–309.
20. Canada, Royal Commission on Dominion–Provincial Relations, *Recommendations*, Book II (Ottawa: King's Printer, 1940), 75.
21. Maurice Yeates, *Main Street: Windsor to Quebec City* (Toronto: Macmillan, 1975).
22. Helpful references for this period are T.N. Brewis, *Regional Economic Policies in Canada*, (Toronto: Macmillan, 1969); Helen Buckley and Eva Tihanyi, *Canadian Policies for Rural Adjustment* (Saskatoon: Canadian Centre for Community Studies, 1966); and Economic Council of Canada, *The Challenge of Growth and Change*, 5th Annual Review (Ottawa: Queen's Printer, 1968).
23. Cf. John Perry, *Inventory of Regional Planning Administration in Canada* (Toronto: Intergovernmental Committee on Urban and Regional Research, 1974).
24. Gertler, *Regional Planning in Canada*, 71–85.
25. James W. Wilson, *People in the Way* (Toronto: University of Toronto Press, 1973).
26. A. Paul Pross, *Planning and Development: A Case Study of Two Nova Scotia Communities* (Halifax: Dalhousie University Institute of Public Affairs, 1975).
27. Gouvernement du Québec, *Guide explicatif de la loi sur l'amanagement et l'urbanisme* (Québec: Ministère des Affaires municipales, 1980).
28. Jean Cermakian, "Geographic Research and the Regional Planning Process in Quebec: A New Challenge," *Proceedings of the New England St. Lawrence Valley Geographical Society* (1984).
29. As described in Doug Aberley, "How to Map Your Bioregion: A Primer for Community Activists," in Aberley, *Boundaries of Home*, 71–129.
30. W. Donald McTaggart, "Bioregionalism and Regional Geography: Place, People, and Networks," *The Canadian Geographer* 37:4 (1993), 307–319.
31. Stephen Frankel, "Old Theories in New Places? Environmental Determinism and Bioregionalism," *Professional Geographer* 46:3 (1994), 289–295.
32. David L.A. Gordon and Ken Tamminga, "Large-scale Traditional Neighbourhood Development and Pre-emptive Ecosystems Planning: The Markham Experience, 1989–2001," *Journal of Urban Design* 7:3 (2002), 321–340.
33. Richard Scott, "Canada's Capital Greenbelt: Reinventing a 1950s Plan," *Plan Canada* 36:5 (September 1996), 19–21.
34. Ian Wight, "Framing the New Urbanism with a New Eco-regionalism," *Plan Canada* 36:1 (January 1996), 21–23.
35. Stan Schwartzenberger, "A Region of Watersheds," *Plan Canada* 40:5 (November–December 2000), 23.
36. Michel Dupras, "The Necessity of Integrated Water Management in Quebec," *Plan Canada* 40:5 (November–December 2000), 24–25.
37. Cf. Charles Hostovsky, David Miller, and Cathy Keddy, "The Natural Environment Systems Strategy: Protecting Ottawa-Carleton's Ecological Areas," *Plan Canada* 35:6 (November 1995), 26–29.
38. Hans Blumenfeld, "Metropolitan Area Planning," in Paul D. Spreiregen, ed., *The Modern Metropolis* (Montreal: Harvest House, 1967), 79–83.
39. Year 2001 data from Statistics Canada website, available online at www.statcan.ca.
40. A good description of metropolitan conditions just after World War II is found in Albert Rose, *Problems of Canadian City Growth* (Ottawa: Community Planning Association of Canada, 1950).
41. Hans Blumenfeld, "Some Lessons for Regional Planning from the Experience of the Metropolitan Toronto Planning Board," in Spreiregen, *Modern Metropolis*, 88–92.
42. See, for example, Anne Golden, "The Agony and the Ecstacy," *Plan Canada* 38:6 (1998), 22–25; and John Sewell, "Thanks, Tories, for City Chaos," *NOW*, May 8–14, 1997.
43. Cf. Hodge and Robinson, *Planning Canadian Regions*, 304–307.
44. George Nader, *The Cities of Canada*, vol. 2 (Toronto: Macmillan, 1976), 293.
45. Renate Banik-Schweitzer, "The City as Form and Idea" in Eve Blau and Moniker Platzer, eds., *Shaping the Great City: Modern Architecture in Central Europe* (Munich: Prestel Verlag, 1999), esp. 78–86.
46. William H. Whyte, Jr., ed., *The Exploding Metropolis* (New York: Doubleday, 1958).
47. Blumenfeld, "Metropolitan Area Planning," 82.
48. Municipality of Metropolitan Toronto, *Metroplan Concepts and Objectives* (Toronto, 1976), 2.
49. Patrick Abercrombie, *Greater London Plan 1944* (London: HMO, 1945).
50. Greater Vancouver Regional District, *Livable Region Strategic Plan* (Vancouver, 1996), 2ff.
51. Canada, National Capital Commission, *Tomorrow's Capital* (Ottawa, 1974), 27ff.
52. Ontario, Department of Municipal Affairs, *Choices for a Growing Region*, a report of the Metropolitan Toronto and Region Transportation Study (Toronto, 1967).
53. The 2005 draft regional plan for the Halifax Regional Municipality is available online at www.halifax.ca/regionalplanning/RegionalPlanDraft1.html.
54. The 2004 Master Plan for Montreal is available online at www2.ville.montreal.qc.ca/plan-urbanisme/en/index.shtm.
55. Francois Lapointe and Pierre Dubé, "A Century of Urban Planning and Building in Canada's Capital Region," *Plan Canada* 40:3 (April–May 2000), 18–19.
56. David Gordon, "Weaving a Modern Plan for Canada's Capital: Jacques Greber and the 1950 Plan for the National Capital Region," *Urban History Review* 29:2 (March 2001), 43–61.
57. Ottawa, National Capital Commission, *Greenbelt Master Plan Summary* (Ottawa, 1996), 19.
58. Ibid. See also Scott, "Canada's Capital Greenbelt."
59. City of Ottawa, *Ottawa 20/20: Environmental Strategy* (City of Ottawa, October 2003).
60. Ontario Ministry of Municipal Affairs and Housing, *Greenbelt Plan 2005*, (Toronto: OMMAH, 2005); also available online at www.mah.gov.on.ca/userfiles/HTML/nts_1_16289_1.html.
61. Greater Vancouver Regional District, *Livable Region Strategic Plan.*, 9–10.
62. Greater Vancouver Regional District, *Building a Sustainable Region* (Vancouver, 2002).
63. Hodge and Robinson, *Planning Canadian Regions*, 294–366.
64. Greater Toronto Task Force (Ontario), *Greater Toronto* (Toronto: Publications Ontario, January 1996).
65. Urlan Wannop, *The Regional Imperative: Regional Planning and Governance in Britain, Europe and the United States* (London: Jessica Kingsley, 1995), 364.

Internet Resources

Chapter-Relevant Sites

Planning Canadian Communities
www.planningcanadiancommunities.ca

Vancouver, Livable Region Strategic Plan
www.gvrd.bc.ca/growth/lrsp.htm

Saskatoon, Meewasin Valley Authority
www.meewasin.com

Manitoba Capital Region
www.gov.mb.ca/ia/capreg/index.html

Ontario Ministry of Public Infrastructure Renewal
www.pir.gov.on.ca

Ottawa, National Capital Commission
www.capcan.ca

Nova Scotia, Association of Regional Development Authorities
www.nsarda.ca

New Brunswick, Regional Development Corporation
www.gnb.ca/0096

10

Chapter Ten

Planning Communities in Rural Regions

Planning becomes (n)either easier (n)or more difficult in the small community.... Conditions are simply different and demand different approaches, not big-city hand-me-downs.

James Wilson, 1961

Most planning practices have grown out of a concern about the problems that *cities* experience with rapid and large-scale growth. But there is another type of Canadian community, actually the oldest and most prevalent type, which has its own special planning considerations. These are the small towns and other communities in rural regions, of which there are over 9000 in Canada. They range from good-sized towns of up to 10 000 population (such as Smith's Falls, Ontario, and Estevan, Saskatchewan) down through small towns (such as Clark's Harbour, Nova Scotia, and Luceville, Quebec), small villages and hamlets, outports, crossroads clusters, and nearly 3000 Native reserves and their communities. By contrast, there are only 27 metropolises and 150 small and medium-size cities, all of which began as small towns. Close to seven million people resided in these rural communities in 2001.

Smaller communities, too, have problems of land use, housing, traffic, public facilities, and population change. But because of their small size, the planning problems of towns and villages manifest themselves in ways that are different from those of cities. The land-use dynamics assumed by the land-use control regulations found in most provincial planning acts do not fit the mould of most small towns. Although the planning problems of small towns seem to fit the general categories of larger centres, they differ in scale, intensity, and pace of change. And there are, in addition, special planning problems that do not usually occur in cities. The recognition of these differences in planning small communities is the concern in this chapter and obliges consideration of this question:

- —*What distinguishes the planning needs and planning practice in small communities from those of regions and cities?*

The Context for Rural-Community Planning

Types of Rural Regions and Their Communities

Small towns occur in several different rural situations in Canada, and, within each, planners are presented with special circumstances and planning problems. An early observer posited the notion of "a continuum" of rural situations from the "completely rural" to the "nearly urban."[1] This continuum approach keeps us from considering all non-urban areas as simply *rural*; it better describes the diversity of small towns and rural communities. Looking at rural Canada, five distinctive regional situations emerge:[2]

- Completely rural regions
- Rural recreation regions
- The city's countryside
- Northern resource regions
- Aboriginal rural regions

Completely rural regions in Canada are largely devoted to natural resources development and harvesting. They have been long settled and are characteristically served by small towns and villages. Probably the most readily recognized are towns situated in agricultural regions (such as Cabri, Saskatchewan and St. Mary's, Ontario). Other rural resource regions, such as those settled for fishing and forestry, also have numerous small communities (Fogo Island, Newfoundland, and Port Hardy, B.C.). The primary economic function of small communities in such regions is to provide services to those involved in the prevailing resource development. The universal planning problem for this type of town is the provision of health, transportation, and recreation services for their own population and for those living in the surrounding countryside. In forestry, fishing, and mining regions, many communities also have few housing choices, chronic unemployment, and degraded natural environments.

Rural recreational regions are also resource-dependent, but in this case it is the land and/or water resource(s) that are considered desirable for recreation and tourism, such as in the Laurentides in Quebec, the Bay of Fundy in New Brunswick, and Tofino in B.C. Communities in these regions, many often long-lived and founded to exploit other resources, find themselves coping with vast changes that enlarge their populations, escalate housing prices, transform employment structures, and create social cleavages.

The city's countryside encompasses small communities caught up in urban expansion. They are frequently transformed from their original rural functions as agricultural service centres into "dormitory" towns.[3] Among them are municipally incorporated places such as Manotick, Ontario, and Cochrane, Alberta, which find themselves confronting pollution, conflicting land uses, displacement of population, and spiralling land costs. Equally impacted, and much less equipped to tackle such problems, are the urban fringe areas that are not incorporated. The latter rural–urban fringe areas fall outside recognized municipal jurisdictions and may lack any land-use planning capacity.[4]

Northern resource regions lie north of the rural regions identified above, and they are often simply called "The North." They are the northern parts of all provinces (except the

Maritimes) as well as the northern territories and Nunavut. Resource development dominates the economy (most recently, diamonds and nickel), but their settlements, as well as being small, are very widely dispersed and often not interconnected, with resulting problems for communications, difficulties in supplying public services, and boom-and-bust economies. Noteworthy about these towns is that many were built as "new towns" to facilitate resource exploitation—for example, Shefferville in Quebec, Leaf Rapids in Manitoba, and Kitimat in B.C. (see Figure 1.1, page 10). There are a few sizeable centres, such as Yellowknife, Whitehorse, and Iqaluit, but by far the majority of the communities in these regions are small and their populations mostly, if not entirely, comprise Aboriginal people.

Aboriginal rural regions are located within the northern milieu just described. Again, the populations are sparse and the communities are small, but they are distinguished by the special relationship between their Aboriginal inhabitants and the land, water, air, and animal and fish life.[5] Community conditions often feature extreme poverty, isolation, poor housing, and lack of services, which, given their inhabitants' generally holistic concept of "natural resources," means that special planning approaches are called for with these communities, several of which are described in a later section.

Development Characteristics of Small Communities

Although there is great diversity and individuality among small communities given their regional milieux, there are commonalities that are important for planners to grasp. The discussion below focuses on places with populations of up to 1000 persons, of which there are several thousand in Canada. At this level, it is appropriate to call them "towns" because they begin to develop a distinctive street pattern and distribution of land uses, and selected planning processes and tools become pertinent to employ.

Land-Use Patterns

Small towns differ from one another, as any resident or observer can tell you. But in their physical development, there are a number of common characteristics that distinguish them from large communities and that affects their planning. These relate to their size or area, density, and land-use patterns.

Size. It may seem self-evident to say that small towns occupy very little land area. It is, however, worth noting because this dimension is related to such issues as the provision of streets and public utilities, as well as to the accessibility of different land uses to their users. Even the largest small towns, those having 10 000 or so in population, occupy less than five square kilometres of land. All the land uses would be within 1300 metres of the centre of such a community. The typical small town of 500–1000 in population occupies less than 1.5 square kilometres and has only 150–300 houses.

Density. Despite their small size, small towns are usually not very compact. Development is spread out at gross densities of two to five dwellings per hectare. Only when the population of a small town approaches 5000 is it likely to have a density near that found in city suburbs of 10–15 dwellings per hectare.

Vacant Land. A factor that contributes noticeably to the low density is the large amount of vacant, undeveloped land found in most small towns. It is not uncommon to find as much as one-third of the land to be vacant, and not just on the fringes of the town. This is a reflection of the generally non-competitive land market. There is, indeed, plenty of land for all potential uses.

Commercial Development. The land-use pattern characteristic of most small towns is one of residential areas concentrically arranged around a single commercial area. "Main Street" is both a retail and social focus for the town. The number of establishments is relatively small, and, not infrequently, there are vacant buildings that attest to a slow-growing local market and the competition of urban shopping centres. There is, however, evidence of stabilization, and even some growth, in the commercial base of many remote towns and villages. And, of course, those within the orbit of a metropolis tend to show major expansion of shopping facilities.

Land-Use Diversity. A mixture of diverse land uses characterizes all but the largest small towns. It is not uncommon to find on the principal commercial street stores, houses, service stations, churches, and other uses, all juxtaposed. Even in residential areas, there is a diversity of uses. Yet these land-use patterns seem to function satisfactorily, suggesting that a high degree of tolerance is possible among land uses when the intensity of development is relatively low.

Built Environment Issues

To plan effectively for small towns, one must recognize the issues that arise from their characteristics of small-scale, small-growth, and diverse land-use situations. These are not, it must be stressed, issues one normally addresses in city planning, which is more concerned with such matters as traffic congestion, rezoning for high-rise development, and the stability of residential areas. People in towns and villages tend to see their development in terms of a number of *specific problems* rather than broad issues.

The following list, compiled from various small-town planning studies, illustrates the types of problems with which planners are often confronted:

- There is not an adequate water supply.
- There is no sanitary sewer system.
- Much of the land in the community is not suitable to build upon.
- Existing development is very scattered.
- The railway crossing is dangerous.
- Many buildings are vacant in the town centre.
- The sidewalks need repair.
- There is no suitable means of garbage disposal.
- The cemetery has become overgrown.

City-oriented planners might tend to dismiss these issues as being trivial, and certainly there is little place for them in conventional municipal plans. However, they are significant for small communities, especially in rural regions remote from the city. As a Newfoundland report on the topic noted several decades ago noted, "If the day-to-day problems can be solved with the aid of a plan then . . . the value of planning is established."[6]

Water Supply. One often finds, for example, that communities with 500 or fewer residents have no community-wide water-supply system or sewage system. These services are usually provided by individual households and firms through wells and septic tanks. Even though most such places have low-density development, pollution problems may arise; thus, the common planning approach of increasing the density by infilling on vacant lots either exacerbates these problems or necessitates costly utility systems. A related problem in providing basic water and sewer service is that dwellings are, very often, situated in areas where land is unsuitable for building (steep slopes, rocky, or swampy). Even larger towns with their own waterworks may be at risk if their water aquifers become polluted by surrounding agricultural or industrial uses, as was seen with the *E. coli* tragedy at Walkerton, Ontario, in 2000.[7] The numerous media stories since then remind us of the vulnerability of water supplies in smaller communities, even those close to cities, such as Central Saanich, on Victoria's outskirts.[8] The forced evacuation of the Aboriginal community of Kasechewan on James Bay in 2005 is the most recent reminder of this critical factor in community life.[9]

Traffic and Roads. Small towns have their own forms of traffic and road problems. Within the community, both the unsuitable land and scattered residential development may lead to excessive and costly road building and maintenance. Traffic congestion may also occur in the business district because of a lack of adequate parking space. Inconvenient access for the community to a major highway nearby or the hazards of the highway running right through the community are two other burdens of small towns. Other small communities sometimes find themselves as recreational destinations as compared to their traditional natural resource roles, and this often results in seasonal traffic congestion and parking dilemmas.[10]

Care of the Environment. There are frequent complaints about the lack of care regarding the physical environment of a small town. The disposal of waste materials—household garbage, worn-out vehicles, and so forth—can pose problems both for the maintenance and the location of dumps. Unkempt cemeteries and playgrounds and derelict buildings are other common problems affecting the environment in a small community. That they often defy solution is due as much to anything as to the community's meagre resources.

Social Development Issues

There are several demographic aspects of small towns that often have implications for planning their physical development. The first—the aging population—is more general and affects almost all small towns. The second concerns population cleavages in towns, particularly in the rural–urban fringe. A third issue revolves around lack of population growth and the loss of population, especially the out-migration of younger people. A brief description of these situations will quickly reveal some of the land-use and development impacts associated with them.

Population Aging. Whereas two decades ago people living in small towns often bemoaned the fact that they had little choice but to move to the city when they retired, today they are staying on and other elderly people from the city may be moving there too. Smaller communities in all provinces, and in all locations except northern resource regions, tend to have considerably higher proportions of the elderly (those 65 and older) than do most cities.[11] Moreover, this tendency has been increasing so much that many small centres have population concentrations of the elderly of 25 percent or more. Among British Columbia's small towns, there have appeared two distinct types of retirement communities. One might be termed **indigenous** retirement towns, where the retired population is composed mostly of those who have remained living in the community after retirement, a tendency called aging-in-place.[12] The other might be called **itinerant** retirement towns, because most of the retirees have migrated there. The latter towns tend to be in regions with high natural amenities, and the influx of retirees or near-retirees can be a boon to a community's economy.[13] Regardless of the source of retirees, most small towns present some immediate problems in meeting the needs of older people. Whether it be health care, home-support services, or transportation, there are usually few such services in rural areas and considerable distances to reach those that do exist.[14] Housing accommodation can also be problematic for the elderly living in small towns, mainly because of the lack of variety in housing types. Added to these, the local governments have very limited resources to provide the services and facilities needed by their elderly. Also to be noted in regard to a town's population aging is that this often occurs in conjunction with the out-migration of younger people, which can exacerbate problems of providing public services.

Population Cleavages. Cleavages often occur in the wake of population growth in towns in the rural–urban fringe of cities (and towns in rural recreational regions). The expansion of cities into the countryside often has the result of displacing the rural population that has, sometimes, lived there for generations. The pre-existing social networks of, for example, farmers and traditional service people are changed by new arrivals (e.g., exurbanites, retirees, and "back-to-the-landers").[15] These newcomers, in turn, affect the political dynamics of the community and how it responds to development pressures and the need for additional services.[16] And, not infrequently, as newcomers settle in, they are likely to become the vanguard of resistance to further change. Therefore, one cannot overlook the social dimensions of what may seem to the outsider to be a relatively small change in the community environment.

Population Decline and Migration. The majority of communities in rural regions register little, if any, growth in population from census to census. Just keeping a stable population level is seen as a favourable result by some, while others suffer decline, and only a relative few see an influx of new households. The primary reason is simply that there are few economic reasons for people to stay in these communities, especially younger people, and they migrate to larger centres.[17] The resulting population is often older and poorer, and housing stock tends not to be replenished, much less new housing being created. Services and facilities for social and health care are usually minimal, if they exist at all. Thus, there is often little basis for long-range planning and development, and other ameliorative solutions must be sought.

Economic Development Issues

The economy of a rural community is tied intimately to its natural environment. Each is surrounded by natural resources that contribute all or part of the economic base of the community: the farmland around a prairie town; the forest that surrounds a village in B.C.'s interior; the ocean on which a Newfoundland outport is situated; the northern river basin that provides a territory for trappers and hunters, to name a few. Despite these being the resource bases that led to them being founded, almost all rural communities today are experiencing dramatic changes in their relationship to the surrounding natural environment. Some resources have been depleted (e.g., fish stocks, timber, minerals) and others are being exploited differently due to market and/or technological changes.

Agricultural communities, for example, have been affected by increasing industrialization of farming, with operations becoming larger and more specialized. The tendency toward large-scale, mechanized operations in livestock production, notably hog raising across the Prairie provinces, has led to community-planning conflicts over odours and water quality.[18] The latter situation is reflected in the issues being confronted in Newdale, Manitoba, as described in Planning Issue 10.1. Still other communities are coping with new demands on their resources (e.g., new mines, new recreational uses) that make additional demands on housing stocks and local services. Not least are the impacts on those communities in the city's countryside that, in the face of urban expansion, seek ways to retain traditional agricultural uses.[19]

In short, economic factors largely out of their control confront rural communities with significant change: employment losses, shifts in demand for labour skills, household-income disruptions, business-income uncertainty, and inability of communities to fund needed services and facilities. A recent external trend affecting small communities is restructuring in the retail sector, which has been leading to a retail geography in which there are fewer and larger stores.[20] Given the much smaller scale of rural communities, these changes loom larger not only in terms of development but often in terms of survival. Planners working with these communities cannot long ignore economic and natural-resource issues.

Scope of Rural-Community Planning

Rural-community planning obliges planners to respond to a wide range of issues, as the above discussion indicates. One Canadian rural planner puts it this way:

> While rural planners focus on traditional land use issues, they also concern themselves with the local economy, labour and employment, demographics, resource management and environmental protection.[21]

Figure 10.1 shows the "diversity of issues, strategies and approaches" that defines the scope of the rural planner's task. Clearly, a scope as broad as this calls for a range of skills well beyond that of most planners. Rural-community planners, thus, frequently depend on consultants for the specialty knowledge needed in their work and may utilize them to conduct public meetings dealing with their particular technical area.[22] Nonetheless, planners working with rural communities must be generalists who appreciate the wide array of factors that impinge on rural areas.

Planning Parameters for Small Communities

Dimensions of Physical Development

It is vital to grasp two things about small communities: first, they are not just scaled-down versions of cities, and, second, their development proceeds in distinctive ways from that of cities. Therefore, planners need to have an appropriate perspective from

Figure 10.1 The Scope of Planning for Rural Communities

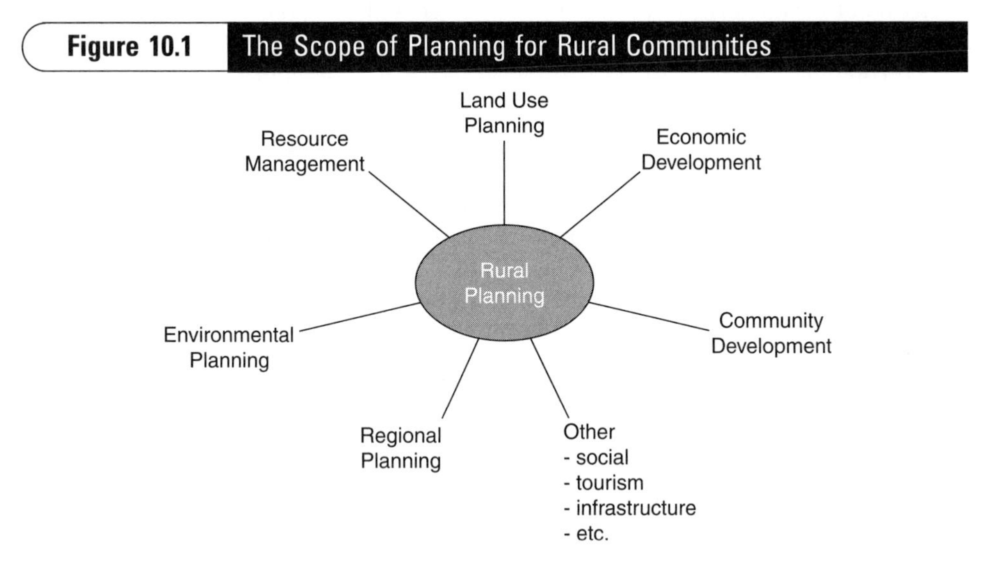

Rural planning is very broad in scope and the planner must take all these facets into account when preparing a plan for a rural community.

Source: Wayne J. Caldwell, "Rural Planning in Canada," *Plan Canada* 45:3 (Fall 2005) 26.

which to assess their development. Four useful dimensions of physical development are:[23]

- *Scale* of development
- *Range* of types of development
- *Intensity* of development
- *Pace* of development

And, for towns with adjacent farmlands, we need to add the dimension of *agricultural land* development. A brief elaboration of each will illustrate the nature of the process of development in most small communities.

Scale. Smallness is an important dimension of several aspects of rural-community development. The level of density is low and the scale of population and of the built-up area is small in towns and villages. So, too, is the scale of change. The additions to a town's housing stock, for example, are likely to be relatively few in number, even though proportionately large for the community. A centre of 500 in population experiencing a growth of 40 percent in dwelling units, as many did in the 1970s and 1980s, would have added only 50 to 60 new houses and required, say, eight hectares of land. This means that the scope of planning problems is usually small, and their solutions should be scaled appropriately. The scale issue is at the crux of controversy in Manitoba and Alberta over the use of "big barns" for use in raising hogs (see Planning Issue 10.1). This is the small town's counterpart to the urban dilemma over "big-box" retail stores, both of which promise local tax revenue and jobs, but at the cost of community transformation.

Range of Types. Almost all small towns have only a small array of different activities and structures. From place to place, the range may vary according to the economic base of the community (i.e., a certain industry may predominate), the climate of the region, or the building materials available. There is seldom, however, a large variety in a town's residential, commercial, industrial, or public land-use areas. This is equally true when new development occurs: it will most likely bring the same kinds of activities and structures as have existed to that point. Thus, the planning implications of most new development have ready precedents. But as we see with the "big barns" mentioned above, the impact of globalization on agricultural communities can be physical as well as economic.

Intensity. Unlike the extensive areas of development of a city, a small town does not have either large areas covered with the same kind of development (such as residential suburbs) or highly concentrated activities (such as a cluster of apartment buildings). In most small centres, the development tends to be in discrete units in a more scattered fashion. Exceptions would be the mill, mine, cannery, transportation terminal, or other industrial enterprise in "single-industry" towns; the new shopping centre in a region-serving town; or the condo complex in a recreation town. The value of this dimension is in determining the effect that new physical development has on a community, such as in the traffic generated, the utilities needed, and the schools and parks required.

Pace. There are two important facets of this dimension in regard to small towns. The first is that it is often misleading to view the pace of development in terms of percentage rates of growth. It is usually better with towns and villages to look at the actual amount of new development, or **absolute growth**. In the example used above of a town experiencing a 40 percent expansion in housing in a decade, not only was the number of

Planning Issue 10.1 Not Hog-Wild on Big Barns?

Winnipeg Free Press
August 28, 2001

Not Hog-Wild on Big Barns?
Bill Redekop and Mia Robson

NEWDALE—A rural municipality plans to be the first jurisdiction in Western Canada to hold a referendum on whether to allow large, industrial hog barns. Residents of the RM of Harrison, about 80 kilometres north of Brandon, will vote Nov. 1 on whether to limit livestock operations to 400 animals.

Council made its decision after a proposal by Premium Pork to build a 2,500-sow barn in the RM. "I got 17 calls from surrounding farms, and they were big farmers, and they said we don't want (the sow barn)," Reeve Tony Novalkowski said yesterday. "The ratepayers have no problem with smaller hog farms. They don't want factory farms."

The sow barn proposal has been withdrawn. It could have added $25,000 to $30,000 in tax revenue per year to the RM's coffers, but it could also cause pollution and hurt the area's tourism, he said. "We've got lots of lakes here (like nearby Sandy Lake) and lots of tourism and we don't want to ruin it," Novalkowski said. He added large livestock operations also bring heavy truck traffic to deliver feed, which causes extra wear and tear on roads. The cost to rebuild a mile of road is $30,000, he said.

Local councils for neighbouring RMs of Rossburn, Strathclair and Riverside have all recently rejected pleas from hog barn opponents to hold a referendum. "I've been to (livestock) hearings in adjoining municipalities where the ratepayers were just about begging council to listen. Well, we've chosen to listen," Novalkowski said.

A spokesman for Agriculture Minister Rosann Wowchuk's office said the province isn't worried the referendum will be precedent-setting. Once the referendum goes ahead, the province will have a better idea of what it is facing, said Ronuk Modha. Modha added referendums aren't binding unless the municipality chooses to make a by-law change, which the RM of Harrison does not plan to do. Even if the referendum vote is against hog barns, it may not prevent one from opening in the municipality in the future, he said.

Novalkowski acknowledged the referendum is not binding but said his council would not go against the expressed will of the people. "If the ratepayers say no, then it's no," he said. "I'm reeve for maybe a term or two, and I can't see myself imposing something like an intensive livestock operation that could be here for 25 or 30 years."

Bruce Robertson, a councilor and owner of the Leisure Inn in Newdale, opposes the referendum. Robertson said the area needs additional business if it is to survive. "These small towns are dying out and if we don't get something soon, we'll have nothing left in Newdale. If we don't get some industry, you won't see the Leisure Inn and a co-op store much longer," he said. Robertson added that some cattle farms in the area are approaching 400 animals in size, and the referendum will prevent them from expanding.

Marchel Hacault, the chairman of the Manitoba Pork Council, said the referendum is being held because the municipality isn't prepared to make a decision. "It's more a reflection of the inability of municipalities to come to terms with what they want to do," Hacault said. He said there isn't a clearly defined decision-making process yet either, and neither the province nor the municipalities are exactly sure who has the ability to make what decisions. "It's a symptom of frustration on both sides," Hacault said.

There are about 1,000 potential voters in the RM. The referendum question will be drafted with help from legal counsel.

Source: Bill Redekop and Mia Robson, "Not Hog-Wild on Big Barns?" *Winnipeg Free Press,* August 28, 2001. Reprinted with permission.

houses built not very large (50–60), but also this was taking place over a 10-year span. The second facet is that seldom is this amount of change spread evenly over the 10 years. As noted with development occurring in discrete projects, all the houses quite likely are built in one or only a few years. Development activity is usually infrequent and irregular in commercial, industrial, and public building as well. Each project must, therefore, be planned on a more or less ad hoc basis, because a continuous level of growth cannot be expected in small towns.

The process of development described within the four dimensions cited above pertains to most, but not all, small towns. There are some that are affected by rapid growth, a high demand for land, and major changes to their character. Resource towns such as Fort McMurray, Alberta, come readily to mind, as do towns on the outskirts of metropolitan centres, such as Markham, near Toronto, and towns in recreation regions like Pemberton near Whistler, B.C. But this type of town and village development is an exception and probably affects no more than a few hundred of the several thousand small centres referred to here.

Agricultural Land. Although more of a contextual dimension than those above, the presence of agricultural land in its setting can be a significant factor in the physical development of a small town. In several provinces, notably Quebec and British Columbia, stringent limits are placed on the conversion of agricultural land to non-agricultural purposes, especially for residential development. In these two provinces, application must be made to a provincial commission for permission to further subdivide "protected agricultural areas," as they are called in Quebec, and "agricultural land reserves," as they are called in B.C.[24] Two major thrusts of this type of intervention are those aimed at (1) assuring current and future supplies of farm products, and (2) preserving rural landscapes that symbolize rural ways of life.[25] Other measures to restrain town expansion onto farmland are generally weaker, such as the "urban-containment boundaries" in Ontario.[26] Small towns in the rural–urban fringe are usually the ones most affected by such regulation because of the urban development pressures on them.

Planning Supports for Small Communities

There are two types of supportive frameworks that many small towns may call upon in their planning. Provincial ministries throughout Canada have a variety of measures to assist small communities. In addition, many rural regions have formal frameworks for regional government or regional planning. There is a great variety in these undertakings, but a brief sampling will indicate some of the sources of planning support to which small towns might turn.

Provincial Level

Provincial ministries may offer technical assistance for planning in small towns and rural areas. This can take the form of providing professional planning services directly to communities, financial assistance to obtain outside advisors, or technical assistance in the form of publications on relevant planning topics. The actual forms and the combinations available differ from province to province. For example, Manitoba has operated community-planning field offices in several regional centres from which staff planners could assist small communities in preparing local-planning instruments and acting as advocates at formal hearings. Land Use Service Centres fulfill much the same role for Prince Edward Island towns, while in New Brunswick, planners are assigned from the ministry to assist designated rural areas and, sometimes, to become the resident staff in regional planning commissions.

Another form of planning assistance that many provinces provide is financial support to promote the preparation of general planning instruments and plans for specific kinds of projects. Several provinces have, for example, a program of "community-planning study grants," which are available only to smaller municipalities. Such programs allow them to obtain the services of consultants to conduct planning studies, to prepare community plans, zoning bylaws, and so on. Aid is also available in most provinces for studies and plans for such specific kinds of projects as business-area revitalization, seasonal cottage areas, and housing-needs statements. Saskatchewan's Main Street Development Program was designed specifically to assist small towns and is typical of these kinds of programs. Most planning assistance involving provincial grants is usually in the form of a shared-cost program, in which the province contributes the major portion of the costs and the municipality, the remainder. Provincial shares usually range from 50 to 80 percent.

Self-help planning manuals are often available from provincial ministries to provide assistance in a wide range of planning needs. Many are designed specifically to help small communities or those that seldom become involved in planning more than a subdivision. The manuals tend to avoid jargon and provide brief, coherent models for planning reports that are well within the capabilities of small towns and townships.

Technical-assistance programs and services such as those described above may play an invaluable role in planning small towns where resources are meagre. Their success, however, depends upon two factors. First, the provincial programs themselves must have continuity and the ability to integrate the various elements of planning a community. This stems from the fact that the planning process for any size of community involves a long-term commitment on the part of the town—the adoption of a community plan is, for example, just the beginning of the next stage of implementing it. In Manitoba, provincial planners are cognizant of pressures on rural communities for "sustainable development" and nowadays urge them to go beyond traditional land-use planning to consider "community facilitation, community economic development, and co-management" of rural resources.[27]

The second factor, and probably the most vital one in ensuring effective rural-community planning, is having the capability of accessing provincial programs. A small town with few resources often lacks skilled personnel able to recognize the value of the programs that are offered; it may lack the funds needed even to participate in shared-cost programs, and its lack of municipal organization and other community institutions might make it incapable of sustained long-term planning. Well-intentioned programs tend to assume that all communities have similar needs and also capabilities. This can result in two problems: communities may not "fit" programs, and many communities may fail to seek solutions out of lack of awareness of how to adapt to the program.

Regional Level

Planning support programs and services at the regional level are the best arrangement for providing assistance to small towns. County planning departments (Ontario and Nova Scotia), and regional district planning departments (British Columbia)—two of the forms currently used in different parts of Canada—comprise all the communities in a defined area. Such agencies can provide continuous planning-staff support for individual communities, as well as for the interrelated services and facilities of the region. Importantly, this type of arrangement makes available to small towns their "own" staff planner, thus approximating the level of staff support needed to nurture a continuous community-planning process.

Planning Tools for Small Communities

The essential point to be made in planning for small communities is that planning approaches should be suited to them, that is, be *small-community* approaches. Most of our planning tools are more suitable for large communities than for small ones.

Planners of small towns should consider whether the tools they propose to use are appropriate to the problems and capabilities of small communities. The planning situation usually features a distinctive set of easily identifiable problems that call for seemingly mundane solutions, rather than an abstract arrangement of land uses. Moreover, plans and other planning instruments need to match the resources and the capabilities of a few hard-pressed and often untrained municipal officials.

In small towns, planning approaches can be simplified yet be appropriate to the situation; small-town residents often recognize this faster than the city-trained planners who try to help them. There is, fortunately, an increasing amount of experience in small-town planning from which examples of appropriate planning tools can be drawn. Experience shows that the community-planning process is essentially the same as for larger communities (as described in Chapter 7), but differs in the tools that are needed to make it effective. Thus, the community plan is a vital component, as are the tools for analysis and implementation. The difference can perhaps be grasped as that between a large and a small wrench in a mechanic's toolbox; each is peculiarly suited to its task. The planner's toolbox in the small-town setting need not contain as many or as wide a range of tools as for city-planning situations. But given the uniqueness of each town and village, care must be taken in their application.

The Community Plan

The community plan for a small town deals with land uses, circulation, recreation facilities, and parks, just as a plan for a large city does, but it can be simpler and more direct both in content and style. Its objectives can address the problems residents would like to solve; it needs fewer analyses and fewer land-use categories; and it can be concise and brief. Experience in Newfoundland demonstrates such principles of appropriateness. As long ago as 1968, the Newfoundland government explored the special planning needs of small communities, using Clarenville (population about 2800) for a prototype plan. The accompanying report advocated that such plans focus on a statement of community problems and possible solutions, noting that "if day-to-day problems can be solved with the aid of a plan then, even if they are of minor importance in themselves, the value of planning is established . . . and there will also be an immediately beneficial effect on the physical environment."[28] The land-use planning solutions suggested in the plan thereby had a clear connection with the felt needs of community members; they were not as abstract as the solutions in many conventional municipal plans.

This concern over the style of presentation is not misplaced, for it is through its presentation that a plan achieves relevance for community members. A community plan is a way of demonstrating consensus, and a small-town plan should heed this fact. Residents of small towns are usually intimately aware of land-ownership patterns and the issues with which the town is concerned, and they will want to see these accurately portrayed. Moreover, there are likely to be few professional planning, administrative, and public-media skills and resources available to interpret the plan if issues arise later. The danger in a poorly presented plan is that it may prove frustrating to officials and citizens; they may either ignore it or only pay lip service to it. A plan for High River, Alberta, in the mid-1970s (population then 4800), is exemplary in its presentation style.[29] Vivid graphic techniques show new facilities, and the five major districts of the town are listed on the town map with their main problems and solutions. Ucluelet on British Columbia's Pacific Coast reiterated the importance of its Official Community Plan (OCP) in a 1998 update because of the loss of traditional fishing and logging industries and the impact of the increasingly popular Pacific Rim National Park just to the north. To quote the planner guiding the OCP process:

> ... the OCP resembled nothing of its predecessor and included nine new Development Permit Areas establishing policies ranging from environmental protection to design guidelines ... [and] involved the entire community as well as a substantial steering committee which fueled its momentum.[30]

Because planning practice in Canada is conducted within an elaborate statutory framework, this seems to have fostered the lengthy municipal plans one commonly finds. It is rarely apparent that these cumbersome, repetitious plans, so often given to legalistic language and jargon, are inappropriate for any size of community, let alone small towns and rural municipalities. This conventional approach has produced a 50-page plan for a town of 400 people in New Brunswick and a 90-page plan for a community of 1000 persons in Ontario, to cite just two examples. One has to wonder who, besides the planners, will read and use them.

Saskatchewan's Planning and Development Act has a special approach to small-town plans. That province now acknowledges that not all municipalities may require a comprehensive municipal plan. Those communities that are very small or have a very low level of development activity may prepare a Basic Planning Statement. These are envisioned as short statements of municipal planning policy that would be sufficient to guide the local council in its (probably infrequent) deliberations over development proposals. The statements would play a role similar to that of a municipal plan in that they are a prerequisite to any intended zoning bylaw.

The plans referred to above are for incorporated small towns; this still leaves to be considered the planning situation of the more than three-quarters of small communities that are not municipalities. These places would, at best, be included within the plans of rural municipalities, townships, or counties. Given that the towns, villages, and hamlets are usually the most important centres in the rural locale, attention should be paid to them within the larger plan. This could mean preparing supplementary plans for unincorporated towns, much as one makes special-area plans for downtown districts, historic neighbourhoods, and parks in large city plans. The County of Huron in Ontario has done this, as has the Municipal District of Rocky View, Alberta, bordering Calgary on the north. The latter district has developed a joint-planning process to prepare plans with several small communities within its boundaries; they are called Intermunicipal Development Plans.[31]

Analytical Tools

With respect to the type of analytical tools that are required, the same general principles apply as for community plans: that is, planners should strive for simple, direct methods of providing the base of knowledge needed for the plan's preparation. Many tools developed for city planning are simply not needed, given the small size of towns and villages. For example, elaborate studies of land use and economic base will probably be irrelevant; similarly, statistical analyses involving correlations and sampling cannot provide reliable results because of the small size of the population. However, a concomitant of the small size that favours thorough understanding of the community is that the amount of data that is usually required can be readily obtained from field surveys and direct interviews. Information about housing, land use, jobs, age structure, and so forth will not be voluminous, and the direct contact can help elicit richer local information than is generally available to city planners. Moreover, many of the necessary studies may be conducted by community members, thus facilitating the task and engaging community interest in the plan-making process.

Community Profile

This tool, as the name suggests, gathers together various data in a single document describing the community's present status, recent trends in population growth and structure, the environment, the local and regional economy, transportation, recreation, and so on. Residents of small places have a fairly unified view of their community; the profile makes use of this view, instead of fragmenting it with a series of separate studies. Furthermore, the community profile lends itself to use as a community self-survey tool. People of all ages, from children in schools to adults in service clubs, can participate in gathering the necessary data. It is helpful for the planner to incorporate issues into the profile that assist residents in identifying community problems and aspirations. Also worth incorporating is a historical overview of the community that provides both a sense of place and the factors that are important to the town's heritage. Historical photographs and maps may be used to advantage in conjunction with those of contemporary situations.

Analyses of Housing and Population

The planner also needs to make analyses and forecasts beyond those that citizens can help provide such as the trends in the size, composition, and location of the community's housing stock and population base. An accurate survey of all households is the only sure way to acquire these data in a reliable form for analysis, although Statistics Canada now provides some small-area data from the quinquennial census. The best analyses and forecasts use the simplest methods. It has been found that forecasts for small towns that use linear extrapolation—the projection of absolute increments in past change—are the most accurate. Because changes in population and housing can vary dramatically from year to year, projections based on percentage changes can be misleading. Nowadays, it should be noted, special attention needs to be given to the age structure of the population given that small communities tend to have higher-than-average elderly populations.

Economic Analyses

Two kinds of economic analysis can prove useful in planning small communities. One is a **locational analysis** of the residents' places of employment and shopping. People in small communities tend to interact over extensive areas, because they can seldom find a full range of employment and shopping opportunities right in their own communities. A knowledge of the extent to which residents commute to work in other places and shop in other centres will indicate the ways in which the town's economy is entwined with that of other communities. Such information could be helpful in forecasting the impact of changes occurring in the region surrounding the town—for example, the opening or closing of a large plant or new shopping centre nearby. Residents' linkages may be gathered by a local survey that asks about trip purposes, mode of travel, distance, and frequency, and then the results mapped and tabulated.

Another useful tool is called **threshold analysis**. Retail and other commercial firms are the economic backbone of most small centres because they provide basic goods and services and are major sources of employment. This tool calculates the amount of population required in the town and its trade area to support additional retail and service firms. It can be used to identify those types of firms that have good prospects and those that have poor prospects. The Province of Ontario provides an online sample of threshold analysis in its report *Marketing Your Downtown*,[32] and Statistics Canada publishes a regular Retail Store Survey.[33]

Implementation Tools

As with other planning tools used in small-community planning, one must consider the problems that implementation tools are intended to solve. How, for example, will zoning work in the small-community context of low density and diverse land-use patterns, given that it "grew up" in the city, where there are large common-use districts and competition among land uses? In light of the meagre resources in small communities for administering any kind of regulations, it is important to avoid implementation tools that require continual and demanding administration. If one were to specify the needed characteristics of plan-implementation tools that would be most fitting for small towns, two of the most important would be *flexibility* and *adaptability*. Planners need to assess whether these qualities can be achieved in the tools they consider applying.

Making Zoning Appropriate

Zoning as commonly employed derives its usefulness from its ability to parallel the development patterns of cities, where land uses tend to sort themselves into districts of similar uses and types of structures. Only when population exceeds 2500 does a small town's development pattern begin to show evidence of homogeneous districts of different land uses. Even then, these districts are not large (perhaps less than 40–80 hectares), and there are likely to be only a few of them (such as one each for the business area, the residential area, and the industrial area). Moreover, they will each probably accommodate a variety of uses. A general or comprehensive zoning bylaw would be too cumbersome in this kind of situation. However, it may be pertinent to frame regulations for any special district where land development is more volatile and subject to conflicts with adjoining areas. There is often a need in small towns to control development along highways, watercourses, or in the business district.

Performance Zoning

The diverse and sporadic development process in small towns tends to accommodate diverse uses adjacent, or in close proximity to, one another that might be considered problematic in urban situations. Conventional zoning tries to anticipate the activity effects of various classes of land uses, thereby predetermining proximate allowable uses. But, in small towns, it is possible, and indeed commendable, to use a more flexible case-by-case approach that deals with the *effects*, or impact, of a proposed land use rather than with its *functional use*. For example, it is probably more important to know whether the proposed use will generate a lot of traffic, have abnormal hours of operation, create extensive noise or odours, or produce dangerous effluents, than to understand the use itself. A type of zoning originally used in industrial districts in cities, performance zoning, is widely recommended for use in small communities either independently or along with conventional land-use zoning.[34] This tool, sometimes referred to as "flexible zoning," works by establishing performance standards against which proposed uses and their location are judged. Performance standards can include traffic generation, noise, lighting levels, storm-water runoff, loss of wildlife or vegetation, or even architectural style, according to one manual.[35]

Development Control

Development control, in which each proposal for a new or changed land use within a specified area (e.g., a business district or historic zone) must receive permission to proceed, is analogous to performance zoning. That is, a **development permit** must be obtained showing that specified planning criteria have been met. These criteria may, and often do, include protection of the environment and certain design specifications, as well as normal zoning limits on height, coverage, parking, and other details. The new zoning bylaw of, again, Ucluelet, B.C., includes nine Development Permit areas, and the bylaw allows for various land uses "to co-exist" that assure the continuation of the coastal marine community atmosphere.[36] Provision is also made for receiving development proposals that are pre-planned using the **comprehensive development zoning** approach (see Chapter 13).

Cluster Zoning

A fairly recent planning tool for rural areas, especially in the rural–urban fringe, is one that blends both zoning regulation and subdivision control. Since expansion of many towns often means impinging on adjacent agricultural land, it is vital to try to reduce the impact that such expansion could bring through normal large-lot subdivision. The technique of cluster zoning, or open-space zoning as it is sometimes called, groups the new development on a part of the property while the remainder is left for farming or other open use.[37] The "cluster" notion comes from the concept of having new development on several adjacent parcels be grouped together. This technique leaves larger blocks of open space, thereby maintaining a rural character to the landscape and also making farms more viable.[38]

Land Subdivision

The subdivision of vacant land for new building lots and structures can pose special technical and resource questions for a small community. What may appear to be a boon in terms of new development and an expanded tax base may turn out to be a drain on financial resources and cause future problems for the community if not considered thoroughly at the start. All too often, the sites chosen for houses are on lands with poor buildability—steep slopes, floodplains, poor drainage, or subject to erosion. If the community is called upon later to provide adequate road access, public utilities, and proper drainage, the costs may be very high. Thus, subdivision proposals should be scrutinized for their relation to the development pattern envisioned in the community plan, for the provision of roads, parks, and utilities, and for the treatment of unique and hazardous topography.

Many technical and engineering questions arise in regard to subdivisions, and a small community may not have sufficient expertise on its own staff to deal with them. In many provinces, the final authority for approving subdivisions rests with the province or with the regional government, and advice will usually be available from these upper levels. However, it is still advisable for the town to be involved in reviewing such plans as thoroughly as possible. It may well be able to convene a review committee from among knowledgeable citizens. Such a group, or the council, could make good use of available provincial handbooks on the principles of good subdivision design when assessing proposals.

Lastly, a variant on regulating land subdivision in agricultural areas is using a preemptive strategy to remove the financial incentives for breaking up parcels of land for housing and so on. In areas where maintaining viable agriculture is an important consideration, a process of **purchasing the development rights** of farm land can be used. In the especially valuable tender fruit belt of the Niagara Peninsula, for example, the Ontario government established a program to induce farmers to place a covenant on their land that restricts the land from being used other than for fruit growing. In return, they received a stipulated cash payment from the province and the regional municipality.[39] Thus, the farmer still owns the land and can continue to farm it but the right of the property owner to otherwise develop cannot be exercised. An easement, as it is properly called, preventing non-agricultural development, has been purchased. Private land trusts often use this kind of device to protect scenic and historic resources. In British Columbia and Quebec, essentially the same thing occurred when agricultural land reserves were established by government fiat. In these instances no compensation was paid because it was considered in the larger public interest not to do so.

Figure 10.2 Rural Land-Use Concept Plan, Langley, B.C.

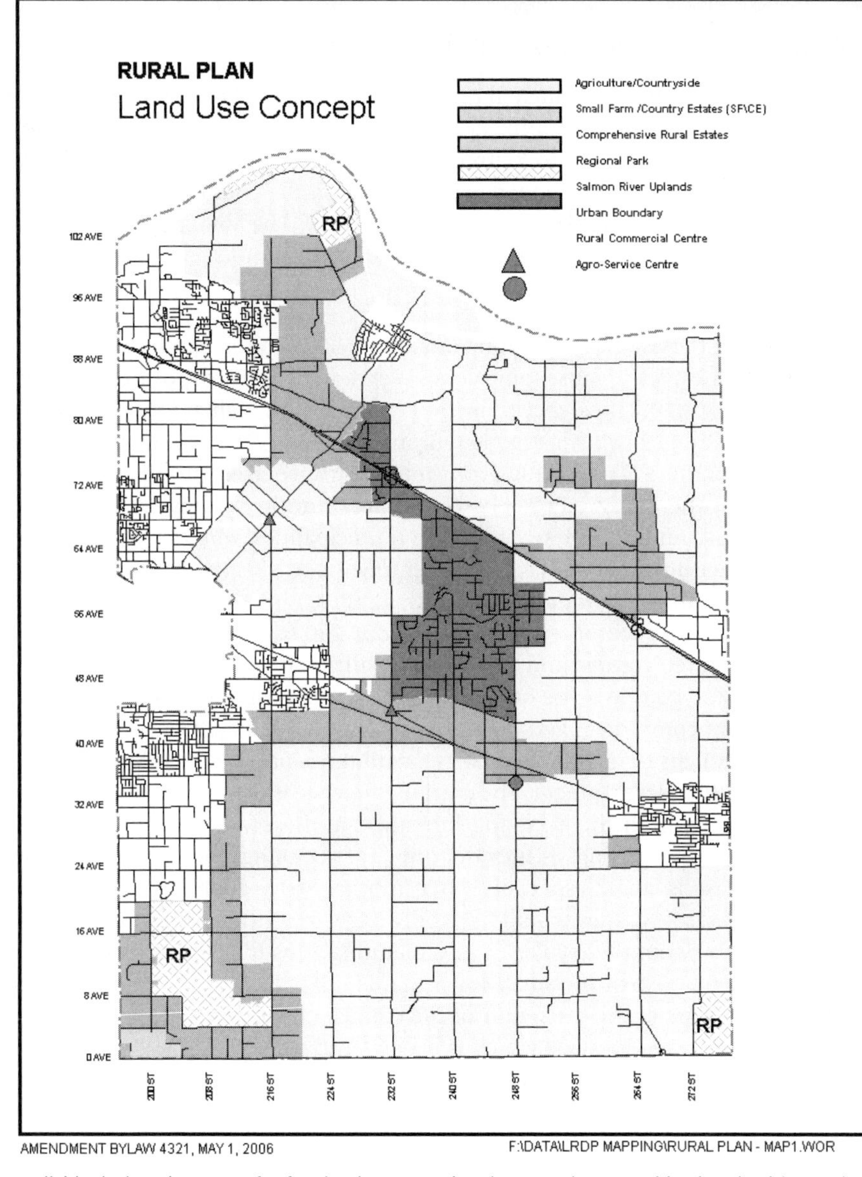

By explicitly designating areas for farming in community plans, rural communities in urbanizing regions can sustain their agricultural resources.

Source: Courtesy Township of Langley, B.C.

Planning and Governing Small Communities

Community planning, regardless of the size of the community, is an integral part of the way a community governs itself and takes responsibility for its future. In the Canadian social and political milieu, local (municipal) government is the cornerstone of such community undertakings. All provinces have legislation that enables a municipality to make plans, regulate land uses, and make expenditures regarding future development. However, this provincial legislation is, by its very nature, meant to apply universally to all municipalities in the province, regardless of size, location, or resources, a situation often unfavourable to small communities.

Local Government Structures and Resources

Small towns exist within governmental milieux that put special constraints on their performance of the community-planning activity. An essential question in rural regions is how to allocate resources and governmental authority in areas where people live in widely separated centres and in low-density settings in the countryside. The issues are those of determining at what population size a centre will be able to govern itself effectively, and over what area the governmental jurisdiction should prevail.

Rural Municipalities

Each province has provisions for granting municipal status to towns and villages, usually based on population size. Thus, some small places may be **incorporated**, and others not. In Saskatchewan, the incidence of incorporation is very high among towns and villages, including hamlets, whereas in New Brunswick and Nova Scotia, only the larger towns are incorporated. To provide government for dispersed rural populations, all provinces (except the Atlantic Provinces) have **area municipalities**, known variously as townships, rural municipalities, and district municipalities, where settlement is widespread. These latter units often include smaller towns and villages that are not incorporated. They tend to have a common size, for example, and in Ontario, Quebec, and the Prairie provinces, they encompass an area of about 260 square kilometres.

Both incorporated centres and area municipalities are local, self-governing units with structures analogous to those of larger communities: that is, each has its own council and constituent committees and boards, perhaps reduced in scale but with the same powers as city councils to raise taxes, make expenditures, pass land-use regulations, and make plans for future development. The crucial difference is the much-reduced scale and quality of resources, both financial and human, that these small municipalities have available to them. Comprising, as they do, only a few hundred or a few thousand persons, their tax base is small, especially when monies are needed for new or renewed physical facilities. As well, they would have only a few technical and administrative persons on staff to carry out the regulatory and planning tasks the community might wish to undertake. In the case of the small town that is unincorporated, both the financial and staff resources are at the discretion of the council of the township or rural municipality in which they are located (which is usually also small).

It is only partly true, it may be argued, that small towns and townships require much less in the way of resources and administrative structures because their needs are likely to be fewer. There often are fixed costs in providing such town facilities as a water or sewer system, streetlights, or sidewalks, and the modest local resources may not allow the community even to begin development of this kind. Increasingly, the province assigns to municipalities such technical tasks as planning, or environmental protection, or energy conservation, which are administratively infeasible in small towns and townships. In these situations, either the task does not get done or the small community must rely on outside help from consultants, a regional government, the province, or all three. Rural local governments, generally, are in a dependent position in regard to achieving their own goals for planning and development.

Regional Government

Another structural form for providing governmental needs to rural regions is regional government. The county (in eastern Canada, except Newfoundland) and the regional district or county (in western Canada) are the various provincial counterparts. They may provide services and facilities—hospitals, parks, planning—directly to all communities because of the costs involved or the overriding regional need. They may, and often do, provide technical assistance to constituent communities; this is usually the case in planning. It should be noted that for most small communities the regional or county planning agency is not an alternative. The regional districts in B.C. may provide planning services to constituent communities, but in provinces to the east, there is no consistent system of county or regional planning, and most small towns must rely on their own resources.

Planning Resources

It is assumed in all provincial planning acts that all (incorporated) communities are capable of establishing and maintaining a workable planning function. But, as already indicated, the job of making community plans and land-use regulations and enforcing them in a small town falls to a small group of people. Typically, towns with populations below 2000 people have fewer than five employees, and it is on them that the burden falls to make technical and administrative planning judgments. The key personnel on a small town staff are the town clerk, building inspector, and roads superintendent. Seldom does a town below 5000 population have a municipal engineer, much less its own professional staff planner, although many employ consultants on an ongoing basis.

To sum up, in the majority of small towns, resources are simply too meagre to permit a satisfactory planning process. Nevertheless, just as with large communities, both the plan-making and the plan-implementation phases must be undertaken. Some mechanisms exist to assist small communities to make plans and land-use regulations. These mechanisms include grants, which can be used to hire consultants, and technical assistance programs from provincial ministries and regional governments. However, the job of plan implementation almost always falls to the community alone. This includes both the enforcement of land-use regulations and the programming of capital expenditures. Meanwhile, the intricacies of planning are increasing. Not only are provincial planning regulations and programs undergoing frequent change, but there are also new forms of legislation that impinge on such planning interests and resources as environmental assessment, farmland preservation, and coastal zone management. The small communities, which exist within these dependent situations, are often left to seek out mechanisms to complement their needs and compensate for their meagre resources.

Planning Aboriginal and Northern Communities

Among the most encouraging situations arising in Canadian community planning over the past decade or so are the advances being made in planning Aboriginal communities. They are encouraging for two reasons: one is the recognition of the often-dire needs in these communities that are finally being tackled; the other is the recognition by planners of the need to seek and use planning approaches that are appropriate to Aboriginal communities, not least that they are community-driven.[40] The concept of integrated community-based planning for Aboriginal communities emerged in the 1980s. It replaced the previous top-down master plan program administered for Native communities by the federal Department of Indian Affairs and Northern Development. The new planning approach seeks to consider "all aspects of life and livelihood and their interrelationships in the community."[41] Evidence of these efforts is seen in Nunavut, the Northwest Territories, Labrador, and the Maritime provinces.

One such project is the planning and building of Natuashish in Labrador to replace the older Davis Inlet (Utshimassits); the latter had come to national attention after a series of tragic events. In 1997, the people of the community, the Mishuau Innu, began a process of designing and building themselves a new community.[42] The approach gave the Innu control of every aspect of the plan. Planners visited elders in their homes, travelled to outpost camps, and asked children in school for their ideas. Family trees were drawn up to determine family relationships that would guide the location of extended-family groupings of housing, which the community wanted. Community members were trained in electrical wiring, plumbing, and carpentry so they could assist in the building. Everyone was consulted, in their own language if wanted, about the location of the school (at the foot of a hill for sledding, but not near the commercial area), about the type of housing (typical clapboard style with a basement), and so forth.[43]

Regional approaches covering several communities and the development of their resource base using similar approaches has occurred in the Gwich'in Settlement Area in the Northwest Territories. Here the land-use planning is for separate communities and the areas between them with the aim of developing a framework of "co-management" of resources—shared resource decision-making—among the communities and the territorial government.[44] A similar project was undertaken in Nunavut involving 26 communities to assess their facilities for elders and youths. As with the Natuashish new town in Labrador, the planners followed pragmatic principles that respected the Aboriginal context from the start:

- Ask questions. Planners are not cultural experts. The experts in a community are the people in the community.
- Enter into a new world view and respect history. People in a community may have answers and solutions to problems that are sustainable in the culture of that community.
- Assume nothing. Community relationships may be invisible to the planner.
- Maintain relationships. Relationships with elders and youth bring accountability and should be maintained after a project is complete.
- Observe body language. Facial expressions and hand language can help gauge the energy level of participants, acknowledge areas of frustration, and determine if your message is being understood.[45]

Planners for Iqaluit, the new capital of Nunavut, a much larger place (6000 people) than those discussed above, also found the need to adapt planning principles from "the South" to northern needs.[46] Faced with goals of "protecting the Arctic way of life and reflecting the Inuit cultural heritage," meant, among other things, permitting shacks on the beach to store equipment, and providing snowmobile trails, safe places for sled dogs, and working areas for drying skins. The response of these planners to the ongoing realities of Inuit life goes far in acknowledging and reinforcing the meaning of *place* in the eyes of the residents.[47] Other planners at Dalhousie University have fashioned a "First Nations Community Planning Model" and worked with several communities in the Atlantic provinces in preparing community plans.[48] A step-by-step planning workbook was prepared to assist communities such as Pictou Landing, Nova Scotia; Abegweit, Prince Edward Island; and Metepenagiag, New Brunswick, in creating their own plans.

Finally, plans for northern communities must acknowledge the climate. Simply providing basic services can be difficult in the Arctic, sometimes requiring special systems like the above-ground "utilidors" in Inuvik. Careful community design can help create "livable winter cities by providing shelter from winds, trapping sunlight, and avoiding snow drifts.[49]

Reflections

Despite their small size, the planning for small communities, both Aboriginal and non-Aboriginal, demands that a broader range of concerns be part of the planning process than does the planning for cities. Urban community planning focuses on the use of land and its regulation, but planners for smaller communities have that concern, as well as those over the natural resource base, the environment, and the cultural milieu. In addition, what has begun to be achieved in small-community planning over the past two decades is an approach that recognizes the crucial connection between the land base and the people of the community.[50] This has led directly to involving the citizens of small towns, from Tofino to Iqaluit, in debating and deciding the merits of planning issues. In Huron County, in southwestern Ontario, the integrated approach has guided planners of small communities to go beyond traditional physical planning; to function more as an "informed facilitator and community enabler" with the objective "to empower the local community with additional responsibility for its own future."[51] This principle and the ones cited above suggest a planning style that is less about the technical aspects of planning or about a received methodology than about developing a *relationship* with people in the community. They could, and should, apply just as well to urban community planning. Thus, as we move to a fuller discussion of citizen participation in planning in the two chapters that follow, it will be helpful to reflect on the question:

- ***In what ways is citizen participation in planning similar or different in small towns and cities?***

Endnotes

1. Alan J. Hahn, "Planning in Rural Areas," *Journal of the American Institute of Planners* 36 (January 1970), 44–49.
2. Gerald Hodge and Ira M. Robinson, *Planning Canadian Regions.* (Vancouver: UBC Press, 2001), 140.
3. C.R. Bryant, L.H. Russwurm, and A.G. McLellan, *The City's Countryside: Land and its Management in the Rural–Urban Fringe* (Longmans New York:, 1982).
4. John F. Meligrama, "Developing a Planning Strategy and Vision for Rural–Urban Fringe Areas: A Case Study of British Columbia," *Canadian Journal of Urban Research* 12:1 (2003), 119–141.
5. Jackie Wolfe, "The Native Canadian Experience with Integrated Community Planning: Promise and Problems," in F. Dykeman, ed., *Integrated Rural Planning and Development* (Sackville NB: Mount Allison University Small Town and Rural Research Program, 1988), 213–234.
6. Newfoundland, Department of Municipal Affairs, *Planning for Smaller Towns* (St. John's: Project Planning Associates, 1968), 23.
7. Dennis O'Connor, *Report of the Walkerton Inquiry* (Toronto: Ontario Ministry of the Attorney General, 2002).
8. *Victoria Times Colonist*, "Medical Officer Seeks Further Test of Wells," July 20, 2001, B1.
9. CBC News, "Water Treatment Plant Too Small," October 27, 2005.
10. Kevin S. Hanna, "Planning for Sustainability: Experience in Two Contrasting Communities," *Journal of the American Planning Association* 71:1 (Winter 2005), 27–40
11. Gerald Hodge, *The Elderly in Canada's Small Towns* (Vancouver: University of British Columbia Centre for Human Settlements, 1987), Occasional Paper 43, 14; recent tendencies are confirmed in Gerald Hodge, *The Geography of Aging in Canada.* (Vancouver: UBC Press, forthcoming 2007).
12. Gerald Hodge, *Seniors in Small Town British Columbia: Demographic Tendencies and Trends, 1961–1986* (Vancouver: Simon Fraser University Gerontology Research Centre and University of British Columbia Centre for Human Settlements, 1991), 10.
13. Raymond Chipeniuk, "Planning for Rural Amenity Migration," *Plan Canada* 45:1 (Spring 2005), 15–17.
14. Gerald Hodge, *Managing an Aging Population in Rural Canada: The Role and Response of Local Government* (Toronto: ICURR Press, 1993).
15. Gerald Walker, "Networks and Politics in the Fringe," in Michael Bunce and Michael Troughton, eds., *The Pressures of Change in Rural Canada*, Geographical Monograph No. 14 (Toronto: York University Department of Geography, 1984), 202–214.
16. Greg Halseth, "Community and Land-Use Planning Debate: An Example from Rural British Columbia," *Environment and Planning A* 28 (1996), 1279–1298.
17. Neil Rothwell et al., "Migration To and From Rural and Small Town Canada," *Rural and Small Town Analysis Bulletin* 3:6, Cat. no. 21-006-XIE (Ottawa: Statistics Canada, March 2002).
18. Wayne J. Caldwell and Michael Toombs, "Rural Planning, the Community and Large Livestock Facilities," *Plan Canada* 39:5 (November 1999), 27–29.
19. Paul Crawford, "Preserving Rural Character in an Urban Region: Rural Planning in the Township of Langley," *Plan Canada* 33:2 (March 1993), 16–23.
20. Alexander C. Vias, "Bigger Stores, More Stores, or No Stores: Paths of Retail Restructuring in Rural America," *Journal of Rural Studies* 20:3 (July 2004), 303–318; and Jim Simmons and Ken Jones, "Growth and Change in the Location of Commercial Activities in Canada: with Special Attention to Smaller Urban Places," *Progress in Planning* 60:1 (July 2003), 55–74.
21. Wayne J. Caldwell, "Rural Planning in Canada," *Plan Canada* 45:3 (Autumn 2005), 25–28.
22. Hanna, "Planning for Sustainability," 23.
23. Gerald Hodge, "Planning for Small Communities," A Report to the Ontario Planning Act Review Committee, Background Paper No. 5 (Toronto, 1978).
24. Cf. Evelyne Power Reid and Maurice Yeates, "Bill 90—An Act to Protect Agricultural Land: An Assessment of Its Success in Laprairie County, Quebec," *Urban Geography* 12:4 (1991), 295–309; and Christopher Bryant and Thomas Johnston, *Agriculture in the City's Countryside* (Toronto: University of Toronto Press, 1992), esp. 137–189.
25. Barry E. Smith and Susan Hald, "The Rural–Urban Connection: Growing Together in Greater Vancouver," *Plan Canada* 44:1 (Spring 2004), 36–39.
26. Hugh J. Gayler, "Planning Reform in Ontario and Its Implications for Urban Containment and Agricultural Land Use," *Small Town* 26:4 (January–February 1996), 4–13.
27. Peter Mah, "Changing the Dynamics of Rural Planning: A Rural Manitoba Planning Perspective," *Plan Canada* 38:2 (March 1998), 25–29.

28. Newfoundland, Planning for Smaller Towns.
29. Alberta, Task Force on Urbanization and the Future, *High River, Alberta* (Edmonton, 1973).
30. Felice Mazzoni, "Ucluelet: The Little Town That Could," *PIBC News* 43:2 (April 2001), 12–13.
31. Descriptions of these plans are available online at www.gov.mdrockyview.ab.ca/planningservices.
32. Available online at www.reddi.mah.gov.on.ca.
33. Statistics Canada, *Retail Store Survey 2003*, Cat. no. 3F0022XIE (Statistics Canada Ottawa:, March 30, 2005).
34. Judith Getzels and Charles Thurow, eds., *Rural and Small Town Planning* (Chicago: American Planning Association, 1979), 89–95; and Lane Kendig, *Performance Zoning* (Chicago: Planners Press, 1980).
35. Urban Land Institute, *Flexible Zoning: How it Works* (Washington, 1988).
36. Mazzoni, "Ucluelet."
37. Thomas L. Daniels, "Where Does Cluster Zoning Fit in Farmland Protection?" *Journal of the American Planning Association* 63:1 (Winter 1997), 129–133.
38. Randall Arendt, *Rural by Design* (Chicago: American Planning Association, 1994).
39. Corwin Cambray and Laurie McNab, "Agricultural Easements and the Niagara Tender Fruit Belt," *Plan Canada* 35:2 (March 1995), 37.
40. Donald Aubrey, "Principles for Successful Community Planning in Northern Native Communities," *Plan Canada* 39:3 (July–August 1999), 12–15.
41. Peter Boothroyd, "To Set Their Own Course: Indian Band Planning and Indian Affairs," Paper Prepared for the B.C. Region, Indian and Inuit Affairs Canada, 1984; and Margaret Jones, *The Community Is Quite Capable* (Guelph, ON: University of Guelph School of Rural Planning and Development, 1985).
42. Kashetan Rich, Joseph M. Rich, Paul Wilkinsonm and Chris Lowe, "Location Vocation, Natuashish: Planning a New Aboriginal Community," *Plan Canada* 37:6 (December 1997), 16–17.
43. This story is told, compellingly, by Margo Pfeiff, "Out of Davis Inlet," *Canadian Geographic* 123:1 (January 2003), 43–48.
44. Hillarie Greening and Neida Gonzales, "From Theory to Practice: Land Use Planning in the Gwich'in Settlement Area," *Plan Canada* 39:3 (July–August 1999), 16–18; and David Witty, "The Practice Behind the Theory: Co-management as a Community Development Tool," *Plan Canada* 34:1 (January 1994), 22–27.
45. Nalini Naidoo, "Principle of Natuashish in Labrador," *Alberta Planning Digest* (March 2002), 14–15. Reprinted with permission.
46. Pamela Sweet, "Sustainable Development in Northern Urban Areas," *Plan Canada* 44:4 (Winter 2004), 41–43.
47. Erik Borre Nilsen, "Rethinking Place in Planning: Opportunities in Northern and Aboriginal Planning in Nunavut, Canada," *Canadian Journal of Urban Research* 14:1 (2005), 22–38; and John Peters, "Aboriginal Perspectives on Planning in Canada—Decolonizing the Process: A Discussion with Four Aboriginal Practitioners," *Plan Canada* 43:2 (Summer 2003), 39–41.
48. Cities and Environment Unit, Faculty of Architecture and Planning, Dalhousie University, *First Nations Community Planning Model*, 2nd ed. (Halifax: Dalhousie University, 2003).
49. Norman Pressman, *Shaping Cities for Winter Climatic Comfort and Sustainable Design* (Prince George, BC: Winter Cities Association, 2004).
50. Floyd Dykeman, "A Return to the Past for a Rural Community-Based Planning and Action Program for the Future: A Challenge for Planners," in F. Dykeman, ed., *Integrated Rural Planning and Development* (Sackville, NB: Mount Allison University Small Town and Rural Research Program, 1988), 147–166.
51. Wayne J. Caldwell, "Rural Canada: Designing a Desirable Future," *Plan Canada* 32:4 (September 1992), 24–29.

Internet Resources

Chapter-Relevant Sites

Planning Canadian Communities
www.planningcanadiancommunities.ca

Rural and Small Town Programme, Mt. Allison University
www.mta.ca/rstp

Main Street Revitalization Initiative
www.centralhastings.ca

Community Economic Development in the Crowsnest Pass 1985–1996
www.ucalgary.ca/EV/designresearch/projects/2001/CEDRO/cedro/cip_acupp_css/crowsnest

Municipal District of Rocky View, Alberta
www.gov.mdrockyview.ab.ca

Rural Cape Breton District Planning Commission
www.rcbplan.ns.ca

Newfoundland, Department of Rural Development
www.intrd.gov.nl.ca/intrd

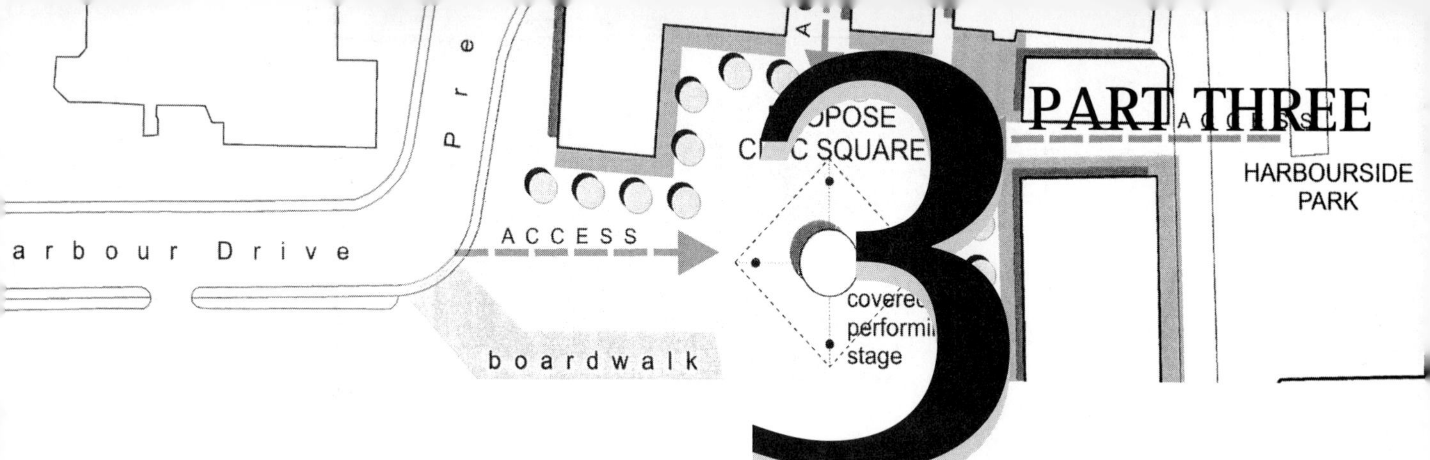

PART THREE

3

Participants and Participation in Community Plan-Making

INTRODUCTION

Probably no other instance of decision making in a community involves so many and such a wide array of citizens as does community planning. Making community plans demands democratic participation before, during, and after the plan is made. This can, and should, involve a wide range of participants, from ordinary citizens to planners, politicians, developers, interest groups, and associated professionals. There are formal structures that provide opportunities for involvement in various phases of making and reviewing plans. But there is as much or more participation that occurs in informal settings that also contributes to community plan-making and amending.

In the end, participation is about community members communicating with one another about what should be included in a plan so that it will best enable the community to reach its goals. The effectiveness of a community's plan is more a function of the participation that occurs in planning decisions than of any other factor. Planning acts may prescribe a planning process, but it is people, individually and in groups, who make it a reality and who give it texture. This requires blending the myriad values of participants and integrating the numerous roles they play in this extensive community dialogue.

The image above is from the Downtown St. John's Strategy for Economic Development and Heritage Preservation *project, St. John's Newfoundland and Labrador, which received the Canadian Institute of Planners' Award for Planning Excellence, Category of Implementation, 2002.*

11

Chapter Eleven

Deciding Upon the Community's Plan

Effective planning of human settlements . . . will come to depend more on human relations in the process of arriving at decisions than it will on the planner's science and art of preparing plans.

Harry Lash, 1976

Community planning has been referred to as a social activity many times in preceding chapters, for it is the process of a community of people deciding upon its future environment. Community planning is not just a matter of planning *for* a community; it is equally a matter of planning *by* and *with* a community. Thus, community planning is a process orchestrated around the aspirations and needs of a diverse array of people—citizens, firms, groups, institutions, developers, politicians, planners, the fortunate and the less fortunate, landowners, tenants, long-time residents and newcomers, and others—who make up the community.

It is now time to consider *who* participates in plan-making, to recognize *who* makes up "the community" that makes decisions concerning the plan and its various tools of implementation. That is the purpose of this chapter and the next. Here, the decisions made in the course of making and adopting a community plan and the complementary processes of amending the plan and implementing it are described. The principal participants in this process are arrayed against the series of technical and procedural steps that need be taken, and the parts they play and the roles they represent at the different stages are indicated. In the next chapter the dynamics and texture of participation are examined. As this exploration of the social, human side of community planning unfolds, consider this question:

- —***How is the process of community planning affected by the amount, nature, and quality of participation?***

The Decision Sequence in Municipal Plan-Making

When a community sets out to make a plan for its future built environment (or renew, or amend it and many of its associated tools) it initiates a decision-making *process* (see also Chapter 7). Further, it is a process that has a certain structure, a certain logic, and steps that must be followed. Some of these steps are spelled out in legislation, and are prescribed; these are the ***formal*** steps and they are considered first because they must be taken. Other steps in the process develop out of community convention and norms, and still others emerge as part of professional planning practice and the interplay of interests in the community; these are the ***informal*** steps, and they are considered second.

The Formal Steps

The community-planning process in Canada is highly formalized. A provincial planning act prescribes for all its incorporated communities a number of specific steps that must be taken, as well as the participants and their responsibilities at the various steps. The formal plan-making process does not contain all the same steps as in the ideal case of community plan-making described in Chapter 7 and, as well, contains other steps that are necessary for an institutionalized, governing process. A comparison of these two processes is shown in Figure 11.1.

Although the two processes have the same end of producing a plan to guide a community's development, their means for achieving it differ in three significant ways. First, the ideal planning process is a concept that does not specify participants, or specific decisions, or legal responsibilities. It is more like a reminder to a community of what to keep in mind when striving to prepare a plan. Second, the municipal plan-making process is concerned with defining the roles of public and private interests in the community with regard to the development of land. When the built environment might be changed either by private initiative or government action, many interests become involved besides the interests of those who directly initiate change. There are the interests of other landowners, neighbourhood residents, other public agencies, and of the community as a whole. Thus, decisions must be specified so that all interested parties (i.e., stakeholders) will be aware of their rights and responsibilities. Third, the formal process is concerned with establishing the balance between provincial and municipal authority in regard to private property rights. Since responsibility for these rights, when they are delegated to a municipality (as they are under the planning act), rests ultimately with the province, this shift in responsibility must be specified. The municipal plan-making process as shown is, therefore, a composite of steps found in various planning acts; actual steps may differ from province to province.[1] However slight these differences, they only heighten the concern over specifying steps and responsibilities. The seven main steps in the formal plan-making process are described below in terms of the decisions that are made, the interests involved, and the planning objectives of the decision.

1. Decide to Prepare a Plan (or Amend It)

Since a municipality's (or other local body's) plan becomes an official statement of policy, an explicit decision by the local council is required to initiate plan-making. Thereby all interests in the community, and those with an interest in the development of land in the community, are informed that the process is to begin. Such a decision to prepare a plan may arise from a sense in the community that development problems exist and/or that initiatives for major new projects are underway and that these should be dealt with in a more comprehensive, long-term, rational framework. The municipal council makes

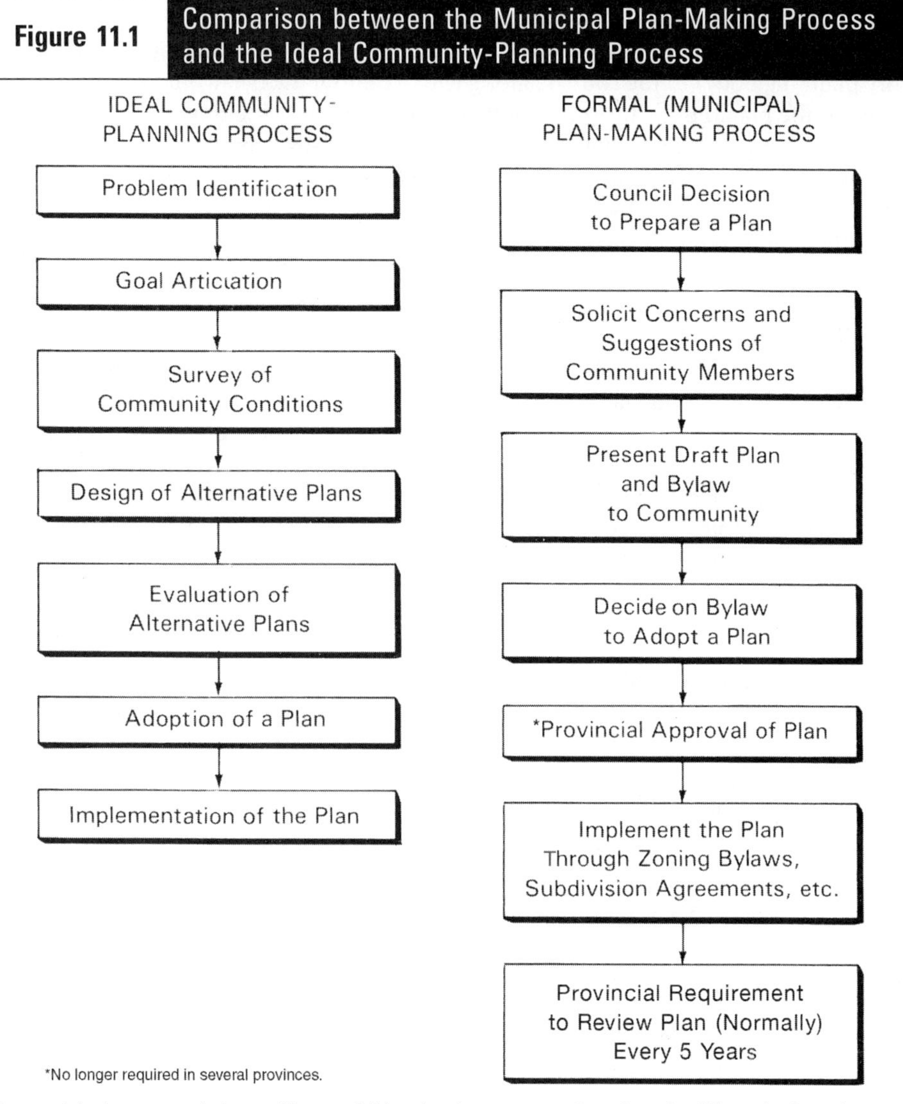

The municipal process of plan-making explicitly takes into account the roles of public and private interests that participate in the development of the community's land base.

this decision, which may have been prompted by planning staff, other public officials, an advisory planning body, developers, and/or public interest groups.

2. Solicit Concerns and Suggestions of the Community

Recognizing that there will be diverse opinions in the community concerning the future environment, most provinces mandate public participation at this step so that these views can be solicited. The object is to acknowledge the right of democratic involvement of community members in the identification of goals for the built environment. This input is usually obtained through one or a few public meetings held in the community, but may also include open houses, visioning processes, design charettes, and surveys.[2] No official decision is forthcoming from this consultation but the conclusions drawn from it usually become input for planning staff and other officials drafting plans. The community's planners and/or the local planning advisory committee normally conduct these consultations.

3. Present the Draft Plan

A proposed plan is submitted by the plan-makers (the professional staff and/or advisory planning body) to the council, along with a draft of the bylaw that would officially sanction the plan. Public hearings are then authorized to be held by the council (or its advisory body) as required by provincial legislation. This provides community members with an opportunity to evaluate the plan's proposals. Normally, only one plan alternative is presented. In most provinces, provincial ministries, federal agencies, and other public bodies with an interest in the plan's outcome are also given the opportunity to evaluate the draft plan. The local council makes the decision to set this part of the process in motion by giving first reading to a bylaw that incorporates the draft plan and allows it to go forward for debate. Presentations by citizens, developers, and other groups at public hearings and other submissions can then be made. These are documented and become official input into subsequent deliberations. The planning staff, other officials, and the planning advisory body may modify the plan in light of presentations at official public hearings before recommending the draft plan to the council. It should also be noted that such formal hearings are usually late in the process and can often become adversarial and result in little real consultation. Except in environmental assessment, planning legislation does not encourage the development and evaluation of alternative plans in the manner proposed by rational–comprehensive planning theory.

4. Decide on the Plan Bylaw

A community plan is adopted by the local council by approving a bylaw that makes the plan an official statement of policy about the community's built environment. This is a formal, legislative procedure requiring a formal vote by councillors on the second and third readings of the bylaw. In this way, the plan becomes "official" and legally binding on the council and all parties with interests in land development.

5. Province Approves the Plan

Most provinces require that a community plan be submitted for review and approval to the provincial government department responsible for local planning matters. The plan is checked for the adequacy of its planning approach and its consistency with provincial government policies. The plan is usually not considered official, or the bylaw in effect, until it has been approved and signed by the minister. In several provinces this step is now a formality.

6. Implement the Plan Using Various Tools

The passage of a zoning bylaw, the acceptance of a subdivision plan proposal, and the renewal or amendment of the community plan tend to follow a set of steps similar to the process for adoption of a first-time plan. Each initiative of this sort is regarded as a supplementary process of plan-making, in which the aims of the plan are reaffirmed, refined, and possibly revised. Indeed, provinces normally require that implementation tools conform to the community plan. Initiatives for amendments often originate with private land-development interests, individual landowners, or corporations. Vetting of such proposals involves planning staff, the planning advisory body, and, sometimes, provincial agencies (especially in land subdivisions that impinge on provincial highways). Again, they are approved by a vote of the local council on an enacting bylaw (see Chapters 13 and 14).

7. Review the Plan

It has become standard practice for provinces to require that each municipality (or similar jurisdiction) review its community plan on a regular basis, commonly every five years. The stated purpose is to update the plan with relation to demographic, economic, and physical changes in the community and revise policies where needed. Left unstated is the simple need for the local council and its citizens to re-acquaint themselves with this key document and reaffirm its policies. Review of the plan entails the same steps as in the creation of the original plan by the planning staff, advisory body, and provincial agencies. The process involves public consultations and formal public hearings and, lastly, approval of an enacting bylaw by the local council. Concomitantly, implementation tools, such as a zoning bylaw, that are based upon the community plan are required to be revised accordingly following the plan review.

Overview of the Formal Process

This set of formal steps is what gives structure to the planning process in a Canadian community. It acts to ensure that the rights of property owners are properly considered, the democratic right of community members to participate is respected, and that the community plan achieves legal, binding status. It also prescribes who may participate at each step and their responsibilities vis-à-vis the decision to be made. There are, essentially, only three sets of participants in the formal process:

- members of the municipal council;
- members of the general public; and
- provincial government officials.

Of these three, only the councillors and the provincial officials are charged with making binding decisions, of giving final approval to the plan. Although the citizenry participate, their role is only advisory. Another facet of these steps is that each consumes a relatively small amount of time in the total planning process. For example, there is the time taken by the council to debate the bylaw that enacts the plan, the time for one or a few public meetings, and the time for provincial review of the plan. These steps are, in effect, the points of convergence of various technical and consultative planning efforts in the total planning process, at which decisions need to be taken. These are the ***visible*** steps in the process, but in the interstices of these formal procedures, between the prescribed steps, much planning occurs, many more participants are involved, and the substance of the plan is gradually refined, as is discussed in the next section.

The Informal Steps

The formal steps described above deal mostly with the *procedure,* authority, and responsibility of participation—with getting the plan ready for adoption. But, in the course of accomplishing this, other important initiatives are undertaken (preceding and following the formal steps) that deal with the *substance* of the plan—with its content and style, as well as with its acceptance in the community. These latter undertakings are technical, consultative, and deliberative in nature and are concerned with the development of a community consensus about the need for a plan, the articulation of community goals based on community concerns, the survey of community conditions, and the design and evaluation of alternative plans. They consume the most time in the entire process, but are often the least visible to those outside the local government's planning apparatus. British planner Paul Cloke refers to them as being "obscure" because many occur in planning offices as planners gather and analyze data, in meeting rooms during negotiations with developers, in communications between local and provincial officials, and so on.[3] They occur between and during the overt (formal) decision-making steps. They do not constitute a progressive set of steps as much as they are part of the ongoing context of plan-making.

Some of these less visible facets have been made intentionally more visible in, for example, public participation programs, focus groups, and panel discussions. Some are made more visible through such events as street demonstrations (see Figure 7.3, page 185). Regardless of mode, the informal "steps" are crucial to the planning process; for it is through them the plan is actually formulated. Without the output from these steps, no formal decisions could be made. Below, they are considered in their two broad formats: (1) technical, and (2) consultative or deliberative.

Technical Steps

A large part of the planning process, it will be remembered from Chapter 7, is technical and involves, for example, the studies and analyses that have to be done by the professional planners. In this phase, the professional planners, municipal administrators, and elected officials are the leading participants. There is no doubt that they are able to influence much of the approach and content of the plan. Where this happens unduly, it can lead to a skeptical rejection of the objectivity and openness of the plan-making process and the charge that it is wholly "political." While much of the technical work has to be done within the planning office, it is a challenge to keep the plan-making process from becoming overly obscure. This challenge falls on the planner and the politician alike. More will be said about this later in the chapter.

The planner may call upon the skills of a wide array of professionals, including others employed by the municipality, consultants, and staff of other public agencies. This is the phase when analyses of economic and social factors must be done, and when the implications for public utilities, roads, and parks become evident, not to mention schools, libraries, and health facilities. As a result, individuals and groups outside the formal planning framework are drawn into the planning process. Except in the smallest municipalities, the planner will need to involve, at a minimum, the chief administrative officer, the public works director, and the director of parks and recreation. Some communities ensure this participation by forming a "technical planning committee," with the planner chairing it. It is important that the technical advice of specialists on city staff be available to the community, and also that there be a large measure of commitment to the emerging plan by those officials who will have a prominent role in implementing it. Sharing the results of planning studies with the public through published reports and/or media presentations are ways in which the process may be kept more visible.

Consultation Steps

Plan-making involves considerable consultation on community goals, needs, and development policies. Those consulted include the general public, public-interest groups, community organizations, and other public agencies. There are usually several forms of consultation. For individuals and groups, there may be an organized public participation program conducted by the community's planners through public meetings, open houses, questionnaires, and so on. Such public agencies as school boards, public utilities, and other levels of governments may be asked to react to planning proposals. Most communities establish an advisory planning board or committee to channel the various reactions into the plan-making process. Such a committee is usually given a mandate by the municipal council to seek public reactions and to resolve these, along with the planner's advice, into a plan that can then be formally debated.

It will quickly be realized that this informal part of plan-making is a complex process of gathering information and giving advice. The information provided by individual specialists, agencies, development interests, citizen groups, and the general public must be compounded until workable planning propositions are reached. This requires a high degree of social cooperation among the participants. Moreover, each participant brings to the consultations his or her own values, knowledge, and motivations, as well as hopes, fears, and criteria for judgment. These perspectives may be personal, professional, or group views, or some combination of the three. Suffice it to say that, at this point, the planner needs to have an acute awareness of the social relations inherent in his or her local planning process, as well as a personal frame of reference regarding the social *intervention* that plan-making constitutes for the various participants.

While cooperation is necessary for community plan-making, the task will not be completed, or if completed may not be fully accepted, unless there is consensus among participants about the goals and methods of the plan. Achieving consensus on planning proposals is a delicate task, but one that has become considerably more practicable, if not demanded, in recent years.[4] The formal planning deliberations discussed above tend to start with planning solutions that are debated in some prescribed (i.e., parliamentary) procedure, whereas consensus techniques start with identifying options and criteria for choice and end with mutually agreed-upon solutions. These techniques are often invoked when there are multiple interests involved in reaching a planned solution to a, sometimes, contentious problem. The siting of waste-treatment facilities at a citywide level or of group homes at a neighbourhood level are two such examples.[5] The Islands Trust, which is a form of regional government for more than a dozen of B.C.'s Gulf Islands, uses this approach in developing community plans: a broad-based committee develops the planning options and criteria before they are submitted to the formal deliberative process. In Prince Edward Island, the Island Regulatory and Appeals Commission promotes the use of consensus building, especially "to enhance the process in dealing with conflict situations."[6]

Participants in consensual processes are usually referred to as **stakeholders**. In practice, this means individuals or groups "with something significant to gain or lose as a result of the deliberations."[7] Choosing stakeholders as participants is, in principle, a matter of being more *inclusive* with respect to different interests in the community and, in practice, of being aware that "the public" in public participation includes a diverse array of publics.[8] The latter includes those often overlooked or inaudible in public debates, such as women, cultural minorities, the elderly, and youth.[9] It also means recognizing the value of a participant's practical experience—his or her local knowledge—and not just formal education or professional background.[10] Consensus building and perspectives on participation are described more fully in the next chapter.

Dispute Resolution. In carrying out their role, planners frequently find themselves having to balance competing demands of politicians, other municipal departments, developers, and public-advocacy groups. Sometimes, these demands are not readily reconcilable through technical or design analyses, and even political bodies may be stymied. Usually, this is because the demands are rooted in differences in basic attitudes and values. Dispute-resolution methods are increasingly being used by planners to reach constructive solutions without going to costly litigation or to appeal tribunals.[11] Two such methods being used are **negotiation** (where advocates for each side seek a mutual solution) and **mediation** (where a neutral third party, such as a planner, assists each side in finding an acceptable solution). Planners trained in these methods find that they can help solve problems, enhance mutual understanding, and strengthen the planning process. Outside professionals may also be utilized by the planners to facilitate problem solving between stakeholders.[12]

A number of cities in Canada have established mediation mechanisms for resolving difficult planning problems, notably Kitchener-Waterloo, Ontario, and Kamloops, B.C.[13] The City of Calgary, which has been operating a planning-mediation program since 1998, provides a good example of how this mode works. It is important to understand that mediation is used to resolve disputes between applicants for planning department approvals (say, a developer) and opposing parties (often nearby residents) to a project, and not between the planning department and applicants. Mediators contracted by the city's Planning Mediation Program work to help the parties to achieve their different goals, not to determine who is "right" or "wrong." The program has resolved conflicts over bylaw enforcement, design issues, conflicting land uses, group homes, restaurant noise, and neighbourhood industrial uses.[14] Mediation does not always succeed in resolving planning disputes, and courts may have to be involved, but it does succeed in many cases. Further, it is an approach that, along with other non-confrontational techniques, cultivates consensus building around planning issues, as we will see below and in the next chapter.

Making Supplementary Plans

Community planning does not stop with the formulation of the overall community plan. Once the latter step is completed, it is usually followed immediately by a number of supplementary plans to bring the community plan to fruition. The most prominent of these is the zoning plan, which, again, addresses the entire land area of the community. Larger cities may also prepare secondary plans to institute detailed land-use policies for individual neighbourhoods. The citywide plan can become a policy framework for environmental, transportation, and open-space systems if the details are in the secondary plans.[15] More specific plans that require decisions are the plans of subdivision for areas of undeveloped land and replotting schemes to consolidate and modernize scattered subdivision plans. Then there is the review and updating of the community plan itself.

All of these planning efforts supplement the community plan and refine its goals and policies. Supplementary plans are usually required by the provincial planning act to conform to the overall plan. Further, the decision-making sequence is formal and very similar to that for broad, general plan-making. Public information meetings are held to solicit views about the draft plan; a planning advisory committee deliberates on the planning proposals, taking into account the views of the public, and makes a recommendation to the municipal council; the council debates the proposal in the form of an implementing bylaw; and, lastly, provincial approval is sought. For supplementary plan-making, the community's planners assume the role of proponent. The array of participants is also much the same as in general plan-making.

A major exception in the array of participants occurs in planning for specific projects, as with a secondary plan for a large suburban greenfield site, a new subdivision, an application for rezoning, or for a development permit. In these cases, the draft plan is usually prepared by consultants and put forward by the individuals wishing to modify the sites. The proponents in these instances are commonly referred to as "developers," and, in secondary plans, subdivisions, rezoning, and development permit applications, they play a prominent role in propounding the merits of the plan in which they have an interest. It is the latter type of planning situation that is most likely to elicit controversy and conflict and call forth opponents to the project. Opponents may include competing landowners, neighbourhood groups, functional-interest groups, and other public agencies. This is the kind of planning situation, as noted above, that may generate the need for mediation or other dispute resolution.

Responding to Development Initiatives

The community-planning process must be responsive to ongoing initiatives for urban development, as well as for the promotion of its own long-range goals. With regard to development on private land—the bulk of all community land—the initiatives that will result in new buildings or redeveloped older buildings come from a wide range of private sources. Implementation of much of the community plan, thus, is effectively a function of decisions by diverse decision-makers. And it is only in the course of these private decisions being taken that the vagaries of land development become apparent.

Three facets of land development affect the process of plan-making or, more specifically, call for modification of the plan's policies and also affect who participates in the planning decisions:

1. Inherent differences among pieces of property that could affect the type of development;
2. The ways different developers view the potential for land development on specific sites; and
3. The different ways other landowners or the public view development initiatives.

None of these potential situations can be ignored. To do so would be not only unrealistic but could also deny "natural justice" to participants. Formal planning frameworks thus provide the means by which these various differences can be reconciled.

Differences in Developability

Zoning bylaws treat each property within a zone as equally developable. However, occasionally one or a few properties may differ from their neighbours in terms of their topography, size, shape, or location, such that the property cannot be developed as required by the zoning bylaw without causing their owners hardship. For such cases, provision is made to vary, or adjust, the zoning regulations; this is called a **variance** (see Chapter 13). A landowner who feels that her or his property requires a variance may make an appeal to a special committee established for this purpose. In some provinces, these committees are called a **Committee of Adjustment**; in others, a **Zoning Appeal Board.** They are appointed by the municipal council, and are usually composed of three members.

The decision sequence and the set of participants in the process of applying for a variance are distinct from other planning situations. The process is initiated by a landowner through an application to the appeal committee. Other landowners in the same vicinity (usually within a 100-metre radius) are advised that the application has been made and that they also may make submissions (pro and con) at a scheduled hearing before the committee. A hearing (which has legal standing) is held, and the committee can decide to allow or disallow the application. The committee's decision can usually be appealed to a provincial planning appeal body.

It should be noted that in the case of variances, neither the municipal council nor the advisory planning committee plays a specified role. The zoning appeal process is considered to be largely a judicial rather than a planning process. It is not unknown for decisions of appeal committees to have major planning consequences, such that the municipal council may feel disposed to appeal the ruling of its committee. The most contentious appeals are for variances in the permitted land uses, which usually arise because a desired use is neither specifically allowed nor disallowed. A very common case is the request to carry on an occupation in one's home in a residential zone. Variances may also be requested because the shape and/or topography of a property does not offer an adequate site for building.

Differences in Developers' Perceptions

The land-use designations in a community plan and zoning bylaw are the planners' *estimate* of the development potential of properties at particular locations. But not until a developer takes the initiative to build a new building or refurbish an old one on a site can the accuracy of that estimate be determined. Further, it is difficult to imagine all the uses that could be appropriate for a site, even within generally acceptable uses. For example, the properties adjoining an intersection may appear to be appropriate for retail commercial uses, such as a shopping plaza. The same properties may, however, appeal to a developer as a good location for a motel or an office building. In this kind of case all the possible uses are commercial. But take another example in which the plan envisions, and the zoning allows, a site to be developed for residential use, and a developer comes forward with a proposal that a shopping centre be built there. In order to accommodate these sorts of possibilities, planning regulations provide for applications to amend the zoning bylaw and, possibly, the community plan. The proponent of the project, the developer, usually initiates this process and remains a key participant throughout.

The examination of proposals to amend the plan and zoning bylaw is analogous to their original formulation. This is because it is felt that the plan or zoning bylaw should always represent a solid commitment by the community and, therefore, should not be modified without proper study and extensive deliberation. Proposed development projects requiring amendments to existing planning policy, thus, receive (1) technical scrutiny by the planning staff and other officials, (2) a formal public hearing to receive citizen comments, (3) debate by the municipal council, and (4) final approval by the province.

In contrast to original plan-making, the process of amending a plan or bylaw is usually initiated by someone external to the local government, such as a developer. The developer must make the case with various public bodies for changing the already agreed-upon planning policy. This, then, casts the staff planner in a different role, for in such instances, she or he is required to appraise the merits of the proposed development and advise the municipal council as to the desirability of modifying the plan or bylaw. Here the planner is expected to act in the public interest and not on behalf of the developer or any other interests. Much has been written on the delicate balance of roles of various participants in these planning actions. There is probably no initiative in planning that can generate as much controversy as one to change existing planning policy. Further, the required public participation and consultation processes are frequently looked upon with suspicion by developers and politicians as slowing down a development considered "good for the economy." This calls for considerable skill in designing processes that involve all affected interests effectively in the decision-making and do not inhibit investment. Many communities require developers to discuss their project with affected groups and seek their support before initiating a re-zoning. The same applies in many places regarding applications for development permits.

Different Reactions to Development

An essential feature of community planning is the consideration given to how development on one site might affect surrounding properties, as well as the community as a whole (if the project is large). It is also in the nature of community planning that there is the right of appeal of planning decisions. This serves not only the democratic rights of citizens in the community but also the rights of natural justice of the property owners, who are directly affected because their property rights may be infringed upon by the decision.

It is something of a truism in planning that reactions to planning decisions become more intense the closer a proposal for development appears about to become an actual project. Thus, if we take the example in the preceding section, in which a developer proposes a project that differs from what is allowed at a location under the plan and bylaw, we have one of the classic situations in which strong reaction emerges in the planning process. The reactions may, of course, be varied and may come from diverse and sometimes unexpected sources. Our planning statutes allow for these reactions to be channelled formally into the plan-making process.

There are three formal avenues by which reactions to development proposals can be expressed. The first mechanism is **public information meetings,** at which applications for rezoning and amendments to the community plan are presented and explained. The second is the formal, legislated mechanism known as the **public hearing,** which is held prior to the decision by an approving body such as a municipal council. The third mechanism is the filing of **formal appeals** against the planning decision of the council with a provincial appeal board, provincial ministry or cabinet, or a court. Each of these has specific functions, protocols, and limitations for dealing with reactions to a development proposal.

Public Information Meeting. This venue offers an opportunity for a proposed project or plan and bylaw amendments to be described to the public and for them to raise questions about it. This forum serves to allow the community's planners (and developers who are able to make presentations) to assess the amount and type of reaction to a proposed project. In light of that, they will be able to modify (or not) the formal planning policy and the bylaw that the council will debate so that it better reflects the interests of affected property owners as well as the mood of the community. No formal votes or decisions are made at such meetings.

Increasingly, communities are employing other methods to expand the opportunities for citizens to express their reactions to development projects and proposals to amend a plan or zoning bylaw. One that is used frequently is the "open house," where the proposals are displayed and community members can drop in to view them, ask questions, and, sometimes, be surveyed as to their preferences. This format tends to be used early in the planning process and serves to educate people about the project or planning proposal. It makes subsequent information meetings and hearings more fruitful

Public Hearing. This is the setting that is mandated in provincial legislation for presenting the draft proposal for a project or plan or bylaw amendment to the public. Views, both pro and con, are heard and recorded for the purpose of informing the local decision-makers and other reviewers and appeal bodies. Members of the council, or other appropriate body, proceed then to a formal debate on whether to approve or not approve the proposal (see Planning Issue 11.1).

Appeal Process. The decision of the municipal council (or other body), if objected to, may be appealed. The objection must be heard in a judicial setting (unless the appeal board considers it trivial or simply dilatory). Those who seek to file an objection must be prepared to enter into the legalistic setting and, in some cases, to pay a fee to do so. Lawyers usually need to be retained, and this tends to limit appeals to those individuals, firms, or groups that can afford to appeal. There is generally provision for individual citizens or groups of citizens to be heard, but the format is so highly structured that, unless their representations are well organized, they tend not to have much effect. Appeals are most often undertaken either by a developer whose proposal has been denied or by organized citizens' groups objecting to a municipal planning decision. A common sort of appeal is that made by a citizens' group against a private development project or a project of a local or regional government, such as the location of an expressway, a waste-disposal site, or an airport (see also Chapter 8).

Who Gets Involved and How

The field of participants in the various processes of plan-making in a community is indeed large. Potentially, every person or household may be involved, along with the local government, the provincial government, community planners and other professionals, business interests, land developers, and appeal bodies. This wide range of participants, so characteristic of modern democratic communities, contrasts with planning in cities of the past. The planning of these, described in the first few chapters of this book, was the responsibility of a few people—it is often possible to identify who in those cities planned or decided upon this or that element—and they usually had some official capacity. Not so today, for the "planners" now are more diverse than ever.

Who then are the community's "planners" nowadays? They might easily be inferred from the decision processes described above. A list of participants, in order of their likelihood to become involved, would look like this:

- Politicians (members of the local council, board, or committee)
- Planning advisory board or commission members
- Professional staff planners and/or consultants
- Private developers and development interests
- The citizenry, in all its facets, from individuals to public advocacy groups
- Other professionals (e.g., lawyers, engineers, mediators)
- Extra-local agencies (e.g., regional governments, provincial ministries)
- Appeal bodies

Implied in this list is a hierarchy of participants, starting with those who are either responsible for or representative of community interests. The latter tend to be more prominent in community-planning processes, and may even seem to dominate them, but others may also play important roles in distinctive ways. Indeed, each individual in whatever category on the list comes with a different background, understanding, and set of values. Achieving a viable planning process in a community requires that a planner have an appreciation of and an ability to work with such a complex array of participants.

Key Processes and Their Participants

It should be clear by this point that community planning is not a monolithic process, but rather a set of sub-processes, each concerned with different phases of decision making. Each of these sub-processes has its own distinctive set of participants. The listing of participants above may be better appreciated within the context of the following three sub-processes, which reflect the basic decision sequences in community planning:

1. the process of plan-making;
2. the process of plan implementation; and
3. the process of plan clarification (see Figure 11.2).

In Plan-Making. In this pivotal phase of planning it is understandable that the key participants are community members and community advisors. Although in formal terms the local councillors initiate the making of an overall plan (and the zoning bylaw), the stimulus may arise from the public and/or the planners employed by the community. Further, once this phase has begun, all three of these major participants become highly dependent on one another. The planner, acting as advisor to the council, will need ideas and comments from the public in order to put forth politically acceptable proposals. (This very special relationship between politicians, planners, and the public is highlighted in the next chapter.)

A few other points about the participants in this phase are important to note. First, the *politicians* on the council are representatives, as well as members, of the public. Second, the *public participants* comprise many elements—individuals and organized groups within the community, as well as those representing their personal interests and those espousing community interests. All can claim, with some justification, to reflect the

Figure 11.2 Key Participants in the Three Phases of Community Planning

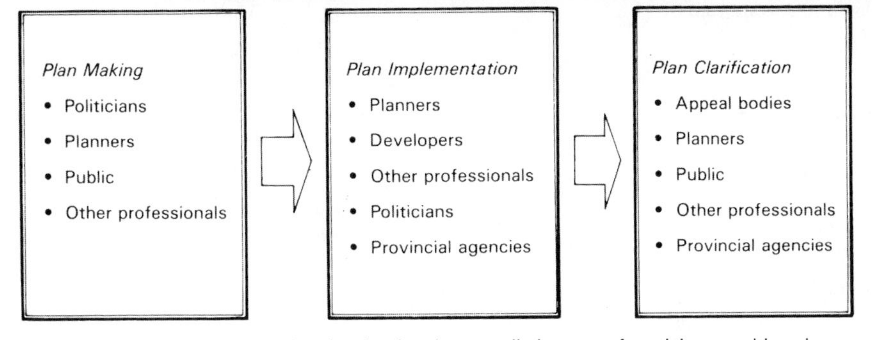

Each of the three phases of community planning has its own distinct set of participants, although some are involved in more than one phase. Within each phase, there is considerable interdependency among participants in their aims and decisions.

interests of the community. Third, *other officials of local government* are involved along with the politicians, planners, and the public. In roles such as city manager, engineer, solicitor, parks director, or traffic engineer, they can affect the form and content of planning proposals, and also, as advisors to the council, can influence the debate about the plan. Fourth, the *planning advisory board* can be a major mediating force, both through its citizen membership and through its review of planning proposals before they go to the council.

A fifth, and important, point is that the plan-making phase itself is increasingly being opened up to larger numbers and a greater diversity of citizens. This is especially the case in what is often referred as the "pre-planning" phase, where goals need to be articulated, citizen preferences elicited, and alternative futures discussed. Often, planners use an array of techniques to obtain public input and ensure the participation of people who do not usually participate, such as youth, the elderly, and members of multicultural communities.[16] Stronger and more broadly acceptable plans emerge, it has been found, when such initiatives are taken early, when a wide array of different people are involved, and when several different techniques of participation are employed.[17]

In Plan Implementation. The planning decisions in this phase have mostly to do with responding to development initiatives. These may come both from private land developers and from public agencies, including the local government itself. The community's staff planner or consultant is primarily responsible for determining whether these initiatives correspond to the intentions of the plan and/or bylaw. If they do, the next steps tend to be routine, with various approvals for development being granted. In many communities, development proposals are subject to further public scrutiny at information meetings or open houses before approval. Plans showing the siting of new structures must be submitted to a site-plan or development-permit committee for approval. The latter committee is usually composed of council members. Many other communities administer development proposals through special departments, sometimes called development departments, or through their building-inspection departments.

When private development initiatives do not conform to the community's planning policies and regulations, the developer has the option to petition council to amend the plan and/or bylaw. This sets in motion formal processes in which the planner is responsible for orchestrating the various steps. The developer appears at each formal hearing to advocate acceptance of the amendments. The public, individuals and public advocacy groups, may also attend these hearings and make submissions. The council will usually ask the planner to give an opinion on the proposed development. If the decision of the

Planning Issue 11.1 Residents Worry about School-Area Growth

St. John's Telegram
May 1, 2003

Residents Worry about School-Area Growth
Sandy Bierworth

There's a lot of new development in the Town of Paradise. And while the most residents are excited about the ongoing growth, many are concerned that at least one proposed subdivision may be putting children at risk and harming the environment.

The town held a public commissioner's hearing at the Town Hall Wednesday night to hear objections and concerns relating to two proposed developments. One is for an extension to the Sunset Gardens residential subdivision that would see about 70 new homes built, the other is for a new subdivision of about 80 homes that would be built near the Paradise elementary school.

William Parsons, commissioner under the Urban and Rural Planning Act, chaired the meeting, and Alton Glenn, Paradise town planner, helped explain the proposed amendments that would allow the two developments to proceed.

Concerned Citizens

Several members of the Paradise Concerned Residents Committee (PCRC) voiced their concerns.

Committee member Kevin Green presented Parsons with a petition of more than 700 signatures which listed the concerns the committee has with the proposed developments, including road infrastructure, clear-cutting of trees, recreational infrastructure, and the safety of children attending the school.

Dan Bobbett, a member of the PCRC and co-chair of the Paradise Elementary School Council, said the proposed development would turn the edge of the school's parking lot into the main roadway for the new subdivision.

"Right now, Mallow Drive ends in the school parking lot," he said. "This development would turn it into a major collector road. Surely the safety of children should take priority."

He suggested that if the development does go ahead, a different entrance be constructed for the school and that "proper sidewalks be built for children who are walking to school. Or better yet, move Mallow Drive away from the school."

He added that nearly 500 children attend Paradise elementary, which goes from kindergarten to Grade 6.

Fence Not Enough

PCRC member Stan Moores said that while there is a fence around the school playground, it isn't enough.

"The fence will keep the kids in, but it won't keep the cars out if something happens, like slippery roads or a drunk driver," he said.

Valerie Bartlett, also on the PCRC, outlined the committee's environmental concerns.

She said the Sunset Gardens development would leave little open space to be used for parkland, which "leaves nothing for future generations to enjoy."

She is also concerned about the trail system students, teachers and parents at Paradise elementary have been working on for years.

"The road behind the school would ruin plans to develop the trail there," she said. "If this development does go through, what we're asking is that it not be linked with the existing trail system, for safety reasons."

Parsons will submit his recommendations to the Paradise council within six weeks. Council will review his report and either accept, reject or modify his recommendations in its decision.

Source: Sandy Bierworth, "Residents Worry about School-Area Growth," *St. John's Telegram*, May 1, 2003, A1. Reprinted with permission.

council is to accept the amendments, provincial officials are then given the opportunity to comment upon and accept or reject them.

In Plan Clarification. There are a variety of reasons that the intent of the plan or the provisions of the zoning bylaw may need to be clarified, as in the simple case of variance or the more complex situation where a developer feels a project should be allowed within the plan and the council disagrees. The appeal processes thus serve to clarify the current thinking with regard to planning in the community. The appeal bodies are a focal point for participation in this phase, and their procedures define the participants. With variances, for example, participants usually are closely associated with the property, such as the property owner, owners of neighbouring properties, or other neighbourhood residents. However, appeals of council decisions involve a wider array of participants: the community's planner, other professionals appearing as expert witnesses (either for the community or for the appellant), and members of the public. In most instances, the decision on the appeal is reviewed by provincial officials, and sometimes even further appealed—to an appeals tribunal, a court, or the provincial cabinet.

The Milieu of Roles

Knowing *who* is involved in the various phases of planning is one side—the objective side—of decision making. The other side is qualitative—it is the *how* side; that is, how the various participants respond, react, behave, perform, and so forth in the actual decision making. (The cartoon below reflects on this qualitative side of participation.) The quality of the involvement in the planning process is affected by the nature and extent of the participation of those involved. The *formal* agents (councillors, planning advisory board members, and provincial officials) have defined places and functions in plan-making: they *have to* participate. Others act in an *individual* capacity (professional planners, other professionals, private developers, members of the public) and *choose* to participate. And some act in *both* capacities, in particular the politicians and professional planners, especially when the latter are on the staff of the local government. This mixture of participants means that we can expect that the people involved will act from a variety of personal and professional backgrounds and possibly have different perceptions of the aims of planning.

Those who participate in the various processes of planning do so much as actors play roles in the theatre or in a film. But planning processes are complex dramas, with each participant possibly playing several roles. While some of the roles are prescribed by the planning statute or regulation, many are not. Further, each participant plays at least one formal and one informal role. The formal roles come from legal and institutional definitions of the participant, while informal roles arise from cultural

Source: Hellman cartoon from *Built Environment* 22:4 (1996), "Theory and Practice in Urban Design," reproduced by permission of Louis Hellman and Alexandrine Press.

norms of the situation, as well as from the personality and behaviour of the participant. For example, the local councillor's formal role is to debate and decide on the bylaw that implements a plan, a role that no other participant shares. A citizen is ascribed a role at public meetings to comment about the plan. The local planner has a formal role, not by legal definition but by professional status and contract, to advise the planning board and council on the technical matters pertaining to the plan.

A public-participation program over the future development of an important site in the heart of Winnipeg, which was examined for the roles assumed by various participants, provides a helpful view of the "players" in such situations.[18] In this case, the public-participation process became as much an issue as the proposed development itself, with participants playing roles that ranged from "proponents" (who supported the process) to "challengers" (who felt the process was manipulative). Between these extremes were roles such as "positive explainers" (who supported the project but not always the process), "negative explainers" (who felt the need for changes in the process), and "observers" (who did not take a position about the process). Proponents included consultants to the agency charged with the development, agency board members, provincial officials, and local politicians—essentially those who had decision-making power. Challengers were mostly activists, along with one local politician—essentially individuals outside the decision-making arena. This research concluded that public-participation programs are "primarily shaped by the values of the powerholders."

Each participant also "acts out" informal roles. A municipal councillor may, for example, champion development interests in debates because the district he or she represents is populated by voters known to profess this orientation. The way in which the councillor plays out this role is further shaped by his or her own personality, which may range from that of an aggressive participant in council debates to that of a contemplative, behind-the-scenes participant. Beside these facets of a councillor's role-playing, the person's actual behaviour comes into play; that is, how he or she reacts to crises, to political pressure, to personal stress, to criticism and misunderstanding. A thoughtful self-evaluation by a councillor of his role in a fractious planning situation in a Midwest U.S. city allows a rare view "behind the scenes," as the following quote reveals how personal predispositions affect participants:[19]

> The most emotionally intense experiences I had ... occurred in those situations where people felt that their worries, fears, hope, dreams, or visions were cavalierly disregarded.... Whenever I saw, heard, or read such intemperate dismissals, I could feel my blood boil and my own anger beginning to build.

There are, of course, parallels for citizens and planners who also have their roles shaped by personal factors. A citizen participant with the professional qualifications of, say, a lawyer is likely to pursue things differently from a retired schoolteacher or blue-collar worker. And, if a planner is personally convinced of the need for citizen participation in planning, he or she will respond differently from one who sees the planner's role as strictly technical and neutral. Regardless, the planner may find it difficult to play the often-ambiguous role of being both the proponent of change (in, say, density) and the facilitator of public debate over the matter. In one reported case, in Tofino, B.C., the planner proposed the use of consultants to carry out the public-participation program for this very reason.[20]

Community Planning versus Corporate Planning

It should be evident by now in our discussion of community plan-making that *there is no single entity or person in control of the decision-making process.* The local council has a central role in making planning decisions, but it depends upon advice from others and reactions from still others, and its decisions are often subject to appeal. The planner is involved in almost all phases of plan-making, but is not in a position to direct the process of carrying out the plan.

One observer calls this diffuse undertaking of community planning a process in "social cooperation."[21] This is an apt term for this diffuse community activity, for community planning involves the processing of information, ideas, and reactions among a diverse group of participants who, it is likely, do not fully share each other's values about the development of the community. In community planning there must be a sharing of information and the opportunity to refine proposals and bring new alternatives into the discussion. It has an essential need for a large measure of feedback so that goals and courses of action may be reexamined and, where necessary, modified. In short, community planning is an ungainly process, one that has been the subject of much discussion. There are those who would streamline it using corporate planning as their model. And there are those who wish to clarify the community-planning process so that it is seen as one of political choice. Both these perspectives deserve examination in order to better understand community plan-making.

The Limits of the Corporate-Planning Model

For many decades, going back to the 1920s and earlier, there has been considerable pressure to model community-planning decision making after the process followed in private business firms. Early initiatives in this regard sought to "keep politics out of planning" and to "get on with making decisions." More recent initiatives, recognizing that community planning is, essentially, a process of political choice, seek rather to improve the information on which councillors base their choices and to organize the process of implementing decisions.

Much of this latter effort focuses on improving budgeting and financial planning. Various new approaches to budgeting that are being introduced into local government decision making aim to improve the link between the service and programs being provided and the goals of the political decision-makers. Performance budgeting (which looks at the output obtained by expenditures), program budgeting (which looks at the output of groupings of expenditures), and zero-base budgeting (which requires all operating units annually to justify expenditures against community goals) are all adding significantly to the effectiveness of local governments.[22] It is through the budget that actions of officials are controlled to achieve policy objectives.

Some proponents of improving local government performance and planning advocate the use of corporate-planning models, which aim at a rational selection of effective means to obtain predetermined ends. Plunkett and Betts, for example, propound a four-stage model: (1) *policy planning,* for the purpose of defining priorities and selecting objectives; (2) *action planning,* for the purpose of establishing program alternatives and budgets; (3) *operations,* which involves carrying out program activities; and (4) *feedback and review,* which involves assessing program impacts.[23] At first glance, this paradigm seems to mirror the rational–comprehensive model of community planning. However, almost all corporate models start with the assumption that the ends (goals) are given and that the need is to activate an administrative structure to achieve these designated ends. They further assume that the tasks of the organization can be subdivided and delegated to facilitate their implementation.

A typical corporate-planning model is that offered by Redman, illustrated in Figure 11.3.[24] It is a four-stage model that includes a feedback loop for plan modification and relates the tasks of the chief executive officer(s), the managers, and the department heads. This model puts into operation the sequence of decision making implied in the typical organization charts made for firms and governments: from chief executive to staff officers to line departments. It is a *hierarchical* model in which administrative responsibilities may be specified progressively from the highest to the lowest levels of the organization. It further assumes that problems or tasks may be handled simultaneously and independently from each other.

The corporate-planning model is suited to some facets of local government operations and is used in many larger Canadian communities, such as Edmonton and Ottawa, at the broad, citywide level. The municipality in these cases is referred to as "the corporation," and the work of the planning department is expected to help fulfill the entire municipal corporate plan (see Figure 8.3).[25] Nonetheless, there are several important ways in which the corporate model is not consonant with physical planning for a community. A rather obvious difference is in the time horizon of corporate planning, which is much shorter (3–4 years) than that of most community-planning activities (10–20 years). Further, there is the assumption in corporate planning that the goals will be established by the top executive of the organization and that these will be agreeable to all other participants. In community planning, by contrast, goals are a product of extensive deliberation at all levels, as well as subject to debate between levels. Another difference is that most corporate planning assumes a hierarchy of responsibilities between participants, but such a division of labour does not exist within community planning. Lastly, the corporate model assumes that tasks can be divided in such a way as to allow participants to act independently. But an essential feature of community planning is that it has always been "collaborative" and is becoming more so.[26] Although it is tempting to want to adapt the efficient means of decision making in corporations to community planning, it is clear that the operations and planning aims in the two realms are fundamentally different.

Figure 11.3 Model of the Planning Process Used in Corporations

PHASE	FUNCTION	RESPONSIBILITY	TIME HORIZON
1. Strategic planning	Establishes objectives, strategies, goals, and policies to govern acquisi tions, use, and disposition of resources, and provides resources to business units.	Executive management	Long-term 4+ years
2. Management control	Establishes objectives and strategies relating to implementation of strategic plan, and allocates resources.	Managers	Long- and short-term 1 month to 3 years
3. Operational control	Develops programs to utilize resources effectively and efficiently.	Department heads	Short-term 1 month to 1 year
4. Plan modification	Assesses performance of resource use.	Executive management	Short-term

This model is essentially hierarchical and allows responsibilities for achieving objectives to be specified from the highest to the lowest levels of the organization. Such a division of labour does not usually exist within community planning.

Source: Louis N. Redman, "The Planning Process," *Managerial Planning* 31:6 (May–June 1983), 24–25.

Community Planning and Political Choices

One of the main reasons community-planning decision making is not easily adaptable to corporate or other organizational planning models is its inherent involvement with *plural* political choices, with politics.[27] Or, put another way, community planning is a mode of decision making peculiar to the needs of diverse interests to decide upon their community's future. There are, therefore, a multiplicity of goals to be satisfied, representing a plurality of interests. There is no single hierarchical structure for channelling initiatives or commands. What regularity there is comes from the ever-present competition between economic and social interests.

The Local Political Economy

Before the advent of community-planning institutions early in the 20th century, decisions about the future development of a city or town had to be resolved by one of two mechanisms: either the **economic marketplace** or the **political arena**. These mechanisms still exist, of course, and play a prominent role, but are now joined by the approach of community planning. It will help in understanding the special character of decision making in community planning to look at these counterpart mechanisms and their relationship to planning as shown in Figure 11.4. It can be seen that (1) the scope of both the political arena and the economic marketplace overlap; (2) the planning realm must cope with both these influences; and (3) planning directly affects only a small proportion of community decision making.

Marketplace Decisions. The built environment of a community results from choices made either by private landowners and project developers or by public agencies. The former build the homes, apartment buildings, shopping centres, and factories, while the latter build the roads, schools, parks, utility lines, and other public buildings. The decisions of private developers are largely motivated by economic considerations—How much will a home cost? How much profit can be realized by erecting an office building? Such decisions are articulated by mechanisms of the economic marketplace such as the residential real estate market or the commercial and industrial land markets. The marketplace in Canadian society is not controlled by any central mechanism, but is open for the participation of all who have the financial resources and the will to risk them. Decisions are made on the basis of financial resources that are bid for houses, properties, and so forth, and on the asking price for them. Transactions of this sort are made all the time in the typical Canadian community, the outcomes of which determine much of the character of the built environment—its appearance, location, and stability. The essence of the marketplace mechanism is the contention that the development of a community's land is most likely to occur, and in the best fashion, through the efforts of buyers and sellers who most fully appreciate their own interests—this is the so-called neoclassical economic model.

Political Arena Decisions. The political arena exists to make those decisions about the development of the community that affect the entire populace. On the one hand are those decisions about facilities and services whose provision is the responsibility of the community, such as the street system and the public utilities. On the other hand are those decisions about steps that are taken to protect the interests of all citizens, such as building and traffic regulations, and waste-disposal services. Provincial statutes provide local councils with the authority to raise money through various taxes to pay for facilities and services and to pass regulations in the general interest of the community. Further, the council's decisions frequently affect development in the private sector, such

Figure 11.4 The Place of Planning in Community Decision Making

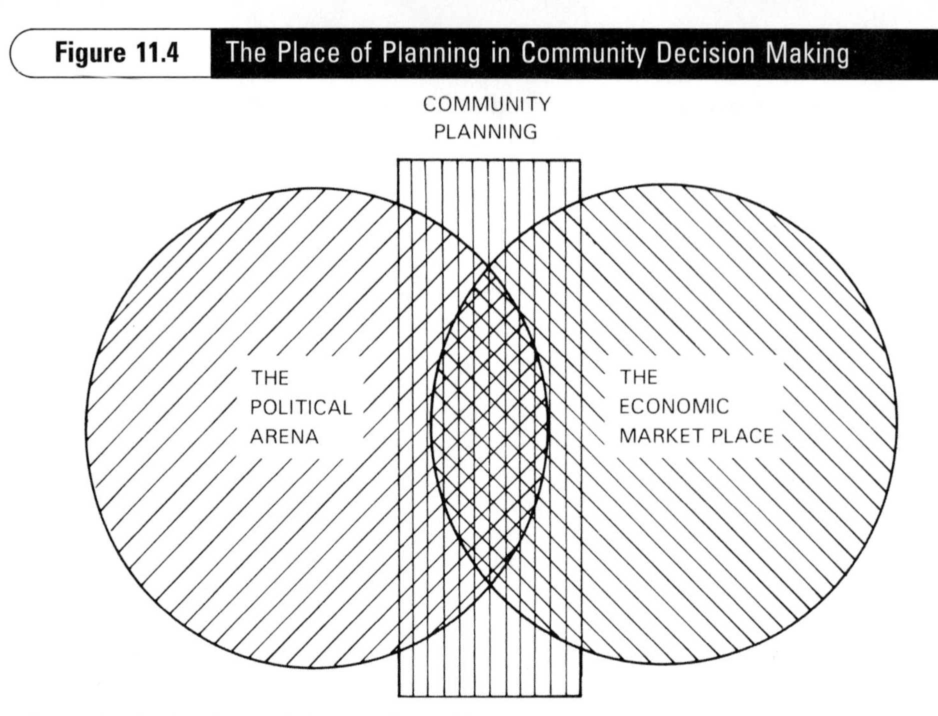

Community planning plays a role in reconciling political and economic interests affecting the development of the physical environment in the interest of the community as a whole.

as decisions about the basic infrastructure and regulations that might constrain the quality, location, and/or pace of development.

Voting. It is important to note the inherent differences in the form and basis of decision making in the political arena and the marketplace. Both, it might be said, involve *voting.* In the council's milieu, the voting is done by a group of people that must publicly declare its position. Councillors are elected to represent the values and interests of the community, so that when they decide upon matters affecting physical development, their vote is expected to be for the general good. In the economic marketplace, the voting is done with money and the transactions are conducted in private, with only the interests of the participants at stake. In both arenas, the decisions are similar in that they tend to concentrate on short-term considerations: concluding a deal, passing the annual budget, establishing a regulation.

We call this overall community milieu, using the older tradition of the social sciences, a **political economy**. The decisions of local councillors are seldom made without either explicit or implicit reference to economic consequences: What will be the impact on jobs with or without this new project? Will we drive investors away if we don't approve this project? And, as has been seen since the beginnings of formal community planning in the century just past, those involved in developing land and trading real estate have taken an explicit interest in the workings of local government—running for council and getting appointed to planning boards, not to mention lobbying. Indeed, some politicians may have an inherent interest in furthering marketplace solutions to community-planning issues. Conversely, many councillors have been elected in recent years on "reform" platforms that espouse more social content in decisions—for example, housing programs for the poor and protection of the natural environment. The task of local government is to find the appropriate balance between the two realms, between the competing values of each. Community planning is the governmental activity that attempts this reconciliation. This intermediate position imparts special characteristics and tensions to the practice of planning.

Planning, Politics, and Power

Community planning grew up and persists as a distinct mode of decision making because neither of the two traditional modes deals effectively with all the concerns that arise when deciding upon the future environment. Developers usually have their eyes on short-term economic gains from a particular project. And politicians often have their eyes on an upcoming election. The approach of community planning is to employ a more pervasive rationality than either of the other two modes in those situations where future-oriented, community-wide interests are at stake. But in pursuing this course, planning is subject to the vagaries of political power and choices, for community planning is an integral part of the political life and machinery of the municipality wherever it is in use.[28] This means that the planners play a political role, an attribute that planners themselves have often had difficulty accepting. For many planners "politics has meant conflict," to quote Forester.[29] And, he continues, "Conflict has meant irrationality...[and] loss of control." Thus, it is virtually impossible for planners, or any others who become involved in planning decisions (e.g., citizens, businesspeople), to avoid the politics entwined in community planning. To quote Forester again:

> We often have to interpret what a goal, policy, regulation or by-law means. Once we do that, knowing that multiple and conflicting interpretations are always possible (some favouring some people, others favouring others) we're right back to politics.

This is nowhere more evident, nowadays, than in environmental issues. Whether it concerns protecting the water supply in Wawa, Ontario, or ridding Sydney, Nova Scotia, of pollutants, these issues are highly politicized. Friedmann has said in this regard: "No matter what position planners take in an argument concerning environmental policy they are certain to antagonize important segments of the population."[30] Hence, as both Forester and Friedmann conclude, planners must establish *relationships* with key participants in order to function effectively. This is due largely to the fact that working in a political milieu means coming to terms with power—the power of others as well as one's own power. Planners in Canada have little political power in the conventional sense (i.e., they are able only to give advice) and cannot control political outcomes. But planners are able to wield considerable influence on those who can (the politicians) and those who try to influence politicians (developers, citizens, business interests, and others).[31] Much more will be said about the role of planners and other participants in the next chapter.

Community planning, seen this way, does not operate independently of either economic or political modes and their criteria, for planning does not entirely replace either of the modes and must draw upon them both, thereby reducing the extent to which it can adhere to its own rational–comprehensive ideals. Those involved in community planning cannot help but bring their own biases and interests into the process. Some would argue, since not all segments of the community are likely to be able to participate, that planning outcomes tend to favour the interests of the more powerful over those of the weak and disadvantaged.[32] From this perspective, community planning is a political activity in all respects, because every planning alternative tries to balance the several sides of a development issue. Plans are, therefore, not just neutral technical solutions. And when planners advocate a planning solution, they are in effect taking a political stand that will affect different groups in different ways.

The potential for planning decisions to be discriminatory is very real. When a stable neighbourhood is divided by an urban expressway, or tenants with low incomes have their rooming houses demolished for high-rise apartments, or small businesses on downtown streets are forced to compete with modish shopping malls or "big-box" stores, it is clear that there are always *some who gain* and *some who lose* from a planning decision. These sorts of situations create tensions in the planning process and affect the behaviour of participants. The long-standing utilitarian ethic of planners—to produce the greatest good for the greatest number—cannot, it seems, ever be fully met.

A graphic illustration of this kind of planner's dilemma is found in the situation that faced the Victoria Park neighbourhood on the southeast edge of downtown Calgary. In 1992, the Calgary Stampede Board proposed an expansion of the Stampede Grounds over the entire 32 hectares of Victoria Park, wiping out a 1000-person older community that had undergone some gentrification. The main justification offered by the board was the need to sustain the economic contribution of the Stampede to the local economy, reputed to make up about 20 percent of all tourist revenues.[33] The city's planners supported the board's plan and, further, argued that the community was marginal and not entitled to protection, even though their previous redevelopment plan had tried to do so. They chose to support the position of the "developer" (albeit a public developer in this case), weighing the whole city's presumed economic gain against the needs of the citizens of Victoria Park, who would lose their homes and community. The reconciliation of such positions in the realm of community planning is never easy. Thus, the milieu in which planners frequently find themselves working has a moral as well as a political side to it.

Reflections

Although we establish structures that offer opportunities for involvement—such as consultation, debate, deliberation, and appeal—in planning matters, we cannot know the nature and extent of that involvement. For, at its best, community planning is a process in social cooperation. It requires the involvement of many participants of many different kinds, and demands their interaction with one another to make, implement, and clarify plans. It will always be somewhat ungainly, time-consuming, and unpredictable, not least because some participants have formal roles they are required to play according to provincial legislation, while many other roles emerge and are played out in the myriad off-the-record deliberations that comprise community plan-making and amending. The latter are the invisible or obscure phases of plan-making that Paul Cloke has shrewdly noted.[34]

The involvement in planning may also differ from one community to another, both in quantity and quality. This arises from the simple fact that the composition of persons and groups and the cultural norms of the community determine who becomes involved in plan-making in any of its phases. The presence (or absence) of persons with certain skills or of specific interest groups can affect the quality of involvement in community planning. As well, each community may differ in its inclination to become involved in planning issues and in the intensity of that involvement. Beyond these "chance" factors are the personal and professional preferences of the plan-makers themselves (both planners and politicians) toward the kind and amount of involvement they wish to encourage. The numbers, composition, and performance of participants, as well as the effectiveness of their participation, may be enhanced or dampened by the access that is afforded to citizens and other stakeholders and by the processes employed, as recent research has shown.[35]

Moreover, there is concern that participation processes constrain the access of many citizens, especially of those who are considered marginal because of their ethnicity, gender, age, or income.[36] The search for more effective and inclusive processes in planning has become more energetic in the past decade, along with pressure not to judge the level of citizen participation by numbers alone, but also by the diversity, and the inclusiveness, of participants.[37] As the next chapter discusses in detail, modes of participation can be expected to be further adapted and expanded in the future, and the pressure to continue on planners and others who are responsible for making this happen. In anticipation of this discussion, consider:

- —***How is plan-making affected by who is involved, and how they are involved as participants?***

Endnotes

1. A useful review of the formal processes in use in each province is found in R. Audet and A. Lettenaff, *Land Planning Framework of Canada: An Overview,* Working Paper No. 28 (Ottawa: Lands Directorate, Environment Canada, September 1983).
2. For a description of one of the most comprehensive public consultation processes see Ann McAfee, "Vancouver CityPlan: People Participating in Planning," *Plan Canada* 35:3 (May 1995), 15–16.
3. Paul J. Cloke, *An Introduction to Rural Settlement Planning* (London: Methuen, 1983), 3.
4. Judith E. Innes, "Planning through Consensus Building," *Journal of the American Planning Association* 62:4 (Autumn 1996), 460–472; and Patsy Healey, *Collaborative Planning: Shaping Places in Fragmented Societies* (Vancouver: UBC Press, 1997).
5. John Andrew, "Examining the Claims of Environmental ADR: Evidence From Waste Management Conflicts in Ontario and Massachusetts," *Journal of Planning Education and Research* 21:2 (2001), 166–183.
6. John L. Blakney, "Citizens' Bane," *Plan Canada* 37:3 (May 1997), 12–17.
7. Karen S. Christensen, "Teaching Savvy," *Journal of Planning Education and Research* 12 (1993), 202–212.
8. Ananya Roy, "A 'Public" Muse: On Planning Convictions and Feminist Contentions," *Journal of Planning Education and Research* 21 (2001), 109–126.
9. Cf. Rae Bridgman, "Criteria for Best Practices in Building Child-Friendly Cities: Involving Young People in Urban Planning and Design," *Canadian Journal of Urban Research* 13:2 (Winter 2004), 337–346; and Penelope Gurstein, "Gender Sensitive Community Planning: A Case Study of the Planning Ourselves In Project," *Canadian Journal of Urban Research* 5:2 (December 1996), 199–219.
10. Ann Van Herzele, "Local Knowledge in Action: Valuing Non-professional Reasoning in the Planning Process," *Journal of Planning Education and Research* 23:2 (2004), 197–212.
11. Richard B. McLagan, "Custom Negotiation and Mediation: Updating Our Planning Toolkit," *Plan Canada* 36:4 (July 1996), 26–27.
12. Philip Dack, "Mediation for Land Use Decision-Making," *Plan Canada* 41:1 (March 2001), 10–12.
13. Rand Diehl, "Resolving Community Development Disputes: The Kamloops Experience," *Plan Canada* 35:5 (September 1995), 30–34.
14. Dack, "Mediation."
15. The most recent comprehensive land-use plans for Montreal (2005), Toronto (2003), Ottawa (2003), and Vancouver (1996) have taken this general approach.
16. Note the Vancouver experience in McAfee, "Vancouver CityPlan."
17. Raymond J. Burby, "Making Plans That Matter: Citizen Involvement and Government Action," *Journal of the American Planning Association* 69:1 (Winter 2003), 33–49.
18. Beth Sanders, "A View from the Forks: Coming to Terms with Perceptions of Public Participation," *Plan Canada* 38:2 (March 1998), 30–32.
19. J.A. Throgmorton, "On the Virtues of Skilful Meandering: Acting as a Skilled-Voice-in-the-Flow of Persuasive Argumentation," *Journal of the American Planning Association* 66:4 (Autumn 2000), 367–383.
20. Kevin S. Hanna, "Planning for Sustainability: Experience in Two Contrasting Communities," *Journal of the American Planning Association* 71:1 (Winter 2005), 27–40.
21. Rolf-Richard Grauhan, "Notes on the Structure of Planning Administration," in Andreas Faludi, ed., *A Reader in Planning Theory* (Oxford: Pergamon Press, 1973), 297–316.
22. Cf. T.J. Plunkett and G.M. Betts, *The Management of Canadian Urban Government* (Queen's University Institute of Local Government Kingston:, 1978), 230–249; and C.R. Tindal and S.N. Tindal, *Local Government in Canada,* 2nd ed. (Toronto: McGraw-Hill Ryerson, 1984), 207–224.
23. Plunkett and Betts, *Management of Canadian Urban Government,* 247.
24. Louis N. Redman, "The Planning Process," *Managerial Planning* 31:6 (May–June 1983), 24–40.
25. Gord Jackson and Mary Ann McConnell-Boehm, "Plan Edmonton: A Plan and a Process," *Plan Canada* 39:5 (November 1999), 17–19.
26. Among the expansive literature on this topic see Patsy Healey, *Collaborative Planning: Shaping Places in Fragmented Societies* (Vancouver: UBC Press, 1997).
27. Art Cowie, "Politics and Planning: Ten Lessons from an Old Campaigner," *Plan Canada* 43:3 (Autumn 2003), 18–20.
28. Ian Wight, "Mediating the Politics of Place: Negotiating Our Professional and Personal Selves," *Plan Canada* 43:3 (Autumn 2003), 21–23.
29. John Forester, "Politics, Power, Ethics and Practice: Abiding Problems for the Future of Planning," *Plan Canada* 26:9 (December 1986), 224–227; and, among many other publications, John Forester, *Planning in the Face of Power* (Berkeley: University of California Press, 1989).
30. John Friedmann, "Planning, Politics, and the Environment," *Journal of the American Planning Association* 59:3 (Summer 1989), 334–338.
31. Jana Carp, "Wit, Style, and Substance: How Planners Shape Public Participation," *Journal of Planning Education and Research* 23:3 (2004), 242–254.
32. Two Canadian planners noted this issue more than two decades ago: Matthew Kiernan, "Ideology and the Precarious Future of the Canadian Planning Profession," *Plan Canada* 22:1 (March 1982), 14–24; and T.I. Gunton, "The Role of the Professional Planner," *Canadian Public Administration* 27:3 (Fall 1984), 399–417.
33. Barton Reid, "The Death of Victoria Park Neighbourhood, the State of Urban Reform and the Battle of Mythologies in Calgary," *City Magazine* 13:1 (Winter 1991–1992), 36–42.
34. Cloke, *Introduction to Rural Settlement Planning.*
35. Samuel D. Brody, David R. Godschalk, and Raymond J. Burby, "Mandating Citizen Participation in Plan Making: Six Strategic Planning Choices," *Journal of the American Planning Association* 69:3 (Winter 2003), 245–264.
36. Among others see Leonie Sandercock, *Towards Cosmopolis: Planning for Multicultural Cities* (New York: Wiley, 1998).
37. Two examples are Beth Moore Milroy, "Some Thoughts About Difference and Pluralism," *Planning Theory* 7–8 (1992), 33–38; and Peter Marris, "Planning and Civil Society in the Twenty-first Century," in Mike Douglass and John Friedmann, eds., *Cities for Citizens* (New York: Wiley, 1998), 9–17.

Internet Resources

Chapter-Relevant Sites

Planning Canadian Communities
www.planningcanadiancommunities.ca

A list and profiles of Canada's municipalities.
www.munimall.net/municipalities/municipalities

City and Town Official website.
www.citytown.info/Canada

Community Visions Program, Vancouver
www.city.vancouver.bc.ca/commsvcs/planning/cityplan/Visions

Focus Edmonton
www.focusedmonton.ca

Public Participation, City of Toronto
www.toronto.ca/torontoplan/participation.htm

Regional Plan Process, Halifax
www.halifax.ca/regionalplanning

12

Chapter Twelve

The Texture of Participation in Community Planning

Challenging public decisions could be viewed as a form of Kafkaesque baseball. Citizen groups are always the visiting team in their own home town.

Linda Christianson-Ruffman, 1977

The effectiveness of planning in a community, in the end, is more a function of the participation in planning decisions than of any other factors. Planning acts may prescribe a planning process, but it is people, individually and in groups, who make it a reality and who give it texture. A well-designed plan and a thoughtfully drafted zoning bylaw must be deliberated, communicated about, approved, and applied by members and officials of the community. The process they engage in amounts to a "flow of argumentation," in which the various participants attempt to persuade others of the merits or otherwise of the proposed plan, project, or regulation.[1] Mirroring the significant shift in the thinking about planning practice in recent years, Judith Innes says bluntly: "What planners do most of the time is talk and interact."[2] This observation applies equally to *all* participants in community planning: their participation carries with it the need, indeed the demand, to communicate, to become involved in dialogue with other participants.

This chapter examines the texture, the shape, the rhythm, and the pace of participation to see how the community participates in community planning. For this, one needs to move beyond the various formats for participation, especially those that are prescribed, and consider its style and quality, that is, to consider the nature and quality of "the dialogue." In this examination of the special place of politicians, professional planners, the public, and developers in community plan-making, the following question should be the focus for this chapter:

•—***How does "the community" actually participate in community planning?***

Linking the Public, Politicians, and Planners

Within the varied milieu of decision making for community planning there are several key community participants whose behaviour is crucial to an effective planning process. Politicians, professional planners, the public, and developers make up this group and significant relationships exist among them, as we shall see. But, first, the key relationship between the members of the community, the members of the local governing body, and the professional planners is explored.

The effectiveness with which this particular triad—public, politician, and planner—can work together will largely determine the success of the planning process in a community. Their situation is not unlike that of the distinctive Russian sleigh that is pulled by three horses (*troika*) whose energies must be balanced to achieve both forward motion and the desired direction of the sleigh. So, too, the citizenry and the municipal planners and councillors are dependent upon one another in the process to attain a plan that embodies an acceptable direction for the future of the community. The perspective within planning practice and theory of how to achieve a harmonious and productive blend of these participants has undergone considerable change, especially in the past two decades. Prior to 1960, for example, community planning was a relatively sedate activity. Politicians received and accepted the "expert" advice of the planners, while the public seemed willing to let their elected representatives judge the appropriateness of planning proposals. However, the disruptions to neighbourhoods caused by the planning responses to urban growth and deterioration—especially large redevelopment and expressway projects—changed the way these relationships worked. Citizens confronted politicians over proposed projects and rezoning, and the politicians sought more workable solutions from the planners. In the 1990s, many of these kinds of confrontations, and newer variations on them, continued.[3] Although many Canadian communities have extended their public-participation efforts, effective community participation requires still more.[4]

Two lessons for planning emerge from this experience. The first is that community planning is a *political* activity that makes choices among values and affects different segments of the community in different ways. Thus, individuals, groups, firms, and others are entitled to a voice in the decision making consistent with planning's venerable democratic values and more recent concerns with civil society. The second is that participation in planning decisions is not just a one-way process of communication. There needs to be a *dialogue* among all three interests. In some communities, planners and politicians have learned these lessons only grudgingly. Where they have been absorbed, the planning process features a high degree of interaction between the public, the politician, and the planner—a "six-sided triangle" of communication, as Harry Lash once called it.[5] The six-sided triangle refers to the six important links of communication in community planning: each side of the triangle is a double-headed arrow. This is a profound statement that each interest (public, politician, or planner) must have the potential of dialogue with each of the other two. Anything less and the effectiveness of the planning process is diminished (see Figure 12.1).

But achieving a good working relationship among all three sets of participants is not a simple matter. Personal and professional expectations, mutual trust, and communications skills and preferences are at the heart of making the six-sided triangle work effectively. Planners must consider whether their role is to be that of a neutral technical advisor to the politician or that of an advocate of particular positions held within the community. Politicians must consider how and whether to invite citizens into the decision-making process, to share some of their power. And citizens must consider how to work in a constructive way with planners and politicians who make a place for them

Figure 12.1 The Six-Sided Triangle of Planning Participation

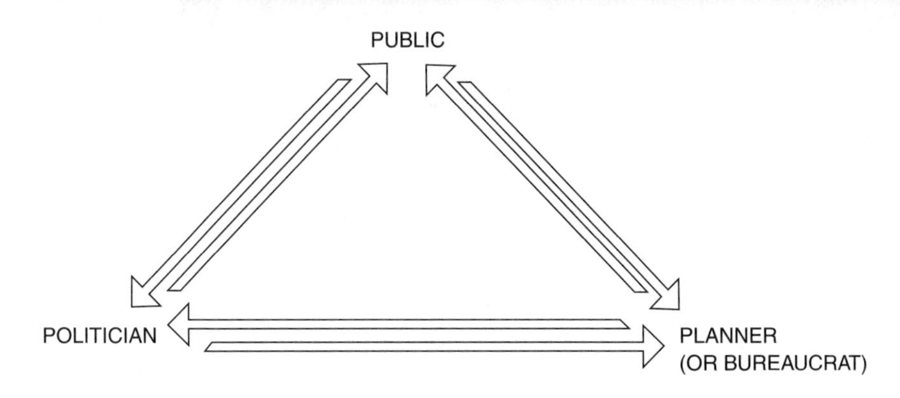

The effectiveness of the community-planning process is largely determined by the degree to which this triad—public, politician, and planner—can work together. It may be called a process of social cooperation.

in the planning process. These are the main threads that weave the texture in this important part of the planning process.

Linking Politicians and Planners

The relationship between the politician and the professional planner is complex and, in some ways, paradoxical. It has several aspects, one of which is that of employer (politician) and employee (planner), because the planner is either on the staff hired directly by the local governing body or is hired as a consultant to the council. Another aspect is that of expert (planner) and client (politician), because the planner comes to the situation with certain socially sanctioned skills from his or her education and professional experience. In the planning process, however, they are each dependent upon the other—indeed, interdependent—since both seek the similar end of coordinating diverse interests in the community and achieving a good quality of environment generally.[6] Yet in this relationship tensions exist, and politicians and planners can find themselves in conflict over who should give leadership in plan-making.

The Politician. The job of the politician is not well understood. Nominally, elected officials are there to represent the interests of the people in the community. But those interests are diverse and often compete with one another. Thus, the job of "representing" becomes one of reconciling and integrating the many competing demands that citizens, groups, firms, and other officials place on the local councillor. This role of the politician has been called that of "broker-mediator."[7] The politician also has to provide leadership as to what direction the community's development should be taking, so that the competing private interests are merged into a general, community-wide interest.

The politician's job of providing leadership and resolving conflicts has some incompatible aspects. If there is conflict around an issue—for example, the closing of a neighbourhood school or allowing a group home—the politician must try to hear as many of the competing views as possible and not make up her or his mind too early. This uncommitted stance may be scorned as failing to show leadership. But when the politician decides on a possible solution to an issue too early in the process, he or she may then be rebuked for being biased toward one interest. A planner who was active in local environmental matters and then elected as a city councillor in a U.S. city, recounted that he was portrayed as having "an underlying agenda" reflecting that of a special interest group.[8] As Lash similarly observed about politicians faced with the same dilemma:

> On any given issue in the community, he/she cannot *lead* unless he/she is *committed* to a solution, but if there is serious conflict around an issue, he/she cannot *resolve* it if he/she is *committed*.[9]

A related dilemma for the politician revolves around sharing his or her power to make decisions with others in the process, or what might be perceived as "giving up" power to others.

The politician generally tries to bring together two sets of interests. One set comprises the views held by members of the community. The other comprises the advice given by the community's planners and other technical advisors, which is taken up more fully below. But first it is important to point out a distinctive facet of community interests that the politician is obliged to respect, and incorporate into her or his understanding and deliberations—the perspective of the developer. Those individuals and firms involved in promoting land and property development are usually also members of the community. They include, as well as the landowner(s), local builders, architects, appraisers, lawyers, engineers, realtors, and bankers (see Figure 12.1 and further discussion below). These latter "citizens" represent a legitimate view about their projects being in the community's interest in a capitalist economy such as ours. It is also a view that may be contrary to that held by other citizens and, thus, add to the work of the politician.

The Planner. In reconciling the diverse interests in the community, the politician is very dependent upon the local planning staff. The issues often require an almost immediate appreciation of technical matters, and it is the duty of the planner to present all the relevant information and analyses to the politician. Yet even exercising such seemingly straightforward tasks involves the planner in the political realm by virtue of the choices about the type and quality of information to present, thereby complicating the interdependence of the two.[10] And when it comes to making a recommendation of one or another course of action, the planner then becomes a competitor for the politician's attention, indeed a stakeholder in the planning process. This situation raises at least two ethical dilemmas for the planner regarding the option he or she advocates in the political debate. On the one hand, if the planner's position seems to support that of the politician, then citizens may lose confidence in the planner's impartiality. And, on the other hand, too-vociferous support for citizen positions on an issue may be seen by the politician as being at odds with those of the planner's "employer." And to the politician is left the difficult task of bringing together the judgment of the "expert" and the "will of the people."

At some point in the process, the relationship between the politician and the planner may shift from employer–employee to client–expert. The ease with which this shift of roles occurs depends a great deal upon the planner's concept of leadership in community planning. The planner's essential contribution to the planning process is to be able to prepare a plan for a physical development situation that integrates a multitude of public and private interests and concerns. Through training and experience, the planner is in a position of "intellectual leadership" in plan-making.[11] Yet conflict can arise between the planner and the politician if the planner thinks herself or himself best fit to serve the community needs and assumes a measure of political leadership by acting strategically. But the planner's training seldom equips her or him for political leadership, a factor that is increasingly stressed as planning processes become more interactive.[12]

Tensions between planners and politicians is correlated with the relationship that each has with the other set of participants—the public. In playing out their respective roles, the politician and the planner have both an obligation and a need to involve the larger public in order that both can say they hold planning positions reflecting the community at large. (Just as with the politician, the planner has close contact with developer–members of the public, including involvement in a process of mutual learning, as we saw in Chapter 7, about policies, regulations, economic realities, and so on.) The competition for the public's (including developers') attention simply indicates the high degree of interdependence of the three sets of participants in the community-planning process. It can also be seen that there are few formal specifications for making these relationships work.

Linking with the Public

There is a long democratic tradition of the community making plans for its own future through its locally elected councillors. There is likewise a growing tradition of public participation, where public-spirited citizens and groups advocate the need for a community plan. However, neither of these traditions provide for the direct participation of the public in actually making a plan. When the public is included in making decisions about the plan for the future community environment (as is the expectation nowadays), the relationships among all three sets of participants are affected. Both the politician and the planner must re-define their roles. With citizen involvement, the politician may no longer assume that he or she provides the best reflection of the views of the public in the planning process; the planner may no longer assume that he or she serves the public by serving only the politician.

Involving citizens in plan-making constitutes a sharing of power over the content and implementation of the plan and over the process of getting to the plan. There have been available, for some time, mandated, formal avenues of communication between the public and local government officials, such as public meetings and hearings. These permit citizens to be heard, but not necessarily in a comprehensive or continuing way or with any guarantee that their concerns are being taken into account. Effective participation implies some direct citizen input into decisions that could influence subsequent developments in the plan. This often leads to the apprehension among citizens that they will be used for the politicians' own ends if they participate. As for planners, citizens often feel "talked down to," or ignored by them, or hampered in obtaining information. Many citizens may feel they are not listened to at all, as numerous feminist critics have pointed out with respect to the participation of women and of persons from marginal groups in planning processes.[13] (A later section discusses this more fully, along with some of the new approaches for enhancing public participation.) These tensions are an inevitable part of any political process in which those with power are being asked to share it. Further, the planning process does not assign power to citizens, only the right to be consulted. Even though provision may be made for including citizens in plan-making decisions, the final responsibility rests with the politicians and planners to see that effective participation occurs.

Despite the conflicts and difficulties, the relationships between citizens, politicians, and planners are essential to the planning process. The reasons are quite simple. The citizens are a primary source of information about the problems that are being experienced by the community, about the impacts of proposed solutions, and about the values and aspirations of community members. The politicians and planners, on the other hand, know the resources that are available to solve problems, the limits of knowledge about project impacts, and the institutional and procedural avenues that must be observed.

"We want real community participation in this decision. Plan A is too expensive. Plan C is inefficient. Now, which plan do you prefer?"

The politician is the key actor in determining the form participation will take and whether it will be positive, for it is the politician's ultimate decision-making prerogative, with regard to issues and options, that would be opened up to scrutiny. The politician must come to believe that public participation will enhance his or her ability to resolve conflict and provide leadership; in other words, to know the community better. When the public is included, there are sometimes risks for the planners: "expert" opinions may have to be justified to "non-expert" citizens, uncertainties may have to be acknowledged, and information may have to be shared. And, while the politician may make the ultimate decision about participation, it is the planner who shapes the participation process and controls much of the flow of information.[14]

In short, another party is brought into the decision making, and this always complicates the resolution of issues. As the six-sided triangle shows, there are *three times* as many interactions to consider if citizens participate than if the planner and politician are the only participants.

The triangle of public–politician–planner can lead, as does any three-sided relationship, to unequal coalitions and resulting stress. If the politicians come to rely more on the citizens' views than on those of the planner, the latter may feel that his or her technical competence is being devalued.[15] Politicians may come to distrust their staff planners, fearing that they will cater to citizen groups espousing different political ideas. Citizens, for their part, often learn that they cannot expect their recommendations always to be accepted by politicians and planners. Indeed, each of the participants has much to learn, not only about the others, but also about the planning process and its possible outcomes when seen through others' eyes. The latter point is often referred to as *social learning* and reflects both the need for broad learning to occur and the mutual obligations (i.e., "to attend, share, listen, initiate creative contributions, and identify and mediate problems") of all participants in the process.[16]

Roles of the Planner

In many ways, the community-planning process centres upon the professional planner. Although not necessarily the one to initiate the process, the planner soon becomes responsible for sustaining it, for shaping it so that plans, policies, and programs emerge to guide future physical development. It can be said that it is in the planner's own interest to carry out the task. Nevertheless, it is a demanding task, involving interaction with all the other participants in the planning process. The planner, it must be emphasized, plays a multifaceted role in a political milieu, a role that is gaining even more facets as the milieu itself shifts. Early on, the scope of the planner's task was identified as involving four basic roles:

1. the planner as leader (or representative) of the planning agency,
2. the planner as representative of the planning profession,
3. the planner as political innovator, and
4. the planner as citizen educator.[17]

The past two decades have revealed additional roles and elaborations of these four roles as planning theory and ideology have evolved. Canadian planner Tom Gunton identified eight variations of planners' roles, for example.[18] The broader perspective will be filled out as we proceed, but the four basic roles listed above are a good starting point. Also, it needs to be noted that each of the roles may be played differently, depending upon the individual planner's personal disposition and the community setting. Different communities may call for the emphasis to be placed more on one role rather than another, and in some cases other participants might substitute for the planner.

As Planning Agency Leader/Member. The planner must be concerned with the organizational base from which he or she operates. This involves the agency's status and relationships with community groups, governmental organizations, and developers as well as its staffing and morale. The planning agency must be accepted as an essential part of the community's governmental machinery if the goals of planning are to be achieved. Thus, the planner must develop the confidence of elected and appointed officials and the public about the necessity of planning, along with that of private developers.[19] This means developing relationships of trust, cooperation, and encouragement with each of these groups and, just as important, cultivating channels of communication through which to obtain information for planning, and directing attention to planning problems.

The effectiveness of the planner is closely linked with the credibility of the planning agency. The social and political relations of the planning agency itself, by which it is either welcomed or excluded in promoting planning solutions with developers, other agencies, or the public, must be perceived and nurtured by the planner as a routine part of the task. Allies as well as enemies, cooperators as well as antagonists, must be identified. Planners thus need skills that go beyond the technical knowledge of planning; that is, as one observer notes, the need "to go beyond technical expertise to organizational and political savvy."[20] They need to be able to work with others and develop trust and support for the agency and its views. This is especially cogent for planning professionals whose agency's development proposals will not be carried out by them but by others (e.g., engineers, lawyers, architects). And planners must, in multicultural cities and towns, not only become more sensitive to ethnic group needs but also learn to consider issues of inclusiveness and the norms of practice.[21]

Not to be overlooked in this particular role is the bureaucratic milieu in which the staff planner works. Municipalities and other forms of government that employ planners have hierarchical structures that assure that major decisions are taken by elected officials and tactical decisions within the agency are taken by department heads. But, as Filion notes, this hierarchical form of decision making "slows the planning process and reduces planners' capacity to make commitments."[22] So, being a "leader" for some significant planning or development reform may be a challenge for the planner.

As Technical Advisor. The planner brings to the community-planning job the values and standards of the planning profession. By virtue of education and training, she or he learns how to approach planning situations, each of which is unique in some respect. It is the planner's knowledge of theory and practice that helps to bridge the gap between current planning needs and circumstances and the longer-run need of preparing viable plans. The community looks to the planner to provide these skills and experience to organize the particular planning task. In doing so, the planner employs approaches that are consistent with the outlook and practice of the planning profession, and this requires both a continual monitoring of these mores (from professional literature and conferences, for example) and ability for self-reflection about his or her practice.

It will be recalled that planning practice involves both technical and social-organizing skills, which both contain a political component. On the one hand, the planner uses accepted technical skills in research and design. Although apparently neutral, decisions about what data to gather, which analytical model to use, and what to make public involves a judgement about what is most likely to further plan-making. Thus, even in carrying out technical tasks, as Sandercock says, "there is no way to avoid being political."[23] On the other hand, the planner must also organize the process that will be followed, including who will participate, when, and to what extent. In the latter case, the planner generally will invoke the prevailing and professionally accepted values of governmental responsibility, democratic planning, public participation, and so forth. Thus, nowadays, for example, planners generally advocate widespread public participation to make planning more accessible and inclusive. This may draw objections from politicians who would like a narrower context for planning decisions. It is a delicate task for the planner to balance the value of being a neutral technical advisor with the social values of the profession. Indeed, some would say that this is impossible to attain because all of the planner's tasks are political to some degree.

As Political Innovator. The planner pursues the acceptance of planning ideas and proposals by those persons in the community with the influence and authority to act on them. All planning activity involves changes in the community, either in the short or long run, and requires approval by the governing body. Even where the changes are slight, there will be some political pressure to resist or delay the change. Affected landowners may protest; politicians may be reluctant to commit capital expenditures; other department heads may have competing proposals. Getting planning proposals accepted often means changing the political climate in the community toward new ideas. The planner, therefore, is frequently called upon to promote proposals in such a way as to bring about political innovation.

In general, in every community there is a reasonably well-defined group whose support is crucial to those political innovations that would interest the planner. They are sometimes referred to as the "influentials," and may comprise elected politicians and appointed officials, as well as others outside the local government whose views are often sought. In some situations, the influence of such special groups as chambers of commerce, real estate interests, environmental organizations, property owners, and tenant associations may also have to be taken into account in order to gain acceptance of planning proposals. The planner will usually try to identify all actors with a specific interest in a proposal and predict their reaction to it. Then the planner will have to consider what steps to take to improve the chances that the plan or other proposals will be accepted. These steps may include the timing of proposals, the structure of participation, the use of outside advisors, the establishment of advisory committees (elite or otherwise), selected pre-release consultations, and even overt political pressure. For example, the evolution of Vancouver's innovative development-approval and design-review systems required two decades of effort by the planning commissioners to negotiate the roles of politicians, the public, outside advisors, and professional staff.[24] This innovator role implies a possible need for the planner to adapt to political realities. There are limits to this adaptation, however, as modifications to plans may begin to conflict with the planner's professional standards, or the integrity of the planning agency may start to be threatened.

Lastly, a dominant view of the planning profession is that the planner's primary obligation is to serve the public interest.[25] This is related to the notion that community members share certain underlying personal and group interests with regard to the subjects that planning deals with, as well as the way it deals with them. For example, it may be argued that it is in the interest of the whole community to acquire parkland today to ensure that it is available for future generations and, similarly, that means of public participation are in everyone's interests to ensure that issues are properly aired. The planner adopts this stance in order to fulfill the mandate of serving all the community and not only the interests of one group. But it is not an easy stance to defend because it depends upon deciding *who defines the public interest.* It has been found that many planners tend to choose only from a limited array of possible participants, even when public participation is mandated. They have tended to favour business groups, elected officials, development groups, and local government departments and far less frequently target groups representing disadvantaged people, ethnic minorities, or the elderly.[26] Such practices indicate who the planners see as relevant stakeholders.

Politicians often argue that voters have sanctioned them and their views of the public interest; many individual citizens and groups argue that their grassroots views truly reflect the public interest; and planners may argue that their comprehensive view of the community provides the basis for such a definition. Clearly, there is ample room for conflict among participants in whichever definition the planner adopts. Quite simply, there is no ready mechanism for determining the public interest or, as some claim, can there ever be.[27] Probably most important in this frequently contentious situation is that the planner not argue for his or her own purposes, but for openness in hearing all interests and facing all issues.[28]

As Citizen Educator. The planner seeks to affect the basic attitudes and values of the community at large regarding the benefits and consequences of planning. The planner is obliged to do this, given his or her commitment to the broader community interest. The extent to which planning is undertaken will depend upon the degree of tolerance for new ideas in the community. Communities vary in this regard, depending upon their previous experience with planning and development, their cultural milieu, and their perception of the resources that might be needed. The planner needs, therefore, to be aware of the factors affecting attitudes toward planning and also of the avenues available to enlarge the area of acceptance in the particular community.

This is doubly important, for example, in planning with Aboriginal communities, as discussed in Chapter 10.[29] The planner needs, therefore, to be aware of the factors affecting attitudes toward planning and also of the avenues available to enlarge the area of acceptance in the particular community.

Specific techniques include publications about planning (including translation into other languages), dissemination of information through the media, personal contacts by the planning staff, speeches, public meetings, neighbourhood drop-in centres, and provision of planning services and information to citizen groups. Some planners have become quite inventive in seeking widespread participation. In the mid-1990s,

Vancouver's planners, when preparing CityPlan, the new citywide plan, created the opportunity for small groups of people to discuss planning issues. Each group, called a CityCircle, was supplied information kits and a facilitator and the language of the community was used when requested.[30] Over 200 such groups formed; they made submissions that were put into an "Ideas Book," which was then used in a citywide Ideas Fair. It is estimated that 20 000 citizens (out of 400 000) took part in this process. Calgary followed a similar process in its GoPlan transportation planning.[31]

Enhanced Planners' Roles. The four roles discussed above provide only the broad institutional parameters of the planner's participation in the planning process. Within this institutional context, there are a number of specific roles that a planner might play. In a planning agency requiring several planners, some will function as plan-makers and others as researchers, regulators, or managers. The scope of their contacts with other participants outside the planning agency will differ, and so too their possible perception of the outcome of the process. Some will come in contact more with the public, with developers, with elected officials, or with other agencies. Increasingly, in recent years, it has been noted that planners practise in relation to a variety of special-interest groups, settings, and regulatory systems, and may be seen as technical staff, evaluators, and advocates.[32] Or, as Gunton observed, planners may, and may need to, modify their role such that it becomes "social reformer," "advocate," "referee," or "social learner, or all four," and, possibly, Guttenberg's suggested role of "social inventor" as well.[33]

In turn, each planner brings to the particular role individual differences in outlook that affect the way in which the role is actually played. Research into planning practice indicates that the individual planner tends to fashion his or her own concept of how the role should be played.[34] Planners thus "frame their role" according to their personal views of the problem at hand and their preferred courses of action in solving problems, especially in conjunction with others. Planning practice is still evolving, both in relation to changes in the role that planning plays in our society and in relation to planners' own understanding of planning processes. The most recent shift involves the elaboration of participatory processes in plan-making from the formal consultation approach, sometimes referred to as "notice and hearing," to more open, inclusive "collaboration" among plan-makers.[35]

The planner has a key role in the latter approach—one might call it the role of **consensus builder**—in identifying stakeholders, choosing venues for deliberations, and providing information for discussion. This requires the planner to develop skills in facilitation because collaborative processes depend upon using processes of consensus rather than formal hierarchical authority.[36] Implicit in this role, as one observer notes, is the need "to learn to listen and how to communicate in ways that allow others to hear and enable others to speak for themselves."[37] The import of these suggestions is to indicate that the traditional perspective on what a planner does, or could do, is clearly too limited. The milieu of the planner, as we have noted several times, is more diverse and complex, both in terms of participants and subject matter. More will be said of this later in the chapter.

The Developer as Participant

A very large part of community planning is involved with anticipating and responding to the initiatives of persons and firms outside the governmental milieu, who are referred to as **developers.** This is a generic term referring to those individuals, corporations, and other commercial groups (and may also include institutional developers such as school boards and churches) who make the decisions to convert raw land to urban use and/or to convert an already existing use to a different use (called redevelopment). The developer's role, broadly, is to satisfy the demand for new space for such establishments as homes, factories, stores, offices, and schools. This is a key role in the overall process of creating the future built environment. Thus, the activity of the developer could be called **community-building**, which, in turn, is closely entwined with community planning.

The developer is the central actor in the process by which a piece of land is turned into buildings and other usable spaces. It is a complex process that usually involves a number of persons and firms—land assemblers, builders, site planners, mortgage lenders, and realtors among them. However, it is the developer (an individual, firm, or public agency) who makes the decisions about whether a project will go ahead and when. The process of development occurs in a series of four discernible stages, each of which comprises its own set of participants: (1) land acquisition, (2) site preparation, (3) project production, and (4) project marketing (see Figure 12.2).[38]

The stage at which a developer usually becomes involved in the planning process is **site preparation**—that is, when it has been decided to proceed with a project to build anew or renew. The next steps will require some physical and legal changes to the parcel of land. Physically, the developer may need to demolish existing buildings, undertake excavations, and arrange for provision of roads, sewers, and water supply. Legal changes may be needed in the form of official approvals for subdivision, zoning, and building construction. Most of these physical and legal changes are under the control of public bodies. Municipalities may exercise various kinds of development control as well as require financial contracts to provide basic utilities, for which they may make development charges (see Chapter 14).

Each application for approval by the developer involves personal interaction with public officials, especially planners. What ensues is a complex **negotiation–bargaining** situation. The developer seeks approvals that will keep the project plans intact. The municipality seeks, where necessary, to modify any aspects of a project that will require undue expenditures of public funds and/or prevent infringement on public rights of access. The planners may also try to persuade the developer to include more amenities in the project for the occupants and the public, such as more open space. In large cities and in large projects in particular, the bargaining between developers and planners has become very complex, with various tradeoffs being made.[39] Planners may offer bonuses, such as additional height on office building projects, in return for the provision of a public plaza; or land may be given by the developer in return for closing a public right-of-way; or a heritage building may be retained in return for other concessions. Even citizens' groups may become involved in such bargaining, as when a large office project in Toronto was stalled in order to obtain money from the developer for low-income housing.[40]

The often-intense negotiations between developers and planners in the site-preparation stage indicate the strong mutuality of the two participants. One observer has termed it a "symbiotic interrelationship" in which the following occurs:

1. Planners prepare plans basically intended to modify, but heavily influenced by and building upon, what developers already do; and
2. Developers make their development decisions based on their interactions with planners and their knowledge of what planners will accept.[41]

In situations involving the development of raw land, most of the bargaining will be done by the developer and the planner, along with other municipal officials.[42] In redevelopment situations where the local populace has a strong interest in maintaining the status quo, a good deal of the bargaining may be subject to public scrutiny and public input as well. Indeed, many local governments require that developers discuss

Figure 12.2 Participants in the Development of an Urban Project

STEPS TAKEN BY DEVELOPER	PARTICIPANTS (A) ACTIVE	(B) PASSIVE	(C) CONSULTANTS	(D) INTERMEDIARIES	(E) REGULATORS
1. Land Acquisition	-landowner(s) -land developer	-lenders	-appraisers -lawyers	-realtors -lawyers -assembly agents	
2. Site preparation	-land developer	-landowner(s) -lenders [blank]	-planners -engineers -architects -lawyers		-planning board -council -ministry -appeal body
3. Project production	-builder	-land developer -mortgage lender -mortgage lender	-architects	-mortgage brokers	-building inspectors
4. Project marketing	-builder -consumer	-mortgage lender	-lawyers	-realtors -lawyers	

Many different individuals, groups, and professions participate in bringing a project to fruition, and their roles differ according to the stage of development: (a) active participants have a financial interest in the land and are directly involved in improving its value; (b) passive participants also have a financial interest in the land and/or improvements, but are not actively involved in the development; (c) consultants are called in by active participants to advise on technical and legal aspects; (d) intermediaries act as a liaison between active and passive participants; (e) regulators represent the public's interests, and their approval is required for the project to go ahead.

Source: Simon B. Chamberlain, *Aspects of Developer Behaviour in the Land Development Process*, Research Paper No. 56 (Toronto: University of Toronto Centre for Urban and Community Studies, 1972).

their plans with neighbourhood residents and organizations and attempt to reach consensus on their projects before initiating formal proceedings for rezoning, development control, or subdivision approval. This, as we shall discuss further below, adds greatly to the "transparency" of the development process, the lack of which has brought much valid criticism from community members.

Involving the Community in Planning

Community planning developed with strong democratic principles; however, the participation of citizens in community planning was not always as common as it is today in Canada. It has gone through several stages and continues to change.

In the decades leading up to World War I, the impetus for planning frequently arose from elite community groups, such as boards of trade and arts organizations. In the 1920s, using the provisions of new planning acts, many municipalities established advisory "town-planning" commissions whose members were frequently selected from the real estate and construction industries or other lines of commerce. The Community Planning Association of Canada (CPAC), established in the 1940s, functioned as a broad-based citizen pressure group, outside government, to promote the advantages of planning with citizens, local and senior governments, and the business community. By the 1960s and into the 1970s, when planning was quite pervasive, proposals by planners began to be questioned, often vociferously, by ordinary citizens and neighbourhood groups. Terms such as "citizen activism," "participatory democracy," and "advocacy planning" were coined; participation was demanded by citizens. This marked a major shift in the evolution of community participation in planning. It could not be turned back, but it did not proceed easily, and calls for its expansion still resound.

There are two basic dimensions for assessing public participation in community planning. The first deals with the *depth* of participation, which is the degree to which the power to make decisions regarding approval of plans is shared with members of the public. The second deals with the *breadth* of participation, which is the extent of the citizenry included in plan-making. A discussion of each follows.

Degrees of Participation

Citizen participation is not a unitary concept; that is, power may be shared with citizens to different degrees. It may vary according to the needs of the decision situation as well as the disposition of those in control of making decisions. In its most modest form, the "notice and hearing" format noted earlier, citizens are informed of planning proposals by a formal notice and told of a public hearing they may attend and participate in. More power is shared when, for example, citizen advisory committees are employed in the planning process. Still greater power is shared when citizens are delegated to make plans. Again, legal and practical considerations may limit the sharing of planning power, but participation may also be deliberately constrained by those holding power so that it becomes nothing more than an empty ritual.

The different degrees of citizen participation are readily discerned in Sherry Arnstein's classic "ladder of citizen participation," with each higher rung corresponding to a greater degree to which citizens could share power in planning decisions (see Figure 12.3).[43] There are eight rungs on the ladder in three broad categories: (1) "Contrived Participation," or ways of avoiding sharing any planning power, are the two bottom rungs; (2) "Token Participation" shares a bare minimum of power with citizens over the next three rungs above; and (3) "Power Sharing," with citizens in increasingly greater degrees of power, constitutes the upper three rungs. In other words, the greater the depth (or height) of participation, the more the power to make decisions about plans is shared.

Contrived Participation

1. **Manipulation** may be practised when participation is organized to "educate" and persuade citizens to support already decided-upon plans and programs. Citizen committees with no mandate, even to give advice, may simply act as "public-relations" vehicles for plans and planners (see cartoon on page 304).
2. **Therapy** refers to the practice of engaging citizens in diversionary activities that will "cure" them of their concerns over basic flaws and injustices. This rung is seldom encountered in community planning in full-blown form, but there may be gradations of it in "workshops" provided for citizen members of planning boards.

Token Participation

3. **Informing** is the first level at which the planning process is opened up to citizens. Information is supplied to them on the nature of the planning task, its schedule,

Figure 12.3 Ladder of Citizen Participation

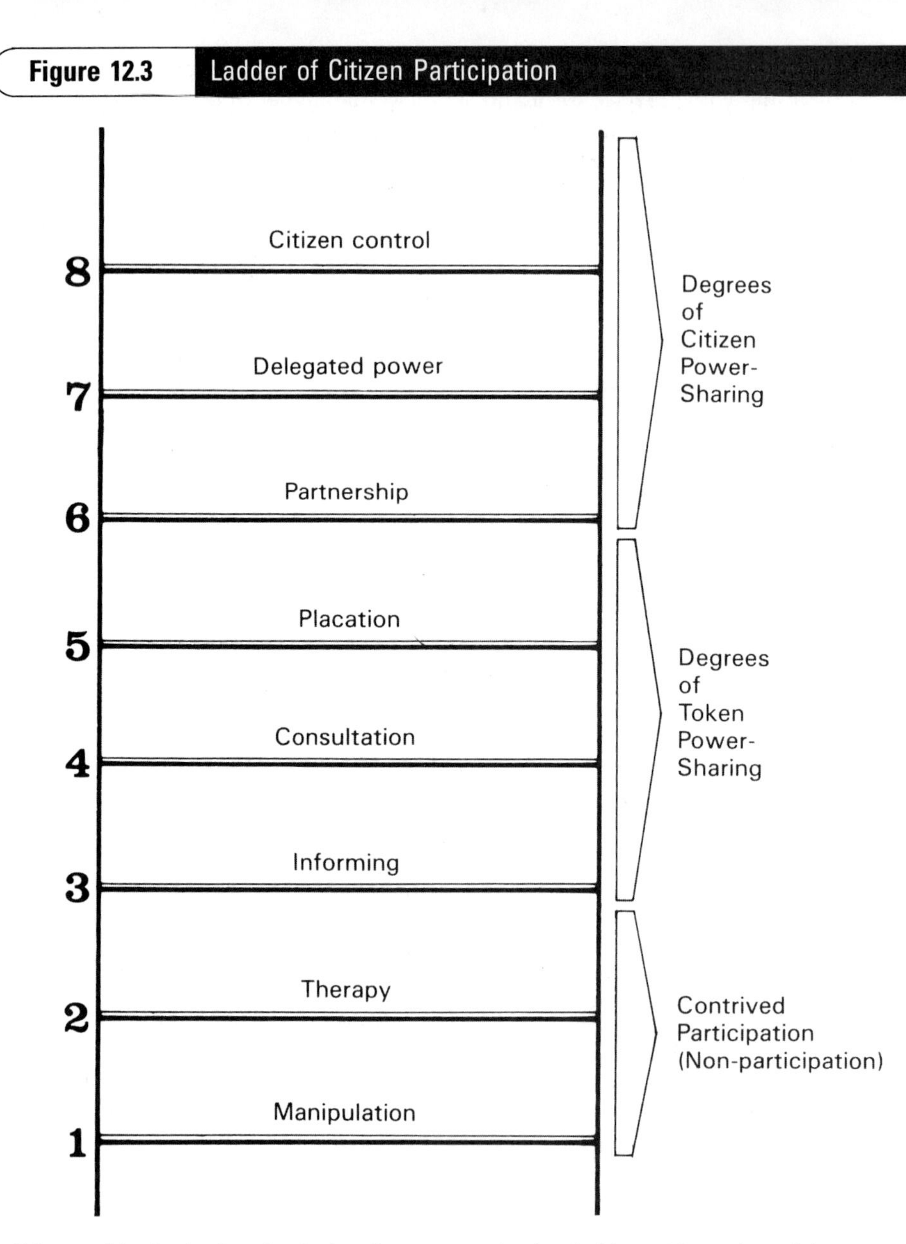

Citizen participation involves the sharing of power over planning decisions with members of the community. This may vary according to the needs of the decision situation and the disposition of those in control of planning decisions to share their power.

Source: Adapted from Sherry R. Arnstein, "A Ladder of Citizen Participation," *Journal of the American Institute of Planners* 35 (July 1969), 216–224. Adapted by permission of the Journal.

and the role of citizens. At its best, a full description is provided to each household and responses of citizens are sought and facilitated. At its worst, this stage will feature only one-way communication, such as by means of the use of printed or news media formats, combined with legalistic and technical jargon.

4. **Consultation** occurs when explicit means are used to obtain the views of citizens, such as through attitude surveys and public meetings. However, while at this level citizens are given the opportunity to be heard, they are not assured of being adequately understood. The design of questionnaires and their interpretation may require some forms of feedback and additional dialogue with citizens, for example.

A recent, and encouraging, example on this level is the "Picket Fence Project" of departments of parks and recreation in three municipalities in Vancouver's eastern suburbs to survey the sense of community held by residents.[44]

5. **Placation** refers to forms of participation in which citizens are given the opportunity to be heard, say, on "advisory committees" but with no certainty their advice will be heeded. One such effort at this level of participation was the system of nine citizen policy committees used to advise the governing board of the Greater Vancouver Regional District in the preparation of its 1970s metropolitan planning policies. Committees were provided with technical assistance by staff planners; but, in the end, the politicians only "received" their reports and did not debate them, thereby relegating the citizens to a token advisory role.[45]

Citizen Power Sharing

6. **Partnership** involves an agreement to share responsibilities for planning through joint policy boards or committees. Often used to resolve impasses over a specific project, such mechanisms allow citizens to influence the outcome of the plan by their votes (on preliminary if not final versions). Partnerships may also be used on a broad community plan such as occurred with the development of Vancouver's City-Plan, where citizens made choices all the way to the draft plan presented to city council.[46] Similarly, the City of Red Deer, Alberta, in 1998, fostered an extensive partnering for revitalization of its Riverside Meadows neighbourhood.[47]
7. **Delegated power** gives citizens dominant decision-making responsibility over a plan or program, usually from the outset. The traditional planning board or commission, with a majority of appointed citizen members and legislative authority to prepare the community plan, is an ideal example of this level of participation. Some communities have delegated to neighbourhood committees the task of making neighbourhood plans and provided them with a professional planner to assist them (as with several neighbourhoods in Vancouver). This approach was also used in some older urban renewal programs.
8. **Citizen control** is a level at which citizens govern a program or project in all its policy and managerial aspects. Although this is not likely in regard to preparing a community-wide plan, instances of citizen control are found in cooperative and other citizen nonprofit housing developments throughout the country. Community groups that obtain grants from senior governments for recreation facilities also operate at this level, as did a group in Kingston, Ontario, that designed and constructed a public waterfront walkway. It should be noted that the municipality must retain ultimate control of it because of its responsibility for public safety. A variation of this level of participation occurred in Calgary, where a citizen's organization took the initiative of developing community-sustainability indicators and preparing a "State of Our City" reports analyzing 36 indicators.[48]

Toward Greater Inclusivity

It is commonplace, today, to talk about the diversity of a community's population, meaning that many differences exist among its citizens based on age, gender, economic status, literacy, and cultural background, to cite several of the most obvious factors. Yet public-participation programs in community plan-making seldom broach these differences. Even though the depth of public participation has increased to higher rungs on Arnstein's ladder in many places, the same progress cannot be said for increasing its breadth. While more attention is now paid to neighbourhood differences, differences between "neighbours" are not usually considered. To be sure, the public *is* being involved in these programs and in greater numbers. But, in most instances, the public is seen as no more than an undifferentiated human category.

When any part of the public is not included, intentionally or otherwise, the effect is to deny their citizenship and right to engage in planning and developing their own community. Research and critiques have been emerging for some time that point toward the failure both to acknowledge differences among citizens and to engage those excluded in plan-making.[49] To rectify this situation is, of course, no small task for, as Sandercock says, the public is a very diverse collectivity comprising "multiple publics."[50] The excluded "publics" that have received the most attention are women, youth, the elderly, the physically impaired, and those from culturally different backgrounds. Each is discussed briefly below.

Women. One of the groups most excluded from public participation processes are women, as feminist critics have frequently noted.[51] This exclusion can come from at least two sides. First, it may happen because the process is not made physically accessible to women (the time of day of meetings, the need for child care, the availability of transportation, etc.). Second, the planning process may also fail to take into account women's knowledge of the city, or their experience from their daily lives.[52] Planners have tended to rely on expert, technical data about land use that can result in so-called objective, rational, universal knowledge. Thus, the knowledge that comes from the personal experience of women—such as safety on the streets, difficulties with mass transit, or scarcity of affordable housing for a single-parent family—has had little or no place in the planning process. In the eyes of some observers, the knowledge that planners do use, therefore, has a male/professional bias.[53] Including women's perspectives in planning calls for dealing with **aspatial** concerns, those human things that one can't always put on a typical planning map. A Vancouver project that involved women in the planning of their own neighbourhood alerted planners about the need for things such as crosswalks, street lighting, community meeting spaces, "latch-key" programs, recycling, and a crisis centre.[54]

Youth. Also not included in most definitions of "the public" are young people who are, indeed, the inheritors of the community form and character that is decided in plan-making. Again, as with women, there is extensive research on the ways that children and older youth experience the city and engage with it to define place.[55] But to attract youth participants requires programs that are aimed at them and that give them decision-making power along with adults. Planners can benefit from working with and through organizations that have youth components, such as community centres and schools.[56]

The Elderly. Sixty decades after the baby boom of 1946–65, the dominant age group in almost all Canadian communities has become seniors (those 65 and older) and near-seniors. The community needs of this "public" have barely been acknowledged by planners or their presence invited into public-participation programs.[57] When they do participate it will be necessary to recognize that they, too, are diverse, with a range of incomes, skills, and physical impairments, as well as age differences that span one-third of a century. To include seniors in plan-making, planners will need to accommodate this growing segment of the population by holding meetings in neighbourhood settings, during daylight hours, in venues that have adequate acoustics; provide special transportation for physically impaired seniors; and use graphic and printed materials that are easily grasped by older people with diminished vision, as well as by elders who are accustomed to other languages and symbols.[58]

Physically Impaired. Over 12 percent of Canadians experience some form of physical impairment that limits their participation in daily activities.[59] The result, as several geographers have pointed out, is to significantly affect the way people with such limitations experience and navigate built environments to the extent of often being excluded from the activities of other citizens.[60] To be included in public-participation efforts will demand that planners convene these activities so that they are accessible to the physically, visually, and hearing impaired.

Ethnic Populations. Canada is a country built on immigration. However, recent trends have changed the composition of the immigrant stream. Seventy percent of all new immigrants are "visible minorities," with the bulk of new immigrants, both visible minorities and others, destined for the cities.[61] During the 1990s, 94 percent of immigrants went to live in the country's 27 metropolitan areas. Of these, over three-quarters now reside in just three census metropolitan areas: Toronto, Vancouver, and Montreal. One result of such flows is the emergence of numerous "visible minority neighborhoods," or "ethnic enclaves," in these cities.[62] Further, these enclaves are frequently not ethnically homogenous but home to several visible minority populations. Thus, planners are confronted with the challenge of working with people who may see the world from quite different perspectives, and will need to adapt their participatory planning practice.[63] Among the responses open to planners are to translate information into the language(s) of the neighbourhood, as is already done in several cities; to augment planning staffs with people who understand ethnic-community concerns; and to use modes of public consultation with which the community is most able to articulate its aspirations.[64]

Immediate reflection suggests many interconnections among these members of the population, such as elderly women, physically impaired youth, and elderly male immigrants. This, in turn, indicates the need to reform public-participation programs so that they recognize the implicit breadth of the citizenry and use categories that are porous. Further, there are a number of cogent issues regarding community planning that arise from this critique. A primary one is to expand the perspective on the substance of planning from mainly land-use and spatial relations to include social and personal relations; that is, to consider relations that come out a perspective of "home" and its role in people's lives, rather than just the physical attribute of "housing"; relations that come out of a "community" perspective, rather than just a physical "neighbourhood" perspective. This means accommodating a broader, less routine, more subjective approach into planning practice with the concomitant issue of about how to obtain needed "data" in such settings.

Using Public Input Effectively

The importance of community planning almost guarantees that citizens will participate in it. And, so much so, if they are not accorded a viable role, they may unceremoniously insist upon it, because people nowadays are reluctant to give up control over their local environments. Nevertheless, how to include citizens effectively is yet another question. It is a process that involves a good deal of mutual trust and confidence among all participants, characteristics that cannot be legislated or easily institutionalized. Further, there are a number of aspects about citizen involvement in a public-participation program that are frequently raised that can prove to be contentious, if not disruptive. The following aspects need to be identified and better understood:

> **Representativeness:** About those citizens who participate in the name of a larger citizen constituency, in person or on a committee, it is often asked: Who and how many do they represent?

Experience shows that the citizens who actively participate are bound to be small in number and proportion of the population. But counting heads will not ascertain whether they are an isolated activist faction or a good cross-section of the public.

Regardless of whom or how many they represent, those who do get involved bring the views of truly interested citizens, and these are valid in and of themselves. The issue of representativeness is possibly more crucial when it comes to selecting a few citizens to sit on committees or targeting groups to be involved. For those making the appointments, there will be questions of the completeness of the representation; for those appointed, there is the matter of to whom they are accountable.

Accountability: To whom are citizen participants accountable? This question often arises when citizen participation is used because citizens cannot be held legislatively, professionally, or legally responsible for their decisions.

This should not, however, be invoked to restrict public participation, since final authority in municipal matters always rests with the elected council. It may simply require a limitation to participation no higher than the partnership level on the participation ladder. Regardless, it should be axiomatic when involving citizens in plan-making to be clear at the outset "about who is making what decisions."[65]

Responsibility: Citizens have a duty to act responsibly when they become involved. Despite that principle, instances of neighbourhood "protectionism," often broadly referred to as NIMBY-ism ("not-in-my-backyard"), occur to distort responsible discussion.

It is known that citizen participants are able and willing to consider the broader needs of the community when they are accorded a meaningful role in the planning process; that is, actually helping to make the decisions. NIMBY-ism is frequently a sign of citizens taking responsibility for their own neighbourhood when it may seem that such a voice has not been allowed for.

This can also indicate that more active, inclusive participatory methods, such as consensus building, are required (see below).

Access: Who may participate in the planning process? This question is related to one of the fundamentals of democracy, i.e., that all citizens have access to the processes of governing and planning their communities.

Related to this issue is whether there are barriers that prevent any citizens from participating either from the procedures being used or out of neglect to include them, or both. Clearly, the public cannot be involved in all levels of planning decision, so the nature of public access must be made clear (and unfettered where it can occur). One unfortunate instance of exclusion was the Alberta Energy Board's use of a hall for a public hearing in Calgary (in 2001) that had capacity for less than half of the 300 people who wished to attend, and the hearing had to be called off. Although rescheduled for the next day, this probably meant that some people who had taken time off work may not have been able to attend at the new time.[66]

Transparency: How are planning decisions being made, and who is making them? It is also a vital facet of democracy that the processes of decision making not be obscured from the public—even those processes that cannot be shared with them. (This could be said to be the citizens' right to accountability from their government.)

Further, citizen participants will want to be assured that any limits placed on their role are not just a form of co-optation, an attempt to make the participation appear to be broad-based.

Dialogue: Who is communicating with whom in plan-making? If planners are merely providing information *to* public participants, the communication is a one-way process, not a dialogue, not real involvement.

Public participation must be set up in such a way as to ensure that the viewpoints of the various publics in a community have an equal chance of being heard, as well as safeguarding their right to receive relevant information and to be able to collaborate in the planning.[67] In short, participation has the most chance of being effective in the public's eyes if there is both the appearance and experience that the individual has been listened to and has the potential to influence planning decisions.

Planning and Consensus Building

The success of a community's plan, in the end, will be determined by the extent to which individuals and groups in the community agree with it, especially when it comes time to implement it. Many a plan remains on the shelf because key interests in the community did not have a direct role in framing it and will not support it. Planners and politicians have witnessed controversies over planning proposals becoming protracted, and have seen citizens become disenchanted with planning itself. A much more aware citizenry regarding planning matters, as is the case now in most communities, can be a blessing to plan-makers if the citizenry is able to participate actively, or to be a burden if not. The need to confront these unsettling conditions, fortuitously, has spawned a variety of ways to address complex and controversial public issues that involve multiple interests. Methods of negotiation and mediation referred to in earlier chapters grew out of the need to resolve disputes that occurred during implementation of a plan or other public policy.

Refinements to these methods have led to broader, more democratic approaches to achieve agreement and support among the diverse interests one finds in a community. In general, these are called **consensual**, or **consensus-building**, approaches and they aim to reach mutually beneficial agreements among participants about planning issues.[68] Most of these methods have two essential features. The first concerns process, the second concerns the array of participants. The consensus-building process works *horizontally*, with discussion, not votes, on principles and criteria that would provide a framework for reaching agreement on solutions to issues. This contrasts the usual, formal, vertical approach in which, typically, "solutions" (i.e., motions) are debated using parliamentary procedures. The latter generates and heightens differences among participants and does not ensure that all viewpoints will be expressed and explored. The former, consensus building, provides for knowledge, experience, ideas, and concerns to be shared among equal participants and accumulated as part of the solution. A consensus is thus achieved in which all participants have been partners and share a concern over the solution—say, the community plan—and its future disposition.[69]

The second major feature of the consensus-building approach is the extensive array of participants who are involved, most often referred to as **stakeholders**. Indeed, consensual approaches are assumed to work best when the discourse is broad and not limited by professional rank or social or electoral status. So, if all groups or individuals "with something significant to gain or lose by the deliberation" are included as stakeholders, the process has the highest chance of success. Agreement is facilitated and implementation more assured because participants are, essentially, the authors of the planning decision.

There are now many instances of consensus building being used to resolve planning issues, and the following instance can serve to demonstrate its value. In Port Colborne, Ontario (population about 18 000), in the early 1990s, a planning process was established that ensured that anyone who was interested could become "a fully-participating member of the planning process."[70] Further, the planning exercise was not the product of one single body in the community but rather evolved from a partnership of groups. The result, in their words, was that such a process "enhances communications and mutual awareness among its stakeholders." Another outcome was that this consensual process permitted both difficult issues (such as the allocation of scarce resources) and a wider range of issues to be addressed than in a typical plan for the built environment for a community. The Port Colborne plan, for example, comprises themes such as economic development, physical infrastructure, social development, and commercial revitalization.

The interest in consensus building methods arises from the recognition that planning, for all intents, is a "communicative" activity—it's about the talking and interaction that is carried on among participants.[71] In short, it is about carrying on a *dialogue*. The need for consensus building is based on many less-than-salutary instances of planning dialogue. But it is more than a mitigation of the problems of participation, to which extensive reference has already been made. It represents a fundamental shift toward a more democratic, pluralist approach to public participation in community planning.

NIMBY and Participation

A not untypical public meeting in Vancouver in the early 1990s heard objections from residents of the city's East End to city council's plans to locate a plant *in their neighbourhood* for recovering recyclable materials from city garbage. Skeptics at city hall, in the media, and in other neighbourhoods probably classed this as just another case of NIMBY (not-in-my-backyard). Community objections to planning decisions that are perceived as damaging neighbourhood quality (through noise, traffic congestion, and decreased property values) are frequently heard today. However, NIMBY is not new. It was documented in the early 1950s by Meyerson and Banfield in their pioneering study of planning for public housing in Chicago.[72] They found that Chicagoans, in general, favoured the establishment of housing projects for low-income people. But the Housing Authority was rebuffed by many neighbourhoods in its search for sites. Such a reaction recalls, in many ways, demands by affluent property owners, decades earlier, for protective zoning regulations in Canadian cities.

Regardless of its history, NIMBY has become more frequent and frustrating to planners, politicians, and developers. NIMBY is blamed for costly delays in formulating plans and bylaws, and for holding up construction projects. While this is frequently true when viewed from the city level, from the neighbourhood perspective it is the *residents* who have to live with the outcome of the decision. Projects visualized in plans become actual buildings, traffic, and so forth in someone's neighbourhood. The need for a LULU (locally-undesirable-land-use), such as a recycling plant or group home, may be accepted by the whole community as meeting its value for a better environment. And the affected neighbourhood may even share this value, but it may also have other values that are unique and just as important to it. NIMBY is, thus, a new name for one of the basic tensions in the planning process (see Chapter 7) between neighbourhood and city values (see Planning Issue 12.1).

Why NIMBY? At its heart, NIMBY is the desire to alert planners and politicians to local concerns, and to bring local knowledge, which is often invaluable, into the process. On some occasions, NIMBY is the reaction of a neighbourhood or a community to not having been informed, or only at the last minute, about a planning proposal affecting it. To long-time Vancouver civic activist Jim Green, "NIMBY-ism arises from a perceived feeling of powerlessness. It is often the result of the failure of those in power to allow a democratic planning process."[73] Politicians may make hasty decisions about a project (for or against the neighbourhood) or make none at all. Or they may plan a new round of meetings, which further protracts a process that is already perceived to accord citizens only a token role. In many planning situations, citizens see that they are allowed no

Planning Issue 12.1 The Power of NIMBYism

The Toronto Star
October 16, 2005

The Power of NIMBYism
Catherine Porter

TORONTO—NIMBY. Meaning "Not In My Back Yard."

It's a term Margaret Smith has long grown used to.

After successfully stopping the city's plan to build a separated streetcar lane down most of St. Clair Ave. W. last week, the leader of the "Save Our St. Clair" campaign was hoping for a more generous label—like "right," maybe.

"It's very difficult for people in a local community to be heard and to make their point," says the bureaucrat-cum-community activist. "When we tried to do that, we were labeled anti-transit, NIMBY, anti-change.

"Labeling people is just a way of dismissing their concerns and not listening to them."

But the label isn't going away. If anything, since the ruling by the Divisional Court that sided with "Save Our St. Clair" against the streetcar lane, Smith has been accused of something worse than being a NIMBY. That's being a successful NIMBY.

Environmentalists, transit riders, and many political thinkers have called the group selfish and small-minded, fringe players who took the city hostage. They say the ruling is a disaster that sets a dangerous precedent.

It's already the case. NIMBYism has become more than a fringe tactic; it appears to be a mindset that has settled across Toronto neighbourhoods.

People want to run their air conditioners all summer long. But the prospect of a generating station in their neighbourhood sent Newmarket residents into a frenzy. People want a subway line to the airport. But when people in Weston heard of the plan to run the rail through their community, they packed a public meeting in protest.

"It does beg the question as to how a city is supposed to proceed with any agenda, either positive or negative, when communities have the ability to stop what's meant to happen," says Mark Guslits, chief development officer for Toronto Community Housing and the former manager of the city's affordable housing file.

"At what point," he asks, "are elected officials and the civil service able to be given enough confidence to make certain decisions for the benefit of all, rather than just specifically zoning into 'the some'?"

The message behind all the attention is that NIMBYism is dangerous. It reflects thoughtless, knee-jerk reactions that go against the common good.

But what if the opposite is true? Could NIMBYism be good for our city?

Margaret Smith compares her own fight with another recent community struggle. "When the Toronto Islanders beat the airport link," she asks, "did people say they were NIMBYs? No, they were validated by council's decision."

There's much debate as to when the term NIMBY was coined. But the Random House Historical Dictionary of Slang attributes the first official use to Walton Rodger of the American Nuclear Society, in the early 1980s, to describe the syndrome of immediately rejecting almost any large construction project. NIMBY has since spawned a whole family of terms, some pejorative, some not. These include NIABY (Not In Anyone's Back Yard) and the more recent Banana (Build Almost Nothing Anywhere Near Anyone).

"I never use the term NIMBY," says David Crombie, the city's former mayor and now president and chief executive of the Canadian Urban Institute. "Whether it's NIMBY or not is entirely subjective and usually directly related to your position on the substance of the matter."

Opposition to a local halfway house, for instance, might be sheer prejudice. But would you regard opposition to a landfill or nuclear waste yard in the same light?

"People could be right, even if they're a minority," Crombie says.

Source: Catherine Porter, "The Power of Nimbysim." Reprinted with permission. Torstar Syndication Services. From an article originally appearing in the *Toronto Star,* October 16, 2005.

higher than the middle rung of Arnstein's "ladder of participation"; that is, they are only *consulted*.

The motivations of NIMBY are diverse, sometimes altruistic (e.g., about the environment) and sometimes ill-informed and selfish (e.g., about lifestyle). Examples of the latter include the tribulations of a Winnipeg clergyman who tried for four years to find a site for a seminary to train Aboriginal people. Urban and rural municipalities alike refused the necessary land-use permission, citing increased traffic and lowered property values.[74] This story is shared by many who have sought to establish group homes in Canadian communities, even though the effect on neighbouring land values has been shown to be almost nonexistent.[75] Paradoxically, the opposite situation can occur, as it has in an old downtown neighbourhood in Toronto that attracted middle-class residents wanting to gentrify the area. Several drop-in agencies for many of the city's destitute that already existed in the neighbourhood found themselves and their homeless clients subject to a campaign to shut them down.[76] The Winnipeg and Toronto cases illustrate a situation of competing claims on neighbourhood land from contrasting value positions.

NIMBY and the Planning Process. On a basic level, NIMBY protests are a form of public participation, and public participation satisfies one of planning's primary values—**democratic participation**. That these protests have grown more prominent is, no doubt, a reflection of the more open planning processes of the past two decades. It is also an indication of a further "feedback loop," where many citizens, experiencing the constraints on public participation when they do get involved, opt for direct ways to be better heard. The result is often no action being taken and the planning problem being left in limbo. That neighbourhood groups have taken to NIMBY with such alacrity, and become so adept at it, suggests that citizens strongly desire to be *partners* in making planning decisions affecting their "turf." Planning processes in their present form often seem to invite protest and stalemate and heighten neighbourhood/city/region value conflicts. Often, there is only a limited technical rationale for resolving such conflicts over LULUs through the normal land-use or environmental-assessment approvals processes. The issues are moral ones as much as anything and planners, who have a heritage of social justice in their professional ethos, should continue to address them with consensus building and alternative dispute-resolution techniques, despite their limited success to date.[77]

Reflections

As desirable and necessary as it may be, public participation may not occur in the scope and form hoped for; in fact, it may not happen at all. Participation is a voluntary act; it can be promoted but not guaranteed. Given their experience, the poor and cultural or ethnic minorities may feel powerless to influence decisions. Middle-class citizens may choose not to participate out of cynicism or apathy. Deliberate efforts to promote participation have had only limited success, especially where the plans are for the distant future. People respond much more readily to changes in their immediate environment than to large, remote projects or abstract plans. Moreover, organized groups of citizens, it has been known for some time, tend to avoid devoting much time to pre-planning exercises, preferring to save their resources for more immediate decision-making stages.[78]

The vagaries of citizen participation reflect the complexity of the actual choices to be made in planning for the future built environment of a community. Irksome though it may sometimes be to other participants, the advent of direct involvement of citizens has revealed the poverty of the notion of one correct plan and the suppression of relevant value positions that often accompanies it. At the same time, it helps reveal the wider range of possibilities for action that exist in reality, thereby enriching the substantive side of planning as well. Probably most important, many community members are involved in decisions about their environment. Notably, this includes women, who provide much of the initiative and leadership in citizen-participation efforts. Through citizen participation, planning our cities and towns becomes much more a *community* planning process, and raises this question:

- ***How can we continue to expand participation in planning, and what modifications might be needed to the roles of various participants?***

Endnotes

1. J.A. Throgmorton, "On the Virtues of Skillful Meandering: Acting as a Skilled-Voice-in-the-Flow of Persuasive Argumentation," *Journal of the American Planning Association* 66:4 (Autumn 2000), 367–383.
2. Judith E. Innes, "Information in Communicative Planning," *Journal of the American Planning Association* 64:1 (Winter 1998), 52–63.
3. Leonie Sandercock, "The Death of Modernist Planning: Radical Praxis for a Postmodern Age," in M. Douglass and J. Friedmann, eds., *Cities for Citizens* (New York: John Wiley, 1998), 163–184.
4. Andrei Nicoli, "The Twenty-First Century is Here: Is Anybody Home? Community Participation and the Role of Local Government," *Plan Canada* 41:1 (January–February–March 2001), 21–23.
5. The innovative approach of Harry Lash and his staff at the Greater Vancouver Regional District in the 1970s foreshadowed today's shift toward participatory planning decision making and consensus building. See Harry Lash, *Planning in a Human Way,* Cat. no. SU32-3 (Ottawa: Ministry of State for Urban Affairs and Macmillan Canada, 1976), 9–13. The following paragraphs in this section draw heavily upon Lash's monograph, especially Chapters 3 and 5.
6. Recognizing this interdependence is crucial according to David E. Booher and Judith E. Innes, "Network Power in Collaborative Planning," *Journal of Planning Education and Research* 21:3 (2002), 221–236.
7. Norman Beckman, "The Planner as Bureaucrat," *Journal of the American Institute of Planners* 30:4 (November 1964), 323–327.
8. Throgmorton, "On the Virtues of Skillful Meandering."
9. Lash, *Planning in a Human Way,* 75.
10. Leonie Sandercock, "Towards a Planning Imagination for the 21st Century," *Journal of the American Planning Association* 70:2 (Spring 2004), 133–141.
11. Melvin M. Webber, "Comprehensive Planning and Social Responsibility," *Journal of the American Institute of Planners* 29 (November 1963), 267–273.
12. Karen S. Christensen, "Teaching Savvy," *Journal of Planning Education and Research,* 12 (1993), 202–212.
13. Cf. Susan Fainstein, "Planning in a Different Voice," *Planning Theory* 7:8 (1992), 27–31; and Penelope Gurstein, "Gender Sensitive Community Planning: A Case Study of the Planning Ourselves In Project," *Canadian Journal of Urban Research* 5:2 (December 1996), 199–219.
14. Jana Carp, "Wit, Style, and Substance: How Planners Shape Public Participation," *Journal of Planning Education and Research* 23:3 (2004), 242–254.
15. Michael Seelig and Julie Seelig, "CityPlan: Participation or Abdication?" *Plan Canada* 37:3 (May 1997), 18–22.
16. Carp, "Wit, Style, and Substance."
17. These basic categories derive from Robert T. Daland and John A. Parker, "Roles of the Planner in Urban Development," in F. Stuart Chapin, Jr., and Shirley Weiss, eds., *Urban Growth Dynamics* (New York: Krieger Publishing, 1962), esp. 190–196.
18. Thomas Gunton, "The Role of the Professional Planner," *Canadian Public Administration* 27:3 (Fall 1984), 399–417.
19. John Forester, "Know Your Organizations: Planning and the Reproduction of Social and Political Relations," *Plan Canada* 22:1 (March 1982), 3–13.
20. Christensen, "Teaching Savvy."
21. Mohammad A. Qadeer, "Pluralistic Planning for Canadian Cities: The Canadian Practice," *Journal of the American Planning Association* 63:4 (Autumn 1997), 481–494.
22. Pierre Filion, "The Weight of the System," *Plan Canada* 37:1 (January 1997), 11–18.
23. Sandercock, "Towards a Planning Imagination for the 21st Century."
24. John Punter, *The Vancouver Achievement: Urban Planning and Design* (Vancouver: UBC Press 2003), Chapters 8 and 9.
25. The extent to which this value pervades the Canadian planning profession may be seen in the survey results of John Page and Reg Lang, *Canadian Planners in Profile* (Toronto: York University Faculty of Environmental Studies, 1977), a report to the Canadian Institute of Planners.
26. Samuel D. Brody et al., "Mandating Citizen Participation in Plan Making: Six Strategic Planning Choices," *Journal of the American Planning Association* 69:3 (Winter 2003), 245–264.
27. James Tully, *Strange Multiplicity: Constitutionalism in an Age of Diversity* (Cambridge: Cambridge University Press, 1995).
28. Matthew Kiernan, "Ideology and the Precarious Future of the Canadian Planning Profession," *Plan Canada* 22:1 (March 1982), 14–24.
29. John Peters, "Aboriginal Perspectives on Planning in Canada—Decolonizing the Process: A Discussion with Four Aboriginal Practitioners," *Plan Canada* 43:2 (Summer 2003), 39–41.
30. Ann McAfee, "Vancouver CityPlan: People Participating in Planning," *Plan Canada* 35:3 (May 1995), 15–16; and Barton Reid, "Looking into the Future: Vancouver's City Plan Attempts Innovation," *City Magazine* 14:2 (Spring 1993), 11–12.
31. Calgary, City Planning Department, *Public Participation in the Planning Process* (Calgary, 1993).
32. Donald A. Schon, "Some of What a Planner Knows," *Journal of the American Planning Association* 48 (Summer 1982), 351–364.
33. Gunton, "The Role of the Professional Planner," and Albert Z. Guttenberg, *The Language of Planning* (Urbana and Chicago: University of Illinois Press, 1993), xiv.
34. Schon, "Some of What a Planner Knows."
35. Cf. Richard D. Margerum, "Collaborative Planning: Building Consensus and Building a Distinct Model for Practice," *Journal of Planning Education and Research* 21 (March 2002), 237–253; and Patsy Healey, *Collaborative Planning: Shaping Places in Fragmented Society* (Vancouver: UBC Press, 1997).
36. Richard D. Margerum, "Evaluating Collaborative Planning: Implications from an Empirical Analysis of Growth Management," *Journal of the American Planning Association* 68:2 (Spring 2002), 179–173.

37. Booher and Innes, "Network Power in Collaborative Planning."

38. This classification is derived from two sources: Simon B. Chamberlain, *Aspects of Developer Behaviour in the Land Development Process*, Research Paper No. 56 (Toronto: University of Toronto Centre for Urban and Community Studies, 1972); and Urban Land Institute, *Residential Development Handbook* (Washington, 1978).

39. A penetrating review of such negotiations and their consequences for city development in Canada is found in James Lorimer, *The Developers* (Toronto: James Lorimer, 1978).

40. The validity of this tactic is questioned in Stanley Makuch, "Planning or Blackmail?" *Plan Canada* 25:1 (March 1985), 8–9.

41. Chamberlain, *Aspects of Developer Behaviour*, 45.

42. An excellent description of bargaining between developers and municipal planners in Scarborough, Ontario, is given in Hok Lin Leung, "Mutual Learning in Developmental Control," *Plan Canada* 27:2 (April 1987), 44–55.

43. Sherry R. Arnstein, "A Ladder of Citizen Participation," *Journal of the American Institute of Planners* 35:3 (July 1969), 216–224.

44. Cherie Enns and Jennifer Wilson, "The Picket Fence Project," *Plan Canada* 39:4 (September–October 1999), 12–15.

45. Lash, *Planning in a Human Way*, 35.

46. McAfee, "Vancouver CityPlan."

47. Neale Smith and Nancy Hackett, "Partnering for Neighbourhood Revitalization: Riverside Meadows, Red Deer," *Plan Canada* 41:1 (January–February–March 2001), 13–15.

48. Noel Keough, "Calgary's Citizen-Led Community Sustainability Indicators Project," *Plan Canada* 43:1 (Spring 2003), 35–36.

49. Ananya Roy, "A 'Public' Muse: On Planning Convictions and Feminist Contentions," *Journal of Planning Education and Research* 21:2 (2001), 109–126.

50. Leonie Sandercock, *Towards Cosmopolis: Planning for Multicultural Cities* (New York: Wiley, 1998).

51. One of the earliest critiques is J. Leavitt, "Feminist Advocacy Planning in the 1980s," in Barry Checkoway, ed., *Strategic Perspective in Planning Practice* (Lexington, MA: Lexington Books, 1986); two later valuable sources are Leonie Sandercock and Ann Forsyth, "Gender: A New Agenda for Planning Theory," *Journal of the American Planning Association* 58:1 (1992), 49–59; and Clara Greed, *Women and Planning: Creating Gendered Realities* (London: Routledge, 1994).

52. Fainstein, "Planning in a Different Voice"; and Mary Gail Snyder, "Feminist Theory and Planning Theory," *Berkeley Planning Journal* 10 (1995), 91–106.

53. Sue Hendler (with Helen Harrison), "Theorizing Canadian Planning History: Women, Gender, and Feminist Perspectives," in Kristine B. Miranne and Alma H. Young, eds., *Gendering the City: Women, Boundaries and Visions of Urban Life* (Lanham, UK: Rowan & Littlefield, 2000), 139–156.

54. Karen Hemmingson and Leslie Kemp, "Planning Ourselves In: Exploring Women's Involvement in the Community Planning Process," *City Magazine* 14:4/15:1 (Fall–Winter 1993), 14–17.

55. Cf. the review in Rae Bridgman, "Criteria for Best Practices in Building Child-Friendly Cities: Involving Young People in Urban Planning and Design," *Canadian Journal of Urban Research* 13:2 (Winter 2004), 337–346.

56. Penelope Gurstein et al., "Youth Participation in Planning: Strategies for Social Action," *Canadian Journal of Urban Research* 12:2 (Winter 2003), 249–274.

57. Gerald Hodge, *The Geography of Aging* (Vancouver: UBC Press, forthcoming 2007).

58. Deborah A. Howe et al., *Planning for an Aging Society*, Planning Advisory Service Report No. 451 (Chicago: American Planning Association, 1994), 47–50.

59. Statistics Canada, *A Profile of Disability in Canada, 2001*, Cat. no. 89-577-XIE (Ottawa: Statistics Canada, 2002), Table 2.

60. David Sibley, *Geographies of Exclusion* (London: Routledge, 1995); Brendon Gleeson,. *Geographies of Disability* (London: Routledge, 1999); and Hester Parr and Ruth Butler, "New Geographies of Illness, Impairment and Disability," in H. Parr and R. Butler, eds., *Mind and Body Spaces* (London: Routledge, 1999), 1–24.

61. Statistics Canada, *Canada's Ethnocultural Portrait: The Changing Mosaic, 2001*, Census Analysis Series, Cat. no. 96F0030XIE2001008 (Ottawa: Statistics Canada, 2003), 39.

62. Feng Hou and Garnet Picot, "Visible Minority Neighborhoods in Toronto, Montreal, and Vancouver," *Canadian Social Trends* 72 (2004), 8–13.

63. Karen Umemoto, "Walking in Another's Shoes: Epistemological Challenges in Participatory Planning," *Journal of Planning Education and Research* 21:1 (2001), 17–31.

64. Mohammad A. Qadeer, "Pluralistic Planning for Multicultural Cities: The Canadian Practice," *Journal of the American Planning Association* 63:4 (Autumn 1997), 481–494.

65. Michael Hibbard and Susan Lurie, "Saving Land but Losing Ground: Challenges to Community Planning in the Era of Participation," *Journal of Planning Education and Research* 20 (Winter 2000), 187–195.

66. Maria Canton, "Energy Board Halts Power Plant Hearing," *Calgary Herald*, August 14, 2001.

67. A recent study of participation in Canadian planning makes this point: Kevin S. Hanna, "The Paradox of Participation and the Hidden Role of Information," *Journal of the American Planning Association* 66:4 (Autumn 2000), 398–410; and Innes, "Information in Communicative Planning."

68. An excellent overview is provided in Judith Innes, "Planning through Consensus Building: A New View of the Comprehensive Planning Ideal," *Journal of the American Planning Association* 62:4 (Autumn 1996), 460–472.

69. The consensus method is laid out in L. Susskind et al., *Consensus Building Handbook* (Thousand Oaks, CA: Sage, 1999).

70. Manfred Fast, "Communities Can Make It Happen: Forging the Port Colborne Ontario Strategic Plan," *Small Town* 26:1 (July–August 1995), 10–15.

71. Judith E. Innes and David Booher, "Consensus Building as Role Playing and Bricolage: Toward a Theory of Collaborative Planning," *Journal of the American Planning Association* 65:1 (Winter 1999), 9–26.

72. Martin Meyerson and Edward C. Banfield, *Politics, Planning and the Public Interest* (Glencoe, IL: Free Press, 1955).

73. Jim Green, "Regional Government, Revolution and NIMBYism," *City Magazine* 13:2 (Spring 1992), 29.

74. Geoffrey York, "Native Seminary Meets Only Rejection," *The Globe and Mail*, Toronto, January 29, 1990, A1.

75. Tom Goodale and Sherry Wickware, "Group Homes and Property Values in Residential Areas," *Plan Canada* 19:2 (June 1979), 154–163.

76. Margaret Philip, "Toronto's Destitute Wear Out Welcome, *The Globe and Mail*, July 18, 1997, A6.

77. John Andrew, "Examining the Claims of Environmental ADR: Evidence From Waste Management Conflicts in Ontario and Massachusetts," *Journal of Planning Education and Research* 21:2 (2001), 166–183.

78. Gerald Hodge and Patricia Hodge, "Citizen Participation in Environmental Planning in Eastern Ontario," *Plan Canada* 19:2 (March 1979), 22–29.

Chapter-Relevant Sites

Planning Canadian Communities
www.planningcanadiancommunities.ca

CMHC Gateway for Housing Professionals and Community Groups
www.cmhc.ca/en/inpr

The Citizen's Handbook, A Guide to Building Community
www.vcn.bc.ca/citizens-handbook

Georgia Strait Alliance—Caring for our Coastal Waters
www.georgiastrait.org

GVRD Homelessness Plan
www.gvrd.bc.ca/homelessness/

Joint Centre of Excellence for Research on Immigration and Settlement
www.ceris.metropolis.net

Metropolitan Action Committee on Violence Against Women and Children
www.metrac.org

National Charrette Institute
www.charretteinstitute.org

PART FOUR

4

BARRINGTON STREET

Implementing Community Plans

INTRODUCTION

Turning the aspirations of a community plan into reality requires tools that can guide the many decision makers with an interest in the outcome of the community's built environment. These tools are more than just complementary to the community plan; they are an integral part of the planning process of attaining the community's goals. In essence, a plan without tools for implementation is probably destined to gather dust on the shelf. The tools of planning—zoning, subdivision control, capital budgets, and so on—are very familiar. People encounter the effects of these tools more often than they come in contact with the plan itself. Indeed, they sometimes have negative connotations in people's minds because they can be used to restrict development or change on privately owned property.

Achieving the aims of a community plan is a matter of combining land-use regulations and public-policy instruments to influence the decisions of private developers and other public agencies. They are complementary: the former are reactive to development initiatives, constraining the choices of land developers, while the latter can be proactive in influencing the decisions of private developers and public agencies alike. Thus, the adoption of a community plan is the beginning, not the end, of the need for choices about community-building.

The image above is from the Halifax Waterfront Open Space and Development Plan *project, Halifax, Nova Scotia, which received the Canadian Institute of Planners' Award for Planning Excellence, Category of Overall Presentation, 2002.*

13

Chapter Thirteen

Land-Use Regulation Tools for Plan Implementation

Discretion is an essential element of planning. Somebody has to say yes or no to a request for permission to develop land at a certain time and in a certain way.

Anthony Adamson, 1956

For a community to prepare a plan is commendable, indeed essential, in order to set about attaining the built environment it aspires to. But turning those aspirations into reality requires tools that can guide the many decision-makers with an interest in the outcome of the community's space. The planner's tools are more than just complementary to the community plan; they are an integral part of the planning process of attaining the plan's goals. The "archway" of planning shown in Figure 8.2 could not exist without both plan and tools. In short, a plan without tools for implementation is probably destined to gather dust on the shelf, and planning tools without a plan add up to meaningless or even arbitrary regulation.

This chapter and the next examine the tools and approaches used in implementing the community plan. In many ways, the tools of planning—zoning, subdivision control, capital budgets, and so on—are very familiar. People encounter the effects of these tools more often than they come in contact with the plan itself. Not infrequently, the planning tools are seen as the end product of planning rather than its means. And where planning tools are also land-use regulations, they often have negative connotations in people's minds because they can be used to restrict development or change on privately owned property. Sometimes they are believed to have more power over the plan's implementation than they actually carry. It is important, therefore, to grasp the nature and capabilities of each of these planning tools.

The planning tools we examine in this chapter are those that are used to guide development on private land. That is, they are the planner's traditional tools for regulating land use. These tools each have a specific role and capacity in implementing a community plan. Thus, as they are reviewed here, an overriding question should be kept in mind:

- —*What potential do the planning tools have to implement a community plan and what are their limitations?*

The Main Tasks of Plan Implementation

The ultimate aim of plan-makers is to establish the conditions that will attain the goals of the plan. Those conditions derive from a combination of policy directives, legal instruments, administrative practices, and means of promoting community participation in planning. Implementing a community plan is, thus, accomplished using a variety of tools and approaches. While making a community plan concludes at a particular point in time (at least until it needs to be amended), plan implementation is a *continuous* process. Since the tools of planning deal with a continuous, functioning community, they must be able to respond to the type and character of present-day development initiatives and to the decisions that will be made as the future unfolds.

Planning tools act as an *interface* between the policies of the plan and the aims of those who make decisions that transform the physical environment. The latter include all those who own some land within the community—the average residential homeowner, who may have no plans for developing his or her property, as well as those who are in the business of land and building development. Also included are the staffs in public agencies who make decisions to invest public funds and construct roads, schools, parks, parking garages, and so on. All have the potential for making decisions that could affect the future built environment. Considering this array of participants involved in land-use decisions, two basic needs, or dimensions, of plan implementation emerge. The need to:

1. **Guide development on private land.** This, in turn, breaks down into two sub-dimensions:
   a. presently developed areas, and
   b. vacant or undeveloped areas.
2. **Coordinate public development efforts.** This applies especially to capital investments.

Around these two dimensions have developed the best-known and most refined planning tools, such as **zoning,** or "districting," as it used to be called, and **development control** or site-plan control; these are used to guide development on presently developed private land. **Subdivision control**, another traditional tool, is used to guide development on large parcels of vacant land. These familiar tools are designed for the *direct* guidance of the individual, group, business, or public agency that wishes to make decisions about the use of land in the community. This guidance is aimed at encouraging certain land use to occur in particular locations while constraining or limiting that development. To do this, the tools are usually backed by statutory powers vested in the municipality by the province.

There are also tools that work to influence development decisions *indirectly*. Indeed, a good deal of planning implementation is involved with more than regulating land use. Considerable attention and time must be devoted, for example, to guiding persons who are considering development of their land and who look to officials in other municipal departments and agencies to explain planning policies. Promoting citizen participation is also important to test public acceptance of plans and policies. These latter tasks have not generated formal, institutionalized tools. Certain implementation *approaches* have emerged to obtain public input and use alternative organizational arrangements for municipal planning operations. The approaches often differ in form from community to community (with the exception of provincially prescribed public hearings). Thus, communities develop individual planning styles with which to achieve the goals of the community plan. For example, Vancouver's approach to downtown development is quite different from that used in Toronto or Montreal, incorporating extensive design review and staff approval delegated by City Council.[1]

One final point about plan implementation and the tools associated with it concerns the outcome—the physical result—of decisions taken in the name of planning. The community plan is intended to encompass *all* community land-use decisions, present and future, in order to achieve some envisioned built environment. In a planned new town, the physical results may approximate the design promoted in the overall plan. But in the typical Canadian city or town, change in the existing environment is not subject to such direct design. The regulations for guiding development on private land are usually invoked only when the persons or groups with interests in the land signal their intention to develop or redevelop their properties.

Planning Tools for Already-Developed Lands

Planning tools that are typically used in the developed portion of a community include **zoning, development control** (site-plan control); and **redevelopment plans.** The developed part of a community is the portion that, in general, comprises relatively small parcels of land that are already built upon or could be developed. There are usually a few large parcels of land occupied by such institutions as hospitals, factories, or parks. These are part of the developed area, too. But large parcels of land in use as farms or for other resource extraction, or those that are simply vacant and not divided into small parcels, are not considered developed land. The latter types are usually found on the fringe of the community and may, indeed, become developed in the future, but their future use is still to be determined at the time of its subdivision into smaller parcels. Here again is another distinction that should be kept in mind when considering such planning tools as zoning: these tools are designed for land for which the intended use has already been determined.

Zoning

Origins and Nature

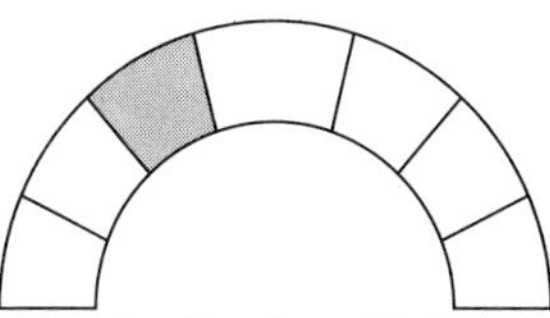
Zoning (land-use) Bylaw

Zoning grew out of the early observation about city development that similar uses tend to congregate in areas separate from other uses. Further, when congregated in their distinctive areas, the land uses and activities—industry, commerce, residences—seem to perform their respective functions more effectively than when intermingled. German town planners in the latter half of the 19th century employed these observations to devise ways of organizing the growth of industrial cities to ensure efficiency for the factories and amenable workers' housing.[2] Municipalities in all countries experiencing industrialization in this period, including Canada, sought to regulate the siting of buildings and the provision of basic services for safety and health reasons. The German approach was, however, to differentiate the regulations according to the needs of the uses located in (or planned for) their respective *districts*. The practice became known as "districting" and, later, as "zoning," especially in North America.

Besides observing the functional differences in land use (e.g., many industries have noxious effluents, commercial areas generate a lot of traffic), early planners also perceived aesthetic differences among land-use districts: the height of buildings, the space between them, their setback from the road, and the road pattern that best suited the use. Not least, early planners noted that differences between land uses reflected differences in *density* of land use. In residential districts, there were differences, for example, in the number of dwellings on each parcel of land, as reflected in the height and bulk of buildings, the size of dwelling units, and the amount of open space around them. Analogous observations were made in regard to industrial and commercial land uses and their structures. In short, districting principles lent themselves to consideration of the physical design of communities to achieve the arrangements and densities of land uses a community might prefer.

Zoning can be applied simultaneously to individual properties or large areas, which may account for its longevity among planning tools. When zoning was established, it was a major departure in the exercise of statutory powers by municipalities. Previously, municipalities had used powers for enforcing safety, health, and structural standards on an individual property basis.[3] But since land use occurs on an areal basis, it was necessary to apply the same regulations to all the properties in a similar area. With this sort of tool, planners were more likely to achieve consistency in the types of uses and structures that constituted the physical environment of the distinctive zones or districts of development.

In North America, the first comprehensive zoning bylaw was enacted by the City of New York in 1916. The first such Canadian zoning bylaw was enacted in 1924 for Kitchener, Ontario, and was formulated by planners Thomas Adams and Horace Seymour.[4] A comprehensive zoning bylaw established districts for the entire community. Prior to the above dates there were districting bylaws in some communities that dealt selectively with heights of buildings and the location of noxious uses.[5] For example, in 1904, Ontario cities had the right to control the location, erection, and use of buildings for laundries, butcher shops, stores, and "manufactories."[6] And there were instances of bylaws designed in a discriminatory fashion to limit the residency of specific minority groups to certain areas of the city.[7] As zoning came more widely into practice in the 1920s, it had to meet the test of court challenges, especially with regard to the principle that it must apply *universally* and *uniformly* to all properties within the areas in question. Earlier exclusionary tendencies gave way to a more general rationale of the protection from dramatic shifts in property values due to the mixing of incompatible land uses.

Zoning practice developed along the same general lines in Canada and the United States, although there are significant constitutional differences between the two countries that affect zoning. The basic difference lies in the fact that the U.S. Constitution spells out personal property rights, whereas the Canadian Charter of Rights and Freedoms does not. (Nor did the British North America [BNA] Act.) In the United States, the validity (constitutionality) of zoning bylaws was not confirmed as a power available to municipalities until it had been tested by the U.S. Supreme Court.[8] In Canada, municipalities were deemed, under powers granted to the provinces under the BNA Act and British Common Law, to have the statutory power to regulate land use.[9] The usual concern over zoning practice in the United States is whether the bylaw involves a "taking away" of property rights granted to property owners. The concern in Canada is more general—that is, whether the bylaw is discriminatory in pursuing the public interest. Thus, zoning bylaws in Canada can have much broader scope than those in the United States. This constitutional difference has given rise in Canada to the complementary practices of development control and site-plan control, which allow a municipality to specify certain land-use regulations on a property-by-property basis. More will be said about these later in the chapter.

Substantive Focus

Zoning has a very precise focus. It provides a set of standards regarding

1. the *use* to which a parcel of land may be put, and
2. the *size, type,* and *placement* of buildings on that parcel.

These standards are made explicit in the text of the zoning bylaw, according to the districts in the community to which they apply. An accompanying **map**, or **zoning plan,** as it is sometimes called, specifies the boundaries of each zone and, thus, the properties affected by the different district regulations. The zoning plan is the second most important planning instrument in the community next to the overall plan.

The zoning bylaw may also regulate the density of population from district to district by specifying the number of dwelling units allowed per building and/or the number of individuals that may occupy a dwelling unit. Most zoning bylaws contain provisions covering the external effects of activities carried on in buildings such as through requirements for off-street parking and loading areas, size and placement of signs, and accessory uses and home occupations. There may be the need or desire in a community to specify land-use districts that contain distinctive local features for functional, historical, environmental, or aesthetic reasons, in which case the zoning bylaw may contain provisions covering a number of special districts.

Land-Use Types and Districts. Three basic land-use categories are usually identified in zoning bylaws: residential, commercial, and industrial. In each of these categories, there may be a series of zones, depending upon the nature of the activities, the type of building, the density of development, or the lot size. A variety of residential zones may be distinguished, for example, by the number of dwelling units permitted on the lots that are typical of the zone. If large apartment buildings and large tracts of land are being developed, the zone may be defined by the number of dwelling units per hectare. Among commercial zones, the distinctions may be in terms of the service needs of people: central business, neighbourhood commercial, highway-oriented commercial, or shopping centre. For industrial zones, the differences in standards usually reflect the characteristics of the processes carried on by the firms. For example, industries using heavy machinery or emitting considerable smoke, noise, or odours would usually be in a zone separate from those whose processes had much less external impact. Similarly, those engaged in mostly the shipping, storing, and transfer of goods would likely warrant a separate zone. Figure 13.1 illustrates a typical listing of zones found in the zoning bylaws of a medium-sized city.

For each land-use zone, it is common practice to specify the uses that are *permitted* to locate there. This may necessitate a long list of valid uses in order to include all likely and desirable types. The following list shows the establishments permitted in the Neighbourhood Commercial Zone of the City of Kingston, Ontario:

- Retail stores
- Neighbourhood stores
- Offices for or in connection with a business or profession
- Banks or financial institutions
- Restaurants
- One-family dwellings and two-family dwellings, provided that such dwellings are located within a commercial structure
- Libraries, art galleries, or museums
- Shopping centres

Figure 13.1 Land-Use Districts in Typical Zoning Bylaw for a Medium-Sized City

CLASSIFICATION	PREDOMINANT LAND USE
R1	Single-family residential
R2	One- and two-family residential
R3	Apartment residential (medium density)
R4	High-density residential
C1	Downtown commercial
C2	Neighbourhood commercial
C3	Automobile-oriented commercial
M1	Light industrial
M2	Wholesale and transportation
M3	Heavy industrial
P	Public use (health, education, religious, administration)
R	Rural and agriculture
EN	Environmentally sensitive area

A brief look at the uses permitted in the above zone indicates how, even in this rather ordinary bylaw, an interesting mixture of land uses might be accommodated: dwelling places, offices, restaurants, and stores. Yet, in recent times, the result is a strip mall. Other communities, especially larger ones such as Toronto, Montreal, and Vancouver, have developed zoning provisions that encourage special combinations of mixed uses, thereby considerably blunting the criticism that zoning prevents mixing.[10]

Zoning bylaws must conform to legal standards and must, therefore, be specific in their language. In the listing above, the specified uses are the only ones permitted and any other uses are thereby excluded from being located within the zone. Occasionally, excluded uses may be specified in order to prevent any ambiguity in interpretation. Unspecified uses already present in the zone before the bylaw was enacted are usually "excepted" (or "grandfathered") from the provisions. The list of permitted uses in each zone is thus distinctive, although a small amount of overlap will undoubtedly occur between the lists of permitted uses within the sets of zones in each of the basic residential, commercial, and industrial categories. Banks and restaurants will generally be permitted in all commercial zones, for example.

The land-use zones, and the lists of permitted uses associated with each, in most instances, reflect the types of uses already existing in the community. Where residential neighbourhoods, commercial areas, and industrial districts already exist and are in stable condition, the zoning regulations are formulated so as to continue those uses in their respective areas. This is the basis for the concern of some people that zoning is essentially *protective* of existing uses and *restrictive* of new uses. But zoning may also be used to specify building types in areas that are in the process of new growth or redevelopment from earlier uses. A new and dramatic example of this strategy is the initiative being taken in Toronto to revitalize two old industrial districts close to downtown. The land-use provisions of the zoning regulations were eliminated, leaving only specifications for building height and performance standards, all with the aim of attracting alternative activities to the districts.[11] In this sense, zoning may also be used in a *prospective* way. In Burnaby, an older inner suburb of Vancouver, there is a similar provision in the land-use regulations to accommodate new development, but using a different approach. Planners in Burnaby employ a regulatory device called Comprehensive Development Zoning, or CD zoning that, in effect, creates a unique zoning district for the site of a project. In order for the developer of the site to obtain a CD zone classification, a Comprehensive Development Plan of the given site must support the application.[12] The CD plan is usually accompanied by architectural, landscape, and engineering drawings, illustrations of the proposed project, and more. As with other proposed zoning changes, CD plans are required to go through public hearings and thereby allow public scrutiny; it is also mandatory in most provinces that the resulting zoning be consistent with the municipality's overall community plan.

Building Height, Bulk, and Placement. Just as important as land use in zoning is the resulting effect of the size of buildings constructed on parcels of land, along with the placement of the buildings on the parcels. For example, **height** controls often existed in cities long before comprehensive zoning was accepted (including in ancient Rome). Controls regulating how close buildings could be placed to the street (now called the **setback**) were also long-lived concerns in cities. Zoning bylaws gathered together these various concerns over the built environment and refined and articulated them in conjunction with the aims for land-use districts to further the community environment.

In a typical zoning bylaw, one will find that each land-use zone includes specifications for the maximum height of buildings and the proportion of the parcel of land that may be built upon. From these two dimensions—height and building area—the potential **bulk** of a building may be ascertained. If there are requirements for leaving space around the building, specified as setbacks from the boundary lines of the property, then the exact space within which the building may be sited on the property is identified. This exact space on the parcel combined with the allowed bulk of building defines what is called the **building envelope** (see Figure 13.2).

Figure 13.2 Basic Dimensions for the Placement, Coverage, and Height of Structures on Building Lots (the "Building Envelope")

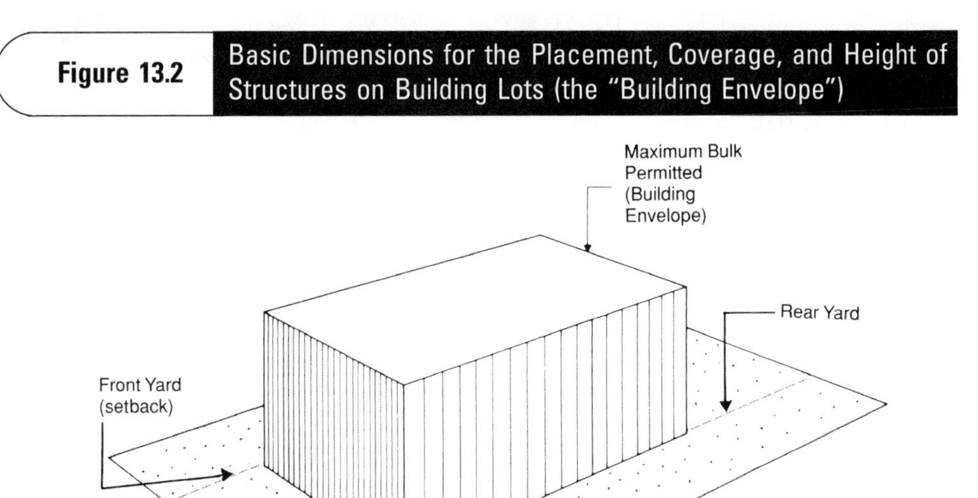

These dimensions determine a "building envelope" and govern the bulk of building allowed on a site.

In each zone there is a distinctive building envelope within which property owners may build. If the building built on a parcel of land completely filled the building envelope, this would be the most intense development that could occur in the zone. In practice, for a number of reasons, this intensity does not always result: for example, owners may not wish their buildings to be as tall as allowed, or may want more space around them, or may not be able to afford to build such a bulky building, or the shape of the building may be different than the building envelope (as with peaked-roof houses in residential zones). Custom is also important in influencing the bulk of buildings. An owner may want his or her building to be in keeping with the bulk of buildings already existing in the district. And existing norms for building bulk, it should be noted, may be a function of long-standing zoning standards.

Figure 13.3 codifies the requirements for building height, lot coverage, and setbacks for a typical set of residential zones in a city. If one were to visit the actual districts in which they apply, it would probably be obvious that the requirements reflect the norm of existing conditions in those districts of the city. For example, older residential districts may have smaller or more variable yard or height situations than newer, more uniform suburban districts. Often, more affluent residential areas will be found to have requirements for large lots and setbacks commensurate with their more spacious character. Until recently, the regulation of building bulk was confined to areas where large office or apartment buildings could be built. But nowadays, in many cities, bulk regulations are being added to zoning bylaws concerned with one- and two-family residential areas. The trend to "mega homes" (sometimes called "monster homes" or "McMansions") caused consternation in many communities because neighbours felt that the new, large houses changed the neighbourhood character. In effect, what was happening was that builders were now using most, if not all, of the allowable building envelope under the zoning regulations, thereby permitting monster homes to dominate the streetscape (see Figure 13.4). Municipalities in the suburbs of Vancouver have revised residential-area zoning regulations by reducing the size of the building envelope and applying a Floor Area Ratio (FAR), usually around 0.3, that would limit the bulk of

Figure 13.3 Typical Building Envelope Standards for Residential Buildings in a Larger City

RESIDENTIAL ZONE	MIN. LOT AREA m^2	MIN. LOT WIDTH m	MAX. LOT COVERAGE	FRONT YARD m	SIDE YARD m	REAR YARD m	FLOOR AREA RATIO %	MAX. HEIGHT m	STOREYS
R1	1000	22.5	35	10.5	4.5	9.0	–	10.5	3
R2	500	15.0	30	7.5	3.0	7.5	–	10.5	3
R3	330	9.0	40	4.5	2.5	7.5	–	10.5	3
R4	200	7.5	50	4.5	1.5	4.5	–	10.5	3
R5	1000	15.0	50	6.0	√	4.5	1.0	12.5	√
R6	1500	15.0	50	4.5	√	√	1.5	25.0	8
R7	1500	15.0	70	4.5	√	√	3.0	√	√
R8	4000	25.0	100	√	√	√	5.0	√	√

√ Indicates variable standard to suit project.

A zoning bylaw specifies the maximum or minimum measurements for the placement, coverage, and height of buildings in each district. In an older city, it is usually necessary to accept the smaller dimensions of lots plotted before the bylaw (e.g., R3 and R4). In larger cities, provision must be made for tall structures (e.g., R7 and R8), with no height limit but constrained in their bulk. New "monster homes" may require Floor Area Ratios in zones R1 to R4.

Figure 13.4 The Increasing Bulk of Recent Dwellings

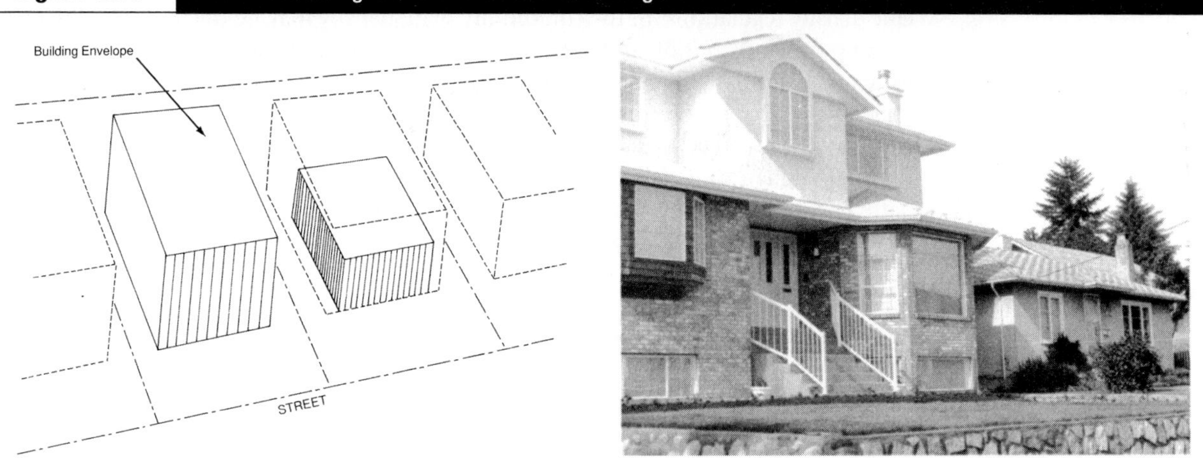

New "monster homes" often dwarf older neighbouring dwellings, even though they are built within the same zone regulations. The new houses take advantage of the entire building envelope.

Source: G. Hodge.

new dwellings. The Floor Area Ratio, sometimes called the Floor Space Index, was discussed in Chapter 6.

In central commercial areas, it is common to permit building on 100 percent of the lot area in recognition of owners' need to extract maximum use from the higher value land. Larger lot coverage is usually allowed in areas where tall apartment and office buildings are located, also because of the higher value of land and buildings. However, in recent decades, in order to reduce the bulk of such buildings and to provide usable open space for residents of apartments or for office employees, provisions are included in zoning bylaws to require more open space at ground level for every increase in height. As well, the overall height of tall buildings may be regulated according to the total area of all floors expressed as a ratio of lot area—that is, according to the Floor Area Ratio of the building. Thus, referring to Figure 13.3, an FAR of 1.0 in an R5 zone that limits lot coverage to 50 percent means that the owner could build a building of two storeys equal in floor area to the lot size, or else build a taller building, also equal in floor area to the lot size, but covering less than 50 percent of the lot.

Since cities are always changing in their physical form, it is not uncommon to see a conventional urban activity change in its physical characteristics. Such is the case with retailing, an almost inherent urban activity. In the past decade or so, new retailing "formats," as they are called, have appeared in the form of bulky, warehouse-type establishments that dwarf existing retail operations. The advent of these so-called "big-box" retailers has led many municipalities to revise their zoning regulations and community plans to accommodate a wider mix of retail building types. The town of Markham, Ontario, for example, has added a "retail-warehouse" designation to its zoning provisions.[13]

Density. The intensity with which the community's land is used is a basic planning factor because it relates to the issue of congestion in the use of public facilities such as water and sewer lines, streets, parks, and so on. Although land uses and building bulk on a parcel of land may be specified, this does not deal with the number of users of the land and buildings. This is especially crucial in residential areas where multiple-family accommodation is provided to ensure that overcrowding does not occur, whether inside or outside of dwellings.

Most zoning bylaws where multiple-family dwellings are common now include specific density regulations. In the community plan, density may be calculated by the number of people per unit of land area. However, the zoning bylaw must be specific enough to relate to individual parcels and buildings. Density specifications in zoning bylaws may take one of two forms, or a combination of both. The first, and most common, is to *specify the maximum number of dwelling units per unit of land area (hectare)*. Various multiple-residence zones may then be created that correspond to existing development patterns or follow new styles and trends in the housing market. At the lower-density end of the scale, zones corresponding to three- and four-storey walk-up apartments may limit development to 25–50 units per hectare, for example. Medium densities (100–200 units per hectare) are usually related to areas in which the buildings would be taller and have elevators, but probably not exceed 10 storeys. High-density zones would be for much taller buildings with densities of 240 units per hectare and more. Of course, what is considered a "high" density or a "low" density is related to local conditions and tastes. (It may help to refer to Figure 6.3, page 152.)

The second method of specifying residential density is according to the *amount of floor area* included in a project site (net density) or in the neighbourhood (gross density). This is sometimes stated as a minimum area (in square metres) in order to ensure that adequate-size apartments are built. Such a regulation is often required in older parts of communities, either where conversion of existing buildings to apartments is likely to occur or where original lot sizes were small. In many urban areas, the bulk of apartment buildings is controlled by a floor area ratio (FAR), which eliminates variations caused by the size of residential units. Such a regulation is usually combined with a limitation placed on the height of buildings.

Other Concerns of Zoning

Parking/Loading. Minimum requirements for off-street parking and loading are common in zoning bylaws. Again, the main objective is to minimize possible congestion in the public streets in the vicinity of a property that might be used more intensively, such as for multiple residences or commercial purposes. Parking standards have been developed through planning experience with various land uses. In residential areas, at least one parking space per dwelling unit is usually needed. However, in higher-density areas this may need to be higher in order to accommodate parking by residents and visitors. In commercial and manufacturing areas, the standards are usually set in terms of the gross floor area (or sometimes by the number of employees) on the assumption of providing space for those who drive their cars to work. Loading facilities located off the street are a common requirement in commercial, manufacturing, and institutional zones, with standards usually set in terms of the number of square metres of the establishment.

Signs. Most zoning bylaws deal with the size, height, and location of signs on a building or property. Signs tend to be discouraged altogether in residential areas; in commercial areas the signs may be regulated to reduce "sign competition" and produce a particular character for the area. The specifications for signs vary from one community to another as a result of local tastes and preferences. Signs that overhang public streets are regulated as well for safety purposes.

Accessory Buildings. Most land uses require, in addition to the principal structure, structures for parking or storage. A garage on the same lot as a single-family house is an accessory use, as is a storage shed on a commercial property and a security guards' building at a manufacturing plant. These are usually quite closely defined in the regulations for each particular zone, including their size and location. The implication is that

Planning Issue 13.1 Sooke's Mayor Calms Fears on Zoning Bylaw

Victoria Times Colonist
July 1, 2006

Sooke's Mayor Calms Fears on Zoning Bylaw: Big Developments Must Follow Rules, Including Hearings
Bill Cleverley

Six-storey buildings aren't going to start sprouting all over Sooke's downtown if a proposed new zoning bylaw is passed, says Mayor Janet Evans.

"We don't anticipate any six-storey buildings going up without complete approval from council and the neighbourhood," Evans said.

Under the new bylaw, any proposed large development would have to go through the regular rezoning process, including public hearings, she said.

Evans made the comments following a public hearing into the bylaw Thursday attended by about 60 residents—the first of its kind for Sooke.

The bylaw would affect land use in the town core and cover everything from property sizes to building heights, parking requirements, ocean views, harbour zoning and even the height of backyard sheds.

With the community's new $22.6-million sewer system and treatment plant—built and operated by Alberta-based utility company EPCOR—up and operating, Sooke no longer requires large setback areas for septic fields in the downtown core.

The need for such fields had kept property owners from easily subdividing properties into smaller single family lots, and served as an obstacle to construction of larger buildings.

Some at the hearing complained that the bylaw would allow six-storey commercial buildings in the Sooke core, up from the current three, which could potentially block water views from some houses.

Evans said building proposals will be dealt with on an individual basis.

"I think the viewscapes were an issue and I agree we have to protect those, for sure," she said.

Others suggested the council has the cart before the horse—that the Official Community Plan should have been updated before introducing a new zoning bylaw.

Sooke director of development services Cheryl Wirsz said a new zoning bylaw is necessary because of the advent of the sewer system.

"It can be seen as backwards, but council needed to get the bylaw done because we weren't ready for the sewer system. So it could be seen as being backwards or it also could be seen as being really slow," she said, given that the most recent community plan review was in 2002.

The bylaw will be scrutinized by a committee for some site-specific revisions, and then return to council for consideration of second and third reading.

It would then be forwarded to the Ministry of Transportation for approval and then back to council to be formally adopted.

Source: Bill Cleverley, "Sooke's Mayor Calms Fears on Zoning Bylaw," *Victoria Times Colonist*, July 1, 2006, B2. Reprinted with permission.

these uses and structures would not normally be permitted in the zone without the principal use to which they are appendages.

Home Occupations. From zoning's earliest days, communities sought to limit commercial activities in residential areas, while usually allowing individuals such as music teachers, insurance agents, doctors, and hairdressers to offer services from their homes. Traffic, safety, and noise were the concerns with respect to some "home occupations," as they are usually called in zoning bylaws. Until the late 20th century, few people carried on business from their homes (exceptions included doctors, accountants, and hairdressers), but this has now dramatically changed. Both economic pressures and the potential of computer and other telecommunications have promoted many new home occupations right across the country, in large cities and small. For example, it was estimated that one-third of Toronto's dwellings in 1993 housed some form of business activity or home occupation.[14] This stimulated the city to amend its zoning bylaw to allow a wide range of "new" home occupations. The exterior of the home would not be able to be changed, nor would exterior signs advertising the business be allowed; there are similar provisions in other cities.

Aesthetics. Although zoning usually strives for compatibility of uses within a zone, many communities also attempt to encourage compatible scale and massing for buildings. However, few explicit specifications of building appearance have been included in zoning bylaws, save those governing the height of buildings. This is largely because zoning grew out of a notion that public intervention in private property should occur only to protect health, safety, and the general welfare of a community.

The regulation of architectural qualities of structures means, of course, being able to define aesthetic norms in legally defensible terms, something not easy to do in such a subjective area. A compromise position used in a number of cities is the designation of particular zones in which the architectural plans of structures are reviewed because the areas are considered of major importance in the "civic design" of the community. One such zone was established for Toronto's University Avenue in the 1920s in the hope of achieving a monumental character to this street leading to the provincial legislative buildings. In the 1950s, Vancouver enacted similar regulations for new structures on its West Georgia Street. The latter approaches act as a separate "zone" superimposed on the bylaw's district regulations. Comprehensive Development Zoning, discussed above, does allow design criteria to be applied to new projects.

The difficulty of formulating aesthetic regulations on a comprehensive basis in zoning bylaws has led to several new variations. Within the jurisdiction given by development-control procedures to municipalities, it is possible to request modifications to façades and exterior materials used on buildings. Because of the recent concern over historically important districts, many provinces have enacted legislation allowing the architectural quality to be closely regulated in such districts. Zoning also can aid in this, as the "heritage main street" district in Markham, Ontario, demonstrates.[15] Both these approaches serve to guide private development on a building-by-building basis, in contrast to zoning bylaws, which strive to develop regulations that apply generally to all buildings and properties. It is because the ideal of comprehensive regulations has significant limits that such supplementary tools have come into existence.

Cultural Diversity. The increasing cultural diversity of Canadian communities has begun to affect land-use planning in general and zoning in particular. New immigrant populations with their different cultural perspectives have frequently encountered difficulty in three facets of the built environment: places of worship, ethnic business enclaves, and housing types.[16] The traditional notion that it is *land use,* not *people,* that is the subject of community planning practice creates an obstacle to changing policies and tools. "People planning" was, with good cause, seen as allowing for human discriminatory practices under the guise of planning, as some early zoning bylaws revealed. This dilemma has been confronted in a number of rezoning applications; a recent Ontario court decision has noted that "land use practices are made by human beings, by people . . . " and, therefore, that social and cultural matters are valid planning and zoning concerns.[17] This is an area of concern that will continue to evolve (see further discussion in Chapter 15).

Recent Zoning Perspectives

It should come as no surprise that a planning tool that has been around for three or four generations would acquire some critics and require some changes. Indeed, zoning has been blamed for many planning ills over the years that were due more to its misapplication than to inherent flaws. But, like all the planning tools described here, zoning has considerable flexibility and has been applied in a variety of ways to suit local conditions and local preferences.

Still, zoning is a simplified view of city development that envisions activities located in discrete districts, housed in distinctive types of structures. It generalizes the pattern of development and then categorizes the activities that take place there in terms of discrete physical forms. This has the result in many communities, especially larger ones, of reducing the diversity of activities that exist in the real world to a limited number of technical categories. The very mixture of activities that gives character to older neighbourhoods may be obscured in the zoning of new residential areas. This is a valid complaint of proponents of **New Urbanism** and **Traditional Neighbourhood Design** (TND).[18] There are now a number of examples across Canada of recent zoning bylaws that promote the latter approach to neighbourhood planning, with a mix of housing types and other uses.[19] Indeed, revisions to zoning bylaws to allow for a **mixture of land uses**, such as retail, housing, office, and home occupations, has now become common across the country to revitalize old commercial and industrial districts.[20]

Mixed-use districts are one response to changing neighbourhood activities. New uses may be accommodated in structures not originally meant for them (such as professional offices in old warehouses or the many new home occupations made possible through telecommuting). Or completely new formats may emerge for an old activity (such as the advent of "big-box" retailing and "ethnic" shopping malls). Planners now allow distinctive land-use demands in zoning regulations to "customize regulations" for individual sites, especially in dealing with large, complex developments as with **Comprehensive Development Zoning** described above.

Development Control

Another way to regulate land use is to institute a permit system, in which each property owner must apply at the time of a proposed development change in order to have the allowable conditions established. This form of regulation is generally referred to as **development control**.

Origins and Nature

Until the 1950s, Canadian planning practice closely followed the primary mode of land-use control in the United States—zoning. In the U.S., the constitutional entrenchment of property rights required that any bylaws that regulate land use be applicable uniformly and universally, and not deprive property owners of any constitutionally guaranteed rights. In Canada, neither the old BNA Act nor the more recent Charter of Rights and Freedoms entrenches property rights. Thus, there is not the same legal basis in Canada for zoning, or any reason to limit land-use regulation to that form of bylaw.[21] Development-control approaches entered Canadian planning practice largely through the efforts of British and British-trained planners who emigrated to Canada after World War II. British planning practice never embraced zoning, preferring to leave responsibility for plan implementation in the hands of local administrators. Development control was considered particularly useful for the period during which a formal community plan was being prepared, especially in fast-growing cities. Each proposal for new development, buildings, or subdivisions thus could be reviewed to ensure consistency with the aims of the emerging plan, and a development permit was issued as warranted. The system was widely used in western Canada from the mid-1950s, but did not enter regular planning practice in Ontario until 1970. While it began as an interim measure to be used during plan preparation, development control has become part of the routine of planning practice, in conjunction with zoning.

Substantive Focus

Development control is known by various names among the provinces: in Alberta, the term is **development permit,** while in Ontario, the term used is **site-plan control.** The approach is, however, similar. Development-control regulation is invoked at the actual time a proposal is made to erect a new building or to significantly expand an existing one usually within predetermined areas of the community (such as a waterfront area, business district, or heritage zone). Importantly, the focus is on the building, rather than its use. Since development control works within the context of zoning, it starts with the premise that the zoning regulations as to the use of the site and the allowable building envelope provide the basic frame of reference. The latter are usually not negotiable at this stage of regulation.

Development control is concerned with the actual placement of the building on the site (within the building envelope), the massing of the building, its relation to surrounding buildings and areas, its relation to streets, and, in some provinces, its appearance. The following extract from the Ontario planning act illustrates the scope of the approach one generally finds in development control. The person or group proposing the development must provide the municipality with

1. plans showing the location of all buildings and structures to be erected and showing the location of all facilities and works (e.g., swimming pools, fences and walls, landscaping, walkways, outside lighting, parking areas); and
2. drawings showing plan, elevation, and cross-section views for each building... sufficient to display:
   a. the massing and conceptual design of the proposed building;
   b. the relationship of the proposed building to adjacent buildings, streets, and exterior areas to which members of the public have access; and
   c. the provision of interior walkways, stairs, elevators, and escalators to which members of the public have access from streets, open spaces, and interior walkways in adjacent buildings.[22]

A development-control bylaw usually requires proponents of a project, which occurs in a specified development-control district, to submit their application and plans for review, especially for commercial, industrial, and apartment projects. Many places have specified their entire community as subject to development control. The applications may be reviewed by a committee or a designated official on whose recommendation a development agreement or contract is drawn up, specifying the conditions that the developer is legally bound to fulfill in the actual construction process.

The original purpose for which development control was introduced into Canadian planning—the bridging of plan-making and zoning—still exists in most provinces. In specific areas, it may be necessary or desirable to reconsider present zoning and planning policies. Development control can be invoked on a limited-term basis under what are often called "interim control bylaws" or, more commonly "holding bylaws." This tool, a hybrid of zoning and development control, must specify the area, the duration of time for which the bylaw applies, and the types of changes that will be allowed in that period. Usually only minor changes will be allowed, according to existing zoning regulations.

The availability of development-control powers opened up avenues of plan implementation beyond those envisioned for it originally. The technique enables a community's plan-makers to deal with developers and their projects on a *case-by-case* basis, rather than on the uniform basis found in zoning. Development control, then, means that a community can enter into *negotiations* with regard to each project and impose individualized conditions for it within the limits of the existing zoning. This approach "provides and encourages flexibility and variety in development that zoning would not allow," says planning lawyer Stanley Makuch, because "different developments create different demands."[23] The advantage of planners being able to directly influence a project's outcome is very attractive. At the same time, it underlines the role of the community plan in providing guidelines for such negotiations. Many larger communities have established site-plan control guidelines to suggest local norms and establish a framework for staff review. This discretionary approach is prominent in much of community planning today.

Redevelopment Plans

In most communities, districts develop in a consistent, amenable way: buildings and uses are compatible, street patterns are regular, and the renewal of properties follows acceptable community norms. But there are two typical instances where this fails to happen: one is in older areas of a community that deteriorate without benefit of rejuvenation; the other is when inadequate or incompatible development takes place in the fringe areas of communities, despite zoning regulations.

There have been many versions of city-rebuilding programs, from the urban renewal of the 1950s to the neighbourhood-improvement programs of the 1970s. The latter were federally financed programs and, although such programs no longer exist, the planning for redevelopment continues (see also Chapter 5). One latter-day example of redevelopment planning is Benny Farm in Montreal, where a post–World War II public-housing site was redesigned and rejuvenated.[24] Another is the Woodward's project in a blighted downtown area in Vancouver, in which an old department store and adjacent buildings are being redeveloped in market housing (condos) and non-market (social) housing, shops, educational faculties, and public spaces.[25]

Central to old and new redevelopment initiatives is the action of designating a specific area for redevelopment and preparing a plan for its rejuvenation. There is an essential difference between this type of planning approach and that of zoning, whereby the community can only encourage development on privately owned parcels. Under a **redevelopment plan,** the community can intervene directly in private property and, indeed, take the initiative for improvement away from the landowner by acquiring land and removing or replacing buildings. Provincial planning acts delegate the power to prepare redevelopment plans to municipalities, under the condition that the plans are conceived within the concept and objectives of the overall community plan.

Terminologies differ among planning acts: in Ontario the redevelopment plan is now called a **community improvement plan,** while in Alberta it is called an **area redevelopment plan.** The scope of a redevelopment plan is well illustrated in the purposes contained in the City of Edmonton's *Planning and Development Handbook.*[26] An Area Redevelopment Plan (ARP) may designate an area for purposes of "preservation or improvement of land and buildings, rehabilitation of buildings, removal of buildings and/or their reconstruction or replacement; and the relocation and rehabilitation of utilities and services." In Edmonton, to continue the example, an ARP may address such topics as urban design, social and community development, transportation, historic preservation, and environmental protection, as well as conventional planning issues of infrastructure and physical-development patterns. These planning prescriptions are very similar to those found in Canada's earliest planning acts. This no doubt reflects a continuing concern that some physical development may be of poor quality from the outset and/or that the land use is now obsolete and the need for public intervention persists to set it on a proper course.

Redevelopment plans are usually thought of in relation to older, blighted, high-density ("slum-type") areas in large cities, but smaller cities may also have areas that require redevelopment. Cornwall, Ontario, faced such a situation when a once-thriving neighbourhood went into decline with the closure of a large factory. Consultants from

McGill University were called in to prepare strategies for urban and architectural rehabilitation to preserve the "Le Village" neighbourhood.[27] Two other redevelopment situations frequently occur in cities. One is on the fringe of a community that may have a combination of scattered country residential acreages, junkyards, and strips of drive-in services of various kinds. The aim of a redevelopment plan would be to achieve more intensive development that is more efficient for providing public utilities and roads, or for aesthetic reasons, or both.

Another situation for redevelopment occurs with the decline of heavy manufacturing industries whose sites, called "brownfields," may have potential to accommodate other uses. These sites are often large, easily cleared and strategically located, say on a waterbody, but due to the nature of the manufacturing processes that were used may also be highly polluted. Well-known examples of this situation include False Creek South in Vancouver, Le Breton Flats in Ottawa, and the area of the tar ponds associated with the former iron and steel mill in Sydney, Nova Scotia.[28] Some communities have successfully redeveloped such sites while many other sites still languish.

Community Design Tools

Conventional Canadian planning practice in the 1970s and 1980s focused upon land use and avoided aesthetic issues. But, as we saw in Chapter 5, concern over community appearance revived in the late 20th century with urban design units added to planning departments in most large cities and in some of the leading suburban municipalities. The urban-design toolkit available for planners varies widely from city to city, but some of the more common policies are **design guidelines, view planes, design review,** and **special districts**. In addition, form-based codes have begun to emerge in some districts.

Surprisingly, some of these urban design approaches can be implemented using standard land-use planning tools, simply re-calibrated to create alternative built environments. For example, zoning bylaws were used to create the skyscraper canyons along Manhattan's avenues, and also to replicate Don Mill's characteristics across Canadian suburbs. Zoning can also be used to implement mixed-use urban precincts, New Urbanist neighbourhoods, and gated communities. The key point is to develop a strong urban-design concept and then prepare tools to implement it. Mixing and matching guidelines and codes from various cities rarely leads to good results.

Design Guidelines

Design guidelines are a combination of text, graphics, and images that explain an urban-design concept aimed at shaping public space or built form (see Figure 13.5). Typically, they address issues such as the design of public spaces and streets, the limit to the location of buildings, and the exterior design of buildings. While public agencies may prepare guidelines for public spaces and streets, other design guidelines have no legal basis, since most provincial planning acts do not allow municipal governments to regulate the external design of private buildings unless they are within a heritage district. However, some municipalities prepare guidelines as an indication of intent, especially after consensus is reached in a public-participation exercise. Design guidelines are sometimes included in secondary plans or business improvement area plans to indicate the standards that will be used to evaluate requests for a rezoning to increase density or to receive public subsides for redevelopment. Finally, both public and private landowners may attach design guidelines to requests for proposals (RFPs) to develop their lands, especially if they have special design objectives for the site.

Figure 13.5 East of Bay (Toronto) Urban Design Guideline, 1990–2005

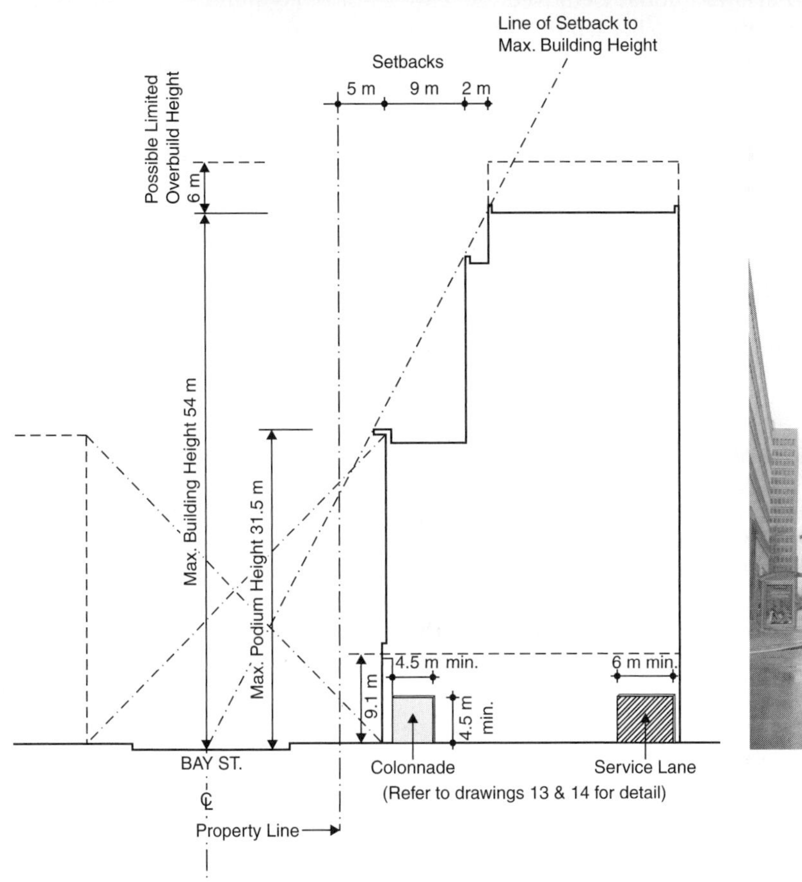

This urban design guideline for the Bay Street canyon in midtown Toronto was consciously modelled on the 60-degree angular plan in the 1916 zoning bylaw that controls building height on the north–south avenues in Manhattan. The image on the right shows the effect of the guideline on built form in 2005.

Source: David Gordon, The Kirkland Partnership.

Design guidelines were mainly used for urban areas in the 1970s and 1980s, often for projects in downtown districts. An example is the guidelines used to shape the "canyon" in Toronto's upper Bay Street, which were modelled on Manhattan's 1916 zoning bylaw (Figure 13.5). Other downtown design guidelines included protection of key vistas with view planes (see below) and access to sunlight.[29] Increasing interest in the design of suburban areas during the 1990s led to guidelines such as Calgary's Sustainable Suburb Design Guidelines.[30] The form-based codes recommended by some New Urbanist designers for American cities such as Austin, Texas, have not yet appeared in most Canadian municipalities, although Markham's Cornell zoning bylaw is a hybrid approach.[31] More recently, Smart Growth has been promoted using guidelines at a regional scale.[32]

View Planes

Throughout the practice of land-use regulation there is always the need to consider the balance between the development rights conveyed to a property owner and the rights of the community, which might be impinged upon by the actual development. Sometimes those latter rights may be infringed upon at some distance from the parcel being developed. A common case in point is the obstruction of views, which people in the community have traditionally enjoyed, by new tall or bulky buildings. Planners have become aware that the character of a community is very often grasped by people through the recurring views they have of landmarks, historic and significant structures, and special landscapes, and have developed legal mechanisms that specify the protection of selected views. Thus, communities can enact bylaws that limit building height and bulk in specific "view-plane" or "view-corridor" areas. **View planes** are three-dimensional and govern both height and breadth of view from a specified area. Halifax, for example, protects traditional views of the harbour and old parts of the city from the area of the Citadel that overlooks the harbour. In Ottawa, the National Capital Commission has regulations governing views of the Peace Tower, but has had only limited success in protecting them. Vancouver's city council decided to allow taller, but not bulkier, buildings in its downtown area in order to maintain existing view corridors to the mountain backdrop.

Design Review

Large public and private land developers sometimes appoint a master architect to review the preliminary designs of structures proposed for individual lots. The results can be commendable, such as the 1950's Modern aesthetic guided by the look of Don Mills (a private development) or the informal industrial look of Vancouver's Granville Island (guided by the Granville Island Trust). Similarly, the National Capital Commission has operated a design review panel since the 1950s for buildings proposed for federal lands in the Ottawa region.

However, few Canadian municipalities engage in a peer review of the detailed design of buildings proposed for private property because provincial planning legislation often does not permit such action. Architects and developers are wary of design review, especially when it is conducted by planners with little design training. The major exception to these trends is Vancouver, where the civic design review panel gradually achieved considerable local support during the 1980s and 1990s. The review panel makes judgments about the quality of proposed building designs and on whether projects meet the guidelines for density bonuses. The panel's members are appointed on recommendation of the design professions, and the planning staff who administer the process are also educated as architects. The results have been praised by international observers,[33] and similar design review panels were recently established in Calgary, Toronto, and Ottawa.

Special Districts

Traditionally, zoning took a simple view of city development and district classifications reflected this. Three general categories of land use were used—residential, commercial, and industrial—and within these only limited numbers of subcategories. Each of the districts—whether single-family, two-family, neighbourhood commercial, or light industrial—had standards for bulk, yards, and uses, standards that were the same no matter where in the city the district was located. As well, these standards ignored differences in character between one such district and another in the same category. Zoning practice now recognizes that many neighbourhoods in cities have unique characteristics or problems. Some may be historically important in the community, and some may be stable while others are not.

Special-district designation, which is now a standard part of zoning, is designed to cope with new and special needs in land use. One of the most common is the historic district. Another is the special designation of districts in which large institutional or public uses occur (such as hospitals, universities, and airports). Other zones to protect and encourage waterfront development are also common. One of the most widely used types of special district is that defined for environmentally sensitive areas, such as those subject to flooding or those having special ecological characteristics. Another widely used special district is the Business Improvement Area (BIA) that is used to promote rejuvenation of older commercial areas. The advent of special districts indicates not only that cities are too complex for the range of responses offered by simple land-use regulations, but also that we may have new perceptions of what is important in land use.

Transfer of Development Rights

As noted earlier in this chapter, zoning defines for each property a building envelope in which the owner has a right to develop a structure of a specified volume or envelope—that is, there are "development rights" for each property. Several decades ago some communities began to allow an owner to sell the development rights on one property to someone who owned land located in an area where the community wished to encourage development, such as in an urban redevelopment area. This policy of **Transfer of Development Rights** (TDR) is now widely practised in large cities (see also Chapter 6).

TDR basically shifts density potential from one area to another. It aims at benefiting three interests. One is the owner, who can sell the development rights on a parcel on which he or she cannot or does not wish to use all of a building envelope, perhaps because there is already a satisfactory building on it. The purchaser, or transferee, of the development rights also may benefit by obtaining sufficient additional density to make a project economically feasible. The community could benefit, sometimes doubly, by achieving development where it had been planned, and by limiting development at the location from which the development rights were transferred. For example, in some cities TDR has saved historically important buildings from demolition and replacement. The process through which such transfers take place is one of negotiation between property owners and city officials. This and other forms of negotiation in the planning process are discussed in Chapter 12.

While TDR can be useful in meeting urban-design objectives in cities, this tool can also be used to implement agricultural or ecological-planning objectives in rural areas. Conservation organizations have purchased the development rights of private wetlands and forests using legal agreements in the past. But municipalities need the power to transfer rights from developable lands to link up green corridors and patches as recommended by the principles of landscape ecology.

Planning Tools for Vacant and Undeveloped Lands

A characteristic of the built environment of cities and towns is that its development usually takes place on individual parcels of land. These parcels are created by splitting up large tracts of land through a formal process known as **land subdivision.** The process is formal because of the need to register and certify land ownership with respect to these parcels. But the process is also part of a larger process of the economic land market—supplying developable parcels of land to the community. In the latter case, land subdivision is a competitive process by which each subdivider aims to attract development to his or her new parcels of land. It should be noted that land subdivision may also occur on those sites approved for redevelopment (as described above) and not just on vacant land.

One of the earliest difficulties confronted by community planners was achieving satisfactory town extensions as cities and towns expanded. Canada's first planning acts recognized that the creation of new parcels of land could have adverse effects on both the nature and direction of the future development of the community. There was experience with subdivisions in which individual parcels had inadequate access from adjoining streets, streets that did not align with or match the size of streets in already-developed areas, and building lots that were not drained properly or were not even of sufficient size and shape to create a pleasant, much less efficient, built environment. Since the community has the responsibility to provide public services and street access to all parcels of land, it has a direct financial interest in the proper subdivision of land for future development. It was the early planning acts that established this community interest to maintain consistent standards among competing land subdividers, and to create the kind of environment that was compatible with the community's aims and within the community's economic capability to maintain. Thus, emphasis was put on the "cost and quality" of the subdivision, as Canada Mortgage and Housing Corporation noted, three decades ago.[34] A contemporary version of these dilemmas occurs with the establishment of "gated communities," wherein private subdivisions are closed off to public access and local street patterns are disrupted.[35] This and other issues of gated communities are discussed in the next chapter.

Subdivision Control

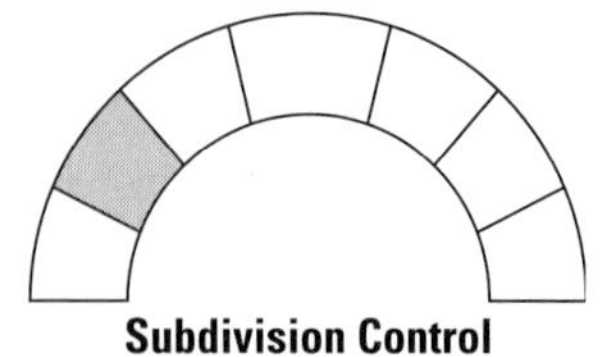
Subdivision Control

The tool developed by planners in response to the above need is known generally as **subdivision control**. In essence, it consists of the authority of the community to approve any plans for splitting up land for development to ensure that such plans meet local standards for health, safety, and convenience. In subdivision control, community approval is tied to the province's power over registration and certification of valid parcels of land for future sale. This enables planning authorities to exercise a high degree of control over the manner in which land can be used for residential building or other purposes, as well as over the time at which the land can be developed.

Provincial planning legislation vests subdivision-control powers in provincial or local authorities, and sometimes in both. The municipality gives "draft" approval and the province grants "final" approval to the registered plan of a subdivision. There is usually a provision stating that registered plans are not required where only one or two small parcels are being created, or where the new parcels are too large to be sold as building lots. Provisions of the latter kind are known as "consents to a land severance" and constitute a special (and not insignificant) form of land subdivision regulation, which will be dealt with in a later section.

Subdivision control has two basic components, one substantive and the other procedural. On the substantive side, this tool attempts to obtain high-quality built environments. It does this by subjecting plans that propose the subdivision of land to an appraisal of their content, according to planning and engineering standards. On the procedural side, subdivision control operates as a monitoring process, with prescribed steps and with respect to all those public bodies having an interest in the outcome of the proposed land subdivision and subsequent development. The latter formal side is necessary because of the constraints that this scrutiny places on the ownership rights of those proposing the subdivision and of the ultimate owners of the subdivided parcels alike.

It can be seen that subdivision control is a *process* type of planning tool (in contrast to zoning, which is legislative). Subdivision-control processes differ in detail from community to community and from province to province, with respect to both provincial planning institutions and concepts of subdivision standards. On the general level, the substantive and procedural components of all the provinces have much in common. We deal now with each of these components briefly, concentrating on residential subdivisions, which constitute the bulk of subdivision activities in most communities.

Subdivision Standards

Central to subdivision control is the examination of the actual design of the proposed subdivision and an appraisal of the standards and dimensions used in its layout. A subdivision plan, as a Newfoundland planning manual states, "is in itself a small planning scheme."[36] It is thus scrutinized according to planning criteria pertaining to the form and density of housing, street systems, open space, and essential community services. Since a subdivision is likely to involve modification of the land surface, various engineering criteria pertaining to drainage, road construction, and the installation of public utilities are also invoked in appraising a plan of subdivision.

Most provinces, in their planning legislation, establish the general factors that a proposed plan must take into consideration. In the Ontario planning act, a plan of subdivision must indicate the legal boundaries of all parcels, the location and widths of streets, the intended use of parcels, natural features of the site, physical features such as railways and highways, and the availability of water supplies and other municipal services. Such a plan, obviously, contains a good deal of detail and must be accurately drawn; subdivision plans are usually required to be certified by a licensed land surveyor as to the boundaries of the overall site and the individual parcels. This is called the **draft plan**, since it may still undergo revisions before final approval is given. The draft plan is also required to consider subdivisions, streets, and land uses on adjoining lands, as well as any zoning or other land-use controls pertaining to the site.

In addition to these basic formal requirements of the planning legislation, most provinces and many communities have manuals or handbooks that set forth design standards for subdivisions. They are part of a continuing effort to promote good-quality subdivision design, an outcome that is easier to advocate than to achieve. According to an Ontario review of subdivision experience, "much contemporary subdivision design is still poor to mediocre in quality" and there is a "pervasive utilitarianism" about most subdivision planning.[37] There are two reasons for much of the lacklustre design. The first is that, as with other aspects of civic design, subdivision plans call for special design skills and an awareness of what constitutes a good residential environment. Such skills are unfortunately not plentiful among many surveyors, engineers, and planners who lay out subdivisions: thus the need for guidelines and manuals. The second reason is the tendency of many subdividers to concentrate on the yield of lots from the site, often without regard for the natural topography, amenities, and rational circulation system. By the 1990s, many municipalities were working with developers to increase the gross density of conventional suburban development by reducing lot sizes and street widths and increasing the mix of townhouses and apartments. More compact development assisted in achieving many objectives for Smart Growth and sustainable development.

An example of the value of striving for good subdivision design can be seen in a typical situation illustrated for a 4.23-hectare site in Surrey, B.C. Figure 13.6(c) shows the design of a similar sized conventional suburban subdivision in an adjacent area. From the developer's perspective, the second plan, Figure 13.6(d), has a higher yield of units, and lower costs per unit for utilities. These factors should compensate for the lower number of big-lot single homes, which would sell quickly. From a sustainable-development perspective, the second plan is much better. It provides a greater variety of unit types, higher gross density (26 vs. 10 units per hectare) and the same proportion of impervious surfaces. Further analysis indicates building the entire East Clayton neighbourhood according to these standards (which was underway in 2006) should also reduce greenhouse gas emissions due to lower car ownership and more walking and cycling for non-work trips.[38]

Other factors that will be taken into account in appraising a subdivision plan include concerns over energy conservation, access for pedestrians, and availability of public transit. Subdivision design can, for example, provide layouts that allow homes to be positioned to maximize transit use,[39] decrease energy consumption, and employ natural features and vegetation to reduce the effects of adverse climatic conditions. London, Ontario, requires non-light-polluting, low-energy streetlights in new

Figure 13.6 Sustainable Subdivision Design, East Clayton Neighbourhood, Surrey, B.C.

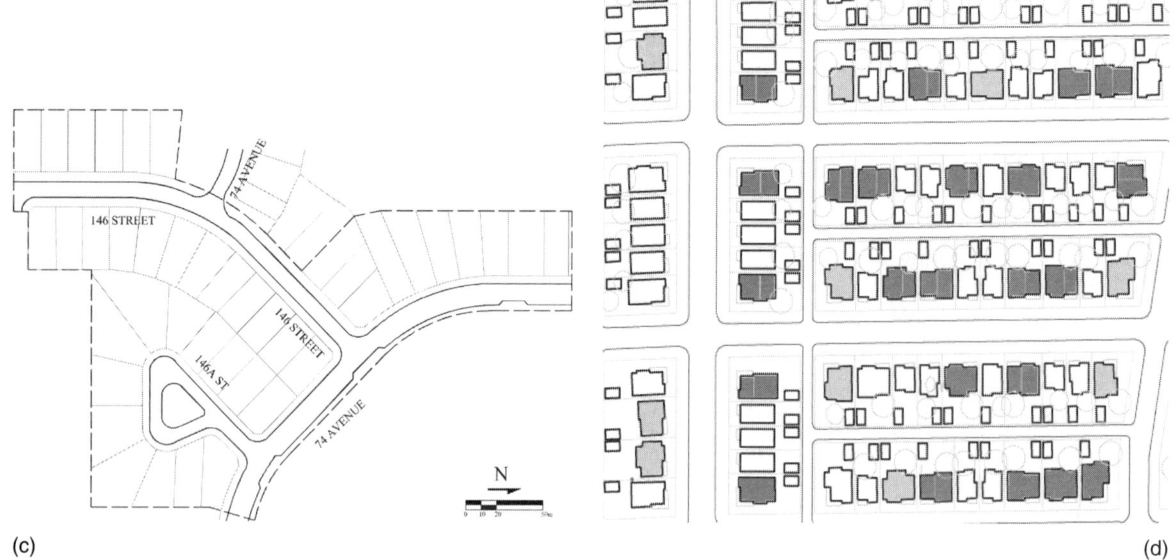

	CONVENTIONAL	SUSTAINABLE
Site Area	4.27 ha	4.23 ha
Total Dwelling Units	41	111
Gross Density	9.6 d.u./ha	26.2 d.u./ha
Site Permeability	50%	49.7%
Total Infrastructure Cost	$23 521 per unit	$11 005 per unit

Source: Patrick Condon, "Case Study: Status Quo Standards Versus an Alternative Standard, East Clayton," *Technical Bulletin,* December 2, 2000, University of British Columbia. Reprinted with permission from the author.

subdivisions, for example. Good subdivision design also stresses the provision of good-quality routes for pedestrians and cyclists. There should be direct and safe connections between dwellings and local activity centres (school, stores, library, church), and recreation areas and public transit stops. And wherever possible, pathways should have only moderate grades, so that they can be used by the elderly and those in wheelchairs. The latter criteria reflect planners' concerns that were first articulated in Perry's Neighbourhood Unit of the 1920s (see Chapter 2) and more recently reiterated by advocates of New Urbanism and Smart Growth (see Chapter 5).

There are clearly many factors that contribute to the ultimate quality of the residential environment that will result from a subdivision plan. No simple list can possibly encompass all of them; however, a set of questions that planners use in reviewing residential proposals will provide a basic overview of concerns (see Figure 13.7). Before leaving this discussion, the reader should note that where a subdivision plan is intended to accommodate commercial or industrial uses, analogous subdivision standards are applied in appraising it.

Subdivision Application Procedures

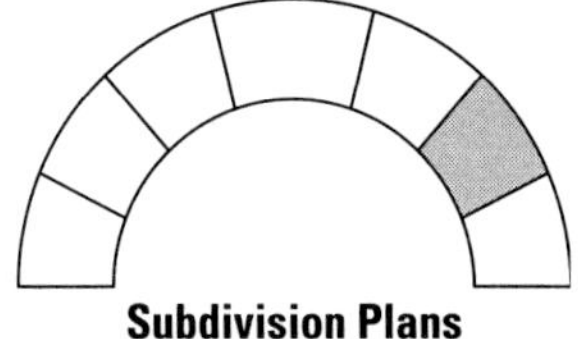

Subdivision control is a system that prescribes the way the subdivider applies for a permit that allows a tract of land to be split up and parcels sold off. Because of legal constraints, the province defines a procedure to be followed in applying for approval of the plan for a subdivision, as well as the prerequisites for such approval. Procedures differ among provinces, largely as a result of where the final authority for approving a plan is lodged. Some have reserved this power for the provincial minister in charge or for appointed officials. Others have delegated subdivision-approval powers to municipalities; still others involve such intermediate-level bodies as regional planning commissions or metropolitan governments. There are, however, five features common to all subdivision control procedures:

1. **Specified process.** In order to protect the rights of owners to subdivide their land, the process of applying for approval is specified. It indicates the form the application must take, the steps the plan's review will follow, and often the maximum amount of time the process will take.
2. **Plan circulation.** Since a proposed subdivision of land may affect the interests and activities of a wide array of agencies, a draft plan is circulated broadly for comment and recommendations. Typically, all provincial government departments with an interest (such as transportation, housing, energy, agriculture, health, and environment) are asked to review a plan. Also, public and private utilities, transportation companies, and special-purpose bodies are included in the review process, along with various departments of the local government.
3. **Conditions for approval.** The subdivider is considered responsible for the provision of the roads, parks, and public utilities necessary to serve the subdivision. Since they ultimately come under community ownership (except in gated communities), it is necessary to specify how they shall be paid for and the standards to which they will be built.
4. **Subdivision agreement.** In order to ensure that the basic services are provided for the subdivision, a contract or subdivision agreement must usually be signed between the subdivider and the municipality, and is often registered against the deed to the property. These agreements cover such matters as staging of development, provision of services, road standards, minimum construction and material standards, conveyance of lands for parks, and demolition and removal of existing buildings.
5. **Final plan.** The plan of subdivision is not operative until it has received final approval. Such approval is given only after any conditions imposed on the draft plan

Figure 13.7 Planning Checklist for Proposed Residential Subdivisions

REVIEW EACH SUBDIVISION PLAN WITH THESE QUESTIONS IN MIND:

Community Context

- ☐ Does the street layout conform to the existing street pattern?
- ☐ Does the plan relate well to existing community facilities?
- ☐ Does the plan conform to the policies of the community plan?
- ☐ Does the plan provide adequately for the education, recreation, and shopping needs of residents?

Residential Needs

- ☐ Do the proposed dwelling types suit community needs?
- ☐ Do dwelling arrangements ensure privacy?
- ☐ Will dwellings be affected by excessive noise, dust, or fumes?
- ☐ Is the size of lots adequate for the dwellings?

Streets and Parking

- ☐ Are the grades, widths, and intersections of streets adequate?
- ☐ Are parking provisions adequate?
- ☐ Is pedestrian access separated from the street?

Public Services and Utilities

- ☐ Has drainage been carefully considered?
- ☐ Are water and sewage facilities adequate and conveniently located?
- ☐ Is street lighting adequate?

Environmental Considerations

- ☐ Does the arrangement of lots and dwellings make the best use of the climate?
- ☐ Have trees been left to stabilize the soil?
- ☐ Has provision been made so that storm-water runoff does not pollute other water bodies?
- ☐ Has the amount of paved area been minimized?

Aesthetic Considerations

- ☐ Do dwelling arrangements result in attractive streetscapes?
- ☐ Will street furniture be provided?
- ☐ Are natural features of the site incorporated in the design?

from the review process have been fulfilled and a subdivision agreement exists to secure the necessary services and standards. At this point, the plan may be registered and will be binding on all future landowners. No additional demands can be made on the developer of the land.

The subdivision plan-review process is the community's opportunity to have a direct effect on the outcome of the development of a portion of its built environment. There are, as well, two vital elements the community should be looking for in reviewing plans of subdivision, regardless of whether it is the final approving authority. The first is to determine the proposal's *compatibility* with the aims and design envisioned for the area in the community's overall plan. In conjunction with this, the community reviews the plan's compliance with zoning regulations that already apply to the area, as well as the possibility that amendments to both the zoning bylaw and community plan may be required to accommodate the subdivision. The second facet that the community will need to be concerned with is whether the subdivision is *premature* or not. An approved subdivision makes demands on the community's resources, in that services must be provided, roads and parks maintained, and so on, whenever the land is built upon and people come to live there. It is therefore in the community's interest to know the timing or staging of the actual development. In most provinces, a community can establish conditions regarding the staging of development on the site to coincide with its own plans for investments in services as well as with its desires to achieve a particular spatial pattern of development. For example, where urban growth boundaries are in effect, the subdivision's conformance with them will need to be checked.[40] In the larger sense, subdivision control becomes part of an integrated planning process, rather than being an isolated act affecting only one proponent and one area.

Consents and Severances

A special form of land subdivision, particularly for rural areas, is the severing of one or two small parcels of land from a large tract such as a farm. A subdivision plan is normally not required for this purpose, but since it is necessary to register the new parcel and convey title properly, a formal process is required. It is known as **granting consent to the land severance,** or, in popular terms, simply as a "consent" or "severance."

This process was, until recent decades, mostly used to allow farmers to create a homestead site for themselves or for a close relative also involved in the farming or other rural activity, and to create occasional lots for summer cottages. However, with the increased demand for year-round country residences and cottages after 1950, obtaining a consent became a source of income to rural landowners. It also became a source of problems for rural communities by creating scattered and ribbon development with consequent difficulties and costs of providing services and road maintenance. In recreational regions, increased numbers of severances for cottages often led to the pollution of lakes. Indeed, the ills associated with excessive use of rural severances led to the popularization of the term "sprawl" to characterize development in fringe areas of cities and in rural areas and, more recently, to efforts to promote Smart Growth (which will be discussed in the next chapter).

Consents must be formally approved, often by the same approving body that processes subdivision plans. A consent is usually required to meet the same conditions as a subdivision plan with regard to dedicating land for streets, road widening, and parks. However, beyond these basic requirements, provincial planning legislation is noticeably silent on the planning criteria that should be considered. Some efforts were made by local approving authorities in Ontario in the 1970s to establish criteria for consents—for example, conformity to community plans, avoidance of high-grade agricultural lands, and adequacy of waste disposal to prevent pollution of nearby bodies of water. The results can be characterized as inconsistent, even in these areas.

An Ontario report calls consents "the Achilles' heel of planning."[41] And a probe of the issue in New Brunswick cites the lack of a planning system that could provide for "suitable consideration of the overall implications" of individual land-severance proposals.[42] While individual applications for a consent may be justified, it is the accumulation of consent decisions that needs to be considered. Two provinces, Quebec and British Columbia, appear to have made some headway in reducing rural consents, especially where agricultural land is concerned. British Columbia established its Agricultural

Land Reserves in 1973, and Quebec established its "protected areas" under its Commission de protection du territoire agricole in 1978. Although there has not been a halt to the subdivision of agricultural land as was intended by these government interventions, it is clear that it has been greatly reduced.[43]

Replotting Schemes

A special form of subdivision planning that has received much attention from planners in western Canada has to do with salvaging subdivisions that attracted little or no development. The practice is called **replotting,** because it involves redesigning the street system, the pattern of building lots, and the open space of a pre-existing legal subdivision. In many communities in the past, the promise of rapid growth often led to considerable subdivision of land in excess of actual growth (see Chapter 4). These subdivisions usually predated planning and subdivision-control practices and, not infrequently, were inefficient and unimaginative gridiron designs bearing little relation to topographic features or existing community patterns. Many of these older, undeveloped subdivisions were in fringe areas, and, as communities expanded, especially after World War II, they were found to be incompatible with present-day planning and subdivision standards.

Provisions were incorporated into planning legislation to allow communities to, in effect, redesign these old subdivisions. A replotting scheme in Alberta involves three steps.[44] First, the existing subdivision must be cancelled and all parcels consolidated. Second, a new design is formulated by the community and registered. Third, the newly subdivided land is redistributed among affected landowners. Replotting offers a number of advantages to a community: more efficient street patterns and utility service, more pleasing residential settings, and higher tax returns from land that was previously under-utilized.

Condominium Subdivisions

A type of subdivision of property that is relatively new in the realm of planning is that of condominium ownership, or "strata-title," as it is called in some provinces. Essentially, it denotes conveying ownership to a housing unit without conveying title to the land on which it sits. The latter, the site, is held in common by the owners of all the dwelling units sharing the site. The typical forms of **condominiums** are row housing and apartment buildings, although this also occurs in commercial and industrial development projects. The main advantages are the sharing of common community facilities and having no responsibilities for site maintenance, which is provided by a corporation formed by the owners.

Because creating a condominium project results in splitting up a larger piece of property among several owners, most provinces have developed means to ensure property rights and consideration of planning standards. Condominium legislation has been developed to cope with the special features of ownership, and existing subdivision-control legislation and regulations are used to allow examination of the planning implications of the project.

Condominium development may take the form of either a brand new building(s) or transference of the single ownership of an existing building to shared ownership. The latter approach, **condominium conversion** of existing multiple-unit rental buildings, has received special attention from planners because it neither creates any additional dwelling units nor assures present residents of the continuation of their housing. While conversion is popular with owners of older apartment buildings, it has frequently proved contentious when tenants are forced to move as buildings are upgraded to attract higher-income purchasers. Elderly and/or low-income tenants are often caught in such conversions and must try to find accommodation in a rental housing market that, in many cases, is already tight. Since condominium conversion may actually reduce the stock of rental units in a community, many cities have enacted conversion policies that limit conversions where the rental vacancy rate is abnormally low and where present tenants are not going to be the subsequent condominium purchasers in the project. The experience in Montreal is a good example of the difficulties that surround this issue.[45] With over three-quarters of city households being renters, it searched for 20 years for a solution. "Benchmark" rents were finally established in 1993 so that conversions could not occur in apartment buildings with rents below this level, the purpose being to protect low-income tenants, including university and college students. Even this can produce a dilemma for municipalities that often enjoy higher tax returns from the refurbished apartments.

Fine-Tuning Regulatory Tools

The main tools of land-use regulation used by planners—zoning and subdivision control—are founded on the principle of being universally and uniformly applicable in regard to the properties and the landowners in the community. In other words, the regulations that will affect the development of a property apply equally to all similar properties and are known to the owner beforehand. This means that land-use regulations are formulated in *general* terms to cover normal configurations of site, location, and property characteristics. But among the numerous parcels of land and kinds of projects to which these must apply, it is evident that not all will have similar characteristics. Moreover, the regulations may cause hardship for some owners, or result in anomalous land-use arrangements for the community. Thus, a number of planning tools have been developed to respond to special circumstances of properties and land uses not adequately covered by the general regulations. The most common of these tools for fine-tuning what occurs in the built environment are described briefly below. In effect, they refine the impact of zoning.

Variances

Zoning bylaws cannot cover all the distinctive situations that might affect each of the parcels of land within a zone. There may be unusual conditions of topography, size, shape, and location that adversely affect the development of a parcel in the manner envisioned in the regulations. For example, a parcel with a steep slope may not allow, without undue cost, a building to be sited to meet setback requirements, or the specific land uses applying to a property in a commercial zone may not cover a similar use that is being proposed. To avoid creating situations of special hardship, planning legislation allows adjustments in bylaw provisions to be considered in regard to a *particular parcel* of land. One type of adjustment is called a **variance**, since it allows the provisions of the zoning bylaw to vary from its stated terms.

Variances introduce a needed degree of flexibility in the application of a zoning bylaw, which otherwise might require a formal amendment. The variance does, however, operate in a delicate area of judgment, wherein it must be decided whether a *major* hardship or difficulty is created by the regulations and, moreover, whether in relaxing the zoning standards there might be adverse effects on adjoining properties and the community environment. Since the variance that is allowed is expected both to create only minimal effects on surrounding properties and to convey no special advantage for the applicant, it is usually referred to as a "minor variance." Criteria defining what qualifies are minor have never been well established, which has led, at worst, to numerous abuses of the variance and, at best, to controversy over many decisions.

The granting of a variance, because it affects property rights and also involves an appeal from the requirements of legislation, is a formal process spelled out in provincial legislation. It is a process that usually devolves to the municipality in the form of a special appeal body. The names of these appeal bodies differ from province to province. In Manitoba, the appeal body is called the Variation Board; in Saskatchewan, the Board of Zoning Appeal; and in Ontario, the Committee of Adjustment. These bodies are usually required to give notice of the application for a variance to all property owners within the vicinity and to provide reasons for their decision, which is ordinarily final and binding. In some communities, the municipal planning staff coordinates the activities of the appeal body, but in most no connection with local planning efforts is attempted or, possibly, even recognized.

Non-Conforming Uses

A zoning bylaw is essentially future-oriented, but in already-developed areas, its standards are not likely to fit all the uses and structures that existed before it came into effect. The concept of natural justice prevents the violation of property rights that existed before the bylaw was passed. Thus, there may be uses and structures that do not conform to the standards in the new regulations but that will legally be allowed to continue. It is customary to refer to them as **legal non-conforming uses**. There are actually three forms of pre-existing conditions that may not conform to the current bylaw. One is the *use* to which a property is put, such as a store in a residential neighbourhood; another is a *structure* that may be situated so as to not provide the required yard space; and the last is a *lot* that is smaller than would now be required.

Non-conforming use status copes with anomalies between the standards of earlier development and the standards now favoured by the community in its current bylaw. However, if the community wants to achieve its new standards, then there must be a way of bringing the earlier development into conformity with them. Thus, most zoning bylaws have provisions that prohibit or limit any expansion of the non-conforming structure and limit the non-conforming use to what it was at the time of the new bylaw, as well as provisions for rebuilding in the event of the building's destruction by, say, fire. In this way, the aim of the community to have the non-conforming uses replaced in time is balanced against vested property rights. The record of non-conformities disappearing is not salutary in most cities, and there are many instances of applications to enlarge such uses. An application for a change in a non-conforming use is often handled by the same zoning-appeal body that deals with variances, and the procedures are much the same.

Planned Unit Development

Many communities have included in their zoning bylaws a means by which the requirements of normal zoning districts may be modified and more innovative standards incorporated in the plan for a project. This approach, which is generally called **Planned Unit Development** (PUD), offers the developer an option to build within a set of requirements established especially for the project, rather than in strict conformity with existing regulations. The incentive for the developer in PUD may lie in achieving a few more dwelling units, in being able to mix several kinds of dwelling units and even some retail uses, or in producing a better quality and hence more marketable project. The community, of course, also wants the better design and offers, through PUD, to relax such standards as height, yard size, lot size, and dwelling type in return for working along with the developer to achieve a mutually acceptable project that has been planned as a unit.

The most noticeable result of PUD is that more common open space is available to residents than would occur if normal zoning regulations were followed. This is often obtained by foregoing the requirements for side yards, which are usually under-used, reducing the width of lots, and allowing dwellings to be built with no intervening space. This approach is called **zero-lot-line** and offers the opportunity to stagger, cluster, and group dwellings instead of using the traditional form of subdivision with standard setbacks. There may also be noticeable savings in land costs for each dwelling, thereby providing more affordable housing. This, along with efficient use of utilities, increased transit demand, and reduced automobile usage are attributes we now find in the Smart Growth strategy for planning. The Alberta Planning Act of 1977, for example, quite explicitly promoted PUD by permitting municipalities to apply to have an Innovative Residential Developments Area established. The success of PUD depends on negotiations between municipal staff and the developer. The Comprehensive Development Zoning approach used by Burnaby, B.C., and referred to earlier in the chapter, works essentially the same way as PUD by requiring that a comprehensive plan detailing the project be submitted prior to rezoning.

Rezoning and Plan Amendments

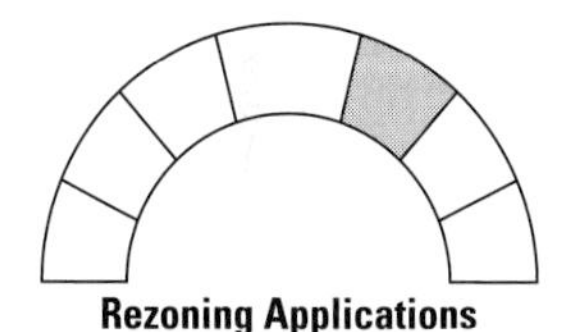

The outright revision of a zoning bylaw, or of a community plan, for that matter, can be seen as a means of adjusting to special needs and changes in the community environment. The reason for not considering revision sooner in this discussion of the myriad of adjustment tools is that it represents a more far-reaching step, involving a broader public interest. In the opinion of most people involved in conceiving and administering plans, this is a step that should not be taken lightly. Provincial legislatures require formal steps to be taken for an **amendment,** steps that are much the same as those for the original enactment of a zoning bylaw or community plan.

The extra statutory hurdles in the amending process, as compared to the other forms of adjustment, reflect the need to provide security in the continuance of major commitments to planning decisions. In the case of zoning bylaws, particularly those of long standing, property owners come to rely on the district standards and have a right to expect their continuance unless a major change can be justified on planning grounds. (Indeed, it is not unreasonable to call amendments "major variances.") For the community plan, the rationale for amendment should be more substantive, such as a major change in growth patterns, the introduction of new forms of transportation such as a rapid transit line, the establishment of large, new land uses, or the disposition of the community to alter its course of development. A change in a community plan often necessitates an amendment to the community's zoning bylaw, and a proposed bylaw amendment may signal the need to consider a plan change. Good planning practice makes these two initiatives interdependent, and some provinces require that the implications for each be considered simultaneously. Amendments of bylaws and plan changes should be possible when the planning considerations are substantial and the community as a whole is in agreement with making the change.

Reflections

Land-use regulation definitely falls on the process side of planning. The various statutory controls described above are themselves not plans of desired development, but rather the means by which the community's planning objectives are linked to the development process—that is, linked to the aims, inclinations, and decisions of those individuals, firms, and organizations that may wish to develop land. Mostly, land-use regulations are intended to shape the plans of prospective developers in the private sector, but they also affect public bodies, including the community itself.

In their basic forms, like zoning and subdivision control, land-use regulations are negative in their approach. They function, essentially, to establish minimum standards for development. But while they can be effective in preventing development considered undesirable, these regulations are relatively powerless in obtaining desirable development. In other words, they inform the developer of what is minimally acceptable, but the actual decision to develop land at or above these standards remains with the developer. A few new approaches, such as Planned Unit Development, Transfer of Development Rights, and Comprehensive Development Zoning, offer *positive* incentives to developers to work with planners in return for relaxing burdensome aspects of the regulations. Development-control approaches are another way in which the community's planners can actively work with developers to secure the most amenable projects.

Land-use regulation is a process in which land developers respond to the standards set by community officials. In this sense, it is a reactive process for the community's planners: they must await the initiatives of the developers. Further, the community's "planners" in this realm are diverse. They include the professional planners on staff (or consultants), planning boards, zoning boards of appeal, the local council, regional or district planners, and provincial planning ministries, as well as other government bodies and public utilities. All of these may be "actors" in one area or another of land-use regulation, depending upon the prescribed procedures. It is important to appreciate that each actor is able to influence the outcome of the planning goals of the community; in effect, each participates in implementing the community plan. Thus, the need for promoting active roles for participants to secure mutual learning (i.e., mutual respect and appreciation) is self-evident.[46]

Making land-use regulation effective in the attainment of a community's planning goals is a complex task. Coordination among the diverse actors is essential to assure that the various tools are used in a consistent fashion. The two main components of this task are (1) the presence of an overall community plan, and (2) an organizational focus for planning within the community. The plan, or at least a clear statement of planning objectives, provides the benchmark against which the various actors can assess the impact of the decisions they must make. But there must also be an explicit commitment within the community to the plan by way of some person or group whose responsibility it is to uphold it. This could be the director of planning, a planning committee, the council, or some combination of all these. The approach that needs to be taken will involve both monitoring all activities in the regulation of land use (such as subdivision proposals, applications for variance, etc.), and influencing the deliberations of various bodies (through representations, analyses, etc.).

Land-use regulation, as noted at the outset of this chapter, is only one of the major tasks involved in implementing a community plan. There are other areas in which the community influences the outcome of development in its built environment by policies and programs it institutes such as through the capital investments it makes for roads, and so forth. These policy tools are described in the next chapter and generate consideration of this question:

- —***In what ways can land-use regulation tools be coordinated with investment policy and program tools to enhance implementation of the community plan?***

Endnotes

1. John Punter, *The Vancouver Achievement: Urban Planning and Design* (Vancouver: UBC Press, 2003).
2. Thomas H. Logan, "The Americanization of German Zoning," *Journal of the American Institute of Planners* 42:4 (October 1976), 377–385.
3. Raphael Fischler, "Health, Safety, and the General Welfare: Markets, Politics, and Social Science in Early Land-use Regulation and Community Design," *Journal of Urban History* 24:6 (1998), 675–719.
4. Elizabeth Bloomfield, "Reshaping the Urban Landscape: Town Planning Effects in Kitchener/Waterloo, 1912–1926," in G.A. Stelter and A. Artibise, eds., *Shaping the Urban Landscape* (Ottawa: Carleton University Press, 1982), 256–303.
5. Richard Dennis, "'Zoning' before Zoning: The Regulation of Apartment Housing in Early Twentieth-Century Winnipeg and Toronto," *Planning Perspectives* 15:3 (2000), 267–299; Walter Van Nus, "Towards the City Efficient: The Theory and Practice of Zoning, 1915-1939," in A. Artibise and G. Stelter, eds., *The Useable Urban Past: Planning and Politics in the Modern Canadian City* (Toronto: Macmillan, 1979) 220–246.
6. Ian MacF. Rogers, *Canadian Law of Planning and Zoning* (Toronto: Carswell, 1973), 120.
7. Cf. John C. Weaver, "The Property Industry and Land Use Controls: The Vancouver Experience, 1910–1945," *Plan Canada* 19 (September–December, 1979), 211–225.
8. Cf. Richard F. Babcock, "Zoning," in Frank S. So et al., eds., *The Practice of Local Government Planning* (Washington: International City Managers' Association, 1979), 416–444.
9. Cf. Rogers, *Canadian Law*, 119.
10. A comprehensive review of Canadian experience with mixed-use areas is found in Jill Grant, "Mixed Use in Theory and Practice: Canadian Experience with Implementing a Planning Principle," *Journal of the American Planning Association* 68:1 (Winter 2002), 71–84.
11. Ken Greenberg and Frank Lewinberg, "Reinventing Planning in Toronto," *Plan Canada* 36:3 (May 1996), 26–27.
12. Kenji Ito, "Comprehensive Development Zoning," *Plan Canada* 37:4 (July 1997), 6–11.
13. Brenton Toderian, "Big-Box Retailing: How Are Municipalities Reacting?" *Plan Canada* 36:6 (November 1996), 25–28.
14. Alan Demb, "No Place Like Home: Legalizing Home Occupations in Toronto," *City Magazine* 14:2 (Spring 1993), 8.
15. Toderian, "Big-Box Retailing."
16. Mohammad Qadeer, "Planning Approaches to Ethnic Enclaves," *Ontario Planning Journal* 19:6 (2004), 6–7.
17. See Nancy Smith, "Diversity: The Challenge for Land Use Planning," *Plan Canada* 40:4 (July–August–September 2000), 27–28.
18. See such critiques as Alex Krieger, ed., *Andres Duany and Elizabeth Plater-Zyberk: Towns and Town-Making Principles* (Cambridge, MA: Harvard University Press, 1991); and James Howard Kunstler, *Home from Nowhere* (New York: Simon and Schuster, 1996).
19. For the experience in Markham, Ontario, see David Gordon and Shayne Vipond, "Gross Density and New Urbanism: Comparing Conventional and New Urbanist Suburbs in Markham, Ontario," *Journal of the American Planning Association* 71:1 (Winter 2005), 41–54.
20. Cf. Jill Grant, "Mixed Use in Theory and Practice"; and Andrea Gabor and Frank Lewinberg, "New Urbanism," *Plan Canada* 37:4 (July 1997), 12–17.
21. Earl Levin, "Zoning in Canada," *Plan Canada* 7 (June 1957), 85–90.
22. The Planning Act, R.S.O. 1990, Ch. P.13. But the act specifically forbids the use of site-plan control to regulate "the colour, texture and type of materials, window detail, construction details, architectural detail and interior design of buildings . . . ," sec. s. 41 (4).
23. Stanley Makuch, "Planning or Blackmail?" *Plan Canada* 25:1 (March 1985), 8–9.
24. Mary Lamey, "Vision for Future Pays Off," *The Gazette*, Montreal, October 15, 2005, E1.
25. Helena Grdadolnik, "Woodward's Take Shape: 'Nothing Like it in North America,'" April 4, 2006, *The Tyee*, available online at http://www.thetyee.ca; and Elaine O'Connor, "Woodward's Nears Sellout: Plenty Buy into Downtown Eastside Revitalization," *The Province* (Vancouver, B.C), April 23, 2006, A4.
26. City of Edmonton, *Planning and Development Handbook* (Edmonton, 2000).
27. Avi Friedman and David Krawitz, "Retooling the Historic Neighbourhood of Le Village," *Plan Canada* 42:1 (January–March 2002), 24–26.
28. Tera Camus, "Sydney's Toxic Woes Widespread," *Halifax Herald*, August 6, 2001.
29. Peter Bossleman, Edward Arens, Klaus Dunker, and Robert Wright, "Urban Form and Climate: Case Study, Toronto," *Journal of the American Planning Association* 61:2 (Spring 1995) 226–239.
30. Robin White, "Designing More Sustainable Suburban Communities: Calgary's Approach," *Plan Canada* 36:4 (July 1996), 16–19.
31. Gabor and Lewinberg, "New Urbanism," 12-17; and David Moffat, "New Urbanism's Smart Code," *Places* 16:2 (Spring 2004), 74–77.
32. Brook McIlroy Planning + Urban Design, *Model Urban Design Guidelines* (St. Catherines, ON: Regional Municipality of Niagara, April 2005), available online at http://www.regional.niagara.on.ca/urban-design/pdf/pdfs/1.pdf.
33. John Punter, "From Design Advice to Peer Review: The Role of the Urban Design Panel in Vancouver," *Journal of Urban Design* 8:2 (2003), 113–135.
34. Canada Mortgage and Housing Corporation, *Residential Site Development Advisory Document* (Ottawa, 1981), 2.
35. Jill Grant, Katherine Greene, and Kirstin Maxwell, "The Planning and Policy Implications of Gated Communities," *Canadian Journal of Urban Research* 13:1 (2004), 70–88.
36. Newfoundland, Provincial Planning Office, *Residential Subdivision Design Criteria* (St. John's, 1975), 1.
37. Ontario Economic Council, *Subject to Approval* (Toronto, 1973), 65–66.
38. Patrick Condon, *The Headwaters Project—East Clayton Neighbourhood Concept Plan*, Research Highlights 62488 Ottawa: CMHC:, 2001); and Geoff Gilliard, "Surrey Shifts Sustainability Status Quo with Neighbourhood Concept Plan," *Plan Canada* 43:1 (January 2003), 13–16.
39. IBI Group, *Transit Supportive Land Use Planning Guidelines* (Toronto: Ontario Ministry of Transportation and Ministry of Municipal Affairs, 1992).
40. Cf. Sharon Fletcher, "Coping With Growth in the Regional District of Nanaimo," *Plan Canada* 41:4 (Fall 2001), 16–17.
41. Ontario Economic Council, *Subject to Approval*, 66.
42. Comay Planning Consultants et al., *A Study of Sprawl in New Brunswick* (Toronto, 1980), 96.
43. Evelyne P. Reid and Maurice Yeates, "Bill 90—An Act to Protect Agricultural Land: An Assessment of Its Success in Laprairie County Quebec," *Urban Geography* 12:4 (1991), 295–309; and Christopher Bryant and Thomas Johnston, *Agriculture in the City's Countryside* (Toronto: University of Toronto Press, 1992), 137–189.
44. Alberta, Municipal Affairs, *Planning in Alberta* (Edmonton, 1978), 25–28.
45. Arnold Bennett, "Montreal Debates Condo Conversion Again," *City Magazine* 14:2 (Spring 1993), 10.
46. Hok Lin Leung, "Mutual Learning in Development Control," *Plan Canada* 27:2 (April 1987), 44–55.

Internet Resources

Chapter-Relevant Sites

Planning Canadian Communities
www.planningcanadiancommunities.ca

Citizens Guide to Land-use planning
www.northstar.sierraclub.org/campaigns/open-space/land-use

Lehmans Zoning Trilogy
www.zoningtrilogy.com/links

Visual Enhancement of Zoning Bylaws
www.cip-icu.ca/English/aboutplan/zoning_nwm2.pdf

Zoning, City of London, Ontario
www.london.ca/Planning/zonebylaw.htm

Centre for the Study of Commercial Activity
www.csca.ryerson.ca

14

Chapter Fourteen

Policy Tools for Plan Implementation

The problem is to distinguish between two questions: What would we like to achieve? and How shall we best achieve it?

Earl Levin, 1962

When one examines the built environment of a community it is readily seen that it is a blend of private and public structures, facilities, and spaces. And while private development comprises the largest part of a community's land use, public development may be said to constitute the most strategic part of the land use. The framework of the public roads, utility lines, parks, libraries, fire stations, parking structures, community centres, and so on can be said to articulate the form of the community within which private development takes place. But both are needed to achieve the goals of the community plan.

Achieving the desired blend is a matter of combining land-use regulations and various other public-policy decisions that can influence the decisions of private developers and other public agencies. The previous chapter discussed the planning tools that establish the basic "ground rules" for the development of the properties of private landowners. Those regulatory tools influence and/or constrain the choices that landowners have when they choose to develop or redevelop land. However, neither the effectiveness of the tool nor the nature of the development is apparent until such choices are exercised—that is, until an application for zoning or a subdivision plan is made, and until physical development actually takes place. As important as those regulatory tools are, they are essentially reactive in nature. A community (with municipal status) can, however, be proactive in implementing its planning goals. On the one hand, it can strategically use decisions about the nature, location, and timing of its own development efforts. And, on the other, it has means by which it can influence the decisions of private developers and other public agencies such as school boards, public utilities, and senior levels of government. In short, a local government can strategically use its policy-making powers to achieve its planning goals. Thus, when implementing a community plan the following question needs to be considered:

- —*In what ways can a community use its public-policy prerogatives to influence and guide private and public actions to bring its planning aims to fruition?*

On Guiding a Plan toward Implementation

A community that makes a decision to plan cannot count on the necessary action being taken by landowners, or the local government itself, to make the plan come to fruition within any explicit time frame. While plan-making is aimed at stimulating and guiding the actions of both public and private decision-makers, ***planning*** and ***action*** are distinctive activities. In logical terms, planning is necessary but not sufficient to obtain the actions needed for a planned outcome. This dilemma has been recognized for many decades, but probably no more incisively than by American planner Martin Meyerson in his concept of the "middle-range bridge" for planning.[1] He perceived the dilemma that is posed because of the long-range aims of the community plan and the short-range perspective of most private development actions and public-action programs. To solve this, he advocated various planning tools to link the two. These tools would include, in particular, the budgeting and programming of the community's capital investment in facilities and land, as well as public incentive and support programs to stimulate private action and detailed plans for improving specific areas. His challenge has been taken up in many Canadian communities to develop a number of middle-range planning tools that can allow a community's planners to take the initiative to bring planning, policy, and action closer together.

It is clear by Meyerson's challenge that a community, having prepared a long-range plan, needs to move into a mode of short-range planning and action. Moreover, local governments have a considerable array of tools at their disposal. On the one hand, there are the various land-use regulations, as described in the last chapter, which can be designed to render land-use decisions consistent with the plan. And on the other hand, there is the ability to make and carry out policies toward the built environment both within the local government itself and toward the broader community. Two middle-range approaches—strategic planning and capital improvements programming—are key to strengthening a community's capacity to link long-range planning with short-range action. Each is a further planning endeavour by the community to ensure progress toward implementation of the community plan and are described below.

Community Strategic Planning

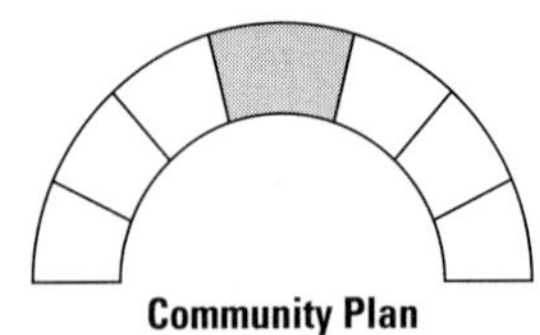

One of the newer middle-range planning tools that many communities now employ is called Community Strategic Planning. Strategic-planning processes, long used in business planning, have been adapted for community use. A strategic plan differs from the overall community plan in several ways: for one, it is cast in a shorter time frame; for another, it is aimed much more at implementing specific objectives; and for yet another, it is more closely oriented to changes in the community's context, or functioning environment.[2] Local government strategic plans typically have a five-year perspective and are reviewed and updated every year so that there is always a five-year planning perspective in place. Strategic plans are usually undertaken in conjunction with a community's overall plan and, thus, provide for a set of actions that comprise a multi-year portion of the long-range plan. Importantly, the community's strategic plan is attuned to changes in social, political, demographic, economic, and natural factors that could affect the plan's attainment and attempts to allow for responses to such changes.

A resource manual for Nova Scotia communities describes a strategic plan as answering the following questions:[3]

- Where are we right now?
- Where do we want to be in the future?
- What strategies will we need to implement to get there?
- What internal and external forces are operating that will hinder or help us achieve our long-term goals?

Strategic planning, as with long-range community planning, involves establishing goals and objectives, and this is frequently done on a broad community basis, as was done in the newly amalgamated City of Kingston in 2000.[4] The small city of Port Colborne, Ontario, conducted a similar process and, according to this community, its strategic plan "has made the journey ahead much clearer."[5] Strategic-planning processes often involve highly collaborative processes involving a wide array of community stakeholders from both inside and outside the local government.[6] In the Nova Scotia approach, a Strategic Planning Team, which needs to include a very broad base of stakeholders, is formed to prepare the strategic plan. Suggested are local elected officials, managers of civic departments, managers of other public agencies (school boards, utilities, transportation, etc.), representatives of the public, the business community, public-interest groups, civic employees, unions, and provincial and federal governments. It is important that all interests that play a significant role in the community are at the table, as for example, a university or major hospital. Such an array respects the fact that many stakeholders outside of the planning department of the community (including those in other local government departments) not only have the means, but often their own plans for utilizing their resources. It can also be appreciated that involving such a variety of people will complicate decision making on the plan, and consensus-building approaches will work best in the give-and-take that is necessary to reach a common ground (see also Chapter 12).

The City of Edmonton has practised strategic planning for several years as a way of implementing its overall plan, Plan Edmonton.[7] The general flow of decision-making in Edmonton is shown in Figure 14.1. A particular aspect of Edmonton's strategic planning is the preparation of a Servicing Concept Design Brief (SCDB). The SCDB is prepared by the civic planners and establishes "a general framework for municipal infrastructure,

Figure 14.1 A Guide to Municipal Decision Making, Edmonton

A community plan, such as Plan Edmonton, has a strategic role to play in the development and refinement of other planning instruments required by the municipality.

servicing, planning, and environmental requirements."[8] The SCDB is prepared for specified areas of the city as designated by the city council to define municipal servicing requirements in advance of new development or renewal. In this way, not only are landowners and developers informed of the city's intent, but the brief also identifies capital projects for the municipality's Capital Priorities Plan. A strategic-planning process is a valuable supplement to preparing the overall community plan and has been taken up in communities from Coquitlam, B.C., to Thunder Bay, Ontario.

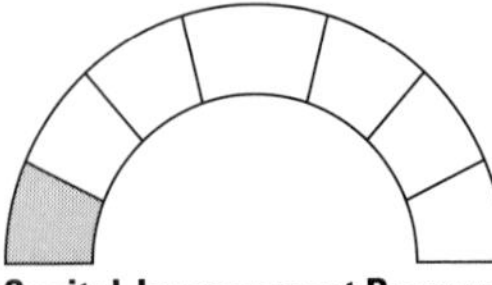

Linking Capital Spending to Local Planning

Each year, local public officials, both elected ones and their staffs, are in the position of making decisions about expenditures that may contribute significantly to the implementation of their community's plan. The most important of these are capital expenditures for physical facilities, for development incentives, and for special-project-area development. Most of these are part of the local government's annual budget and, thus, within the control of a municipal council to directly influence the outcome of much of its own plan. The local council usually sees the annual budgetary process as the most important activity it performs, because its choices allocate the community's normally limited financial resources among various activities, programs, and facilities. However, planning and budgeting are not synonymous, and unless a deliberate effort is made to link the two activities, a lack of coordination in the use of resources can hinder the plan. There is a need to develop procedures that keep the objectives of the plan, and the efforts of planners to implement it, tied to the budgetary process, an approach that is exemplified by the planners, politicians, and civic managers in Edmonton alluded to above.

Coordinating Capital Improvements

There is always a financial cost to a community in achieving its overall development plan. Even in a community that decides its future lies in maintaining the status quo, there will be upkeep and maintenance costs as facilities and streets wear out. In communities that choose to improve their built environments, city facilities will always need to be built, renewed, or upgraded to some degree. Communities that accept the growth of new subdivisions, office complexes, and other developments are destined to expand community facilities at some substantial cost to the local treasury. In each case, the question is not *whether* these sorts of costs will be incurred, but *when*.

The community plan, by projecting the best possible arrangement for development, determines what kind of improvements, or investments, are needed. The plan is, however, only a general guide. Development projects differ as to when they will happen, if private, and as to when they are needed, if public; the plan is not able to anticipate precisely the timing of development. Moreover, the availability of funds for public facilities and projects will determine when the community can afford to make the capital investment. Thus, as the community sets about implementing its plan, it is faced with choices as to which investments it will make from its always limited capital resources, and when it will make them.

This kind of activity, which arranges the expenditures for projects according to a time schedule, is called "capital programming." A form of programming in use for over a half a century in community planning is **Capital Improvements Programming** (CIP). Specifically, CIP is a means of linking physical planning with its long-term perspective and the budgeting of community expenditures with its annual perspective. The basis of CIP is the expenditure of funds for such physical facilities as roads, sewer lines, parking garages, and community centres. These sorts of projects are relatively costly as well as long lasting. Because their financing is often subject to borrowing funds with long periods of amortization, the community must decide on the level of debt that it can support each year. Further, because physical facilities are not easily changed once constructed, it is necessary to consider both the expenditure and the planning ramifications of each project.

Three basic definitions will help provide a context for CIP:[9]

- **Capital Improvements:** new or expanded physical facilities that are relatively large, expensive, and permanent, such as streets, libraries, fire halls, and water mains.
- **Capital Improvements Budget:** the expenditures for physical facilities to which the community is committed in the next fiscal year. This annual budget usually comprises the first year of the multi-year CIP.
- **Capital Improvements Program:** a schedule of proposed expenditures for public physical facilities covering a period of several years into the future, usually five years.

The primary objective of CIP is "to determine that set of projects for each time period which if carried out would provide the greatest 'product' over all future time periods."[10] In order to achieve this objective, CIP involves several important sub-tasks that must be blended together into a viable program to guide capital expenditures. In many respects, preparing the CIP is a process, like that of preparing the overall community plan. Although it deals with more details of development and its time frame is shorter, the CIP must reconcile competing ends into a composite "middle-range" plan of facilities. The cost of the proposed facilities must be within reach of the community's financial resources, and the facilities must contribute toward implementing the overall plan.[11]

The facilities and projects that the CIP will need to consider arise out of the activities of various functional departments and agencies of the community government—the roads department, the parks department, the library board, and others. Each group will have its own set of projects that it believes will enhance the development of the community. Rarely will the community be able to afford all projects at once. Moreover, it is likely that some facilities will need to be constructed before others (e.g., roads will precede a community centre). This leads to two basic decisions about each proposed project:

- What is its importance relative to other projects? and
- What is its desirable sequence relative to other projects?

Steps in the CIP Process

The CIP process involves a number of formal steps conducted by a specified office or department of the local government. Often it is the planning department that is assigned the role of soliciting proposals for capital improvement projects and assigning the priorities for construction. Typically, the CIP process is conducted on two different levels: a five-year program is prepared every four years, with the first year comprising the expenditures in the upcoming budget year (i.e., the capital improvements budget). Then the capital improvements program is reviewed annually to adjust the sequence of projects should major changes occur in financial resources or community needs in the projected time period of the program.

All communities conduct the CIP process according to their own organizational arrangements. The order of steps may not exactly parallel those listed below, but at some point each step will be taken in each community:

1. Analysis of financial resources. Projects in the CIP must be financed out of the community's available resources. Thus, projections of revenues and/or expenditures must determine how much money will be available in the current and subsequent years. The community will have expenditures for projects already built or authorized, usually in the form of bonds that have been issued. New projects requiring the community to borrow funds (issue new bonds) may have to wait until previous projects have been partially or completely paid for. And the operating costs of new and proposed projects after they are built can affect the allocation of revenues between operating and capital budgets in the future.

2. Requests for Proposed Projects. Each department involved in making capital expenditures in the local government submits proposals for the projects that they hope to undertake over the coming five years. Proposals must usually conform to a format that provides the planning department (or other CIP agency) with all the necessary information, enabling it to compare the merits of each project. The basic information to be supplied includes:

- A description of the project, its location, and when it is required;
- The expected cost of the project and when these costs would likely be incurred (possibly over several years if the project is large);
- The justification for the project and its relation to other projects in the same and/or other departments;
- The status of ongoing projects;
- The sources of funding in cases where other-than-local sources might be available; and
- The priority rating in comparison to other departmental projects.

3. Choices Among Functional Groups. The essential aim of CIP is to complete the development envisioned in the community plan for a given capital outlay. There is, however, no analytic system that allows direct choices among separate projects emanating from different departments, as, say, between a new sewer, playground, or fire hall. Also difficult is deciding on the projects according to specific years. It is usually recommended, first of all, to conceive of the five-year CIP as a whole and decide on those functional groups of projects that would best achieve the plan's objectives.[12] For example, projects may be grouped into such categories as transportation, residential development, industrial renewal, health care, recreation, protective choices, or whatever is appropriate for the particular plan and the planning period. Thus, some sewers or streets might be important for industrial renewal areas and others for new subdivisions.

4. Determining the Sequence of Projects. The timing of projects is central to CIP if the community's development sequence is to be both efficient and generally acceptable. Initially, when considering the sequence of projects, it is necessary to exclude those for whose services there will not be a demand until after the five-year CIP period. Then, the projects of each functional group are organized in terms of the order in which they need to be built. At this point, only a sequence of construction, without specific dates, is shown. Some projects need to be built early because (a) they will increase the service performed by existing facilities (e.g., additional sewage-treatment capacity); (b) they could cost much more if delayed (e.g., the acquisition of park space), and (c) they are strategic in influencing the form and pace of private development (e.g., the extension of water mains and arterial roads). There is, as well, the need to consider the advantage of linking such projects as sewer and street construction to avoid repeated road disruptions.

5. Setting the Priority of Projects. Choosing the projects to be built and those that will not be built in the time period is the most crucial step in the CIP process. To determine the priority of projects, it is necessary to go beyond their sequence and rate them according to their relative importance in the community's development. The first criterion is whether they can be achieved within the given capital budget. A second criterion is whether the project is specifically identified in either the comprehensive plan or some functional plan adopted by the community. Beyond these two criteria, the setting of priorities is difficult, because of local preferences and political considerations. Various ranking schemes have been developed to derive a measure of the relative importance of projects.[13] The final determination of which projects should be included combines their sequence and importance positions and, starting from the most important and progressing to the least, they are placed in time sequence until all the capital resources available for the five years are exhausted.

As with all planning decisions, capital improvement programming has a *technical* and a *policy* side. The process described above pertains mostly to the technical side—that is, the process undertaken by the community's staff of planners in analyzing capital needs and formulating a program. Both elected and appointed officials will use their political roles to influence the CIP. Department heads tend to favour projects in their own departmental area more than citywide issues. Elected officials often desire projects that favour particular areas or groups in the community. And often elected officials wish to keep capital commitments vague so as to be able to change their minds and respond to new needs. Thus, the relatively rational CIP process, therefore, cannot be considered apart from its political aspects, which, at their best, enhance the technical judgments and make the CIP more pertinent to community needs and goals.

Beyond these tensions, capital improvements programming has three major advantages:[14]

- It provides a link to the long-range plan by focusing attention on community goals, needs, and financial capacity when considering whether new infrastructure should be constructed and existing infrastructure should be replaced.
- It improves cooperation and communication among operating departments within the local government.
- It clearly indicates the community's investment intentions for several years ahead, so that private developers and other government agencies may better program their investments.

The City of Vancouver calls its CIP a Capital Plan, which it puts forward every three years in a public plebiscite.[15] The City of Edmonton calls its CIP a Capital Priorities Plan. In both instances, as with other communities that prepare a CIP, the process is closely linked to its strategic planning.

Policy Tools for Influencing Development

A local governing body has the special prerogative of making its voice known, and listened to, about how a community's growth and development should proceed. It does this by the *policies* it espouses. A policy decision is like saying "this is the route we should be taking and this is how we should travel it." The decision to make a community plan is a policy choice to improve the built environment using a planning approach rather than leaving future development to chance. Thus, a community plan, upon adoption, is a major statement of policy. And that policy document requires, in turn, further policy choices that reflect preferred actions to achieve its objectives.

Strategic plans and capital improvements programs further refine policy initiatives by pursuing particular objectives, facilities, and programs, or favouring particular areas. Through the latter tools a community may, for example, decide to use its resources to create low-cost housing or refurbish an historic district or attract private investment to an abandoned industrial area that the community plan has recommended. The policies it chooses to pursue also contain the means by which to achieve them. Broadly speaking, a community's policy choices revolve around the use of its resources. A community's resources cover its land resources, financial capacity, and its ability to use statutory instruments; any or all may be strategically important to achieve goals for the built environment. Using these resources may be limited by the provincial government (e.g., by prohibiting lower tax rates to attract business) and sometimes encouraged when the province (or federal government with provincial approval) wishes to achieve results through local government actions, such as area-wide transportation planning.

Financial Development Incentives

Communities often have financial resources available to influence decisions by private landowners and builders. This is done by providing financial incentives through grants and/or low-interest loans to encourage participation in a program favoured by the community. As an example, one of the most widely used incentive programs was the Neighbourhood Improvement Program (NIP) of the mid-1970s. Under NIP, homeowners and communities could receive federal funds in areas that were identified as needing upgrading in housing and community facilities. Homeowners could receive grants and loans to repair and improve their dwellings in such areas. The community could receive grants to assist in acquiring parkland, in improving streets and sewer lines, and in carrying out other activities that would promote residential renewal (see also Chapter 5).

Other similar programs include those to promote the development of low-income housing through municipal non-profit housing corporations, to rehabilitate unused upper storeys of downtown commercial buildings for apartments, and to renovate disused industrial buildings for commercial or residential uses. These programs vary in the way the resources are made available. In some cases, the municipality may offer the incentives directly, while in others the community may be asked to approve the award of funds by a senior government to ensure that the use conforms to local planning objectives.

Financial incentive programs may be strategically important to stimulating private development in previously uneconomic types of uses and in parts of the city avoided by developers. For example, older, often-blighted industrial areas, called brownfields, may be desirable to keep as employment areas in factories or offices. Toronto has such a policy and has developed a program of financial incentives for office and industrial uses that are willing to invest in designated brownfield areas. The program utilizes a tax-increment financing (TIF) approach in which the property owner and the city share, over a 10-year period, the increase in property taxes brought about by the new investment.[16] Further, the city's share of the tax increment is earmarked for public projects (such as parking, roads, landscaping) in the designated area. Other financial incentives for brownfields, or other areas requiring redevelopment, could include property-tax reductions, tax freezes, or cancellation of tax arrears. These tools help bring more certainty in regard to private development and investment that are relatively uncontrollable. Incentive programs may also be linked with capital investments by the community itself for a more concerted effort at rejuvenating the physical environment.

Public–Private Partnerships

A relatively recent approach to funding urban development projects is for a community to enter into partnerships with private developers. They are called public–private partnerships (P3s) and have emerged to blend private resources and expertise with government policies and need for facilities and services.[17] This approach arose in response, on the one hand, to shrinking public financial resources and, on the other, to political pressures for the privatization of services that government had previously provided. These partnerships tend to be focused on specific projects rather than on, for example, downtown renewal or neighbourhood traffic calming. One finds them used in the development of toll highways (such as Highway 407, which skirts the Toronto metropolitan area) and waterfront historic areas (such as Historic Properties in Halifax).

The first generation of Canadian urban public–private partnerships focused on redevelopment of publicly owned land, as described in Chapter 5. Projects included Granville Island and the recent Woodwards project in Vancouver, and Le Vieux Port in Montreal. The public owner would often solicit private-developer interest with a request for proposals (RFP), usually incorporating planning policies and urban-design guidelines for the site. Private developers would purchase or lease the land and agree to follow the appropriate policies.

Almost any public facility that a government body constructed and has operated—including hospitals, bridges, and airports—are now being considered for such partnerships. The essential arrangement is that a private body (a for-profit corporation or non-profit group) provides the capital to build or buy a facility the governmental body wishes to provide to the community or region, and the governmental body establishes standards for its construction, regulates its operation, and often becomes a tenant. The private partner then gets a return on its investment, usually by charging users a toll, as with the Confederation Bridge to Prince Edward Island.

The primary rationale put forward for P3s is that the taxpayers are relieved of a portion of the capital investment costs in the projects. There is still considerable debate over these types of partnerships, however, ranging from concern over their social costs to the failure of governments to maintain existing "free" facilities. Some opponents cite the differential benefits for those citizens able to pay the tolls and those who can't. Others are concerned with the monopoly aspects given to private companies. Yet others claim that the cost of added pollution occasioned by, for example, a toll road is not being recovered by these arrangements.[18] Suffice it to say, such project arrangements are often instituted on grounds other than those established in community plans.

Statutory Tools

There are several other means by which communities can influence the quality of the built environment. Two of the most important, expropriation and building regulations, are embedded in legislation codifying the responsibilities of communities to their residents. Communities are assigned roles, as municipalities, to provide a safe and convenient environment for all their citizens. That is, the municipality has a formal (legal) responsibility to act on behalf of the entire community to make sure that the decisions made by individual landowners not only do not impinge on their neighbours, but also serve the best interests of the entire community. Building regulations and the power of expropriation stem, respectively, from two constitutional powers that reside in the municipality—the *police power* and the power of *eminent domain*.

Building Regulations and Other Permits

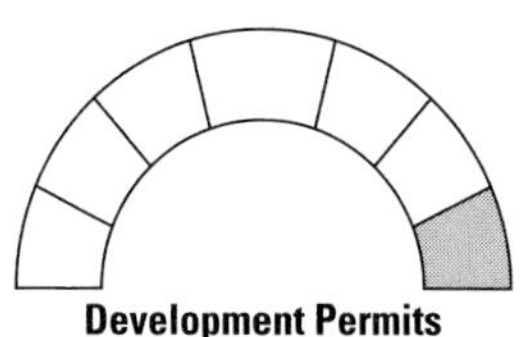

In general, municipalities in Canada have the power to regulate the construction of buildings within their jurisdiction. The reason for this is to ensure that new buildings and the reconstruction of older buildings will not endanger the health, safety, or general welfare of the public. This power to approve building construction derives from a constitutional power—**police power**, in legal terms—delegated to the municipality by the province. The task of the municipality, in essence, is to enforce the provincial building code that sets basic standards for construction. To accomplish this, the municipality employs building inspectors whose job it is to review plans for buildings and to issue building permits if the plans meet basic standards of the code. Building inspectors also monitor the actual construction to ensure adherence to the agreed-upon plans.

The issuance of a building permit represents, in many ways, the final step in developing a building project. It is, therefore, a crucial step for the municipality. It can determine at this point whether the construction plans for the building conform to community-planning expectations of the project. Not infrequently, building plans are found to contravene the setback and height regulations of zoning bylaws. Many communities combine a review of zoning requirements with the building inspector's review of construction plans at this stage. In a related way, the issuance of a building permit is seen as the indication that building on a site is going to proceed shortly thereafter, but there is no guarantee of this. There are usually some sites in a city or town on which development is preferred by the community to occur as soon as possible. Increasingly, communities are tying the approval of zoning amendments and development agreements to either the issuance or the duration of a building permit.

Occupancy permits are often required, especially for residential projects, before use of a building can be allowed. The prime criteria at this stage are usually the meeting of fire and other safety conditions. Since zoning bylaws and/or development agreements may have required the buildings to meet other conditions, such as the provision of open space or walkways, this is another opportunity to ensure that planning conditions have also been met. To some, this may suggest undue constraint on developers and buildings, but it should be remembered that construction requires a legal, contractual arrangement with the municipality in which the steps of issuing building and occupancy permits are specified. There is also the case of development permits that may be needed and their terms satisfied, as discussed in Chapter 13. The municipality should be rigorous in taking these steps and in verifying any related planning provisions, and it need not be dilatory in its procedures.

Expropriation

From another long-lived constitutional power held by Canadian provinces—**eminent domain**—municipalities derive the power to obtain land needed for community purposes. The essence of eminent domain is that the highest property right on the use of land rests ultimately with the Crown, or the community in this case. Specifically, if there are public uses of specified parcels of land that would be of more benefit than those proposed by the private owners, then the community may acquire the land. Normally, most municipalities have been delegated this power by their province, as have such other public agencies as school boards, highway departments, and public utilities.

The term used in Canada for this municipal power is **expropriation**. It is probably most often used for road-widening purposes when an existing right-of-way is not sufficient to accommodate an enlarged thoroughfare. Other instances include a community's need for park space in an already built-up area or space to build a public building or an arterial road. When a community designates a property for public use, it must acquire it from the owner by paying a reasonable price. Usually, the price to be paid will be the current market value as judged by an independent land appraiser. Strictly speaking, the power of expropriation is used only when the public agency cannot acquire the needed property through normal purchase procedures; it then has the power to acquire the property regardless of the willingness of the owner to sell. Municipal councils tend to use their power of expropriation sparingly because of the political ramifications of taking away private property; they try instead to negotiate a purchase. Nevertheless, this power of eminent domain has a valuable place among planning tools, as it ensures acquisition of properties that are of strategic importance to the completion of the plan.

Development Charges and Property Taxes

Municipalities have two fiscal tools for influencing development of private lands. One is the charge applied to the development of a site to provide public infrastructure, called "development-cost charges." The other is the rate of property taxation applied to the land and buildings of a site. Both of these devices impose a cost on landowners and developers that must be included in calculating the cost of a proposed land development project. In general, private-land interests will want these costs minimized while the municipality will want to make sure that its initial and ongoing service costs are met. It follows, therefore, that the size of these costs may affect decisions about the scale and type of project, or even whether it goes ahead at all.[19] Planners also need to be aware of the costs of urban growth and of the effects of these and other fees imposed on new development.

Property taxes are the primary source of income for a municipality and, in most provinces, tax rates are based on the current market value of properties. Separate valuations are made for the land portion and the building portion of a property's value. Thus, vacant land is taxed less than land with buildings, which means that, if a community wished to see a particular site developed, it could reduce the property tax rate on vacant land. It has been shown in a Canadian study that land developers consider differences in property taxes between adjacent municipalities important when deciding on the location of a project.[20] Similarly, development-cost charges, which enable a municipality to have public services for a project paid for by the developer, are another important factor in deciding which municipality to locate a project in. To work, differences in these fiscal tools must occur in similar economic settings, zoning, and administrative-processing practices. Development-cost charges, in particular, may be varied to reflect differences in recouping environmental costs to a community, as well the costs of normal urban infrastructure.

Targeting Special Areas and Functions

Beyond incentive programs to entice private investment to occur in response to the community plan, the community may decide to undertake specific development projects on its own. It may target a specific *area*, such as downtown, an industrial area, a waterfront, or an historic district, the improvement of which would benefit the entire community and attain the objectives of the plan. Or it may target a specific *function*, such as the mass-transit system or hospital services or parks, as a strategic move to achieve the plan. The essential feature of both types of community initiative is the investment of *public funds*, sometimes in conjunction with outside (private or nonprofit) investors. Often, these are accompanied by special land-use regulations devised for the project or area.

Special-Area Initiatives

The targeting of special areas for development offers a strategic tool for achieving planning objectives. Such targeting acknowledges two facets of city development. First, there are differences in the pace of development between various areas at different times (e.g., the suburbs growing faster than downtown). And second, public infrastructure expenditures (for roads, parks, etc.) usually precede private investment. Thus, a substantial investment in public funds within a short period of time may be necessary to stimulate new private development or the renewal of existing uses. Five common initiatives—for business areas, heritage districts, waterfronts, institutional areas, and local ecosystems—will demonstrate the issues and the outcomes of this policy approach.

1. Business Improvement Areas (BIAs). Plans and programs for the revitalization of downtown and neighbourhood business areas are the most common special-area initiatives. Most of the buildings in these areas, in both large cities and small towns, represent extensive private investment, if not the largest concentration of such capital in the community. So, it is usually in the community's interest to protect this investment and maintain the economic and cultural vitality of these business areas. The approach of the

Business Improvement Area was devised in Canada and is used in all provinces. It designates a specific commercial area for which a plan for improvement is made in conjunction with property owners and businesses in the area who form a nonprofit association. These groups are then able to access provincial and local funds to make physical improvements and undertake economic development activities.

BIAs are normally based on a plan that details the objectives for the area. The BIA plan describes the attributes to be achieved (e.g., street patterns, land uses, open spaces, views, streetscape, signage) along with the set of public and private actions that must be taken to attain the plan's objectives. In 2005, there were over 230 BIAs in Ontario, and the province provides a handbook to guide such planning.[21] Other provinces provide similar assistance for all sizes of community from Ladysmith, B.C., to Cornerbrook, Newfoundland.[22] These programs help underwrite capital investments, such as new sidewalks, sewers, and streetlights, and may also offer incentives to private builders and landowners to participate. The community, in turn, is required to make a commitment in budgeting its capital and organizational resources to provide needed improvements such as parking structures, public open spaces, or a civic building to give impetus to the local development.

On a much larger scale are redevelopment programs for central districts of major cities, such as those for the Winnipeg downtown core, *le centreville de* Trois-Rivières, and Market Square in Saint John. In the latter cases, the capital resources of both provincial and federal governments were also required. Their aims may include (as they do in the Winnipeg Core Area Initiative) the development of new housing, heritage conservation, new community facilities, the formation of local economic development corporations, retraining of core-area residents, industrial modernization, and assisting the development of new commercial buildings.[23]

2. Heritage Areas. Many Canadian cities have some heritage-protection policies for older downtown districts, such as Halifax's Historical Properties, Vieux Montreal, Ottawa's ByWard Market, or Vancouver's Gastown, while outstanding heritage areas such as Quebec City's Historic District and Old Town Lunenburg are UNESCO World Heritage sites, with more complex policy controls. In recent years, heritage-conservation areas have spread outside the downtown to coherent residential neighbourhoods and former industrial districts such as Vancouver's Granville Island, Montreal's Lachine Canal, and Toronto's Distillery District. The aims may differ from those found in core-area programs, but the means for achieving them are much the same.[24]

3. Waterfront Enhancement. There is hardly a city in Canada that doesn't have a waterfront on a lake, river, or ocean, and most communities realize its importance to both their heritage and their economy. Planning for waterfront redevelopment has occurred in Montreal and Windsor, in Nelson, B.C., and Yarmouth, Nova Scotia, among others. It has often been used to reclaim a neglected industrial and port area and make it functional once again for business, residence, and recreation. The two basic planning strategies used in waterfront areas are to secure historic preservation and access for the public with both being used often in combination.[25]

For many cities, the waterfront is where the community began, and there is frequently a stock of older historic buildings that are worth conserving and adapting. For both an older port area (e.g., Saint John) and a riverfront scenic area (e.g., Saskatoon), public access is crucial. It is not uncommon for these areas to be cut off from the city by railway lines or highways, and public initiatives can be used to link the area to existing streets and create such things as water's edge promenades and parks. The most successful waterfront projects focus on "the quality of the public space" in the project area.[26] In other words, public-policy initiatives, by way of heritage legislation and capital expenditures on streets and parks, are crucial for waterfront enhancement.

4. Institutional Areas. Many government agencies and public institutions prepare long-range plans, since public agencies are among the largest managers of land and buildings.[27] The federal government's National Capital Commission has prepared regional and institutional plans for the Ottawa area for over a century,[28] and provincial capital plans have been prepared for Regina's Wascana Centre, Yellowknife, and Quebec City, among others.[29] The Department of National Defence is one of Canada's largest landowners and implements plans for military bases and adjacent residential areas, such as the 1950s new town in Oromocto, New Brunswick. In recent years, surplus military bases in urban areas have been redeveloped into new projects such as Garrison Crossing in Chilliwack, B.C., Garrison Woods in Calgary, and Downsview Park in Toronto.

Public institutions such as hospitals, colleges, and universities often have complex long-range plans for their lands. Controversial proposals for redevelopment of former hospital sites in New Westminster (Woodlands), Calgary (The Bridges) and Toronto (Queen Street) have recently won awards for good planning practice. Finally, colleges and universities usually have a master plan for development of their institutions, but several universities with large land banks have also been catalysts for other development. The University of British Columbia and Simon Fraser University are developing large mixed-use communities on their lands, and the Universities of Victoria and Waterloo are developing technology parks to foster spinoffs from their research activities.

5. Local Ecosystems. It has become increasingly common for communities to undertake planning programs devoted to protecting and maintaining natural areas and local ecosystems. These efforts range from the identification and preservation of natural habitat remaining in a community to broad-scale ecosystem restoration.[30] "Ecological rehabilitation," a term now more widely used, addresses two needs, according to a study of ecosystem planning in the Regional District of Waterloo in Ontario:[31]

- To restore natural ecological features and functions in a degraded area; and
- To reconnect or strengthen links between existing natural areas and a wider landscape.

Such planning can involve extensive areas of land in a community that are, by implication, no longer available for built urban development. While it may be argued that ecosystem planning will benefit the entire community, any reduction of land for development will have other difficult and political ramifications, such as raising the cost of the remaining land and thereby housing costs. It can also be seen from the above objectives that a local ecosystem cannot be planned for in isolation from the surrounding region.[32] Not infrequently, local ecosystem planning is combined with planning for larger open spaces and/or environmental sustainability as it was in the Toronto region.[33] Lastly, a common feature of local environmental planning is the involvement of groups of citizens in maintaining the program, as in some British Columbia communities, because ecosystems are growing, changing entities for which ongoing stewardship is needed.[34]

Responding to Ethnic Enclaves

Canada's cities, especially the largest ones, have seen over the years the development of areas of residence and commerce centred on particular ethnic groups. Early examples are the Chinatowns and Little Italys, and the Jewish, Greek, and Portuguese communities formed by the waves of immigrants early in the 20th century. These older ethnic neighbourhoods have been joined by many more in recent decades as a result of increases in the flow of immigration and in its ethnic composition. Immigrants today, and since the 1970s, are predominantly (70 percent) visible minorities.[35] Moreover, they have taken up residence broadly across urban areas and are forming what are called ethnic enclaves.[36] These are particularly evident in Vancouver, Toronto, and Montreal, and not just in inner cities, as in the past, but in inner and outer suburbs such as Surrey, B.C., and Markham, Ontario.

A distinguishing feature of these enclaves is the complex of malls, stores, restaurants, movie houses, places of worship, and professional offices, with their distinctive signage, that serve the nearby ethnic population as well as those living farther afield.[37] The addition of these enclaves to the urban fabric generates two types of community planning issues that will increasingly require attention. The first is the social dimension of providing social and economic opportunities, services, and infrastructure for their residences and businesses. The latter task needs to be done in such a way as to balance the reality of diversity with the promotion of unity.[38] Canadian planner Mohammad Qadeer argues for providing infrastructure and local services that are tailored to these neighbourhoods and yet are equitable with other neighbourhoods. This would involve public schools, community centres and health services, facilities for the elderly, and so on that are linguistically and culturally accessible.[39]

The second issue concerns planning practice in both its procedural and physical aspects. On the one hand, cultural and religious dimensions may arise in the administration of simple matters such as zoning, variances, and development permits. This occurred when a funeral home was allowed to open in a Chinese neighbourhood in Ottawa, which was considered a cultural taboo.[40] The need is to review current planning policies from a multicultural perspective and, if necessary, revise zoning bylaws and development standards accordingly. The physical design side of ethnic enclaves, especially their commercial districts, offers a number of challenges and opportunities. Again, in regard to the cultural character of an enclave, planners have the opportunity to enhance the diversity of a business area through urban-design approaches to street signage, sidewalk designs, and murals, for example.[41] (This subject is discussed further in Chapter 15.)

Meeting Functional Needs

A community plan may aim to improve specific functions within a community. Increased traffic may need to be accommodated, sewage systems may need to be upgraded, and affordable housing may be in short supply. Such needs often occupy large portions of land and/or demand considerable public investment. They usually require a separate functional plan (see Chapter 6) showing location, land acquisition, and the disposition of capital expenditures, as well as other expected outcomes of the project. The following are typical of such projects:

- Rapid-transit systems that aim to relieve current traffic congestion in the downtown area, as with the systems in Calgary, Edmonton, and Vancouver;
- Social-housing projects that provide affordable housing for the elderly and other low-income groups not normally able to compete in the economic marketplace;[42] and
- Regional parks developments that aim to improve recreational opportunities and protect natural environments.

To this list could be added improvements to an airport, a new convention centre, college, or health-care facility, or an expanded freeway system. A demonstration of this is the approach used by the City of Calgary in planning its future transportation system. The Calgary Go-Plan is aimed at guiding capital expenditures on transportation over the next 30 years; the planning process itself is costing several millions of dollars.[43] Often, one of the outcomes from such a functional-planning exercise is to bring into sharper focus the larger questions about the community's future shape. This happened when Calgary proposed transportation corridors that provoked considerable public debate over the urban form, the ecological consequences, and the balance between public transit and automobile transportation, and led to improvements in the plan.

The planning and programming of public projects is undertaken for three main reasons: (1) to benefit present residents; (2) to direct new development in a growing community; and (3) to attract development to a stagnant community. Typically, some combination of these reasons comes into play. A new rapid-transit system, for example, may both relieve downtown auto traffic congestion and direct new development to locales adjacent to new stations that now have improved accessibility; similarly, a new college could both benefit local residents and attract new residents and businesses.

While the functional needs of the community may be spelled out in the community plan, the actual functional plan is often prepared by agencies outside the local government that provides the funding. A provincial government department or Crown corporation may be responsible for rapid-transit and freeway projects, or a metropolitan government may have responsibility for regional parks, hospitals, and airports. An example of this is the siting of treatment and transfer facilities for hazardous wastes in Manitoba communities. This is handled by the provincially owned Manitoba Hazardous Waste Management Corporation (there are similar agencies in other provinces). Not surprisingly, such projects are often called "difficult-to-site" projects. The Manitoba agency has developed a voluntary process in working with communities to assure greater acceptance of these facilities by sharing power and decision-making responsibility and is achieving considerable success.[44]

Some Implications of Targeting

The great advantage of targeting special areas and functional needs is that it can generate tangible results from the planning process in a relatively short span of time. But it must be remembered that such specific planning has the effect of directing the allocation of public spending (and, if successful, private investment) toward one area or project and away from others. Thus, care must be taken not to starve existing projects of public funds and opportunities for private development. As with other middle-range planning tools, special-area and/or project targeting requires that considerable attention be paid to integrating planning and action.

This need is especially evident in view of the diverse interests involved in realizing a refurbished downtown or rapid-transit system. These include departments within a local or regional government, agencies and departments of one or more senior governments, and private development interests. Given the often substantial amounts of resources required, a city may be able to do no more than indicate the desired project or redevelopment in its community plan. The public agency that, for example, can develop a rapid-transit system, may opt to do the project planning itself. Or the private developer with large and/or strategic land holdings may demand concessions in zoning. In such instances, project developers rather than community planners become the ones controlling public policy on the location of facilities and the bulk of buildings. Not infrequently, special development agencies are formed to push a project ahead, and they may come into competition with the community's planners for funds and political favour. This can cause relationships with the host community to turn sour as projects diverge from local planning norms and objectives. Nowhere was this more evident than in the Harbourfront project in Toronto, which required a federal royal commission to seek a reconciliation of local and agency planning aims.[45]

Policies for Managing Community Building

Achieving the vision of the built environment as seen in the community plan requires a process of managing growth and change for as long a period as the plan projects ahead. Community building, and rebuilding, is ongoing and if there is a desire to achieve a preferred outcome vis-à-vis the amount, location, type, and character of the built

Planning Issue 14.1 Meltdown in Mansionville

The Globe and Mail
January 7, 2006

Meltdown in Mansionville
John Lorinc

Like a growing number of affluent Forest Hill residents, Brett Smith, a commercial developer who trained as an urban planner, is stunned by the sheer size of the new houses sprouting up all over the tony mid-Toronto neighbourhood where spacious, stately homes dating to the 1920s and 1930s have become, amazingly, knock-down fodder for speculators.

Citing the case of a lot just north of Forest Hill Village, where the owner conjoined two ravine-side properties and wants to put up a 14,500-square-foot mega-mansion that is 75 feet wide, has three garages facing the street and a three-storey tower peering into the neighbours' backyard, Mr. Smith says: "It's the most ridiculous application I've ever seen, and it didn't raise one flag at City Hall."

Mr. Smith and about 20 of his neighbours hotly opposed the project, which was turned down at the committee of adjustment in October, and are going to the Ontario Municipal Board next month. They bring a new twist to the standard Not-In-My-Backyard argument—the development should be opposed not only because the ersatz chateau is inappropriate, they say, but because it's an assault on the heritage character of their street.

A growing number of residents in the area want to see city council designate Forest Hill as a Heritage Conservation District, a move several older neighbourhoods, including North and South Rosedale, Cabbagetown, Wychwood Park, the Annex and a stretch of Blythwood Road in Lawrence Park, have taken in recent years to slow the pace of the demolitions that fuel real estate speculation and threaten to erase large swaths of Toronto's historic residential architecture.

Heritage expert Catherine Nasmith says a disturbing number of the Forest Hill demolitions involve historic homes built by once-prominent architects, including at least 10 of 22 homes in the area by Eden Smith, one of Toronto's most sought-after designers during the late 19th and early 20th centuries.

"It's a complete disaster. The homes have become nothing more than the value

environment, management and coordination of these factors must ensue. Furthermore, what gets built will be permanently in place for decades to come, so the policy choices regarding growth and change are of strategic importance. The following questions must be asked: How much growth? Where it will be located? What it will look like? How it will function? These questions are recurring ones for planners and the policy-makers they advise. They also arise at all levels of community building, from the neighbourhood level through the district, citywide, and regional levels, and they require policy choices about what would best suit the community. Instances of such choices faced by communities are discussed below, first at the community and regional level, and then on a more intimate scale.

Managing Regional Growth

Two salient characteristics of most larger communities in recent decades are growth in the number of both people and dwellings, and expansion in area. The rapidity as well as the scale of much urban growth has often led to egregious amounts of land being consumed, not infrequently at very low density in a "leapfrog" manner, and environmentally sensitive areas converted to urban uses. These, in turn, leads both to escalating infrastructure costs and uncoordinated patterns of development, not to mention traffic congestion, pollution problems, and a failure to redevelop old neighbourhoods.[46] Responses to these undesirable features have centred on **growth management**—that is, methods to limit and/or direct new growth.

of the land," adds retired broadcaster Donald Duprey.

Two Forest Hill ratepayers groups are now ramping up to push for HCD designation, but none are yet in place. HCD designations require residents to undertake a full inventory of the heritage features of their neighbourhood—an archival sleuthing process that often unearths a mother lode of information about the history of an area, its architectural features, how it was developed and who lived there.

In some cases, as in Cabbagetown, the guidelines can be quite specific (for example, dictating exterior materials and roof slopes), while in others, the guidelines are geared simply toward making it difficult for property owners to get demolition permits. Interior renovations or backyard additions are never subject to HCD rules, which focus on the streetscape and prevailing character.

Michael McClelland, an architect who advised two Rosedale ratepayer associations on their heritage designations, describes HCDs as "vision statements" that go beyond the traditional planning controls of density, height and setback, which he characterizes as "blunt instruments" when it comes to preserving heritage and ambience. "If properly constituted," he says of HCDs, "they're really a valuable tool for good city-building."

That may be true, but HCD designations require hundreds of hours of slogging by volunteers and fund raising drives to pay for studies that the city doesn't fund. (The customary objection from homeowners is that an HCD designation limits their property rights and could depress resale values. But Mr. McClelland cites U.S. studies showing that property values in heritage conservation districts tend to rise, not fall, after the designation is put in place.)

David Townley, president of the South Rosedale Ratepayers Association, which obtained its HCD designation in 2003, insists that his group "doesn't want to be the taste police," but he admits that it is unlikely they would agree to demolitions of homes within the district.

Source: John Lorinc, "Meltdown in Mansionville," *The Globe and Mail*, January 7, 2006, M3. Reprinted with permission from *The Globe and Mail*.

New growth occurring outside a community's boundaries, perhaps in an adjacent community, or simply well beyond existing built-up areas in a large municipality, requires an area-wide or regional perspective to manage it. The basic policy choice that needs to be made is to decide *where* urban development should and should not occur. The latter concern should be prompted by a study of the natural features and an analysis of the landscape that should be conserved and/or reclaimed.[47] With that policy parameter in place, there still remains the matter of where and how much land is needed by ongoing and projected growth and how to contain such growth. **Urban containment boundaries**, which fix the limits on a map, are a widely used approach to this issue. Sufficient land must be provided within these boundaries for continuing as well as expected new development in order to keep land prices stable. In Calgary, a 30-year supply of developable land is allowed for.[48] In addition, the city's policy requires that new growth be contiguous with existing built-up areas. It also tracks the absorption of land being developed and aims to have a three- to five-year supply of land available with municipal infrastructure. York Region in Ontario follows a similar path with a commitment to "firm urban boundaries" and also uses a rigorous system of monitoring regional conditions every three years.[49] In British Columbia, regional districts with growth and expansion problems among their communities may invoke the use of a Regional Growth Strategy. Within this arrangement, all constituent municipalities agree upon the urban containment boundary and a further formal agreement binds each of them and their community plans to it.[50] Alberta communities are strongly encouraged to prepare and adopt inter-municipal development plans with adjoining municipalities for fringe areas lying astride their boundaries that are critical for growth and/or conservation.[51]

A more nuanced approach than simply "containing" growth is encompassed by the notion of **Smart Growth**, which combines urban and suburban planning at the regional scale. Its urban principles include raising residential densities, emphasizing public transit, providing for mixed land uses, revitalizing old neighbourhoods, and allowing for more diverse layouts and street designs.[52] This broad menu of policies is often associated with the movement among planners, politicians, environmental groups, and others to "control sprawl" and develop more compact communities.[53] The concern over the form and quality of suburban development covers a number of issues including prohibiting low-density growth at the suburban margin. Communities are urged to embrace Smart Growth ideas and enact policies to promote them. According to critics, this may be more difficult to achieve than it appears because it requires a coordinated approach both within and between communities and with higher levels of government.[54] Further discussion of the issues embedded in efforts to control sprawl follow in the next chapter.

Shaping Growth at the Community Level

At the community level the issue of managing growth is less about sprawl than it is one of livability in existing and new neighbourhoods. The policy choices range over such matters as designing new subdivisions, redeveloping former industrial sites, accepting gated communities, allowing mixed use, and making streets in old neighbourhoods quieter. A review of responses to these matters in several Canadian communities will help illustrate the use of policy as a tool to implement community plans.

Subdivision Design

The conventional design of subdivisions over the past four to five decades has tended toward curvilinear streets and culs-de-sac bounded by arterial roads. Although efficient in terms of land use and infrastructure, they have not been as successful at providing connectivity to the larger road network, allowing easy orientation, or mixing uses.[55] **New Urbanism** challenged conventional subdivision design, beginning in the 1990s, to revert to a modified, traditional gridiron of streets, to mix housing types, and to produce more compact, less land-consuming residential areas.[56] Many municipalities, from Abbotsford to Winnipeg to Montreal, have changed their development policies to permit New Urbanism, or **Traditional Neighbourhood Design** (TND) as it is sometimes called. Although the results have been mixed in achieving all TND principles, more compact, walkable development has ensued and, presumably, better connectivity.[57]

Another approach to subdivision design being considered by other Canadian communities seeks a variation on the classic gridiron, TND, and cul-de-sac suburban models. It is known as the **Fused Grid** and was developed by Canada Mortgage and Housing Corporation.[58] The goals, according to its designers, are to provide a balance between pedestrian and automobile movement and to create "safe, sociable streets and easy connectivity to community facilities." This policy option has been tested in Stratford, Ontario, and in Winnipeg.[59] Any of the above-mentioned approaches to subdivision design may also come into play at the stage of preparing and reviewing subdivision plans, as discussed in Chapter 13.

Gated Developments

Increasingly, municipalities are being faced with proposals to develop subdivisions that are designed to have access limited to persons living there and with their own internal roads, and so on. Commonly, these are referred to as "gated communities," and they may range in size from a handful to several hundred homes.[60] Studies indicate that well over 300 gated projects exist across the country. The key land-use features of these projects are that their standards for roads, sidewalks, street lighting, and setbacks of houses are unique to the project, and frequently differ from those normally applied in the community. The gated feature prohibits use of the local roads to reach other areas. These conditions are sanctioned by the use of legal instruments such as property covenants that regulate relations with the municipality and activities of residents (i.e., as in a mini-zoning bylaw).

Gated communities are popular alternatives with some segments of the population and deliver built environments that residents find safe, quiet, and friendly.[61] As a policy option it means the municipality must vary standards of development for a specific group of residents that apply universally to all other citizens. There are both advantages and disadvantages for the community. On the positive side for the municipality is the fact that care and maintenance of the street system is the responsibility of project residents (often a condominium association) and property taxes still accrue to the municipality. It can also know that those community residents who live in the project enjoy calm, safe streets. On the negative side there may arise issues such as access for emergency vehicles, and reduced connectivity for pedestrians and automobiles to public sites such as parks and education facilities.

Traffic Calming

One of the assets of gated projects, the control of automobile traffic in a residential area, has also been sought in regular neighbourhoods, particularly those based on a traditional gridiron of streets. At the behest of neighbourhood groups, many municipalities have made a policy decision to reduce the amount and speed of automobile traffic in local areas through the use of various pavement and signage devices and thereby achieve **traffic calming.** This approach got its start in Europe (Germany, the Netherlands, Denmark) in the 1970s and later in Australia.[62] Three basic principles motivate traffic-calming approaches:

1. The function of streets is not to act just as a corridor for automobiles.
2. People have a right to not have their quality of life spoiled by undue traffic caused by automobile use.
3. Trips are a means of accessing some desirable land use or activity, not an end in themselves.[63]

The thrust of these principles contrasts with the traditional planning principle of separating various forms of traffic (autos, cyclists, pedestrians) from one another. Traffic calming invokes public policy in regard to one of the fundamental public responsibilities in a community environment—creating and maintaining public rights-of-way. Various methods are used in traffic calming, including raising the surface of streets at intersections (traffic "bumps"), narrowing streets with trees, building traffic circles, and looping streets.[64] Usually, traffic calming is undertaken on a neighbourhood-by-neighbourhood basis upon the request of the residents.

Reflections

Meyerson's notion of a "middle-range bridge," invoked at the outset of this chapter, has two main supports. The first is the array of land-use regulation tools, described in the last chapter, that undergird so much of planning for the built environment that they are seen by many as synonymous with community planning. That is, it is seen as essentially a regulatory process. As important as regulatory tools and processes are in providing a uniform base for private land-use decisions they are, as was mentioned, primarily *reactive* in nature—they come into play after a proposal has been made to build or rebuild part of the community. The second main support of the middle-range bridge, planning policy tools, described in this chapter is *proactive* in nature. That is, these tools allow the community to take its own steps to achieve the built environment envisioned in the community plan: making strategic capital investments, targeting areas to be rejuvenated, and managing growth and development. Neither of these arrays of tools is more vital than the other in bringing a community plan to fruition, although they may vary in importance in the course of plan implementation—both are needed to support the middle-range bridge.

The policy tools that communities have available to shape both the form and the pace of community-building have much in common with the current approach of embedding natural ecosystem needs early in the planning process. Policy tools, too, can be *preemptive*. They can establish, early on, priorities for the type, location, and timing of private development that is essential for forming and filling out the built environment. They cannot be as precise as are regulatory tools and they involve risks, as with committing public monies for capital investments for facilities and infrastructure. So, policy tools have to be utilized wisely and sensitively with regard to private land and housing markets and this is best done by encompassing them with community strategic plans. The planners in Edmonton framed their municipal plan, Plan Edmonton, knowing that it would have to provide not only land-use policies but also integrate the "business plans" of the municipal corporation and its departments, including city budgets (see Figure 14.1). The planners for York Region, and both Calgary and Ottawa, follow a similar planning trajectory. In short, the community plan is just one plan, albeit the lead one, in a chain of plans to achieve a community's goals.

Given their importance, it is surprising that policy tools are not used more widely. They are adaptable to all sizes of community as demonstrated in Nova Scotia. Further, presenting policy options such as those described above to a community's elected leaders engages them more thoroughly in the ongoing decisions required to implement a community plan. The adoption of a community plan is itself a major policy commitment by local councillors, but it is the beginning, not the end, of the need for choices about community building. Moreover, addressing these issues allows for greater awareness and participation in the planning process, thereby enriching the end result. As the discussion moves to consider future prospects and challenges for community planning, the question below deserves attention:

- ***—How can communities develop more open, integrated planning processes that will allow them to meet challenges yet to come?***

Endnotes

1. Martin Meyerson, "Building the Middle-Range Bridge for Comprehensive Planning," *Journal of the American Institute of Planners* 22 (Spring 1956), 58–64.
2. John M. Bryson and William D. Roering, "Applying Private Sector Strategic Planning in the Public Sector," *Journal of the American Planning Association* 53:1 (Winter 1987), 9–22; and Mark Seasons, "Strategic Planning in the Public Sector Environment: Addressing the Realities," *Plan Canada* 29:6 (November 1989), 9–27.
3. Service Nova Scotia and Municipal Relations, *Local Government Resource Handbook* (Halifax, 2000), 7.
4. It is called the Focus Community Strategic Plan and can be viewed online at www.cityofkingston.ca/cityhall/strategic/index.asp.
5. Manfred Fast, "Community Can Make It Happen: Forging the Port Colborne Ontario Strategic Plan," *Small Town* 26:1 (July–August 1995), 10–15.
6. Richard D. Margerum, "Getting Past Yes: From Capital Creation to Action," *Journal of the American Planning Association* 65:2 (Spring 1999), 181–192.
7. Gord Jackson and Mary Ann McConnell-Boehm, "Plan Edmonton: A Plan and a Process," *Plan Canada* 39:5 (November 1999), 17–19.
8. City of Edmonton, *Planning and Development Handbook* (Edmonton, 2000).
9. Frank S. So and Judith Getzels, eds., *The Practice of Local Government Planning,* (Washington:International City Manager's Association, 1988).
10. Robert Coughlin, "The Capital Programming Problem," *Journal of the American Institute of Planners* 26 (February 1960), 39–48.
11. Government Finance Officers Association, *Capital Improvement Programming: A Guide for Smaller Governments* (Chicago: GFOA, 1996).

12. Coughlin, "The Capital Programming Problem."
13. Robert H. Bowyer, *Capital Improvements Programs*, Planning Advisory Service Bulletin 442 (Chicago: American Planning Association, 1993).
14. Service Nova Scotia, *Local Government Resource Handbook*, Section 6.2.
15. The 2006–2008 Capital Plan may be viewed online at www.city.vancouver.bc.ca/corpsvcs/financial/capital/index.
16. Casey Brendon et al., "Urban Innovations: Financial Tools in Brownfield Revitalization," *Plan Canada* 44:4 (Winter 2004), 26–29.
17. Canadian Council for Public–Private Partnerships, *Canadian PPP Project Directory* (Toronto, 2006).
18. Canadian Centre for Policy Alternatives, BC Office, *Assessing the Record of Public–Private Partnerships* (Vancouver, 2003).
19. Greg Lampert and Marc Denhez, *Levies, Fees, Charges, Taxes, and Transaction Costs on New Housing* (Ottawa: Canada Mortgage and Housing Corporation, 1997).
20. Andrejs Skaburskis and Ray Tomalty, "How Property Taxes and Development Charges Can be Used to Shape Cities," *Plan Canada* 41:1 (January–February–March 2001), 24–30.
21. Ontario, Ministry of Municipal Affairs and Housing, *Business Improvement Areas Handbook* (Toronto: Queen's Printer, 2004).
22. Cf. Francois Leblanc, "La renaissance des centre-villes: Le programme Rues principales," *Plan Canada* 29:5 (September 1989), 8–13.
23. Matthew Kiernan, "Intergovernmental Innovation: Winnipeg's Core Area Initiative," *Plan Canada* 27:1 (March 1987), 23–31.
24. Christaine Lefbure and Eve Wertheimer, "An Indispensable Reference for Heritage Conservation," *Plan Canada* 46:1 (Spring 2006), 41–43.
25. David L.A. Gordon, "Implementing Urban Waterfront Redevelopment," in *Remaking the Urban Waterfront* (Washington, DC: Urban Land Institute, 2004), 80–99.
26. David L.A. Gordon, "Planning, Design and Managing Change in Urban Waterfront Redevelopment," *Town Planning Review* 67:3 (1996), 261–290.
27. Olga Kaganova and James McKellar, eds., *Managing Government Property Assets: International Experiences* (Washington, DC: Urban Institute, 2006).
28. David L.A. Gordon, ed., *Planning Twentieth-Century Capital Cities* (New York: Routledge, 2006).
29. Pierre Dubé and David LA. Gordon, eds., "Capital Cities: Perspective and Convergence," *Plan Canada* (Special Issue) 40:3 (May 2000), 22–40.
30. Ken Tamminga, "Restoring Biodiversity in the Urbanizing Region: Towards Pre-emptive Ecosystems Planning," *Plan Canada* 36:4 (July 1996), 10–15.
31. Soonya P. Quon, Larry R.G. Martin, and Stephen Murphy, "Ecological Rehabilitation: A New Challenge for Planners," *Plan Canada* 39:4 (September–October 1999), 19–21.
32. David L.A. Gordon and Ken Tamminga, "Large-scale Traditional Neighbourhood Development and Pre-emptive Ecosystems Planning: The Markham Experience, 1989–2001," *Journal of Urban Design* 7:3 (2002), 321–340.
33. Eric Pedersen, "The Garrison Creek Linkage Plan: A Model for Developing an Open-Space System," *Plan Canada* 39:5 (November 1999), 20–21.
34. Fern Heitkamp, "Using Stewardship as a Guide for Planning," *Plan Canada* 36:4 (September 1996), 28–32.
35. Statistics Canada, *Canada's Ethnocultural Portrait: The Changing Mosaic*, Cat. no. 96F0030XIE2001008 (Ottawa: Minister of Industry, 2003).
36. Feng Hou, *Recent Immigration and the Formation of Visible Minority Neighbourhoods in Canada's Large Cities*, Analytical Studies Branch Research Paper, Cat. no. 11F0019MIE (Ottawa: Statistics Canada, 2004).
37. Sandeep Kumar and Bonica Leung, "Formation of an Ethnic Enclave: Process and Motivations," *Plan Canada* 45:2 (Summer 2005), 43–45.
38. Mohammad Qadeer and Sandeep Kumar, "Ethnic Enclaves and Social Cohesion," *Canadian Journal of Urban Research* 15:2 (May 2006), 1–17; and Mohammad Qadeer, "Dealing with Ethnic Enclaves Demands Sensitivity and Pragmatism," *The Ontario Planning Journal* 20:1 (2005), 10–11.
39. Mohammad Qadeer, "Ethnic Segregation in Toronto and the New Multiculturalism," Research Bulletin No. 12 (University of Toronto, Centre for Urban and Community Studies, March 2003).
40. Nancy Smith, "Diversity: The Challenge for Land Use Planning," *Plan Canada* 40:4 (July–August–September 2000), 27–28.
41. Sandeep Kumar and George Martin, "A Case for Culturally Responsive Urban Design," *The Ontario Planning Journal* 19:5 (2004), 5–7.
42. An amendment to B.C.'s planning legislation in the early 1990s requires every local government to include a policy statement on affordable housing in its official community plans.
43. Barton Reid, "Go-Plan Looking into the Future," *City Magazine* 16:1 (Spring 1995), 8–10.
44. Alan Richards, "Implementing a Voluntary Process for Difficult-to-Site Projects," *Plan Canada* 36:1 (January 1996), 31–32.
45. Barton Reid, "Harbourfront: Aesthetics vs. Dollars," *City Magazine* 8:3,4 (Fall 1986), 9–10.
46. Anthony Downs, "Smart Growth: Why We Discuss It More Than We Do It," *Journal of the American Planning Association* 71:4 (Autumn 2005), 367–378.
47. Ken Tamminga, "Restoring Biodiversity in the Urbanizing Region: Towards Pre-emptive Ecosystems Planning," *Plan Canada* 36:4 (July 1996), 10–15.
48. Marjorie Young, "Accommodating and Managing Municipal Infrastructure Investments in Calgary," *Plan Canada* 41:4 (October–December 2001), 18–20.
49. Bryan Tuckey, "The Cornerstones of Community-Building," *Plan Canada* 41:4 (October–December 2001), 14–15.
50. Sharon Fletcher and Christine Thomas, "Coping With Growth in the Regional District of Nanaimo BC," *Plan Canada* 41:4 (October–December 2001), 16–17.
51. Alberta, Municipal Affairs, *Land Use Policies* (September 1996), 4.
52. Downs, "Smart Growth."
53. Ray Tomalty and Don Alexander, *Smart Growth in Canada: A Report Card*, Socio-economic Series 05-036 (Ottawa:CMHC, December 2005). .
54. Larry S. Bourne, "The Urban Sprawl Debate: Myths, Realities and Hidden Agendas," *Plan Canada* 41:4 (October–December 2001), 26–28.
55. Ian Wight, "New Urbanism vs. Conventional Suburbanism," *Plan Canada* 35:5 (1995), 3–4.
56. Andres Duany and Elizabeth Plater-Zyberk, "The Second Coming of the American Small Town," *Plan Canada* 32:3 (1992), 6–13.
57. Jill Grant, Planning the Good Community: New Urbanism in Theory and Practice (New York: Routledge, 2006).
58. Canada Mortgage and Housing Corporation, "Applying Fused-Grid Planning in Stratford, Ontario," Research Highlights, Socio-economic Series 04-038 (Ottawa: CMHC, November 2004).
59. Fanis Grammenos, "Stratford Leads the Way to a New Model of Suburban Development," *Plan Canada* 45:1 (Spring 2005), 20–22; and Canada Mortgage and Housing Corporation, "Evaluating Arterial Road Configuration Options for a New Community," Research Highlights, Socio-economic Series 05-008 (Ottawa: CMHC, March 2005).

60. Jill Grant, Katherine Greene, and Kirstin Maxwell, "The Planning and Policy Implications of Gated Communities," *Canadian Journal of Urban Research* 13:1 (Summer 2004), 70–89.

61. Katherine A. Greene and D. Kirsten Maxwell, "Taking Matters Into Their Own Hands: Traffic Control in Canadian Gated Communities," *Plan Canada* 44:2 (Summer 2004), 45–47.

62. A good example from Australia is *Citizens Advocating Responsible Transportation, Traffic Calming: The Solution to Urban Traffic and a New Vision of Neighbourhood Livability* (Angrove Q., Australia, 1989).

63. Ibid., 18–19.

64. J.P. Braaksma and Associates, "Reclaiming the Streets: Setting the Stage for a Traffic Calming Policy in Ottawa," *Report to the City of Ottawa* (1995), 86ff.

Internet Resources

Chapter-Relevant Sites

Planning Canadian Communities
www.planningcanadiancommunities.ca

Iqaluit sustainable subdivision
www.city.iqaluit.nu.ca/:18n/english/plateau.html

Vancouver CityPlan Projects
www.city.vancouver.bc.ca/commsvcs/cityplans/cityplansprojects.htm

Woodward's Project Vancouver
www.vancouver.ca/woodwards

Heritage Planning, City of Windsor
www.citywindsor.ca/000497.asp

Toronto Waterfront Revitalization Corporation
www.towaterfront.ca

Benny Farm, Montreal
www.bennyfarm.org

Montréal Harbourfront
www.havremontreal.qc.ca

Nova Scotia Waterfront Development Corporation
www.wdcl.ca

PART FIVE

The Future of Community Planning in Canada

INTRODUCTION

Community planning is a task that evolves as our society evolves. Cities and towns continue to develop, informed by both their past and their future. The true progress of community planning may be as much a measure of how it deals with existing problems as how it anticipates future conditions. Many of the challenges of the future are the same as those in the past—the appearance of the city, housing new immigrants, disposing of wastes, and traffic congestion—but they come with new faces because of changes in technology, social policy, and attitudes. Other challenges are recent in origin, such as dramatic aging of the population and the demands of achieving sustainability.

Not least, these and other challenges will confront planners with the need to consider their values, moral positions, and their concept of the roles they play as professionals. The ultimate challenge to the ongoing practice of planning is how planners see their role in an increasingly complex and dynamic world.

The image above is from the Special Places E-zine *project, Calgary, Alberta, which received the Canadian Institute of Planners' Award for Planning Excellence, Category of Downtown Planning, 2005.*

15

Chapter Fifteen

Future Challenges for Community Planning in Canada

. . . in the large-scale task of putting cities together in which we all live and work, we have not yet stretched our abilities.

Humphrey Carver, 1955

Not only is community planning the "large-scale task" Carver (above) challenges us with, it is also a task that evolves as society evolves. Cities and towns continue to develop, informed by both their past and their future. The promises of planning's early advocates, as long as a century ago, to deal with such basic problems as providing affordable housing, dealing with traffic congestion, and disposing of human and industrial wastes, still need to be fulfilled, but in the form they present us with today. No small part of the difficulty is due to the changing nature of these problems: housing for the homeless has been added in the last two decades, gridlock is found in the suburbs as well as downtown, consumer and industrial products are more toxic today than ever, and the "face" of immigration, in this country of immigrants, has dramatically changed. New problems arise along with new versions of old problems. Thus, the true progress of community planning is as much a measure of how it deals with existing problems as well as how it anticipates future conditions.

This final chapter discusses a number of salient issues that will challenge Canadian community planning as it moves further into its second century. This will include looking at issues that are still in need of resolution, such as citizen participation, environmental protection, and affordable housing, as well as issues that will become more central in the years ahead, such as the population's aging, its health, and its increasingly multicultural character, and, not least, the future of the suburbs. These various challenges will, in turn, invoke the need for community planners to examine their own premises, presuppositions, and practice embedded in the following question:

- —***How will planning practice need to be modified to provide guidance appropriate to communities in the coming decades?***

Continuing and Emerging Challenges

The future shape of community planning will unfold from its past and also enfold new ingredients. This seemingly obvious statement is indicative of the somewhat paradoxical nature of the activity—it makes plans that promote change in a continually changing world. Those changes, planned and emerging, will challenge community planners both in terms of *substance* (the form and content of plans) and in terms of *process* (making and implementing plans), and often both at the same time. Although the full array of challenges that will face Canadian planners in the future can never be foreseen, the following are already insistent and, in them, one sees the scope of challenges:

- Achieving environmental integrity and sustainability
- Addressing housing, homelessness, poverty
- Responding to the seniors' surge
- Fulfilling citizen participation
- Reflecting multicultural realities
- Retrofitting the suburbs
- Responding to fear and safety
- Planning for healthier communities

Achieving Environmental Integrity and Sustainability

Toward Ecological Planning

Over the course of Canadian planning, a number of issues have been transformed and become more complex, but none more so than environmental protection. The long-standing planning traditions of conservation, park planning, and open-space preservation began to give way, about two decades ago, to concerns over ecosystems and biodiversity. The planners' response to ecological concerns in many ways has been commendable,[1] including among them the North Vancouver's Alpine Area Official Community Plan, Markham's Natural Features Study, and greenways planning for pedestrians and bicycles in Kelowna and Windsor. Nevertheless, local land-use planning is primarily about improving the built environment and achieving livability of a community; the defence of nature has often been left to regional or provincial initiatives. An inherent difficulty in planning for the two environments—natural and built—is that it raises an underlying conflict, as North Vancouver's and other planners have realized, of "protecting the natural environment vs. land development,"[2] or what would come to be called the "property conflict" in debates over sustainable development (see below).

We need to move beyond these dilemmas and see ourselves as part of the natural system of the planet and "re-design ourselves into the system of life."[3] Beyond this, planners need to strive for the "ecological restructuring of urban form and the development of ecological efficiencies in urban systems." The concept of compact urban form has received considerable attention as a model for urban environmental planning, especially the forms to reduce sprawl promoted by the New Urbanists (see Chapters 5 and 6). Although there is much to commend these forms for increasing urban density, more important is that development (compact or otherwise) has to be "in the right place, at the right time, in the right form" so as to not disrupt ecological systems.[4] Planners need to incorporate into their plan-making, from the beginning, an appreciation of the integral connection of human communities and natural ecological processes. One of the ways is to foster partnerships between governments and local conservation groups and neighbourhood organizations to acquire sensitive lands and to steward them.[5] More pointedly planners should undertake:

- —***To make the human-ecological mutuality a guiding principle for land-use planning.***

Sustainability: The Larger Picture

The broad concerns encompassed by the concept of sustainable development influenced environmental planning, after the 1980s, and have since become an idea with worldwide currency. However, sustainability is turning out to be an extremely complex issue; clearly, it reaches beyond simple questions of ecosystems protection and resource conservation. It involves three major facets: economic growth, social development, and environmental protection, or the three Es as they have been called—economy, equity, and environment.[6] A quick glance at these facets of sustainability indicates that even though each of them is integral to the planning of cities, towns, and regions, there is rarely mention of such entities by the World Commission, which defined sustainable development as "development that meets the needs of the present without compromising the ability of future generations to meet their own needs."[7] Yet the building (and planning) of a community constitutes "development," so how can it be made sustainable?

The overarching nature of sustainable development thus has the effect of presenting the planner with three broad goals of community planning that are integral to one another. In other words, the community plan should promote:

- Economic growth and efficiency;
- Environmental protection (natural and built); and
- Social justice and equity.

American planner Scott Campbell suggests that one imagine these as three points of a triangle in which each point, or goal, is linked to the other two.[8] In this form, one can quickly grasp that there are inherent conflicts in attaining each individual planning goal, much less all three. Campbell identifies them as:

> The **property conflict** between economic growth and social justice;
> The **development conflict** between environmental protection and social justice; and
> The **resource conflict** between economic growth and environmental protection.

One of these dilemmas, the "property conflict," has been evident to planners for some time. It goes back to the earliest days of zoning and other public land-use regulation, as well as the social intervention (such as urban renewal) in established neighbourhoods. It has also been evident (although not in these terms) to people in communities suffering the aftermath of major industrial development (i.e., economic growth) causing pollution of their backyards many years later (i.e., social equity). From the pollution in Calgary's Lynnview Ridge (due to a former oil refinery), to Wawa in Ontario (due to iron ore processing), and Sydney, Nova Scotia (due to a steel mill), residents have not only sought solutions but have also been frustrated in their attempts to determine who is responsible for undertaking a solution.[9]

The ideal position for the planner, the sustainability position, is at the centre of the triangle balancing off each of the goals. Even this is not a complete answer, as Godschalk has pointed out, for when planners tout "livability" they posit a competing visionary idea.[10] His resolution to this apparent conflict is to reconstruct the sustainability triangle by adding livability and forming a three-dimensional prism thereby incorporating all the principal values that are needed to broach sustainability in human settlement terms. Further refinement is needed in order to recognize, in actual communities, that sustainability will need to be tackled at different scales from the region to the neighbourhood. For example, the initiatives to design and gather indicators of progress toward sustainability, such as in Hamilton and Calgary,[11] have to recognize that indicators at the regional level will differ from those used at the citywide and neighbourhood levels. In the same vein, what would constitute sustainable housing is something that becomes apparent at the neighbourhood level. One Canadian study has determined that row housing is the most sustainable,[12] and that the "grow home" designed by McGill University's School of Architecture may be an affordable prototype.[13] Thus, scale makes a difference, and regional plans, community plans, and small-area plans will each need to employ different tools and mirror sustainable development need in different ways, hopefully, as an integrated set, an "ecology" of plans.[14] Planners, therefore, should:

• —***Promote dialogue about sustainability and design planning solutions that demonstrate it at all relevant scales.***

Addressing Housing, Homelessness, and Poverty

The problem of providing affordable housing for people of lesser means plagues planners and city councils just as insistently today as it did periodically through the 19th and 20th centuries. Slums, and what to do about them, were central to the concerns of the "utopian" planners of the 19th century, and comprised one of the prime issues for the fledgling planning profession from the beginning of the 20th century and, again, following World War II (see Chapter 5). Besides the persistence of this problem in inner-city areas, one now finds versions of it in suburban neighbourhoods.[15]

Why is this problem so resistant to resolution? The answer, in simple terms, is because ensuring the availability of affordable housing is a complex problem. It is what some call a "wicked problem," because dealing with one aspect can unleash a host of other problems, often of a fundamental nature, that we are not either equipped for or prepared to tackle. The issue of making affordable housing available means going beyond the physical problem of constructing dwellings to encompass both social and economic components, involving as it does the economically and socially vulnerable—the poor, recent immigrants, the elderly, single parents, the physically impaired, and others. Such a situation is vividly revealed in a Toronto study that shows higher-poverty neighbourhoods now predominantly comprise newcomers, visible minorities, and youth as compared to just 20 years ago.[16] Moreover, this is occurring in the city's older inner suburbs that were not originally endowed with strong social and economic infrastructure. A sense of the scale of the Toronto situation is given in the fact that 71 000 households were on the municipal waiting list for affordable housing in 2003.[17]

The problem of low-income families moving to the least expensive areas of a city may be caused by housing-transformation processes such as gentrification, condominium conversion, and redevelopment.

Gentrification. This process occurs when older neighbourhoods with solid housing close to downtown, and currently housing lower-income households, are sought by higher-income households and speculative renovators with the intention of upgrading the dwellings. The usual result includes the loss of rental units, the increase of rents on remaining units, and the displacement of current occupants to other low-cost districts.[18]

Conversion to Condominiums. Owners of older apartment buildings, many times faced with costly renovations, frequently seek either to convert them to condominium ownership or to raze the buildings and erect new condominium apartments. As with gentrification situations, current tenants often have lower incomes (frequently they are also elderly) and are unable to purchase a new unit and are, therefore, displaced. Similarly, conversions usually result in fewer units of higher cost with fewer people being housed.

Redevelopment. A residential area may be razed and replaced as a result of physical deterioration of the area, expansion of adjacent land uses, or space needed for public uses (e.g., highways). The area may not be economically competitive with other "higher" land uses, with the result that the current dwellings are completely withdrawn from the housing market. Most frequently, these are areas that house lower-income households who are then displaced.[19]

Homelessness

Homelessness, no stranger especially to larger cities, became increasingly evident in the 1990s. Not only have the numbers of homeless increased, but they now comprise youth and even children. One report suggests that more than 250 000 live either on the streets or in shelters in Canada. In the Vancouver metropolitan area, for example, the number of homeless was found to have doubled from 2002 to 2005 and those known as "street homeless" had more than tripled. Homelessness occurred in all the area's municipalities, inner city as well as suburbs.[20] Aboriginal people, notably women, also tend to be over-represented among the urban homeless in Vancouver and Prairie cities; indeed the "face" of homelessness varies by community and is no longer a "big-city" problem.

Some of the homeless population may be the outcome of the three processes described above or variations on them. Homelessness on the scale now being seen is a little-understood phenomenon but one can say, with certainty, that the people affected are bound to be poor. Public solutions are sparse and piecemeal; solutions of the homeless themselves range from squatting to the use of charitable shelters and living on the street.[21] These situations are not isolated cases but are one of the inevitable outcomes of urban growth and underfunding of social-housing programs. The economic marketplace is indifferent to these consequences, and the political arena, while it is obliged to listen, is not required to act to redistribute the costs of displacement.[22] Moreover, the social and human consequences are often not evident until the new development thrust is a *fait accompli*, by which time the constituency of the displaced must overcome the community's heady expectations of increased profits from economic development and an increased tax base; another instance of the "property conflict" between economic growth and equity noted above. Planning Issue 15.1 describes such a situation in Vancouver's Downtown Eastside, the poorest neighbourhood in Canada.

It is important to go beyond the images of homelessness often portrayed on TV or in newspapers, for it is a far larger problem than that experienced by those on the street or in shelters, as desperate as is their situation. When homelessness is tackled, as it is in cities large and small across the country, what is revealed is the general lack of affordable housing for many citizens. Take just two community plans for homelessness—London, Ontario, and Greater Vancouver—and both conclude that there will not be success "without an increase in the amount of stable affordable housing."[23] Greater Vancouver's Regional Homelessness Plan recommends a three-pronged approach, of which the first is the provision of affordable housing, along with support services, and income and employment reforms.[24] Both these plans recognize that many citizens, besides those on the street or in shelters, are "at risk of homelessness"; they are described in the Vancouver plan as those people

> living in spaces or situations that do not meet basic health and safety standards, do not provide for security of tenure or personal safety, and are not affordable. This also includes people considered as the invisible homeless, such as individuals who are 'couch surfing' or staying temporarily with family and friends.[25]

In Greater Vancouver, in 2005, the homeless numbered over 126 000 people.

The mundane planning issue of maintaining a supply of good-quality affordable housing turns out to be, in a paradoxical way, the factor that helps bring to fruition the grandiose schemes of development. It permits both the underclass and overclass to participate in the benefits of growth and development. The success of affordable housing ultimately comes down to the vision of the community that the planner has for the shelter and well-being of all segments of the population—of all *citizens*—and, not least, to transmit that vision effectively. Although many of the consequences of city change cannot be anticipated, planners can predict with little difficulty the kind and amount of displacement discussed above. An especially crucial question for the planner regarding the supply of affordable housing is: Morally, where does the planner stand on the issue? Thus, the ongoing challenge for planners is:

- ***To mitigate the social consequences of development proposals that affect the supply and quality of affordable housing.***

Responding to the Seniors' Surge

The rapid increase in the number of elderly people in cities and towns is well known to most Canadians. However, this trend and its implications have received little attention in the planning literature, despite the fact that those 65 and older are the fastest growing age group in our society.[26] In 2001, there were just over four million persons 65 and older, almost 12.6 percent of the total population. This represents a growth in the numbers of elderly of almost 30 percent over that of 1986. But, as substantial as it has been in the past, the seniors' surge still to come will be even greater. When the baby boom cohorts reach senior citizen status beginning in 2011—less than half a decade from now—the numbers of elderly will grow even more spectacularly until they double and account for more than one-quarter of the population at 2031.[27]

But numbers alone are not what make the surge in seniors' population relevant for planners. Not least, the surge of the baby boom cohorts into the ranks of the elderly will greatly diversify an already diverse seniors' population. To those who grew up in the 1920s and 1930s will be added those who grew up in the 1950s and 1960s. The "new" seniors' cohort will bring different social and economic experiences, as well as different perspectives on growing old compared to their older peers. The new cohort will include many foreign-born seniors, especially from Asian immigration since the 1970s, who will age along with the Canadian-born boomers. An even faster increase in the number of Aboriginal elders will be added to this mix of seniors in the same period. And, overall, seniors will be living longer and in far greater numbers. They will be living in almost every Canadian community, large and small, in inner cities or suburbs.[28]

Studies show that the three major needs of the elderly that impinge on the domain of the planner are housing, transportation, and community-support services and, further, that these three needs are closely associated with one another.[29] Regarding housing, seniors require a variety of options that take into account changing circumstances of physical ability, household composition, and income to be able to continue to make choices about their residence—to age in place, if they choose.[30] Collective transportation is also an essential element in the seniors' environment, where one-quarter or more are unlikely to have the ability or desire to drive.[31] Community-support facilities constitute a high priority for seniors' well-being and independence. These facilities should be broad in range (including, for example, seniors' centres, adult day care, and medical clinics). As well, shopping, banking, and restaurant facilities should be easily accessible by transit and/or by foot (preferably within 400–600 metres).[32]

Planning communities that enable seniors to express their independence and enjoy a high quality of life is both a worthy and a complex task. It goes well beyond the location and types of housing and the provision of transportation, although these are major elements, and involves concerns about 'belonging" and "making life meaningful" and, not

Planning Issue 15.1 A Squalid Landmark Is Going Upscale

Special to The Globe and Mail
May 23, 2006

A Squalid Landmark Is Going Upscale: Downtown Eastside Rooming Houses Selling for Millions
Jonathan Woodward Armstrong

VANCOUVER—Tenants who used to live at the Pender Hotel complained about the squalor, the police harassment and the howls of pain from certain rooms when drug deals went wrong.

But what was once one of the most notorious rooming houses in the Downtown Eastside is now at the forefront of a development boom. The Pender was closed and bought for $1.25-million this month by a company planning to build a boutique hotel on the site.

Andrew Fletcher, 46, was one of more than 30 tenants who faced homelessness when they were evicted before the hotel was sold. He says that even though life in the hotel was chaotic—animals roamed the hallways and the rooms, which rented for $325 a month, had no doors—its loss makes it harder for poor people to find a place to live.

"It was awful, but it was home. I left almost my entire life there," said Mr. Fletcher, who depends on social assistance and says he tried many places before he was able to find a new home in the nearby Argyle Hotel.

The Pender is not the only downtown residential hotel to go on the selling block.

After being closed for violations of the fire code, the Burns Block on Hastings Street was emptied of its 18 tenants and is now on the market for $2.5-million.

The asking price is five times what owner Nik Bahrami paid in 2003, a sign of the brisk business landowners believe can be done after the redevelopment success of the old Woodward's building, which is being turned into condos.

City planners are now considering more than two dozen applications for development in the area, and discussions are under way with developers about how to navigate bylaws that restrict the conversion of hotels, said Catherine Wong, a municipal employee who works with the Eastside hotels.

"The reason the market is so hot? It has to be the Woodward's building," she said.

While some people welcome the influx of capital to a traditionally downtrodden neighbourhood, others lament its impact on Vancouver's dwindling supply of low-income housing.

The Woodward's building has changed the business model for housing for poor Eastside residents, said lawyer David Eby of the Pivot Legal Society, a non-profit group that advocates for clients on the margins of society.

In the past, residential hotel owners depended on income from tenants, Mr. Eby said, and would respond if city inspectors threatened to shut them down for violations.

But now those buildings are worth more if they are empty. Any shutdowns are likely to help a sale move forward, saving buyers from the hassle of evicting tenants, he said.

"We're seeing the dark side of Woodward's [success]," Mr. Eby said. "Sure it's revitalizing the area, but it's a double-edged sword."

After threats of a legal battle with Mr. Eby over how the Pender residents were evicted, Mr. Radbourne agreed to pay each tenant $2,000 to leave.

Some Pender tenants left the Eastside altogether. Others, like 57-year-old Brian Sutton, found places to live nearby. Mr. Sutton's new place is $465 a month—a sizable increase from the $325 he paid at the Pender.

"It's tough," said Mr. Sutton, who receives social assistance. "I'm living comfortably, but I can't afford to go out any more. But it's worth it not to see the cops every day."

Source: Jonathan Woodward Armstrong, "A Squalid Landmark is Going Upscale," *The Globe and Mail,* May 23, 2006, S.2. Reprinted with permission from *The Globe and Mail.*

least, safety.[33] There has also to be a concern over details in the environment such, as sidewalk surfaces, the placement of benches, well-lit transit stops, legible signs, and snow clearance.[34] An example of what can be accomplished in this regard is the Urban Braille System to assist the severely visually impaired, the elderly, and the infirm in "navigating" public spaces such as sidewalks and bus stops that has been developed by Hamilton, Ontario, planners.[35] This concern for details needs to include housing features such as the "visitability" initiative of CMHC that aims to promote accessibility to and within homes of seniors so that they may visit and be visited by others.[36]

The planning perspective that will be required should involve a high degree of integration of programs and coordination of decisions of agencies within and outside local governments, a mode of practice that is difficult to achieve and not frequently seen. And popular concepts of urban design such as New Urbanism and Smart Growth should be assessed to ensure that they incorporate the needs for seniors.[37] These are important tasks, for, next to health-care professionals, community planners probably have the greatest influence over the ability of seniors to maintain their independence. The main challenge for community planners is, thus:

- —***To commit to including seniors' needs explicitly in all community plans.***

Fulfilling Citizen Participation

Major strides have been made in the past two decades to engage citizens in the planning of their communities. It is now fairly common practice to consult citizens about their ideas and preferences for future development in large cities and small towns.[38] This progress means that community planning in Canada has reached the middle-rung—*consultation*—on Arnstein's ladder of participation (see Figure 12.3, page 312). However, going further upward into the area of sharing power with citizens in deciding upon plans and programs remains a quest still to be fulfilled.

Although widespread consultation occurs, this achievement is fragile. Offering the opportunity to citizens to participate does not guarantee that all are able to, even if willing. To be truly inclusive means being aware of possible barriers to participation because of language, timing, physical access, cultural custom, and so on for some citizens. It also means utilizing many techniques, from the community-wide information meetings to neighbourhood meetings, citizen panels, and focus groups, and beyond these to interactive cable-TV and Web-based dialogues.[39] Moreover, a good deal of community consultation tends to favour more organized segments of the public and thereby exclude or not provide for an equal voice for those less well prepared. This may require technical and/or financial assistance (e.g., for translation and transportation) to ensure more complete participation.

Citizen participation is constrained, not least, by our legislated means for consultation and public hearings. These remain, essentially, one-way means of communication with no provision either for dialogue or for ensuring citizens have been heard. Based as they are on plans or other proposals already made (often in some detail) there is no role for the citizen but to react; she or he has no entitlement to an answer to her or his questions and assertions.[40] Ideally, it would be preferable if other means of two-way communication were allowed for in legislation. But, in lieu of that, the best course is to determine the public's pulse well before the public hearing and use collaborative means to deal with contentious issues.

Language is one of the confounding issues in citizen participation for both citizens and planners, regardless of the degree of inclusiveness. Words written down or spoken, by professionals or citizens, often convey such different meanings that little meaningful communication actually takes place.[41] As one planner notes:

> The core of planning activity involves bringing the different interests of the private and public sectors into a certain balance and therefore planners need to speak several "languages"... [but] it appears over the past few decades that planners have, to a certain extent, lost their ability and readiness to *listen* and to *speak*.[42]

Other uses of language in planning can also be problematic, such as the too-easy use of words such as "activist" to label those who persist in making their points, much less having them listened to.[43] The same occurs with the use of "NIMBY" whenever citizens question the siting of development projects in their neighbourhoods. Opposition can take several forms, but only when opponents approve of the type of project but reject it being located in their neighbourhood is it true NIMBY. Furthermore, attitudes frequently change when details of a project become known and/or it is shown that the plans, when compared with local knowledge, are faulty.[44] Lastly, there is the fashionable tendency to refer to citizens as "customers," apparently to parallel the positioning of municipalities as businesses proceeding in a business-like way.[45] Referring to citizens as customers also obscures the issue of citizens' rights rather than just being passive recipients of services; it also reinforces an all-too-frequent "ambivalence" on the part of planners and politicians about the idea of *citizen* participation.[46] Some of this wariness may be attributable to the often high level of emotions citizens bring to meetings, emotions bred frequently from the intensity of their attachment to their "place," their neighbourhood.[47] The people who participate, and those who might, are citizens, and they need planning processes that give them more than a place to vent their ideas and feelings, to be more than just consulted. Planning partnerships with citizens are long overdue. Planners are at the crux of this challenge. Thus it is in their interest:

- —***To establish and maintain fair, open, and collaborative participation processes for all citizens.***

Reflecting Multicultural Realities

New immigrants are reshaping the cultural identities of many Canadian communities. Although most communities have faced this change more than once in the past, Canada's recent immigrants come with a wider array of ethnic backgrounds, languages, religious perspectives, and cultural traditions. Since the 1970s, Canadian immigration policy has been aimed at attracting people from non-European countries, with the result that today's immigrant stream comprises mostly (70 percent) visible minorities from a wide variety of Asian, African, and Latin American countries.[48] Moreover, most recent immigrants (over 90 percent) go to live in the country's 27 metropolitan areas; and of these, over three-quarters go to Toronto, Vancouver, and Montreal, where they have formed "visible minority neighborhoods," or "ethnic enclaves."[49] The scale of these changes should not be underestimated, since Canada's three largest cities are now among the world's most multicultural communities.

This raises new challenges for planners everywhere in Canada to broaden their own social perceptions. And new and old citizens alike are challenged to consider the emerging cultural landscape and its implications for planning. The planning issues that emerge with expanding cultural diversity cover the spectrum from housing and transportation to employment, commercial development, institutions, and community services. In other words, the content of the planning agenda is much the same as for other members of the community. However, the understanding—the meaning—of land-use terminology, planning concepts, and participation processes may differ significantly from one ethnic community to another.[50]

As well, incomes of residents in ethnic enclaves may differ, as may national origin, language, religion, family composition, class divisions, housing customs, shopping habits, and mobility.[51] The concept of "neighbourhood," for example, may have different spatial dimensions and land-use composition than that implied in planners' traditional norms. The tragedy of Africville (see Chapter 5) should remain a constant reminder to planners that they recognize "the cultural assumptions" embedded in their current planning practice when dealing with "land use," "activities," and "functions."

Indeed, seemingly technical issues become "cultural issues," and this has been nowhere more obvious than in zoning applications for places of worship.[52] Minarets and temple domes often don't conform to height limits, parking requirements may not be equitable because of seating arrangements used in mosques, and so forth. An apparently simple rezoning application to allow a funeral home in a residential area of Ottawa, a not uncommon land use in other parts of the city, met a storm of protest on cultural grounds concerning the disposition of the dead from new residents of Muslim, Sikh, and Chinese backgrounds.[53] It is easy to say that planners need to be able to cross cultural boundaries in such situations but, as a study of applications for minority places of worship in Montreal shows, it is neighbourhood social and political dynamics that are often central in resolving locational issues.[54]

Nevertheless, these types of challenges, occasioned by the increasing diversity of cultures, provide an opportunity to broaden planning processes, as well as heighten planners' cultural sensitivity and refine their practice. Information can be translated into the language of the neighbourhood, as is already done in several cities. Planning staffs can be augmented to include people who understand ethnic-community concerns. And the mode of public consultation may have to be adapted to the means of communication and the settings with which the community is most comfortable in articulating its aspirations. For example, Vancouver's CityPlan process made a special effort to involve ethno-cultural groups in its visioning process, with some success.[55] One of the foremost observers of this facet of planning, Canadian planner Mohammad Qadeer, notes that while such procedural modifications to practice are important, the aim should be to develop visions of community plans and programs that reflect the inclusion of multicultural realities. He describes this goal as being attained through a series of steps, a "ladder of planning principles supporting multiculturalism."[56]

Qadeer's "ladder" is shown in Figure 15.1. In his view, Canadian planners have reached level 3; generally, that is being sensitive to cultural differences on a case-by-case basis. Some cities such as Vancouver and Toronto now work at higher levels (4–5). The aim is, in his trenchant phrase, to go "beyond sensitivity" to specific group and project needs and enfold a true pluralism in our community planning that would "explicitly recognize promotion of community cultures, religious freedoms and human rights" in community plans.[57]Moving in this direction carries with it this challenge for planners:

- —***To develop visions, plans, programs, and standards that meet the needs of diverse communities within their communities.***

Retrofitting the Suburbs

The growth of cities after World War II was unprecedented, not just in numbers of people and houses but also in the spatial dimensions of urban community life. Traditional urban nodes, already surrounded by some suburbs, saw ever-expanding rings of suburbs added quickly so that, today, most Canadians live in post-1945 suburbs. This dramatic change is often viewed in a pessimistic light, with the suburbs being stereotyped as undifferentiated streetscapes that are inconvenient and energy-inefficient and that

Figure 15.1 A Ladder of Planning Principles Supporting Multiculturalism

Level	Principle
7	A multicultural vision of the development strategy for a city or region
6	Cultural and racial differences reflected in planning policies and acknowledged as bases for equitable treatment
5	Provision of specific public facilities and services for ethnic communities
4	Special District designation for ethnic neighbourhoods and business enclaves
3	Accommodation of diverse needs through amendments and exceptions, case-by-case
2	Inclusionary Planning Process—participation by and representation of multicultural groups on planning committees
1	Facilitation access by diverse communities to the planning department

Planning responses to multiculturalism can range over seven levels, from adapting administrative procedures to redefining goals that inform plans, policies, and programs.

Source: Mohammad A. Qadeer, "Pluralistic Planning for Multicultural Cities: the Canadian Practice," *Journal of the American Planning Association* 63:4 (Autumn 1997), 481–494, Figure 1. Adapted by permission of the *Journal*.

promote social anomie and gender discrimination.[58] Although this decentralization of cities has been apparent throughout history, in the past two decades, this low-density, scattered urban development has been re-coined as "sprawl."[59] Considerable effort has been directed toward eliminating sprawl, most recently under the umbrella of Smart Growth, which argues for more compact (i.e., higher density) communities with better transportation and protection of natural areas.[60] Debate over sprawl has ensued from its supposed ills to how to measure it and even over whether it really exists. One observer states candidly that "urban Canada has relatively little sprawl if this is thought of as 'haphazard, disorganized, poorly serviced and largely unplanned.'"[61]

This difference in viewpoint about the suburbs is as much about attitude as about substance. "Sprawl" is a pejorative, just as "slum" and "urban blight" were before; it is equally hard to define and fails to recognize that the suburbs are where *people* live. Essentially, people make conscious choices to move into these new communities. As suburbs mature, almost all major land uses start to be found there (or nearby) and so residents can, and do, conduct most of their daily lives within these localized "urban realms."[62] Moreover, planning efforts to redevelop older suburbs to accommodate higher densities often encounter the dilemma that residents want to maintain the existing densities.[63]

The debate over sprawl aside, there are two aspects of existing suburban development that need attention. The first of these is that a vast amount of housing is already 50 and more years old and much is considered obsolete. Recognizing this dilemma,

Montreal architect and planner Avi Friedman and his colleagues have been developing approaches that would allow suburbs to evolve gradually to meet current housing needs and become more complete.[64] A comparable effort aimed at retrofitting Quebec City post-war suburbs is being planned by that city's Inter Disciplinary Research Group on Suburbs.[65] The process in both cases would be resident-motivated and gradual, and the transformations would be small scale; its key is the built-in flexibility to allow each neighbourhood to change in its own way at its own pace. The process does, however, depend upon the availability of compatible land-use planning tools and regulations, thus posing the following challenge for planners:

- —***To develop appropriate plans and planning tools that will facilitate the transformation and renewal of older suburbs on the residents' terms.***

The second aspect of suburban areas needing attention reflects the aging of their population. The people who moved to the suburbs as young and middle-aged adults have now become seniors, or soon will be, such that more than half of all urban seniors live in the suburbs.[66] The import of this is that most suburbs were designed for families with children and with access, almost exclusively, by automobile. Housing tends to be uniform, single-family detached types, and commercial and service clusters are widely separated. Seniors, by comparison, are less mobile (as many as one-quarter do not have access to an automobile), subject to more physical impairment, and, with increasing age, often require alternative forms of housing. Given their natural desire to continue living in the neighbourhoods they've become accustomed to, suburban seniors face many constraints: stores and doctors' offices are not usually within walking distance, sidewalks might not exist even where transit stops are relatively near, and options for apartment living are rarely available if a spouse should pass on.

The general solution is to retrofit suburban neighbourhoods to suit the needs of their elderly populations, most of whom will want to continue to live there far into old age. The answers are relatively straightforward (as indicated in an earlier section): add sidewalks and good street lighting, add other housing types, implant small-scale clusters of health, social service, and commercial facilities, and devise transportation options that are flexible and appropriate. To do these things will require planners:

- —***To design (or, more likely, redesign) suburban neighbourhoods that facilitate the everyday needs of seniors.***

Responding to Fear and Safety

Many people, in the conduct of their daily lives, find themselves challenged by the physical configuration of streets and spaces that planners have facilitated being built. A major concern held by many city dwellers is the fear of crime and its counterpart, the lack of safety. It is increasingly pointed out that this fear, if not always borne out by actual crime, is abetted by the design of urban spaces.[67] As an example, the simple placement of bus stops can expose passengers to potential criminal behaviour, as when stops are placed adjacent to "negative land uses such as liquor stores, bars . . . pawn shops, etc." or "in front of surface parking lots, vacant buildings, or other dead space."[68] Such sites should be avoided. Fear in city life is only beginning to be understood as a response of people to the likelihood of criminal risk and sources of protection in the urban environments they experience.[69] Thus, planners and urban designers are challenged to find ways to develop urban spaces that address these fears without creating barriers to public use and interaction. This is, of course, a complex social issue that is wrapped up in poverty, social exclusion, anomie, and so on, in many areas that require social-policy approaches for their solution, as we see with homelessness.

Barriers to the use and enjoyment of urban spaces are also a concern of the physically impaired. The design of urban environments often excludes persons with impairments from accessing places and activities in a city—instead of being enabled they become *dis*-abled.[70] Barriers that may constrain, or even preclude, the simple activity of walking include steep steps, poor lighting, slippery surfaces, uneven and cracked sidewalks, and unsafe road crossings.[71] For the elderly, who are a large proportion of the physically impaired, such barriers frequently lead to falls; the latter is the leading cause of injury among Canadian seniors and for their admission to hospital.[72] Moreover, there are many types of impairment (e.g., physical, visual, mental, cultural), as well as many combinations and degrees within each of these, and what is a barrier to one person is not necessarily a barrier to others with a different impairment. Planners will need to learn more about the urban barriers encountered by impaired persons. They cannot assume that architects, engineers, landscape architects, and urban designers of urban projects and spaces appreciate the constraints with which impaired citizens often have to contend. Herewith lays another challenge for planners:

- ***—To ensure that the design of urban spaces does not contribute to fears about crime and safety and lead to disability for physically impaired citizens.***

Promoting Healthier Communities

The ambitious Healthy Communities concept embraced by the Canadian Institute of Planners in 1988, which focused on all aspects of health (physical, mental, emotional, and economic well-being), remains a goal worth pursuing.[73] Clearly, the condition of the built environment impinges on all these aspects as discussion of each of the challenges above has indicated. Whether it be fear of crime, homelessness, environmental degradation, respecting cultural differences, or providing walking milieus for seniors, all relate to human and community health and all involve the choices community planners' make and/or facilitate regarding the built environment. This is, of course, both a momentous and stupendous task to be undertaking and one not easily achieved as the intervening years have shown, for it invokes the need for integration and coordination of many policy areas, government departments, levels of government, and the private and nonprofit sectors.

There is, however, renewed interest in community environments and health that is being stimulated by widespread concerns over obesity and lack of physical activity in the population. Current interest focuses on how urban form, especially street patterns and their walkability, affects physical activity patterns and, thence, better overall health.[74] Evidence has been accumulating that residents walk more in areas with land-use patterns that have more street intersections and shorter blocks, patterns not commonly found in the suburbs.[75] This, in turn, has led to concerns about the overall role of suburban sprawl and its preponderance of low-density, single-use areas with disconnected street networks in discouraging active community environments.[76] As one might have expected, the research has led to a broader view of walking behaviour as being related not only to street patterns but also to safety and fear of crime and, further, to gender and age.[77] And it has also led to a more-nuanced view of community health wherein all the elements—the built and natural environments, the social environments, and individual and collective health—interact with each other in complex and adaptive ways.[78] The significant but limited success of the Healthy Community initiative in Canada (especially in Quebec) now needs to be revisited and revitalized. It thus behooves planners:

- ***—To promote the establishment of Healthy Community projects in their communities and to help sustain them.***

The Sufficiency of Practice

The topics in the foregoing section raise a basic question: *Is planning practice sufficient to meet the kinds of challenges that planners will have to confront?* It can be seen in this array that there are challenges yet to be fully met and those that have only recently emerged within the ever-changing societal milieu and its values. Indeed, one is struck by both the increasing number and diversity of the challenges and, within them, the demands to respond to the needs and capabilities of individuals and groups from diverse parts of the citizenry: demands for inclusion and participation in their community's plan-making. Though many would demur, planners are the lead professionals in this facet of public life in a community; they have the power to conduct plan-making processes. Further, these are not just challenges about the content of community plans, programs, and policies: they are, significantly, about the content and conduct of planning practice.

These and other challenges confront planners today with the need to consider their values, their moral positions, and, not least, their concept of the roles they play as professionals. Indeed, the demands of this new milieu for planners could require them to choose between the two contrasting roles, so trenchantly pointed to by Sandercock, of "anti-hero" or "passionate pilgrim."[79] The former refers to the neutral technical advisor role, so common for planners in institutionalized and bureaucratic settings, while the latter refers to playing a proactive advocacy role on behalf of citizens, a not-uncommon stance among planners in the 1930s and the 1960s. The planning profession, in turn, must consider whether and how it could encompass these distinctive perspectives within its activities. How would planning staffs function? Who would take the lead role in plan-making? How would this affect the content of community plans, or even the need for plans? This concluding section considers such questions with the aim of developing a more self-conscious, reflective approach to planning practice.

Whither Planners and Planning?

Accepting to meet even one of the challenges posed in this chapter will, at a minimum, demand the planner appraise both the *substance* and *style* of his or her current planning practice. The issues, which range across housing, the environment, the elderly and disabled, and multiculturalism, involve not only plans and their making but moral questions as well. There is today far greater community consciousness about both the potential for planning and its limits. This consciousness has given rise to concerns about the livability of communities and, more specifically, about economic opportunities and social justice. These tendencies argue for a planning process that is both more fine-grained and more inclusive. For city-building is a complex project with multiple actors from governments, the private sector, and civil society, "each of whom brings relevant knowledge to the table."[80] The active involvement of more participants will bring into focus the need for planners to work more and more with custom-made plans for particular neighbourhoods, functions, and projects.

Leaders or Followers?

Facing this sort of agenda will require transformations in planners' styles and planning-agency relations. For one, it will demand that planners acknowledge that planning and politics are linked even in the technical work of data gathering and analysis, modelling, forecasting, and so forth. Even there one has to make assumptions about the best course to take and thereby accept only some options with their "political implications and consequences."[81] Beyond this it will demand that planners know how to make their efforts politically effective, including encouraging citizen empowerment.[82] Given the bureaucratic orientation planners have adopted, can such changes be attained? Can (and will) planners share their powers with other participants? Not easily, if some recent reactions of planners are indicative. Many planners are frustrated by ungainly citizen-participation processes and see themselves as becoming subordinated to the lay public, as "abdicating" their role of shaping cities.[83] Planners may not, however, have the luxury of entertaining the issue of leadership in an increasingly complex world that demands a greater variety of expertise and a broader range of involvement. More fruitful would be to assume the role of initiating and guiding the formation of the partnerships among governments, the private sector, and citizen groups that are so necessary to achieving many of today's planning situations.

Short Term versus Long Term

A dilemma for many planners is how to achieve more timely and more certain implementation of plans and policies. From within the profession, there are calls to abandon long-term comprehensive planning in favour of a "short-term, problem-oriented" stance. It would probably be relatively easy for Canadian planners, with their bureaucratic leanings, to shift to such a perspective. While this may result in more rapid implementation of policies, it could lead to a less open planning process at a time when public sentiment is for more, not less, citizen participation. An approach such as strategic planning, which deliberately links a plan and the means for its implementation, might offer a way out of the dilemma, providing, of course, it allows adequate time for participation. In the end, it is necessary to recognize the need for planning at several scales, from the neighbourhood to the province and perhaps beyond, and that these can "only be loosely coordinated" for each addresses different problems within different time frames.[84]

If planners gravitate entirely toward short-term planning, long-term planning will be weakened. The major conceptual argument against long-term planning has been its inability to take into account unanticipated changes within its time frame—for example, the failure of projects to materialize, the emergence of new opportunities, and even disasters. However, the solution is not to substitute short-term planning (where vagaries also occur) but to institute *contingency* or *action* planning. To this one might add regular plan reviews, which planners in many communities today seem to do grudgingly, if at all, capital improvement programs, and short-term programs of planning activity that set out realizable steps. The lessening of planners' concerns for the long-term future also means a lessening of planners' "capacity to deal with it," as Harvey Perloff cogently noted.[85] If not the planners, who will deal with this important dimension? Who will supply the images of the future?

How Do We Know What We Know?

Regardless of the level at which they practice, planners are looked to by others (colleagues, the public) to provide forecasts about future conditions and outcomes. Yet, too often, these forecasts either overstate or understate the situations they seek to portray.[86] While they tend to fade in prominence, into background studies, forecasts have an impact on the amount and source of resources that are mobilized in implementing a plan. Not only must planners work to improve their means of analysis and forecasting, but also be more aware of their use and impact. In a related way, insufficient attention is paid to developing skills in critical appraisal of data and research findings in the training of planners. The dearth of case studies in the planning literature adds further to this and means that there is no easy way for planners to know what works and what doesn't. This is especially important as planning becomes increasingly interactive, involving multiple actors of whom each brings their own rational understanding to the task. These, too, can be knowledge-producing situations but their value may be diminished unless the planner also has the "organizational and political savvy" to take advantage of them.[87] Another form of savvy is for planners to be aware of the presuppositions that guide much practice. (A few of these are examined in Figure 15.2).

Figure 15.2 Ten Things Every Planner Knows...

Every field of activity is based on *presuppositions*. It is important in reading this book, or any other planning document, to know the presuppositions of planners, to insist that they be articulated. Some of the presuppositions listed below may challenge accepted propositions. The aim is to improve the basis on which we do community planning.

1. To Make a Plan Is to Advocate a Position
A plan constitutes a combination of all the elements of land use in a workable relationship. If it is workable, it is defensible. It constitutes one position regarding how the urban pattern could work. Alternative plans constitute other preferred combinations. One should also ask: can a planner truly recommend more than one plan?

2. Planners Cannot Keep All Their Options Open
In the face of the uncertainties that abound in planning it is tempting not to make a plan for the future. Some would contend that this allows them to deal with new events and conditions when they arise. This approach means, of course, that at least one major option—making a firm commitment to attain desired future conditions—has been foreclosed.

3. One Rezoning Does Not Lead to Wholesale Rezoning
By convention, rezoning is considered for one project at a time through variances or the recognition of changed conditions affecting viability of a project. Whether it will be seen by other landowners in the same zone as a precedent depends upon their individual aims and available resources. In any case, a precedent only exists if the zoning authority wishes to treat the particular situation that way. If a plethora of other rezoning applications arise, the zoning regulations probably should be changed for all landowners.

4. Zoning is Necessary, but Not Sufficient for Development
Grandiose development schemes and a sense of urgency usually accompany (re)zoning applications. All too often, no construction follows a successful application. Since zoning confers development value on a property, this usually results in large potential profits being made merely by getting zoning approval. That is as much the aim as actually building something, at this stage of the process.

5. Public Approval Processes Do Not Slow Down Development
Government regulations regarding subdivisions and zoning are an easy target for those desiring to expand development. However, studies have failed to confirm that such regulations have been the culprit, to any appreciable degree, in reducing the pace of development. The development process

It is now widely assumed that planners cannot rely just on technically based professional knowledge. Planning is "communicative." It is about having dialogues with others and through such dialogues will be introduced other ways of knowing from the various participants. With demands for greater inclusion of more participants, this new "knowledge" must have an equal role to that of professional knowledge in planning processes.[88] Planners thus have the opportunity to know more about the situations under deliberation, knowledge that does not meet the same criteria as scientific, technical knowledge, knowledge that comes from participants' own experience, personal stories, the images and representations participants employ in making their points, and from the intuition of participants.[89] As planners collaborate more with others, the richness of knowing a planning problem from many perspectives and, thus, its various possible solutions, will become evident to all participants.

Professional and Personal Ethics

Planners frequently find themselves facing dilemmas in their professional worlds that impinge on their personal beliefs and the standards they apply in their work. Throughout this book, the normative side of community planning has been highlighted, as have the notions of greater inclusiveness and empowerment of all citizens. Implicit in them are value-based choices of the authors that we, in turn, advocate to those involved in community planning, professional planners as well as others. More than this, these choices reflect involves many actors, besides planners, who have the ability to slow a process, not least those providing financial backing.

6. A Shortage of Subdivided Land Does Not Cause the Price of Housing to Rise
The cost of housing is a function of location, cost of commuting, amenities, types of housing, and consumer preferences exercised in many geographic and social submarkets that have few substitutes. It is essentially behavioural. Subdivided land is a physical entity that will create yet another submarket, on the periphery, if it gets built upon. Subdivided land is to housing costs what a sperm bank is to the birth rate. They are different logical types, as Bateson would say.

7. The Home-to-Work Journey Seldom Exceeds 40 Minutes
For over a century, through changes in city size and modes of transportation, the vast majority of commuters have chosen not to travel more than 40 minutes to or from work. This self-imposed limit is achieved through personal adaptations to the urban environment such as changing residences and/or jobs, modifying working hours, and finding new routes and modes.

8. Things Have to Get Better Before They Get Worse
Hans Blumenfeld many years ago observed that traffic congestion in cities seldom improved for very long by adding more capacity. Cities are self-adjusting systems of land use and those seeking access to it thereby generate a certain level of congestion. It won't worsen unless steps are taken to reduce it by adding more lanes to the freeway or improving public transit that make the land use more accessible and more in demand.

9. Rapid Transit Stations Do Not Cause High Density Nodes
Rapid transit stations are located at accessible junctions to serve existing activity centres and/or link with surface transportation. These are usually traditional nodes that have reached a level of development commensurate with their locale. All a rapid transit line does is bring people to the junction more efficiently so that they may be dispersed more efficiently by bus or car to surrounding areas. Dreams of high-rise development around stations are just that. The political will has to be there, the zoning has to be in place, and the developers have to see the potential. Consider the surroundings of so many stations that never see much change at all.

10. Canadian Cities Will Never Have the Densities of New York and Paris
Toronto, Montreal, and Vancouver are not simply at a lesser "stage of development" than New York or Paris, such that they will eventually have the latters' densities of people and buildings. Population densities of cities and their Floor Space indexes are a product of the local, perhaps even national, cultural milieu. They reflect cultural norms on residential development and building bulk. In effect, city growth and density do not have a linear relationship over the long run.

aspects of planners' work that, increasingly, thrust the "value-laden nature" of the planning profession to the fore.[90] They are intended as an invitation to Canadian planners to question their decisions, the directions of their organizations, the tools they use, and the plans they help make in light of the ethical dimensions they employ. For some the question will be the relationship between planning and politics,[91] while for others it will be the question of mediating between "big picture planning" and the "human dignity of individuals."[92] And these are only two ethical issues of many that surround planners' professional lives and invoke responses, knowingly or otherwise, within themselves as individuals acting from within their own belief systems. As Larry Beasley told fellow planners in 2004: "We must face ourselves and know ourselves because what we intend is often just as influential as what we do."[93]

Reflections

A Future of Challenges

Communities and planners face a very different set of conditions today than were encountered in the recent past and, moreover, there can be little doubt that even these conditions will continue to change. We have come to see many issues in their true complexity and have identified more clearly the competing choices, as is the case with the meta-problem of *sustainability*. The active discourse about *gender issues* serves to alert us that other groups and interests need to be included in planning theory *and* practice. And while the planning process in most communities tends to involve more people, in most it is still at the middle rung—consultation—of Arnstein's ladder of participation, and hence there is the consequent, and often valid, complaint that the process is not much more than "public nattering."[94] Add to these the globalization of the economy, fiscal constraints, and the rise of civil society, and there is no doubt that planners face a future of challenges.

Not least will be the challenge to the ongoing practice of planning in terms of how planners see their role in an increasingly complex and dynamic world. The words of several Canadian planners may help in conveying the shape of the planning task ahead, an ungainly one at best:

Jill Grant: "Our role involves exposing issues and options for those who make decisions and to those affected by the decisions. Our professional credibility depends on openness about our assumptions and transparency in process."[95]

Larry Beasley: "Planners must show real leadership and not just be factotums for politicians or only do background studies. We must inspire, mentor and guide. We must tap interests—not just take positions. We must find solidarity with our citizens. . . ."[96]

Ann McAfee: "Experience suggests planners do not have all the 'right' answers. What planners have are techniques to analyze information and assess consequences. Planners have a responsibility to suggest new ways to meet new challenges. What planners do not have the right is to choose, and thereby limit, the options society considers when difficult choices have to be made."[97]

Doug Aberley: "Reforming the practice of planning in this new millennium will pose a far greater challenge than reforming the theories of the discipline. It is not a matter of reorienting the activities of a few planning theorists, but of thousands of working planners who occupy many types of institutional niches."[98]

Art Cowie: "[H]ow many planners do you know who are constantly second-guessing mayors and councils? The roles have to be separated, yet it is the interplay between the two that makes local planning so fascinating and keeps me coming back for more."[99]

In the end, planning will need to adapt itself further if it is to respond effectively to the issues of its second century, not just those described above, but those that are bound to arise as time continues to pass. For change is a "core concept of planning" and, at the very least, something planners need to anticipate.[100] The planner inhabits a milieu of change: there will be those changes she or he advocates along with those that come unbidden from new societal interests and structures. The essential challenge for planners is, thus:

- —***To prepare ourselves to see opportunities for change both in what we know and in what may await us.***

Endnotes

1. Cf. Special issue of *Plan Canada* devoted to conservation, 29:5, Marie Lessard, ed. (September 1989).
2. Desmond Smith, "Local Area Conservation," *Plan Canada* 29: 5 (September 1989), 39–42.
3. Mary-Ellen Tyler, "Ecological Planning in an Age of Myth-Information," *Plan Canada* 40:1 (December 1999–January 2000), 20–21; see also Timothy Beatley, "Preserving Biodiversity: Challenges for Planners,"*Journal of the American Planning Association* 66:1 (Winter 2000), 5–20.
4. Philip R. Berke and Maria Manta Conroy, "Are We Planning for Sustainable Development?" *Journal of the American Planning Association* 66:1 (Winter 2000), 21–32.
5. Jeff Ward, "Not by Government Alone: B.C. Partnerships in Green Planning," *Plan Canada* 41:3 (July–September 2001), 38–39.
6. Philip R. Berke, "Does Sustainable Development Offer a New Direction for Planning? Challenges for the Twenty-First Century," *Journal of Planning Literature* 17:1 (August 2002), 21–36.
7. World Commission on Environment and Development, *Our Common Ground* (Oxford and New York: Oxford University Press, 1987), 8.
8. Scott Campbell, "Green Cities, Growing Cities, Just Cities?: Urban Planning and the Contradictions of Sustainable Development," *Journal of the American Planning Association* 62:3 (Summer 1996), 296–312.
9. Coincidentally, three similar pollution situations were reported during August 2001: Kerry Williamson, "Report Warns of 10-Year Cleanup," *Calgary Herald,* August 11, 2001; Bill Schiller, "Town Plagued by Fears Over Arsenic," *The Toronto Star,* August 26, 2001; and Tera Camus, "Sydney's Toxic Woes Widespread," *Halifax Herald*, August 6, 2001.
10. David R. Godschalk, "Land Use Planning Challenges: Coping with Conflicts in Visions of Sustainable Development and Livable Communities," *Journal of the American Planning Association* 70:1 (Winter 2004), 5–13.
11. Cf. Noel Keough, "Calgary's Citizen-Led Community Sustainability Indicators Project," *Plan Canada* 43:1 (Spring 2003), 36–36.
12. Karen Ramsey, "What is Sustainable Housing?" *Plan Canada* 42:4 (October–December 2002), 25–26.
13. Avi Friedman, *The Grow Home* (Montreal and Kingston: McGill-Queen's University Press, 2001).
14. Godschalk, "Land Use Planning Challenges."
15. Geneviève Vachon, Gian Piero Moretti, and Julie Vaillant, "Retrofitting Quebec City's Postwar Suburbs: The Renovation and Densification of Rental Housing Developments," *Plan Canada* 46:1 (Spring 2006), 12–15.
16. United Way of Greater Toronto, *Poverty by Postal Code: The Geography of Neighbourhood Poverty, City of Toronto 1981–2001* (Toronto, 2004), 54.
17. City of Toronto, *The Toronto Report Card on Homelessness* (Toronto, 2003), available online at www.toronto.ca/homelessness/index.
18. The plight of Toronto neighbourhoods in the 1980s, as described in Leigh Howell, "The Affordable Housing Crisis in Toronto," *City Magazine* 9:1 (Winter 1986), 25–29, has its parallels today.
19. Aldo Santin, "Tenants Facing Eviction Smell City-Hall Scam," *Winnipeg Free Press,* August 26, 2001.
20. Greater Vancouver Regional District, *On Our Streets and in Our Shelters* (Vancouver, 2005), 1.
21. Allison Hanes, "Squatters Feel a Lot of Love," *Montreal Gazette*, August 11, 2001; and Torstar News Service, "Homeless Bounced Like Tennis Ball: Hamilton Mayor," *The Toronto Star*, July 19, 2001.
22. Jonathan Woodward, "Squalid Landmark is Going Upscale," *The Globe and Mail* (Vancouver edition), May 23, 2006, S2.
23. Ginsler and Associates, *Community Plan on Homelessness in London* (Ginsler and Associates Kitchener, ON:, October 2001).
24. Social Planning and Research Council of B.C., *Three Ways to Home: The Regional Homelessness Plan for Greater Vancouver* (Vancouver, 2003).
25. Ibid., 5.
26. Mark Rosenberg and J. Everitt, "Planning for Aging Populations: Inside or Outside the Walls," *Progress in Planning* 56:3 (October 2001), 119–168.
27. Statistics Canada, *Population Projections for Canada, Provinces and Territories 2005–2031,* Cat. no. 91-520-XIE (Ottawa: Statistics Canada, 2005).. Scenario No. 2 is used here.
28. Gerald Hodge, *The Geography of Aging in Canada* (Vancouver: UBC Press, forthcoming 2007).
29. Larry Orr, "An Aging Society: a Municipal Social Planning Perspective," *Plan Canada* 30:4 (July 1990), 42–45.
30. Mary Catherine Mehak, "New Urbanism and Aging in Place," *Plan Canada* 42:1 (January–March 2002), 21–23.
31. David R. Ragland, William A. Satariano, and Kara E. MacLeod, "Reasons Given by Older People for Limitation or Avoidance of Driving," *The Gerontologist* 44:2 (2004), 237–244.
32. James Wilson, "Assessing the Walking Environments of the Elderly," *Plan Canada* 21:4 (1982), 117–121.
33. City of Regina, Regina and District Seniors' Action Plan Steering Committee, *Seniors and Safety Working Paper* (Regina, 2000), 4ff.
34. Gerald Hodge, "The Seniors' Surge: Why Planners Should Care," *Plan Canada* 30:4 (July 1990), 5–12.
35. Sinisa Tomic, "Hamilton Urban Braille System; Urban Design for an Aging Society," *Plan Canada* 43:1 (Spring 2003), 41–43.
36. Joel Casselman, "Visitability—A New Direction for Housing in Canada," *Plan Canada* 46:1 (Spring 2006), 34–37.
37. Deborah Howe, *Aging and Smart Growth: Building Aging-Sensitive Communities* (Miami: Funders Network, 2001), 8; available online at www.fundersnetwork.org.
38. Dick Ebersohn, Kevin Froese, and John Lewis, "ImagineCalgary: A 100-Year Vision for Sustainability," *Plan Canada* 45:4 (Winter 2005), 35–37; and Kevin S. Hanna, "Planning For Sustainability: Experience in Two Contrasting Communities," *Journal of the American Planning Association* 71:1 (Winter 2005), 27–40.
39. Judith E. Innes and David E. Booher, "Reframing Public Participation Strategies for the 21st Century," *Planning Theory and Practice* 5:4 (December 2004), 419–436.
40. Ibid.
41. Heather Campbell, "Talking the Same Words But Speaking Different Languages: The Need for Meaningful Dialogue," *Planning Theory and Practice* 4:4 (December 2003), 389–392.
42. Marta Doeler, "How to Communicate as a Planner," *Planning Theory and Practice* 3:3 (December 2002), 364–365.
43. Ben Bennett, "An Activist's Experience of the Planning Process," *Plan Canada* 43:3 (January–March 2002), 14–16.
44. Maarten Wolsink, "Entanglement of Interests and Motives: Assumptions Behind the MINBY-ism Theory on Facility Siting," *Urban Studies* 31:6 (1994), 851–866.
45. Robert Shipley, "The Sinister Implications of Language: The Difference Between a Citizen and a Customer," *Plan Canada* 43:1 (Spring 2003), 28–30.
46. Innes and Booher, "Reframing Public Participation."
47. Lynne C. Manzo and Douglas D. Perkins, "Finding Common Ground: The Importance of Place Attachment to Community Participation and Planning," *Journal of Planning Literature* 20:4 (May 2006), 335–350.

48. Statistics Canada, *Canada's Ethnocultural Portrait: The Changing Mosaic, 2001,* Census Analysis Series, Cat. no. 96F0030XIE2001008 (Ottawa, 2003), 39.

49. Feng Hou and Garnet Picot, "Visible Minority Neighbourhoods in Toronto, Montreal, and Vancouver," *Canadian Social Trends* 72 (2004), 8–13.

50. Sylvie Grenier, "Urban Planning in a Multicultural Society," *Plan Canada* 41:3 (July–September 2001), 31.

51. Mohammad Qadeer and Sandeep Kumar, "Ethnic Enclaves and Social Cohesion," *Canadian Journal of Urban Research* 15:2 (Summer Supplement 2006), 1–17.

52. Marcia Wallace and Beth Moore Milroy, "Ethno Racial Diversity and Planning Practices in the Greater Toronto Area," *Plan Canada* 41:3 (July–September 2001), 31–34.

53. Nancy Smith, "Diversity: The Challenge for Land Use Planning," *Plan Canada* 40:4 (July–August–September 2000), 27–28.

54. Annick Germain and Julie Elizabeth Gagnon, "Minority Places of Worship and Zoning Dilemmas in Montreal," *Planning Theory & Practice* 4:3 (September 2003), 295–318.

55. Joyce Lee Uyesugi and Robert Shipley, "Visioning Diversity: Planning Vancouver's Multicultural Communities," *International Planning Studies* 10:3,4 (August–November 2005), 305–322.

56. Mohammad Qadeer, "Pluralistic Planning for Multicultural Cities: The Canadian Practice," *Journal of the American Planning Association* 63:4 (Autumn 1997), 481–494.

57. Mohammad Qadeer, "Dealing With Ethnic Enclaves Demands Sensitivity and Pragmatism," *The Ontario Planning Journal* 20:1 (January–February 2005), 10–11.

58. One sample of the lament about suburbs is found in John Sewell, *The Shape of the City: Toronto Struggles with Modern Planning* (Toronto: University of Toronto Press, 1993).

59. Robert Bruegmann, *Sprawl: A Compact History* (Chicago: University of Chicago Press, 2005), 17.

60. Cf. Don Alexander, Ray Tomalty, and Mark Anielski, "The Challenges in Implementing a Smart Growth Agenda: The BC Sprawl Report 2004," *Plan Canada* 45:4 (Winter 2005), 23–26.

61. Larry S. Bourne, "The Urban Sprawl Debate: Myths, Realities and Hidden Agendas," *Plan Canada* 41:4 (October–December 2001), 26–28.

62. John Friedmann, "Urban Communes, Self-Management and the Reconstruction of the Local State," *Journal of Planning Education and Research* 2:1 (Summer 1982), 37–53.

63. Andrejs Skaburskis and Dean Geros, "The Changing Suburb: Burnaby B.C. Revisited," *Plan Canada* 37:2 (March 1997), 37–45.

64. Avi Friedman, *Planning the New Suburbia: Flexibility by Design* (Vancouver: UBC Press, 2001).

65. Vachon et al., "Retrofitting Quebec City's Postwar Suburbs."

66. Gerald Hodge, *The Greying of Canadian Suburbs: Patterns, Pace and Prospects* (Ottawa: CMHC, 1996), 5.

67. C. Hale, "Fear of Crime: a Review of the Literature," *International Review of Victimology* 4 (1996), 79–150.

68. Anastasia Loukaitou Sideris, "Hot Spots of Bus Stop Crime: The Importance of Environmental Attributes," *Journal of the American Planning Association* 65:4 (Autumn 1999), 395–408.

69. Jo Bannister and Nick Fyfe, "Fear and the City," *Urban Studies* 38:5,6 (May 2001), 807–813.

70. Here we are following the distinction made between *impairment,* when a person has a defective limb or organism, and *disability,* when an impaired person is excluded from public space by barriers of various kinds. Cf. Brendon Gleeson, *Geographies of Disability* (London: Routledge, 1999), 25.

71. Anastasia Loukaitou-Sideris, "Is it Safe to Walk? Neighbourhood Security Considerations and Their Effect on Walking," *Journal of Planning Literature* 20:3 (February 2006), 219–232.

72. Paula C. Fletcher and John P. Hirdes, "Risk Factors for Serious Falls Among Community-Based Seniors: Results from the National Population Health Survey," *Canadian Journal on Aging* 21:1 (2002), 103–116.

73. David R. Witty, "Healthy Communities: What have We Learned?" *Plan Canada* 42:4 (October–December 2002), 9–10.

74. Lawrence Frank, Peter O. Engelke, and Thomas L. Schmid, *Health and Community Design: The Impact of the Built Environment on Physical Activity* (Washington, DC: Island Press, 2003).

75. Cf. B. Giles Corti and R. Donovan, "Relative Influences of Individual, Social Environmental, and Physical Environmental Correlates of Walking," *American Journal of Public Health* 93:9 (2003), 1583–1589.

76. David Charles Sloane, "From Congestion to Sprawl: Planning and Health in Historical Context," *Journal of the American Planning Association* 72:1 (Winter 2006), 10–18.

77. Scott Doyle, Alexia Kelly Schwartz, M. Schlossberg, and J. Stockard, "Active Community Environments and Health: The Relationship of Walkable and Safe Communities to Individual Health," *Journal of the American Planning Association* 72:1 (Winter 2006), 19–31; and Loukaitou-Sideris, "Is it Safe to Walk?"

78. Sholom Glouberman et al., "A Framework for Improving Health in Cities: A Discussion Paper," *Journal of Urban Health* 83:2 (2006), 325–338.

79. Leonie Sandercock, "A Portrait of Postmodern Planning: Anti-Hero and/or Passionate Pilgrim," *Plan Canada* 39:3 (May 1999), 12–15.

80. John Friedmann, "Globalization and the Emerging Culture of Planning, *Progress in Planning* 64 (2005), 183–234.

81. Leonie Sandercock, "Toward a Planning Imagination for the 21st Century," *Journal of the American Planning Association* 70:2 (Spring 2004), 133–141.

82. Elizabeth M. Rocha, "A Ladder of Empowerment," *Journal of Planning Education and Research* 17 (1997), 31–44.

83. Michael and Julie Seelig, "CityPlan: Participation or Abdication?" *Plan Canada* 37:2 (May 1997), 18–22.

84. Friedmann, "Globalization."

85. Harvey S. Perloff, *Planning the Post-Industrial City* (Washington: American Planning Association, 1980), 87.

86. Andrejs Skaburskis and Michael B. Teitz, "Forecasts and Outcomes," *Planning Theory and Practice* 4:4 (December 2003), 429–442.

87. Karen S. Christensen, "Teaching Savvy," *Journal of Planning Education and Research* 12 (1993), 202–212.

88. Karen Umemoto, "Walking in Another's Shoes: Epistemological Challenges in Participatory Planning, *Journal of Planning Education and Research* 21 (2001), 17–31.

89. Judith Innes, "Information in Communicative Planning," *Journal of the American Planning Association* 64:1 (Winter 1998), 52–63.

90. Sue Hendler, "It's the Right Thing To Do—Or Is It?" *Plan Canada* 42:2 (April–June 2002), 9–11. Many planning professional associations have adopted some form of a code of ethics. See, for example, the CIP Statement of Values and Code of Professional Practice, available online at www.cip-icu.ca/English/members/practice.htm.

91. Ian Wight, "Mediating the Politics of Place: Negotiating Our Professional and Personal Selves," *Plan Canada* 42:2 (April–June 2002), 21–23.

92. Mark L. Dorfman, "The Dilemma of Big Picture Planning: A Question of Ethics," *Plan Canada* 42:2 (April–June 2002), 12–13.

93. Larry Beasley, "Moving Forward in Canadian Communities: Soliloquy of an Urbanist," *Plan Canada* 44:4 (Winter 2004), 16–19.

94. John Dakin, "Heading for New Minds?" *Plan Canada* 34:2 (July 1994), 93–98.
95. Jill Grant, "Rethinking the Public Interest as a Planning Concept," *Plan Canada* 45:2 (Summer 2005), 48–50.
96. Beasley, "Moving Forward."
97. Ann McAfee, "When Theory Meets Practice: Citizen Participation in Planning," *Plan Canada* 37:3 (May 1997), 18–22.
98. Doug Aberley, "Telling Stories to the Future: Integrative Approaches to Planning," *Plan Canada* 40:1 (December–January 2000), 24–25.
99. Art Cowie, "Politics and Planning: Ten Lessons from an Old Campaigner," *Plan Canada* 43:3 (Autumn 2003), 18–20.
100. Gary Davidson, "Rummaging in the Compost," *Plan Canada* 39:3 (May 1999), 15–16.

Internet Resources

Chapter-Relevant Sites

Planning Canadian Communities
www.planningcanadiancommunities.ca

Active Living Coalition for Older Adults
www.alcoa.ca

Health Canada, Division of Aging and Seniors
www.hc-sc.gc.ca/seniors-aines

National Round Table on the Environment and the Economy
www.nrtee-trnee.ca

Sustainable Development Research Initiative, University of B.C.
www.sdri.ubc.ca

Dockside Green Victoria
www.docksidegreen.ca

McGill affordable homes program
www.homes.mcgill.ca

Affordable Housing, Saskatoon
www.city.saskatoon.sk.ca/org/city_planning/affordable_housing

Regent Park revitalization, Toronto
www.regentparkplan.ca

Student Planning Network
www.planningnetwork.org

Appendix

Highlights of Canadian Community Planning, 1890–2006

Historically, the achievements in Canadian community planning fall into four distinct periods of development. There was, first, a *formative period* in which the perceived problems of cities were allied with planning solutions, and institutional arrangements began to be put in place to facilitate community planning. This period began in the final decade or two of the 19th century and lasted until 1930 in Canada. A second, the *transitional period*, spanning the years from 1930 to 1955, saw steps to build an infrastructure for community planning at the national, provincial, and local levels. The period from 1955 to 1985, the *modern period*, was an era of putting planning ideas into practice on a widespread basis, a period of experimentation and innovation, of trial and error, and, notably, of the acceptance of the approach of community planning. There are indications that community planning has, since the mid-1980s, moved into a new period that could be called the *postmodern period*. Many planning initiatives in this period are moving beyond single-problem formats and linking factors in broader based approaches, as with Healthy Communities, Smart Growth, and sustainable development.

The evolution of community planning in Canada may be seen in the sets of important events constituting each of the four periods shown in the following chronology.

The Formative Period, 1890–1930

1896 Herbert Ames's study of Montreal slums.

1903 Frederick Todd's plan for a park system for Ottawa and Hull.

1904 City Beautiful plan for Prince Rupert, British Columbia.

1906 Toronto Civic Guild of Art prepares City Beautiful plan for city.

1908 Garden Suburb plans prepared for Shaughnessy Heights in Vancouver, Mount Royal in Calgary, etc.

1909 Commission of Conservation formed by the federal government.

1911 City Planning Commissions established in Winnipeg and Calgary.

1912 New Brunswick passes first provincial planning act.

1913 Alberta's Town Planning Act allows municipalities to acquire 5 percent of subdivisions for parks.

1914 Thomas Adams appointed Town Planning Advisor to Commission of Conservation.

1914 Sixth National City Planning Conference held in Toronto.

1914 *Conservation of Life* begins publication for seven years (Thomas Adams, ed.).

1915 Local planning boards required in Nova Scotia.

1915 Plan for National Capital area of Ottawa and Hull prepared by Edward Bennett.

1917 Thomas Adams submits plan for rebuilding the Richmond District in Halifax, which was devastated by a wartime explosion.

1917 Garden City plan prepared by Thomas Adams for new resource town (Temiskaming, Quebec).

1919 Town Planning Institute of Canada founded with 117 members; publishes its own journal.

1921 Commission of Conservation disbanded.

1923 First zoning bylaw adopted for Kitchener and Waterloo; drafted by Thomas Adams and Horace Seymour.

1929 Harland Bartholomew and Horace Seymour's comprehensive plan for Vancouver.

1929 Provincial Planning Office established in Alberta; Horace Seymour first director.

The Transitional Period, 1930–1955

1935 Prairie Farm Rehabilitation Administration established.

1935 Housing Centre formed at University of Toronto by Humphrey Carver and Harry Cassidy.

1938 National Housing Act passed.

1939 National Housing Conference held in Toronto.

1943 Metropolitan Planning Commission for Greater Winnipeg formed; Eric Thrift is director.

1944 Report of the Advisory Committee on Reconstruction (the "Curtis Committee").

1945 Central (now Canada) Mortgage and Housing Corporation established.

1946 Founding of Community Planning Association of Canada; publishes *Community Planning Review* for the next 25 years.

1946 Ontario Planning and Development Act passed.

1947 McGill University establishes the first Canadian planning program

1948 Regent Park North slum-clearance project built in Toronto.

1949 Lower Mainland Regional Planning Board formed; James Wilson is director.

1950 Jacques Gréber prepares plan for Ottawa capital region, featuring a greenbelt.

1951 Plan made for Kitimat, B.C., by Clarence Stein, using the neighbourhood principle.

1951 University of Manitoba establishes first graduate planning degree.

1953 Metropolitan government instituted for the Toronto area and given wide planning powers; Murray Jones is director.

The Modern Period, 1955–1985

1955 Macklin Hancock's plan for Don Mills "New Town" is implemented.

1959 National Capital Commission established to implement Gréber plan and acquire the greenbelt.

1961 "Resources for Tomorrow" Conference held in Montreal.

1961 Agricultural Rehabilitation and Development Act passed.

1964 National Housing Act, 1964, enunciates new housing and urban-renewal policies.

1964 Establishment of le Bureau d'aménagement de l'est du Québec for regional planning in the Gaspé.

1966 Trefann Court urban-renewal citizen protest begins in Toronto.

1967 Mactaquac Regional Development Plan unveiled in New Brunswick by Leonard Gertler's team.

1968 University of Waterloo establishes the first undergraduate planning program.

1970 Ontario government presents plan for 14 250-square kilometre Toronto-Centred Region.

1971 Spadina Expressway (Toronto) ordered stopped.

1972 Federal government organizes Ministry of State for Urban Affairs, based on recommendations by Prof. Harvey Lithwick in his report, *Urban Problems and Prospects.*

1973 B.C. Agricultural Land Commission established.

1974 Niagara Escarpment (Ontario) Plan initiated.

1975 Greater Vancouver Regional District unveils "Proposals for a Livable Region."

1976 United Nations' Habitat Conference held in Vancouver.

1978 Quebec moves to protect agricultural land with its Loi sur le protection du territoire agricole.

1978 In Toronto, plans for Harbourfront and new inner-city housing in the St. Lawrence precinct.

1978 In Vancouver, False Creek redevelopment, including Granville Island's unique industrial-commercial-artistic mix.

1979 Meewasin Valley Authority established to plan and develop 80 kilometres of river banks in the Saskatoon area.

1980 Revitalization plan begun for Market Square in Saint John.

1981 The Core Area Initiative to redevelop downtown Winnipeg agreed to by three levels of government.

1982 Light Rapid Transit systems built in Edmonton and Calgary.

The Postmodern Period, 1985 to the Present

1986 Regional plan for Communauté urbaine de Québec adopted.

1987 Introduction of the City of Sudbury Strategic/Corporate Plan, encompassing physical, economic, human, and organizational development.

1988 Hans Blumenfeld dies in Toronto at age 96.

1988 Healthy Communities Initiative launched by Canadian Institute of Planners.

1990 Waterfront planning for the Toronto region by the Crombie Commission recommends a bioregional perspective.

1992 Montreal's first downtown plan is adopted.

1993 Highly participative planning process in the preparation of Vancouver's CityPlan and Calgary's GoPlan.

1993 Regional planning commissions are disbanded in Alberta.

1996	New Greenbelt Plan for Ottawa embraces ecosystems approach.
1996	City of Toronto modifies zoning regulations in two "reinvestment areas" to promote diversity and adaptation of uses.
1997	Municipal governments within Metropolitan Toronto amalgamate.
1998	Plan Edmonton breaks new ground in combining strategic and financial planning with physical land-use planning.
1999	The regional plan for Ottawa-Carleton links its growth-management policies to ecosystems needs for the region.
2000	New Urbanism concepts influence suburban plans from Langley, B.C. (Murray's Corner), Calgary (McKenzie Towne), Markham (Cornell), and Montreal (Bois Franc).
2001	The First Nations Community Planning Model is developed at Dalhousie University.
2002	The (newly amalgamated) City of Ottawa adopts the 2020 Plan, integrating land use, the natural environment, transportation, infrastructure, culture, economic development, and social plans.
2003	Toronto's Regent Park Revitalization Plan adopted after extensive community consultation.
2004	First Master Plan for the entire island of Montreal adopted.
2005	Dockside Green begins sustainable redevelopment of Victoria's inner harbour.
2005	Ontario Government implements a Greenbelt Plan for the Golden Horseshoe area, bounded by the Niagara Escarpment and Oak Ridges Moraine.
2006	World Planning Congress and United Nations' World Urban Forum held in Vancouver.

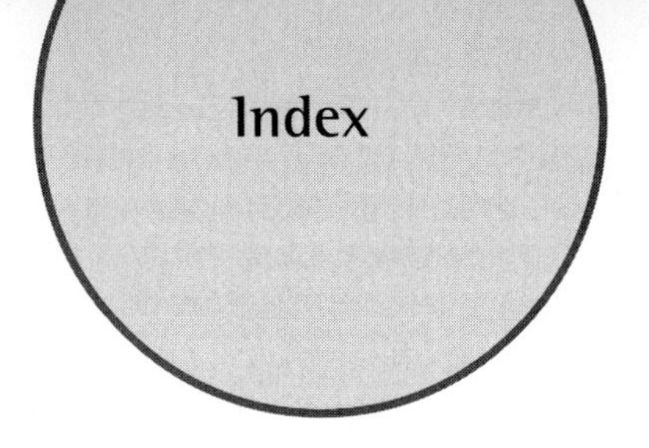

Index

Note: Page references in **bold** type indicate an illustration.

D

E

F

G

H

I

J

K

L

M

N

O

P

Q

R

S

T

U

V

W

Y

Z